The Perception of Speech

The Perception of Speech:
From sound to meaning

Edited by

Brian C.J. Moore
Department of Experimental Psychology,
University of Cambridge,
Downing Street,
Cambridge, UK

Lorraine K. Tyler
Department of Experimental Psychology,
University of Cambridge,
Downing Street,
Cambridge, UK

William D. Marslen-Wilson
MRC Cognition and Brain Sciences Unit,
15 Chaucer Road,
Cambridge, UK

Originating from a Theme Issue first published in Philosophical
Transactions of the Royal Society B: Biological Sciences
http://publishing.royalsociety.org/philtransb

OXFORD
UNIVERSITY PRESS

Great Clarendon Street, Oxford OX2 6DP

Oxford University Press is a department of the University of Oxford.
It furthers the University's objective of excellence in research, scholarship,
and education by publishing worldwide in

Oxford New York

Auckland Cape Town Dar es Salaam Hong Kong Karachi
Kuala Lumpur Madrid Melbourne Mexico City Nairobi
New Delhi Shanghai Taipei Toronto

With offices in

Argentina Austria Brazil Chile Czech Republic France Greece
Guatemala Hungary Italy Japan Poland Portugal Singapore
South Korea Switzerland Thailand Turkey Ukraine Vietnam

Oxford is a registered trade mark of Oxford University Press
in the UK and in certain other countries

Published in the United States
by Oxford University Press Inc., New York

© The Royal Society, 2009

The moral rights of the author have been asserted
Database right Oxford University Press (maker)

First published by Oxford University Press 2009

All rights reserved. No part of this publication may be reproduced,
stored in a retrieval system, or transmitted, in any form or by any means,
without the prior permission in writing of Oxford University Press,
or as expressly permitted by law, or under terms agreed with the appropriate
reprographics rights organization. Enquiries concerning reproduction
outside the scope of the above should be sent to the Rights Department,
Oxford University Press, at the address above

You must not circulate this book in any other binding or cover
and you must impose the same condition on any acquirer

British Library Cataloguing in Publication Data
Data available

Library of Congress Cataloging in Publication Data
Data available

Typeset by Cepha Imaging Private Ltd, Bangalore, India
Printed in China
on acid-free paper by
C & C Offset Printing Co., Ltd.

ISBN 978-0-19-956131-5

1 3 5 7 9 10 8 6 4 2

Contents

List of contributors	vii
1. Introduction *Brian C. J. Moore, Lorraine K. Tyler, and William Marslen-Wilson*	1
2. Neural representation of spectral and temporal information in speech *Eric D. Young*	9
3. Basic auditory processes involved in the analysis of speech sounds *Brian C. J. Moore*	49
4. Acoustic and auditory phonetics: the adaptive design of speech sound systems *Randy L. Diehl*	79
5. Early language acquisition: phonetic and word learning, neural substrates, and a theoretical model *Patricia K. Kuhl*	103
6. The processing of audio-visual speech: empirical and neural bases *Ruth Campbell*	133
7. Listening to speech in the presence of other sounds *C. J. Darwin*	151
8. Functional imaging of the auditory processing applied to speech sounds *Roy D. Patterson and Ingrid S. Johnsrude*	171
9. Fronto-temporal brain systems supporting spoken language comprehension *Lorraine K. Tyler and William Marslen-Wilson*	193
10. The fractionation of spoken language understanding by measuring electrical and magnetic brain signals *Peter Hagoort*	223
11. Speech perception at the interface of neurobiology and linguistics *David Poeppel, William J. Idsardi, and Virginie van Wassenhove*	249
12. Neural specializations for speech and pitch: moving beyond the dichotomies *Robert J. Zatorre and Jackson T. Gandour*	275
13. Language processing in the natural world *Michael K. Tanenhaus and Sarah Brown-Schmidt*	305
Index	337

List of contributors

Sarah Brown-Schmidt Beckman Institute, University of Illinois at Urbana Champaign, Urbana IL, USA

Ruth Campbell Division of Psychology and Language Sciences, University College, London, UK

Christopher J Darwin Department of Psychology, University of Sussex, UK

Randy L Diehl Department of Psychology and Center for Perceptual Systems, University of Texas at Austin, USA

Jackson T Gandour Department of Speech Language & Hearing Sciences, Purdue University, West Lafayette, IN, USA

Peter Hagoort Donders Institute for Brain, Cognition and Behaviour, Radboud University, Nijmegen, The Netherlands; Max Plank Institute for Psycholinguistics

William J Idsardi Department of Linguistics, University of Maryland, College Park, MD, USA

Ingrid S Johnsrude Department of Psychology, Queen's University, Kingston ON, Canada

Patricia K Kuhl Institute for Learning and Brain Sciences, University of Washington, Seattle, WA, USA

William Marslen-Wilson MRC Cognition and Brain Sciences Unit, Cambridge, UK

Brian C J Moore Department of Experimental Psychology, University of Cambridge, UK

Roy D Patterson Centre for the Neural Basis of Hearing, Department of Physiology, Development and Neuroscience, University of Cambridge, UK

David Poeppel Department of Linguistics, University of Maryland, College Park, MD, USA

Michael K Tanenhaus Beverly Petterson Bishop and Charles W. Bishop Professor of Brain and Cognitive Sciences and Linguistics, Department of Brain & Cognitive Sciences, University of Rochester, Rochester NY, USA

Lorraine K Tyler Department of Experimental Psychology, University of Cambridge, UK

Virginie van Wassenhove Division of Biology, California Institute of Technology, Pasadena CA, USA

Eric D Young Department of Biomedical Engineering, Centre for Hearing and Balance, Johns Hopkins University, Baltimore MD, USA

Robert J Zatorre Montreal Neurological Institute, McGill University, Montreal QC, Canada

1

Introduction

Brian C. J. Moore, Lorraine K. Tyler, and William Marslen-Wilson

> Spoken-language communication is arguably the most important activity that distinguishes humans from non-human species. This chapter provides an overview of the chapters that make up this volume on the processes underlying speech communication. The volume includes contributions from researchers with a wide range of specialities within the general area of speech perception and language processing. It also includes contributions from key researchers in neuroanatomy and functional neuro-imaging—in an effort to cut across traditional disciplinary boundaries and foster cross-disciplinary interactions in this important and rapidly developing area of the biological and cognitive sciences.
>
> **Keywords:** Speech perception, Speech production, Psycholinguistics, Audition, Phonetics, Functional imaging, Neuropsychology

1.1. Introduction

Spoken-language communication is arguably the most important activity that distinguishes humans from non-human species. While many animal species communicate and exchange information using sound, humans are unique in the complexity of the information that can be conveyed using speech, and in the range of ideas, thoughts, and emotions that can be expressed.

Despite the importance of speech communication for the entire structure of human society, there are many aspects of the speech communication process that are not fully understood. Research on speech and language is typically carried out by different groups of scientists working on separate aspects of the underlying functional and neural systems. Research from an auditory perspective focusses on the acoustical properties of speech sounds, on the representation of speech sounds in the auditory system, and on how that representation is used to extract phonetic information. Research from psycholinguistic perspectives focusses on the processes by which representations of meaning are extracted from the acoustic–phonetic sequence, and how these are linked to the construction of higher level linguistic interpretation in terms of sentences and discourse. However, there has been relatively little interaction between speech researchers from these two groups.

In addition, there has been a dramatic expansion in recent years of research into the neural bases of auditory and linguistic function. Developments in the neuroanatomy and neurophysiology of the auditory system of non-human primates provide the basis for mapping out the basic organization of the structures and pathways that support the processing of auditory information in the primate brain. Complementary developments in neuro-imaging techniques for visualizing the activity of the intact brain are allowing scientists to probe the dynamic spatio-temporal patterns of neural activity that underlie the representation and processing of speech and language in the human brain.

Despite this ferment of activity across a variety of fields, there has been relatively little interaction between researchers working on these various topics, and perhaps a lack of recognition that they are all participating in the same overall scientific process of understanding how the motor gestures of a speaker are transformed to sounds and how those sounds are mapped onto meaning in the comprehension of spoken language. This volume addresses these issues. It includes contributions from researchers with a wide range of specialities within the general area of speech perception and language processing. It also includes contributions from key researchers in neuroanatomy and functional neuroimaging—in an effort to cut across traditional disciplinary boundaries and foster cross-disciplinary interactions in this important and rapidly developing area of the biological and cognitive sciences.

1.2. Overview of the special issue

The chapter by **Young** describes the representation of speech sounds in the auditory nerve and at higher levels in the central nervous system, focussing especially on vowel sounds. The experimental data are derived mainly from animal models (especially the cat), so some caution is needed in interpreting the results in terms of the human auditory system. However, it seems likely that at least the early stages of auditory processing, as measured in the auditory nerve, are similar across all mammals. A key feature of the representation of sounds is that it is tonotopic; speech signals are decomposed into sinusoidal frequency components or groups of components and different frequency components are represented in different populations of neurones. In other words, the short-term spectrum of the sound is represented in the relative amount of neural activity in neurones that are tuned to different frequencies. This tonotopic organization is preserved throughout the auditory system, although at higher levels in the auditory system there may be multiple 'maps'. Another critical feature of the representation is suppression, a nonlinear process whereby strong neural activity in one group of neurones (all 'tuned' to similar frequencies) suppresses activity in neurones tuned to adjacent frequencies. This suppression is essential for maintaining the representation of the spectral content of sounds over a wide range of sound levels. Spectral features may also be represented in the detailed timing of the neural activity (phase locking), although the role of this 'temporal fine structure' is still controversial. The representation of speech sounds in central auditory neurones is more robust than at the periphery to changes in stimulus intensity and it also becomes more transient. Furthermore, Young argues that it is likely that the form of the representation at the auditory cortex is fundamentally different from the representation at lower levels, in that stimulus features other than the distribution of energy across frequency are analyzed.

The chapter by **Moore** reviews basic aspects of auditory processing that play a role in the perception of speech. Here, the data are mainly derived from perceptual experiments using human listeners. The frequency selectivity of the auditory system refers to the ability to resolve the sinusoidal components in complex sounds, and is closely related to the tonotopic representation described by Young. Moore describes how frequency selectivity can be quantified using masking experiments. The 'auditory filters' inferred from the results can be used to calculate the internal representation of the spectrum of speech sounds in the peripheral auditory system; this representation is called the excitation pattern. The perception of timbre and distinctions in quality between vowels are related to both static and dynamic aspects of the spectra of sounds, as represented in the excitation pattern. The pitch of speech sounds is related to their fundamental frequency, which is in turn related to the rate of vibration of the vocal folds. Moore describes the mechanisms by which the auditory system extracts the pitch of speech sounds and the role that pitch patterns play in speech perception, especially the perception of intonation.

Although some speech sounds, such as vowels, can be characterized in terms of their long-term spectral properties, speech perception in general depends strongly on the dynamic nature of speech sounds, and the way that they change over time. Moore describes the limits of the ability of the auditory system to follow rapid changes, and describes how temporal resolution can be modelled using the concept of a sliding temporal integrator. The combined effects of limited frequency selectivity and limited temporal resolution can be modelled by calculation of the spectro-temporal excitation pattern, which gives good insight into the representation of speech sounds in the auditory system. Moore argues that, for speech presented in quiet, the resolution of the auditory system in frequency and time usually markedly exceeds the resolution necessary for the identification or discrimination of speech sounds, which partly accounts for the robust nature of speech perception. However, people with impaired hearing have reduced frequency selectivity and can hear comfortably over a smaller-than-normal range of sound levels. For such people, speech perception is often much less robust than for normally hearing people.

The chapter by **Diehl** considers further the robust nature of speech perception. For people with normal hearing, speech can be understood even under conditions when there is considerable background noise or reverberation, or when the speech is distorted in a variety of ways. Diehl considers how the acoustical and auditory properties of vowels and consonants help to ensure intelligibility. The properties of speech sounds can be understood by considering the sounds as resulting from a source of sound energy, such as vibration of the vocal folds or turbulence produced by forcing air through a narrow constriction, followed by a filter (the vocal tract) which modifies the spectrum of the source. Diehl describes this 'source-filter' theory, and demonstrates how it can account for the relationship between vocal-tract properties and formant patterns. He points out that certain types of speech sounds (e.g. the resonance patterns or 'formant' frequencies of specific vowel sounds) occur commonly in the languages of the world, while others occur much more rarely. He presents two theories that have been proposed to account for the structure of these 'preferred sound inventories'—quantal theory and dispersion theory.

Quantal theory (Stevens 1989) is based on the fact that non-linearities exist in the mapping between articulatory (i.e. vocal-tract) configurations of talkers and acoustic outputs. For certain regions of articulatory 'space', perturbations in the articulatory parameters result in small changes in the acoustic output, whereas in other regions perturbations of similar size yield large acoustic changes. Given these regions of acoustic stability and instability, quantal theory is based on the idea that preferred sound categories are selected to occupy the stable regions and to be separated by unstable regions. Dispersion theory (Liljencrants & Lindblom 1972), like quantal theory, is based on the idea that that speech sound inventories are structured to maintain perceptual distinctiveness. However, in dispersion theory distinctiveness is viewed as a global property of an entire inventory of sound categories. A vowel or consonant inventory is said to be maximally distinctive if the sounds are maximally dispersed (i.e. separated from each other) in the available 'phonetic space'. Diehl discusses the strengths and limitations of each theory and proposes that certain aspects of the two theories can be unified in a principled way so as to achieve reasonably accurate predictions of the properties of preferred sound inventories.

The chapter by **Kuhl** describes the development of language during the early years of life, and the mechanisms that appear to underlie that development. Particular emphasis is placed on the use of neuroscience techniques to examine language processing in the young brain. These techniques include electroencephalography, event-related potentials, magnetoencephalography, functional magnetic resonance imaging, and near-infrared spectroscopy. Kuhl describes how these techniques can be combined with behavioural studies to clarify how the mechanisms of speech perception and language processing develop. She also describes applications of these techniques to the study of abnormalities in language processing associated with autism spectrum disorder. Infants' speech perception skills show two types of changes towards the end of the first year of life. Firstly, the ability to perceive phonetic distinctions in a non-native language declines. Secondly, skills at making phonetic distinctions in the child's own language improve. The chapter presents recent data showing that both native and non-native phonetic perception skills of infants predict their later language ability, but in opposite directions. Better *native*-language skill at 7 months predicts faster language advancement, whereas better *non-native*-language skill predicts slower advancement. Kuhl suggests that native-language phonetic performance is indicative of commitment of neural circuitry to the native language, while non-native phonetic performance reveals uncommitted neural circuitry. This chapter describes a revised version of a model previously proposed by Kuhl and co-workers, the Native Language Magnet model.

The chapter by **Campbell** emphasizes the fact that speech perception is multi-modal; what we perceive as speech is influenced by what we see on the face of the talker as well as by what is received at the two ears. This is illustrated by the McGurk effect (McGurk & MacDonald 1976), which is produced when a video recording of one utterance is combined with an audio recording of another utterance. What is heard is influenced by what is seen. For example, an acoustic 'mama' paired with a video 'tata' is heard as 'nana'. The influence of vision on speech perception is also illustrated by the fact that, in noisy situations, speech can be understood much better when the face of the talker is visible than when it is invisible (Erber 1974).

Campbell proposes that there are two main ways or 'modes' in which visual information may influence speech perception. The first is a complementary mode, whereby vision provides information more efficiently than hearing for some under-specified parts of the speech stream. For example, the acoustic cues signalling the distinction between 'ba' and 'ga' may be relatively weak and easily masked by background sounds, whereas visually these two sounds are very distinct. The second is a correlated mode, whereby vision partially duplicates auditory information about dynamic articulatory patterning.

Campbell reviews evidence suggesting that these two modes are not reflected in discrete cortical processing systems, but that they reflect somewhat differentiated access to two major streams for the processing of natural language—a 'what' and a 'how' stream. The 'what' stream makes particular use of the inferior occipito-temporal regions of the cortex and of the ventral visual processing stream which can specify image details effectively. It can, therefore, serve as a useful route for complementary visual information to be processed. A major projection of this stream is to association areas in middle and superior temporal cortex. In contrast to this, the 'how' stream for the analysis of auditory speech may be readily accessed by natural visible speech, which is characterized by dynamic features that correspond with those available acoustically. Processing that requires sequential segmental analysis (e.g. identifying syllables or words individually or in lists), will differentially engage this posterior stream. It is in this stream that the correlational structure of seen and heard speech is best reflected. The visual input to these analyses arises primarily in the lateral temporo-occipital regions that track visual movement.

Although the great majority of studies of speech perception have been conducted using speech sounds presented in quiet with little reverberation, speech communication in everyday life often takes place in the presence of background sounds and reverberation. The issues raised by this are considered in the chapter by **Darwin**. He points out that irrelevant background sounds can cause severe problems for computer-based speech recognition algorithms and for people with hearing impairment, but that people with normal hearing are remarkably little affected. A variety of perceptual problems are created by the presence of background sounds. These include: complete or partial masking of some parts of the target speech, the need to decide which parts of the sound 'belong to' each sound source, and the recognition of speech sounds based on partial information. Darwin examines the effectiveness of the cues which can be used to separate target speech from a background of other sounds (including competing speech), focussing particularly on the role of fundamental frequency, onset asynchronies, and binaural cues. At present, human listeners perform far better than any computer system in separating mixtures of sounds. A fuller understanding of how humans do this would have important practical applications.

The chapter by **Patterson** and **Johnsrude** places the study of auditory processing, as applied to speech, squarely in a neuro-biological and neuro-imaging context. Cross-species studies—especially in the macaque—provide a well-developed neuroanatomical and neurophysiological account of the primate auditory processing system. This leads to concrete hypotheses both about the detailed functional architecture of the human system, with sub-cortical auditory processing systems feeding into primary auditory cortex, and about the local and the global connectivity of these areas with other regions of the brain. In this general framework, Patterson and Johnsrude go on to consider some of the basic functional challenges that speech variation presents to the listener, and how these challenges are met in the primate auditory system. Two major sources of variation are differences in pitch and in vocal tract length, which mean, e.g. that the same vowel (in terms of its linguistic label) spoken by a child or by an adult will vary markedly in its acoustic properties.

Patterson and Johnsrude present an innovative account of how adaptive mechanisms, operating before speech analysis can take place, may provide information sufficient to allow the system to normalize for pitch and vocal tract variation. This account combines a computational model of auditory processing with psychophysically constrained neuro-imaging investigations of the spatial locations in auditory-processing areas (in and around Heschl's gyrus) that are particularly sensitive to the relevant acoustic and phonetic contrasts. An important role is played here by magnetoencephalography (MEG), where high temporal resolution is accompanied by significantly improved spatial resolution, relative to the electroencephalogram (EEG). Recent studies using MEG are beginning to tease out the spatio-temporal details of the cortical-processing events underlying the extraction and perception of pitch information.

In a final section, Patterson and Johnsrude address the central question of how the cortical system moves from general auditory processing to potentially voice- and speech-specific processing activities. Research into this question is still in its early stages, but the evidence suggests that the transformation from an auditory signal to speech is localizable, but is not straightforwardly hierarchical. The emergence of a vowel percept, e.g. from the building blocks provided by sub-processes concerned with glottal pulse rate, vocal tract length, and so forth, seems to be distributed across several neural loci, situated around but not directly in core auditory cortex, and with possible links further afield to structures in premotor and motor cortices involved in speech production. This suggests that motor theories of speech perception (Liberman *et al.* 1967; Liberman & Mattingly 1985) may be due for a revival.

The next chapter, by **Zatorre** and **Gandour**, is highly complementary, with its focus on neural specializations for speech and pitch, and provides a balanced and informative account of possible hemispheric differences in these domains. They argue against standard approaches to this issue, which have led to a polarized debate asking whether speech processing is underpinned either by encapsulated, specialized domain-specific mechanisms or whether it piggy-backs on general-purpose neural mechanisms for processing sound which are sensitive to the acoustic features that are present in speech. Zatorre and Gandour propose a more integrated approach, arguing that the brain's response to low-level acoustic features is modulated by linguistic factors, affecting the specificity of hemispheric function.

They outline the considerable evidence that has now accumulated for hemispheric differences in sensitivity to both the spectral and temporal properties of auditory inputs, but go on to argue that these differences can be modulated by the linguistic status of the input. In addition to neuroimaging data on English, they also discuss extensive data from tonal languages, where Gandour and colleagues have been the pioneers in applying neuroimaging techniques to the evaluation of neural contrasts in how pitch is processed as a function of its linguistic role. A clear outcome of these studies, in languages like Mandarin and Thai, is that when linguistically relevant pitch patterns carried by tones cue linguistic differences, activity tends to be left-lateralized, but when they do not, then activity is right-lateralized. Zatorre and Gandour conclude by arguing for an approach to speech processing that recognizes the complexity of hemispheric interactions between general sensory-motor and cognitive processes, modulated by the specific processing demands of different linguistic environments.

The chapter by **Poeppel**, **Idsardi**, and **Wassenhove**, although it covers some of the same ground as the two preceding chapters, addresses the neurobiology of speech in the brain from very different, and more 'external' theoretical perspectives. Poeppel and colleagues revive—and significantly rework—the classic methodological framework put forward by David Marr in the 1980s for the analysis of complex neuro-cognitive systems, and they give linguistic theory equal status with neurobiology and auditory neuroscience in placing fundamental constraints on the representation of speech in the brain. Their key assumption is that speech perception is about the construction of abstract phonological representations, structured in such a way that they can interface with lexical representations as characterized in current linguistic theory. In their Marrist framework, this requirement is related to three levels of scientific description. The highest, computational level refers to the commitment to a representational theory in terms of phonological distinctive features. The two lower levels—the implementational and the algorithmic—describe how the system is organized to generate a linguistically relevant output specified in these terms.

The implementational level centres around the notion of multi-time resolution processing (also considered by Zatorre & Gandour), where speech signals are simultaneously processed on a short (25–80 ms) time scale, and on a longer time scale of roughly syllabic length (around 200 ms), and where there are hemispheric asymmetries associated with these two temporal domains. The output of these processes is the input to an analysis-by-synthesis process—specified at the algorithmic level—that interacts with lexical hypotheses and a partial feature matrix to generate a contextually acceptable lexical outcome. The analysis-by-synthesis approach—linked to current developments in Bayesian methodology and to the notion of a 'forward model'—is well suited to these authors' proposal of a 'phonological primal sketch' at the segmental level. Preliminary, broad-brush hypotheses about feature content can be tested and elaborated relative to stored knowledge about possible lexical analyses.

The next chapter, by **Tyler and Marslen-Wilson**, moves away from the specifics of auditory speech processing to focus on higher levels of the neural language system, combining cognitive accounts of language function with neuro-imaging studies of healthy subjects and patients who have specific language deficits. This research complements standard subtractive analyses of the functional Magnetic resonance imaging (fMRI) data with connectivity analyses in order to better understand the relationship between frontal and temporal regions in the processing of different aspects of language function. These studies develop a general contrast between a core set of morphological and syntactic linguistic functions, likely to be combinatorial in nature, and requiring an intact left hemisphere peri-sylvian language network, with more general processes of semantic and pragmatic interpretation whose neural substrate is more distributed and more bilateral in nature.

The first part of the chapter focusses on the processing of regularly inflected forms in English, as a prominent example of a linguistic process likely to involve the decomposition of a complex linguistic form (such as the past tense *jumped*) into its morphemic components (the stem *jump* and the grammatical affix *–ed*). A growing body of neuropsychological and neuro-imaging evidence points to a decompositional morphemic substrate for lexical processing, that requires an intact fronto-temporal network linking left posterior temporal lobe regions with left inferior frontal cortex (classical Broca's area). The second part of the chapter, focussing on syntactic and semantic processing—and their disruption following left hemisphere damage caused by stroke—confirms the critical dependency of syntactic (but not semantic) processes on a left fronto-temporal network that partially overlaps with the network revealed for morphological processes. Tyler and Marslen-Wilson interpret this overlap as indicating that different linguistic processes are not carried out in neural regions which are functionally specialized. Instead, each language function requires the co-activation in time of multiple regions within the fronto–temporal–parietal system, providing a different perspective on structure–function relations in the human language processing.

Also considering the global structure of the speech comprehension process, **Hagoort's** chapter discusses the speed with which spoken language is processed, focussing on research using EEG, a time-sensitive methodology for probing the moment-by-moment processing of language. As is now well established, spoken word recognition is a remarkably rapid process whereby multiple word candidates are activated on the basis of the sensory input and word recognition occurs when one candidate emerges as having the best fit. This produces a system in which words are identified well before their offset, through a process of activation, selection, and integration with the prior context. Hagoort describes EEG studies which confirm the earliness of word identification and the structure of the system which underpins lexical processing. He complements EEG studies on single-word processing with experiments showing how sentence and discourse contexts modulate the processing of individual words. These experiments show that context speeds up lexical selection, and add to previous findings by relating different aspects of processing to different event-related potential (ERP) components. The EEG data he describes help to develop models of language processing in which the processing of individual words is immediately affected by the discourse and real-world context.

Tanenhaus's chapter continues with the theme of the facilitatory effects of higher level context on lexical processing, but does so in the framework of the 'visual world paradigm', in order to generate a more naturalistic environment in which to study language comprehension. These studies, using eye-movement-monitoring techniques, show that multiple sources of linguistic and visual information are used to constrain the real-time analysis of spoken language processing. In the second part of his chapter, Tanenhaus describes studies which also focus on language use in naturalistic contexts, but here the emphasis is upon natural conversation, on the assumption that language use is typically an interactive process whereby speakers and listeners share common communicative goals. These types of naturalistic context may generate different models of language use compared to those based on more impoverished contexts. Subjects in these studies engage in a referential communicative task while gaze and speech are monitored. The results show that subjects closely coordinate referential domains as the conversation develops. The wider implication of this work is that behavioural context, including attention and intention, affects even basic perceptual processes involved in language processing.

Acknowledgement

This book was originally published as an issue of the *Philosphical Transactions of the Royal Society B: Biological Sciences* (Volume 363; Issue 1493) but has been materially changed and updated.

References

Erber, N. P. 1974 Auditory-visual perception of speech: a survey. In *Visual and Audio-Visual Perception of Speech* (eds H. Birk Nielsen & E. Kampp), pp. 12–30. Stockholm: Almquist & Wiksell.

Liberman, A. M., Cooper, F. S., Shankweiler, D. P. & Studdert-Kennedy, M. 1967 Perception of the speech code. *Psychol. Rev.* **74**, 431–461.

Liberman, A. M. & Mattingly, I. G. 1985 The motor theory of speech perception revised. *Cognition* **21**, 1–36.

Liljencrants, J. & Lindblom, B. 1972 Numerical simulation of vowel quality systems: the role of perceptual contrast. *Language* **48**, 839–862.

McGurk, H. & MacDonald, J. 1976 Hearing lips and seeing voices. *Nature* **264**, 746–748.

Stevens, K. N. 1989 On the quantal nature of speech. *J. Phonetics* **17**, 3–45.

2

Neural representation of spectral and temporal information in speech

Eric D. Young

Speech is the most interesting and one of the most complex sounds dealt with by the auditory system. The neural representation of speech needs to capture those features of the signal on which the brain depends in language communication. Here, we describe the representation of speech in the auditory nerve and in a few sites in the central nervous system from the perspective of the neural coding of important aspects of the signal. The representation is tonotopic – meaning that the speech signal is decomposed by frequency and different frequency components are represented in different populations of neurones. Essential to the representation are the properties of frequency tuning and nonlinear suppression. Tuning creates the decomposition of the signal by frequency, and nonlinear suppression is essential for maintaining the representation across sound levels. The representation changes in central auditory neurones by becoming more robust against changes in stimulus intensity and by becoming more transient. However, it is likely that this spectrotemporal representation – in which neurones basically recapitulate the frequency × time pattern of the speech signal – does not provide an adequate description of the representation at the auditory cortex. In the cortex, the representation is likely more in the form of auditory objects, i.e. the sources of the sounds, as opposed to the sum total of the acoustic signals that they produce.

Keywords: Auditory cortex, Auditory nerve, Cat, Discrimination, Inferior colliculus, Speech, Tonotopic

2.1 Introduction

The general features of the neural representation of speech have been studied for over 25 years (e.g. Sachs & Young 1979; Young & Sachs 1979; Delgutte 1980; Reale & Geisler 1980; Sinex & Geisler 1983; Palmer *et al.* 1986). This is a challenging problem, primarily because of the complexity of the speech signal. Information in speech is encoded in a rapid sequence of different sound segments. The individual segments can be characterized by their frequency spectra, meaning the distribution across frequency of the energy making them up. The spectra change with the speech segment, so the resulting signal is a complex spectrotemporal pattern (Diehl, this volume; Lehiste 1967; Fant 1970); an example is shown in Figure 2.1.

This chapter summarizes some aspects of the progress in understanding the representation of the speech signal in the brain, as reflected in the discharge patterns of single neurones and populations of neurones. The focus is on the peripheral parts of the auditory system, mainly the auditory nerve (AN), but results are shown also for three structures in the central auditory system – the cochlear nucleus, inferior colliculus, and primary auditory cortex. Studies of responses to speech in auditory and other parts of the cortex using imaging and evoked electrical and magnetic signals are considered in other chapters of this book.

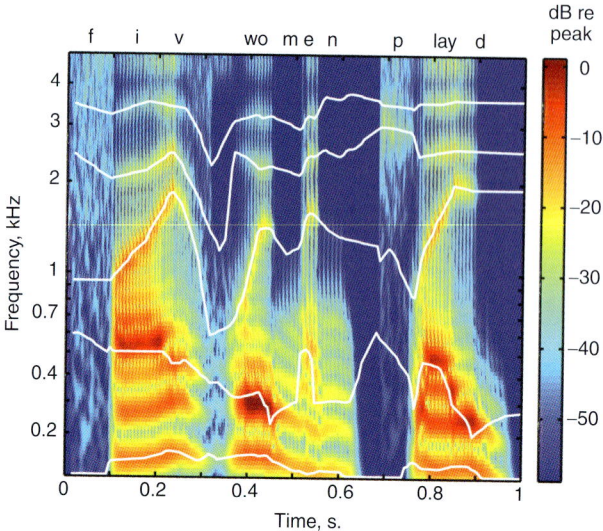

Figure 2.1 Spectrogram of the speech segment "five women played". The plot shows the energy in the stimulus as a function of time (abscissa) and frequency (ordinate). Energy is given by the colour using the scale at right (in dB relative to the peak energy in the signal). The spectrogram was computed using a gammatone filterbank (Slaney, 1993; Patterson *et al.*, 1992) which is a set of linear bandpass filters with bandwidths similar to individual frequency channels in the cochlea. Each horizontal line in the plot is the energy passing through one of these filters. The scale on the y-axis is roughly the layout of frequencies along the human basilar membrane. Thus this display shows an approximation to the cochlear representation of the speech. The stimulus was computer synthesized (courtesy R. McGowan, Sensimetrics Inc.) using the formant frequencies shown by the lines. The lowest line is the glottal repetition rate (the voice pitch) and the next four are the first four formants. At frequencies below 1 kHz, individual harmonics of the signal are visible as prominent horizontal bars; at higher frequencies, the filter bandwidths are wider and the harmonics are not resolved (or separated) by the filters; as a result, the structure of the representation is dominated by the formants.

The problem of the neural representation of the spectrotemporal pattern of speech has generally been simplified either to the representation of the frequency spectrum of stationary speech segments or to the representation of the temporal sequence of energy within a narrow band of frequencies (those to which the neurone under study is sensitive). Here, we consider the spectral representation first, then the representation of temporal stimulus patterns.

The frequency content of speech sounds is largely determined by the placement of the formant frequencies, which are the resonant frequencies of the vocal tract (Fant 1970). As shown in Figure 2.1, the speech signal has peaks of energy at the formant frequencies. For vowels, the formants dominate both the neural responses, described below, and the acoustic properties of the sound (Stevens & House 1961); because of this, much of this chapter is devoted to the representation of the formants. For consonants, the formants are still important; however consonants often vary significantly with time, which means either that the formant frequencies vary with time (formant transitions) or that the sound contains periods of silence bordered by transients in sound energy, or both. Studies of the neural representation of consonants have often focussed on stop consonants, in which both formant transitions and transients are prominent. Examples of those studies will be described further.

The study of auditory neural representations begins from the assumption that the representation is tonotopic, meaning that the speech sound is decomposed into its component frequencies by the basilar membrane, much as is shown in Figure 2.1. Essentially, energy at different frequencies causes displacement of the basilar membrane at different, frequency-specific locations (Robles & Ruggero 2001). Thus the neural elements of the cochlea – the hair cells and auditory nerve (AN) fibres – sense energy at different frequencies, depending on their location along the basilar membrane. As a result, different features of a speech sound, such as the formants, are represented in separate populations of AN fibres, according to their frequencies. Presumably, this separation is important in speech perception because it minimizes the interference between speech components at different frequencies. For example, the second formant (F2) can be masked by the generally more intense first formant (F1); this effect is larger in persons with impaired hearing (reviewed by Moore 1995), where the separation of the formant representations breaks down (Miller *et al*. 1997). Thus, the fact that an auditory neurone responds only to a narrow range of frequencies, called tuning, is the most important factor in the neural representation of speech. However, other properties of cochlear transduction are also important – especially suppression of the responses at one stimulus frequency by energy at another. These points will be illustrated by examples.

For those unfamiliar with neurophysiological studies of the auditory system, a general description of the experimental basis for the results discussed in this book is given in the appendix.

2.2 Tuning in the auditory nerve: the tonotopic representation

The basic representation of stimuli at all levels of the auditory system is tonotopic (Schreiner *et al*. 2000), meaning that different frequencies in the stimulus are analysed separately (although not independently). The initial frequency analysis occurs in the cochlea. The AN conveys a representation of sound to the brain that resembles the outputs of a bank of bandpass filters tuned to different frequencies. For AN fibres, the tuning can be described in the two ways shown in Figure 2.2. Figure 2.2a shows tuning curves for 11 AN fibres from one cat. Each curve shows the threshold sound level of a fibre plotted against the stimulus frequency; threshold here means the sound level producing a criterion increase in the discharge rate (usually 20 spikes/s). Each curve is V-shaped with a minimum threshold at the best frequency (BF) of the fibre.

It is clear that each fibre has a restricted range of frequencies to which it responds and that there are fibres tuned to frequencies (i.e. with BFs) across the audible range of the animal. Assuming that the tuning curves are the (inverted) gain functions of bandpass filters and that the filters are an accurate model of the neurones' input/output characteristics, one expects that each fibre should convey information to the brain about stimulus frequencies near the fibre's BF, thus decomposing the stimulus by frequency. The assumption that such filters are an accurate model for AN fibres is only partly true, in that there are

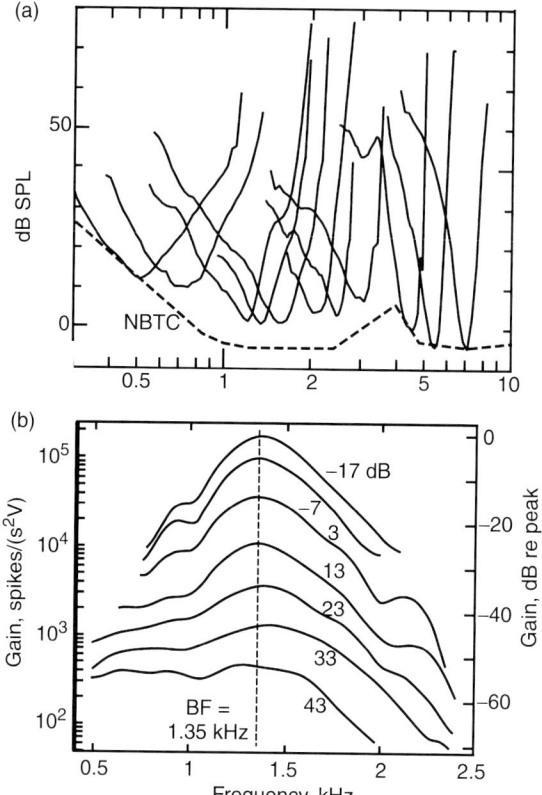

Figure 2.2 (a) Tuning curves of AN fibres from a cat for BFs between 0.5 and 7 kHz. The dashed line labelled 'NBTC' shows the thresholds of the most sensitive fibres in a group of cats. (b) Gain versus frequency functions for an AN fibre measured at different sound levels using the method of reverse correlation. The stimulus was a broadband noise and the functions express the 'gain' of the neurone, meaning its discharge rate divided by the power spectral density of the stimulus. Gain is given in absolute units on the left ordinate and as dB relative to the peak gain on the right ordinate. The units of gain on the left ordinate derive from the way the gain functions are computed; see Johnson (1980a) for a full explanation. The sound levels of the noise are given as spectrum levels next to the curves, as dB re 20 µPa/(Hz)$^{1/2}$, the level in a 1-Hz-wide band. The speech sound levels given in subsequent figures are not comparable to these spectrum levels, because the speech levels are the overall intensity of the stimulus, summed across frequency. The vertical dashed line is the approximate BF.
((a) is redrawn with permission from Miller et al. (1999a); (b) is redrawn with permission from Recio-Spinoso et al. (2005)).

nonlinear effects that complicate cochlear tuning. These include the tuning changes discussed next and the nonlinear interactions which allow energy at one frequency to suppress responses to a different frequency (Sachs 1969; Javel et al. 1978; Javel 1981; Delgutte 1990).

A more accurate measure of suprathreshold frequency selectivity is provided by the gain functions in Figure 2.2b, for a chinchilla AN fibre tuned to 1.35 kHz (Recio-Spinoso et al. 2005). Unlike the tuning curves, these are true gain functions, with units of response divided by stimulus amplitude. The gain curves were computed from the responses to broadband Gaussian noise as the first Wiener kernel (Johnson 1980a), a method similar to reverse correlation (de Boer & de Jongh 1978). Briefly, the data are the responses of the neurone to a broadband noise at the ear. The average waveform of the noise preceding each action potential is computed, resulting in the reverse-correlation or revcor function (Møller 1977; de Boer & de Jongh 1978; Carney & Yin 1988; Lewis & Henry 1994; Recio-Spinoso et al. 2005). The Fourier transform of the revcor is the gain function of the linear filter that approximates the input/output characteristics of the neurone, under the stimulus conditions used to obtain the data. Figure 2.2b shows the gain functions at seven sound levels for this fibre; note that the shape of the gain functions changes with the sound level. Thus, at low noise levels (−17 dB), the gain function is bandpass and is very close to the inverted tuning curve of the neurone; at high noise levels (43 dB), the gain function is broad and low-pass. This behaviour is similar to that of basilar membrane gain functions (Ruggero et al. 1997) and is the basis for models of AN responses that successfully model responses to speech (Carney 1993; Bruce et al. 2003; Holmes et al. 2004).

The gain functions in Figure 2.2b show that AN fibres are less frequency selective at high sound levels. As a result, the neural representation of broadband stimuli, such as speech, changes with sound level. For example, for a vowel, the energy at F1 is typically 10–20 dB more intense than the energy at F2. Thus for neurones with BFs near F2, sharp bandpass filtering is necessary if the neurone is to respond to F2, in preference to F1. Taking the example in Figure 2.2b, if a vowel were presented with F2 at the neurone's BF and F1 an octave lower, the neurone would respond to F2 at low sound levels – where the filtering is sufficiently sharp to attenuate F1 below F2, but would respond to F1 at higher sound levels where the filtering is low-pass and does not attenuate F1 relative to F2. This behaviour is observed experimentally, in that AN fibres with BFs near F2 gradually switch their responses from F2 to F1 at sound levels between 70 and 90 dB SPL (sound pressure level; Wong et al. 1998).

2.3 Analysis of the neural representation

Figure 2.3 displays an example of the responses of an AN fibre to speech, and illustrates the way in which neural representations are constructed from the data. The waveform of the sentence 'Five women played basketball' is shown at low time resolution in Figure 2.3a. The peri-stimulus time (PST) histogram in Figure 2.3b shows the rate of discharge of the fibre in response to the speech, on the same time axis. This histogram was constructed by counting spikes in bins synchronized with the stimulus, as described in the caption. From the PST histogram, it is possible to estimate the strength of response of the neurone to the stimulus (or to any temporal segment of the stimulus), as the average rate of spiking over the appropriate time interval (number of spikes/length of time interval). Clearly, the different syllables activate this neurone to different degrees, presumably because they differ in the amount of energy in the neurone's tuning curve.

A rate representation could be constructed by recording data like these from a population of neurones with different BFs, and then computing the average discharge rate of each neurone during the stimulus segment of interest and plotting those rates versus the BFs of the neurones. For example, the average rate over 0.1–0.23 s could be computed to obtain rate responses to the vowel nucleus in 'five'. Rate representations have generally been computed with low time resolution, over time intervals of 100 ms or more, usually from responses to a stationary stimulus like a steady vowel. However, rate representations

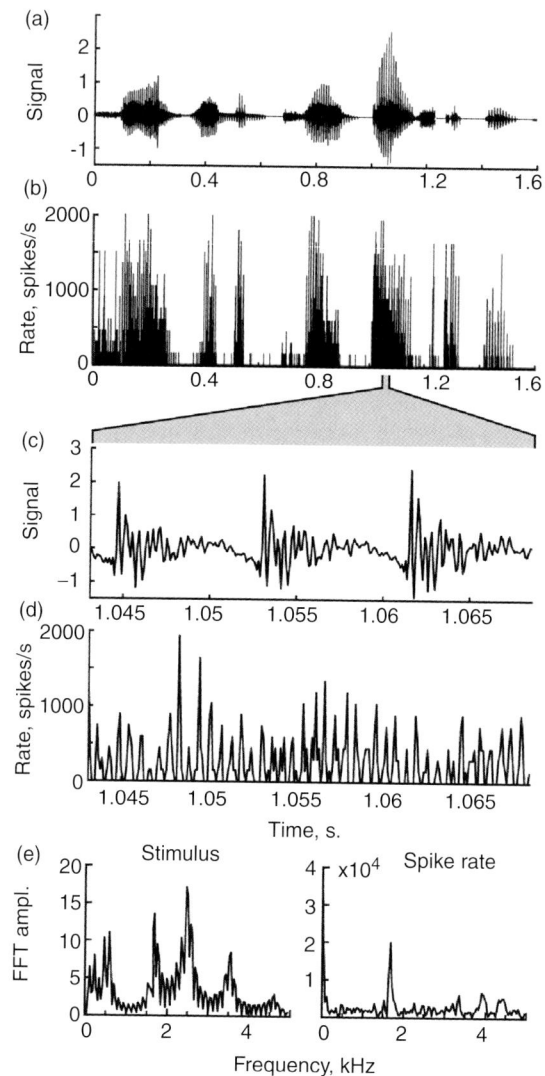

Figure 2.3 (a) The waveform of the sentence "Five women played basketball" is shown at low time resolution. (b) A PST histogram of the responses of an AN fibre (BF = 1.7 kHz) to the stimulus in (a). The PST histogram was constructed by counting the number of spikes that occurred during successive 0.1 ms bins synchronized with the stimulus, over 67 repeats of the stimulus. Although the PST histogram was computed at high resolution, it is displayed at low resolution, so only the overall changes in discharge rate can be seen. (c) and (d) These are the same plots as in (a) and (b), except at higher time resolution. The portions of the low-resolution plots between 1.042 and 1.068 s are shown, as indicated by the schematic between (b) and (c). (e) Magnitude of the discrete Fourier transform of the stimulus segment in c (left plot) and the PST histogram of the response in (d) (right plot). The major peaks in the stimulus plot correspond to F1 (near 0.5 kHz), F2 (near 1.7 kHz), F3 (near 2.5 kHz), and F4 (near 3.5 kHz); the response plot shows only a single peak near 1.7 kHz. Note that both plots have a linear ordinate (i.e. not dB).

can be computed at any time resolution from responses to any stimulus. Examples will be shown in Figure 2.6 for 200 ms and Figure 2.8 for 1 ms resolution.

In the rate representation discussed above, there is no way to know which components of the stimulus (e.g. F1 vs. F2) produced the response. It is possible to gain further information by examining them with higher time resolution, as in Figures 2.3c and 2.3d. The high-resolution rate plot in Figure 2.3d shows that the neurone's activity is not random, but is strongly periodic at a frequency of about 1700 Hz. Although it is not obvious in this figure, the 1700 Hz oscillation is time-locked to a similar oscillation in the stimulus of Figure 2.3c; this is an example of the phase-locking property of auditory neurones (Rose et al. 1967; Young & Sachs 1979; Johnson 1980b; Palmer & Russell 1986), in which spikes occur at a particular phase of the cycle of a periodic stimulus.

By analysing the frequency components that are present in the response and locked to the stimulus, it is possible to infer which components of the stimulus are effective in exciting the neurone. This analysis is shown in Figure 2.3e which shows the magnitude of the Fourier transform of the stimulus at left and the magnitude of the transform of the spike rate at right; these are the transforms of the signals in Figures 2.3c and 2.3d, respectively. Whereas the stimulus contains many significant frequency components, the response is mainly to frequencies near 1.7 kHz – the F2 frequency during this part of the stimulus. Of course, this reflects the tuning of the neurone. On the basis of this analysis, one concludes that this neurone is responding to the F2 energy in the stimulus and, therefore, 'represents' F2.

2.4 Tonotopic representation of vowels

The AN responses to vowels are usually dominated by the formant frequencies (Young & Sachs 1979; Delgutte 1984; Delgutte & Kiang 1984c; Sinex & Geisler 1984; Palmer et al. 1986). Figure 2.4 shows an example of responses to the vowel /ɛ/ as in 'met' at two sound levels (Schilling et al. 1998). The method in Figure 2.3 has been used to analyse the components of the response based on phase-locking. The data consist of the responses to the vowel in a population of several hundred AN fibres. The fibres have been sorted into bins according to BF, shown along the abscissae of the plots. The Fourier transforms of the responses were computed for each fibre and the magnitudes of the transforms were averaged for all the fibres falling into each bin. The box plots show the averages as a function of frequency, plotted along the ordinate. Thus, the plot shows the tonotopic array of neurones along the abscissa and the strength of the response to various frequency components of the vowel along the ordinate.

Notice the concentration of larger boxes at the frequencies of the formants along the ordinate. The largest responses are to F1 (0.5 kHz), with smaller responses to F2 (1.7 kHz) and F3 (2.5 kHz). These responses are centred (along the abscissa) on fibres with BFs near the frequency of the appropriate formant. There are also significant responses to the fundamental frequency of the vowel (100 Hz) among fibres with higher BFs. These neurones have several frequency components of the vowel within their tuning curves, and their BFs are high enough that phase-locking to those near-BF frequency components is weak; as a result, they respond significantly to the envelope of the sum of those components, at the fundamental frequency of the vowel. Responses to other frequency components of the vowel are smaller than those to the formants, except for a response to the second harmonic of F1 (1 kHz). This response has been discussed previously and is

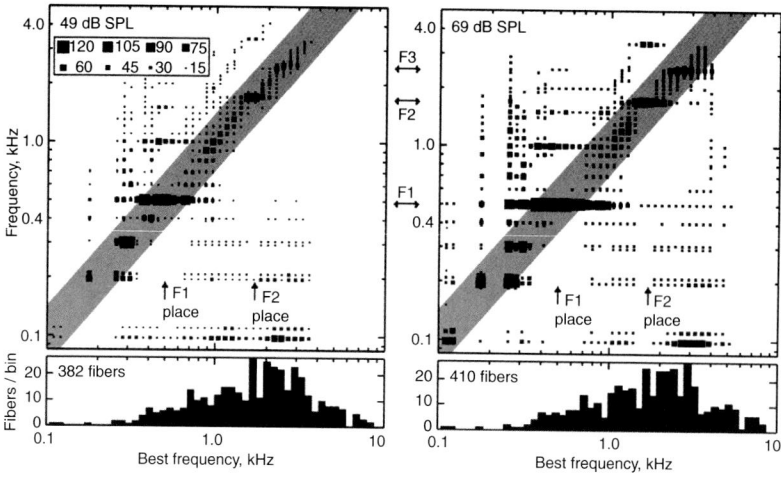

Figure 2.4 Three-dimensional plots of the analysis of phase-locking in a population of AN fibres responding to the vowel /ɛ/. Responses to two sound levels are shown, indicated in the legends. The abscissae show fibre BFs and the ordinates show the frequencies to which the fibres are phase-locked. The spectrum of the vowel is shown in Figure 2.6a (F2 = 1.7 kHz) and the formant frequencies are indicated by the horizontal double-headed arrows between the plots. The formant frequencies are also shown along the BF axis by the vertical arrows. The response strength is indicated by the size of the box, using the scale shown in the inset of the left plot. For example, rates greater than 120 spikes/s are shown by the largest box, rates between 105 and 120 spikes/s by the 2nd largest box, etc. The responses are magnitudes of discrete Fourier transforms of PST histograms, as in Figure 2.3e, averaged across all fibres with BFs within a frequency bin (0.133 octave). The grey stripe shows where the frequency of the response is within 0.5 octave of the BF. Thus, responses within the stripe are tonotopic. The number of fibres in each bin is shown by the histograms below the plots. For this analysis, fibres were combined across 10 normal animals with similar threshold audiograms. (Redrawn with permission from Schilling *et al.* (1998)).

caused by a rectification artefact in the analysis (Young & Sachs 1979); it should be considered a part of the F1 response.

As the sound level increases from 49 to 69 dB SPL, the responses to the formants increase in magnitude and spread along the BF axis (the abscissa) to occupy a larger fraction of the population. This spread is qualitatively consistent with the broadening of fibres' gain functions in Figure 2.2b.

The earlier statement that AN responses to speech are tonotopic is illustrated by the fact that the responses in Figure 2.4 are mostly within the grey stripes. These stripes show where response frequency is near BF (within 0.5 octave). Most importantly, the population of neurones responding to F1 is different than the population responding to F2 and F3. Two mechanisms are important in maintaining the tonotopic representation:

(1) basilar membrane tuning is essential to separate the components initially (Geisler 1989) and
(2) suppression sharpens the representation by restricting the spread of components along the BF axis.

The effect of suppression can be seen in the 69 dB SPL data in Figure 2.4 in that the F1 response amplitude decreases sharply at BFs close to F2. At high sound levels, the responses to F1 and F2 are reciprocal in that, at BFs where the F2 response is large, the F1 response is small. That the reciprocal behaviour is suppression follows from the observation that a tone at the level and frequency of F1 in the 69 dB SPL data, presented by itself, would produce strong responses phase-locked to the F1 frequency among fibres with BFs near F2 (Wong *et al.* 1998). The fact that those responses are not seen in Figure 2.4 implies that they are suppressed by F2 (for the same analysis using two tones see Kim *et al.* 1979). Further support for the importance of suppression comes from the fact that a linear model of AN fibres that does not include suppression does not display the kind of reciprocal responses observed in real fibres (Sinex & Geisler 1984), whereas a non-linear model with suppression does (Deng & Geisler 1987a).

The importance of the cochlear mechanisms that maintain the normal tonotopic representation can be seen by looking at responses in the AN when the cochlea has been damaged by acoustic trauma (Miller *et al.* 1997). For the example in Figure 2.5, acoustic trauma was produced by exposing the animals to an intense narrow band of noise centred at 2 kHz for several hours (Schilling *et al.* 1998). The exposure produced a high-frequency hearing loss, in which the thresholds and tuning curves were within 20 dB of normal for BFs up to about 1 kHz; at higher BFs, there was significant threshold elevation, between 40–60 dB, and tuning curves were abnormal by being broader than usual. In addition, the strength of suppression was decreased, as measured using two-tone interactions (Schmiedt *et al.* 1980; Salvi *et al.* 1982; Miller *et al.* 1997). When the cochleae of animals given similar exposures were examined anatomically, there was a partial loss of outer and inner hair cells and damage to the transduction apparatus in the remaining hair cells (Liberman & Beil 1979; Liberman & Dodds 1984).

Both the broadening of tuning and decrease in suppression reduce the separation of responses to F1 and F2, with the result that the representation of the vowel is degraded

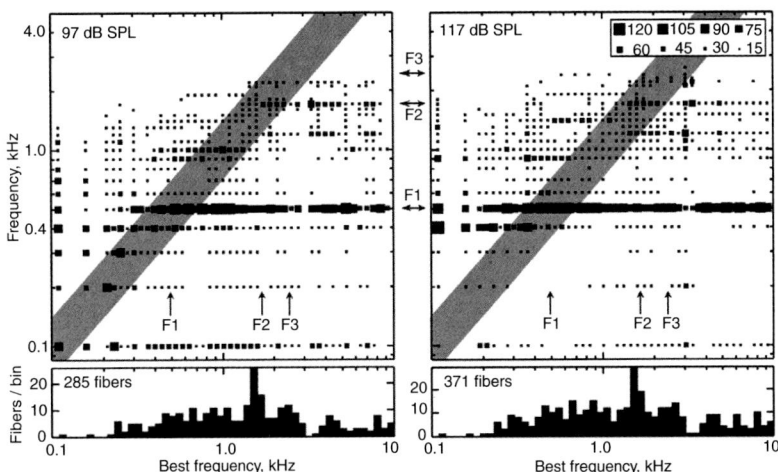

Figure 2.5 Box plots of population responses to the vowel /ɛ/ at two sound levels, as in Figure 2.4. The data are from four animals exposed to acoustic trauma, giving the threshold shift described in the text (see Figure 2 of Miller *et al.* (1997)). The sound levels shown are about the same number of dB above threshold as for Figure 2.4 for BFs near F2 (1.7 kHz). In comparison with Figure 2.4, the responses no longer show F2 and F3 and the formant responses are no longer tonotopic. (Redrawn from Miller *et al.* (1999a) with permission).

(Palmer 1990; Palmer & Moorjani 1993; Miller *et al.* 1997). Two changes are evident in the example in Figure 2.5, in comparison with Figure 2.4. First, responses to F1 dominate the responses of fibres at all BFs tested. Responses to F2 and F3 are hardly seen. This is expected from the degraded tuning, because the broadened tuning curves now do not adequately discriminate between F1 and F2 in neurones with BFs near F2; this is similar to the effect of degraded tuning at high sound levels in Figure 2.2b. Second, there is a wide distribution along the ordinate of phase-locking to frequencies other than the formants, regardless of BF. This broadband phase-locking is consistent with decreased suppression, in the sense that the fibres' responses are not captured by the strong responses to the formants, as they would be in a normal ear.

2.5 Rate representation of spectral shape

The results in Figure 2.4 show that fibres normally respond most strongly for BFs near the formants. This result leads to the simplest neural representation of spectral shape, the 'rate–place profile' (Sachs & Young 1979; Delgutte & Kiang 1984b) – a plot of discharge rate versus BF in the population of AN fibres. Such a profile should have a shape similar to the spectrum of the stimulus. However, there are substantial responses at BFs between the formants in Figure 2.4, and the spread of response along the BF axis as sound level increases suggests that the response should become more uniform across BF at high sound levels. Thus, it is not clear how well a rate–place profile will convey information about the stimulus spectrum. The data in Figures 2.6 and 2.7 will be used to explore this question.

Figure 2.6a shows the spectra of four variants of the vowel /ɛ/ with different F2 frequencies, given in the legend. The question posed in the previous paragraph will be answered here by considering how well the responses to these four stimuli can be differentiated. Rate profiles for a population of AN fibres responding to the F2 = 2 kHz variant are shown in Figure 2.6b (Conley & Keilson 1995); rates are shown for two sound levels (50 and 70 dB SPL), indicated by the different symbols and line weights. The data points show the responses of individual fibres and the lines show moving-window averages of the data points. Clear rate peaks are seen among fibres with BFs equal to F2 at both sound levels and at BFs near F1 at the higher level; presumably, a rate peak at F1 would have been observed at the lower level with sufficient data. No separate peak at F3 is evident. The population response can be said to represent the spectrum of the vowel in the sense that the first two formant frequencies can be estimated from the BF locations of the rate peaks.

An informal measure of the quality of the representation is how well the rate peak at F2 stands out, as measured by its height, the difference in rate between BFs near F2, and the rate minimum for BFs between the formants. The peak height is slightly greater at 50 than at 70 dB SPL in Figure 2.6b. This trend – towards poorer rate representation at higher sound levels – has been reported for several vowels (Sachs & Young 1979).

The F2 peak height depends on the F2 frequency, increasing as F2 moves away from F1. Figure 2.6c shows moving-window averages for the 70 dB SPL stimuli for all four variants of the vowel. For the lowest F2 frequency (1.4 kHz, heavy line), there is only a small rate peak; the size of the rate peak grows, both in absolute and relative terms, as the F2 frequency increases. The increase in the rate peak as the separation between the

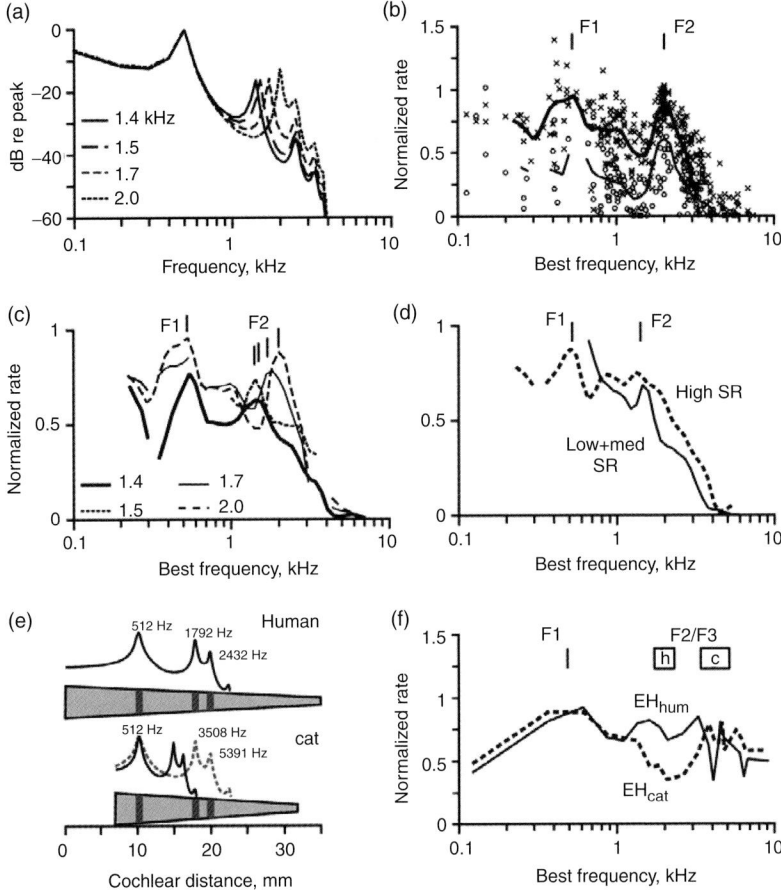

Figure 2.6 Rate profiles of responses to vowels similar to /ɛ/. (a) Spectra of four variants of /ɛ/ with the same F1 and F3, but different F2 frequencies, are given in the legend. For all variants, F1 = 0.5 kHz and F3 = 2.5 kHz. (b) Empirical rate–place representation of the F2 = 2 kHz variant at 50 dB SPL (open circles, light lines) and 70 dB SPL (X's, heavy lines), computed from 200-ms stimulus presentations repeated 50 times. Each point shows a fibre's rate plotted against its BF. Rate is normalized as (rate − SR)/(maxrate − SR), where maxrate is the rate in response to a BF tone 50 dB above threshold and SR is spontaneous rate. The lines are moving-window averages computed from the data points with a 0.15 decade log-triangular window. The formant frequencies are marked at the top of the plot. (c) Moving-window averages of normalized-rate profiles for the four variants of /ɛ/ at 70 dB SPL. The F2 frequencies are given in the legend. The data are sparse at BFs near F1. (d) Normalized-rate profiles for high SR fibres (SR ≥ 20 spikes/s) and low + medium SR fibres (SR < 20 spikes/s) in response to the F2 = 1.7 kHz variant at 70 dB SPL. (e) Schematic comparison of the position of the spectra of /ɛ/ on the human (top) and cat (bottom) basilar membranes. The vowel spectra are plotted with an abscissa that matches the layout of frequencies along the basilar membrane. The formant frequencies are indicated next to the spectra. For the cat, the spectrum is shown twice, once (black) with the usual formant frequencies and once (grey, dashed) with the spectrum spread out to make the physical distance between the formants the same on the cat and human basilar membrane. The schematics show the positions of the formants on the two membranes, for the normal /ɛ/ on the human membrane and the spread /ɛ/ on the cat membrane. (f) Normalized rate profiles for /ɛ/ with its standard formants (EH_{hum}) and with the formants spread as in part e (EH_{cat}). The positions of the formants are marked at the top of the plot; F2 and F3 frequencies are shown as the left and right sides of the small boxes labelled 'h' for human and 'c' for cat. ((a)–(d) are redrawn from Conley and Keilson (1995); (e) is from Keifte & Kluender (2001), and (f) is from Recio *et al.* (2002), all with permission).

frequencies of F1 and F2 increases is probably attributable to a reduction in suppression by F1 of the response to F2 as the frequencies move farther apart. The argument for this point is that the rates of neurones with BFs near F2 change with the sound level of the vowel in the same way as rate in response to a tone at BF = F2 in a two-tone suppression paradigm, with the second tone at F1 (Sachs & Young 1979; Figure 2.13).

Thus, suppression has two countervailing roles: suppression of responses to F1 at the F2 place is important in maintaining the tonotopic representation at high sound levels, but suppression of responses to F2 by F1 reduces the salience of rate peaks in response to F2, as F2 approaches F1. These data suggest that the rate representation of F2 will vary with the vowel and will be stronger in vowels with more separation between the formants.

The peak height also varies in different groups of AN fibres. Fibres vary in their thresholds and dynamic ranges – properties that are correlated with spontaneous discharge rate (SR; Liberman 1978) and the mode of innervation of hair cells (Liberman 1980). Dynamic range means the range of sound levels over which the fibre changes its rate when the input changes in level. Low- and medium-SR fibres generally have higher thresholds and wider dynamic ranges and provide a better rate representation at high sound levels (Sachs & Abbas 1974; Sachs & Young 1979; Winter *et al.* 1990; Yates *et al.* 1990). Figure 2.6d compares the rate representation of the F2 = 1.7 kHz variant of the vowel between fibres with SR < 20 spikes/s ('low + med SR') and fibres with SRs ≥ 20 spikes/s ('high SR'). A small peak is observed for the low- and medium-SR fibres, but not for the high-SR fibres, suggesting that the rate peaks in Figure 2.6c result mainly from activity in the low- and medium-SR populations.

2.6 The question of whether the cat is a good model for the human ear

An important question about the data shown here is whether the human cochlea and the cochleae of animal models differ in ways that are important to studies of speech coding. In simplest terms, one wonders to what extent data like those in Figures 2.4–2.6 need to be corrected in some way before they apply to the neural representation of speech in the human ear. This question has a number of facets, such as the use of anaesthesia in animal experiments, the effects of damage done to neural circuits by the surgery necessary for neurophysiological recording, and the possible differences between the auditory systems of a species that uses language and others that do not. In this section, we will focus on a more elementary question – whether there are differences in frequency representation between human and animal cochleae.

Although the general properties of cochlear physiology are similar across mammals, there are differences between species in the frequency range of hearing and in the layout of frequencies along the length of the cochlea (Fay 1988; Greenwood 1990). For example, human hearing extends from about 20 Hz to 15 kHz in a 35-mm cochlea and cat hearing extends from about 90 Hz to 60 kHz in a 25-mm cochlea. These differences raise the possibility that frequency resolution differs across species. The cochlear frequency maps of laboratory animals (the map of BF onto distance along the basilar membrane) locate any two frequencies closer together, in terms of distance on the basilar membrane, than they would be in the human ear. The extent of crowding for the vowel /ɛ/ is shown in Figure 2.6e, which shows a schematic of the spectrum of /ɛ/ (black solid lines) mapped onto the basilar membrane of the human (above) and the cat cochlea (below; Kiefte & Kluender 2001). The spacing between F1 and F2 is smaller by a factor of about 0.6 on the cat, versus the human, basilar membrane. The difference is larger in other animals commonly used for auditory work.

In order to get an idea of the potential effects of the difference in cochlear maps, Recio *et al.* (2002) assumed that interactions between different frequencies occur over a constant physical distance along the basilar membrane in different species. There is no direct evidence for this assumption, but it is consistent with the fact that auditory filter bandwidths correspond to roughly a constant distance along the basilar membrane within a species (Greenwood 1961, 1990; Moore, this volume). Recio and colleagues synthesized a version of /ɛ/ with modified formant frequencies, so that the formants were the same physical distance apart in the cat cochlea as they are in the human cochlea. The spectrum of the modified vowel is shown by the grey dashed line in Figure 2.6e. Figure 2.6f compares rate profiles from cat AN fibres for two vowels: EH_{hum} is the standard /ɛ/ and EH_{cat} is the modified version. The rate profile obtained with EH_{hum} (solid line) is similar to the profiles in Figures 2.6c and 2.6d, with a small peak at F2 and no sign of a separate peak for F3. The rate profile for EH_{cat} (dashed line) shows clearly defined rate peaks for F2 and F3, and a substantial improvement in the F2 peak height. The increase in peak height is due mainly to a decrease in normalized rate at the rate minimum between F1 and F2 (near 2 kHz). This change is probably due to the narrower relative filter widths (relative to the stimulus spectrum) in EH_{cat} because it is not consistent with suppression effects, which should grow weaker with more widely spaced frequency components.

The demonstration in Figure 2.6f shows that it is important to know the relative bandwidths of tuning curves or of the perceptual auditory filters in the human cochlea versus the cochlea of an animal model. Unfortunately, only psychophysical or indirect physiological measures, such as the compound-action-potential tuning curve (Eggermont 1977), can be obtained in human subjects, which raises a problem of interpretation. The width of an AN tuning curve is a straightforward measure, but the results for other measures of tuning filter widths depend on the methods used to obtain the data (reviewed by Moore 2003). Presumably, these technical problems explain why some studies have obtained psychophysical or indirect physiological filters that match the AN tuning curves (e.g. Evans 1992), whereas others have not (e.g. Pickles 1979; Harrison *et al.* 1981). By the same argument, while most studies have found behavioural filters to be wider in animals than in human subjects, the comparisons are not based on directly comparable methods and it is not clear how that result should be interpreted in terms of the relative widths of AN tuning curves (Ruggero & Temchin 2005).

Recently, Shera *et al.* (2002) published a direct comparison of human, cat, and guinea pig basilar membrane tuning, suggesting that tuning is at least twice as sharp in the human as in the cat or guinea pig cochlea. They inferred tuning width from the group delay of otoacoustic emissions – a measurement that was done in both human and animal subjects. Although this technique may not provide accurate results at the low frequencies that are most important for speech (Shera and Guinan 2003; Siegel *et al.* 2005), the conclusion that tuning is narrower in the human than animal ears was supported by comparable psychophysical measures of tuning in the human cochlea (Oxenham & Shera 2003) and by preliminary direct measures of tuning curves in Old World monkeys (Joris *et al.* 2006). The tuning in the monkey was between that for humans and non-primate laboratory animals, consistent with otoacoustic emission measures. Thus, the current evidence seems to favour the existence of sharper tuning in the human and primate cochlea relative to common laboratory animals. As a result, the differences between the two plots in Figure 2.6f demonstrate how the neural responses to speech measured in laboratory animals may underestimate the quality of the representation in the human auditory system.

2.7 Inferences from discrimination data

The data in Figure 2.6 suggest that the variants of the vowel /ɛ/ whose spectra are shown in Figure 2.6a would be easily discriminated on the basis of AN discharge rates (and, in fact, they are easily discriminated in informal listening tests). However, the scatter in the discharge rates (Figure 2.6b) and the small size of the rate differences as F2 approaches F1 (Figure 2.6c) suggest that it would be profitable to quantify this impression. One approach is to measure the change in discharge rate as F2 changes and to compute from the rate changes the detectability of changes in F2 frequency for various stimuli (Conley & Keilson 1995; May et al. 1996; Miller et al. 1999b). Figure 2.7b shows rate differences between responses to variants of /ɛ/ with F2 = 1.4 and 2 kHz, plotted versus BF. The points are data from individual fibres and the lines are moving-window averages of the points.

The largest rate changes are in fibres with BFs equal to the two F2 frequencies (the vertical dashed lines). The rate changes are larger in fibres with low and medium SRs

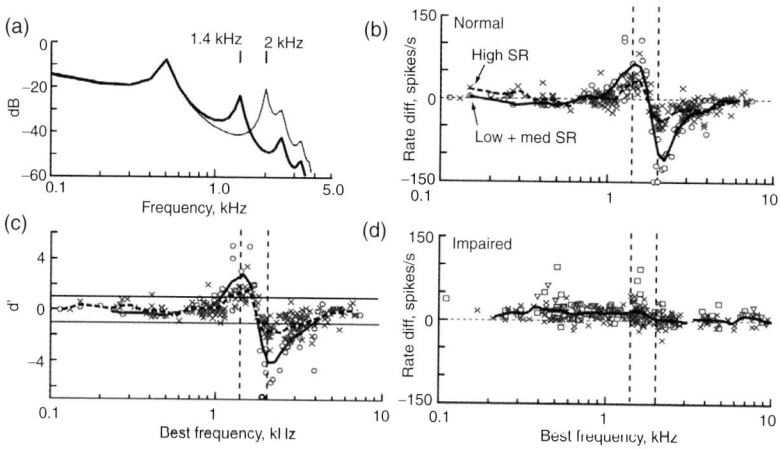

Figure 2.7 (a) The spectra of two variants of the vowel /ɛ/ with F2 frequencies of 1.4 and 2 kHz (the same as in Figure 6a). (b) Rate difference between responses to the two variants presented at 70 dB SPL (rate to the F2 = 1.4 kHz variant minus rate to the F2 = 2 kHz variant), plotted versus BF. SRs are identified by symbol: open circles and solid lines are for SR < 20 spikes/s, X's and dashed lines are for SR≥20 spikes/s. Lines are moving-window averages as in Figure 2.6. Vertical dashed lines show the two F2 frequencies. (c) The data of b replotted as d' versus BF (see text). The horizontal lines show $d' = 1$. (d) Rate differences for the same stimuli from a population of AN fibres studied in cats with a high-frequency threshold shift due to acoustic trauma, the same population as in Figure 2.5. The plot is the same as (b) except that triangles are for SR < 1 spikes/s and squares are for SR between 1 and 20 spikes/s. The moving-window average is for all SRs. The stimulus level was 97 dB SPL.
((a) and (d) redrawn with permission from Miller et al. (1999b); (b) and (c) redrawn from Conley and Keilson (1995)).

(open symbols), consistent with Figure 2.6d. The fact that rate changes are largest in fibres with BFs equal to the formant frequencies is expected from the tonotopic representation of the vowel (Figure 2.4) and provides direct evidence for the statement that these neurones 'represent' F2.

The rate changes provide a reliable representation of F2 frequency to the extent that the rate changes themselves are reliably detectable. Because AN fibres give randomly varying rates, this is a statistical problem; detectability can be measured as the rate change divided by the standard deviation of the rates (Green & Swets 1966), called d'; Conley and Keilson (1995) computed d' as $(\mu_1-\mu_2)/(\sigma_1^2+\sigma_2^2)^{1/2}$, where μj and σj are the mean and standard deviation of the rate in response to vowel j. The d' results in Figure 2.7c show that rate changes in the best fibres (those with BFs equal to the F2 frequencies) easily achieve a d' value of 1, usually taken as the detectability at the discrimination threshold (or 'jnd' for just-noticeable difference).

The data in Figures 2.6b and 2.6c are for the largest F2 difference in the data set, between vowels with F2 frequencies of 1.4 and 2 kHz. Using data from the other vowels studied, Conley and Keilson (1995) showed that the threshold for discriminating the vowels, where $d' = 1$ at the peak of the moving-window average of the rate difference, occurred for a change in F2 frequency between 125 and 240 Hz, depending on SR. Vowel-formant discrimination thresholds for F2 are usually 50–100 Hz in human observers depending on vowel and formant frequency (e.g. Liu & Kewley-Port 2004). The value computed by Conley and Keilson is based on information from a single average AN fibre. If information were combined across different fibres, the jnd would be considerably smaller (about 1 Hz). Thus, the psychophysical performance on vowel formant discrimination could easily be achieved from rate responses of AN fibres.

To emphasize the point made in Figure 2.5, that the representation of F2 is lost following acoustic trauma, Figure 2.7d shows rate differences for the population of impaired fibres from Figure 2.5. There is minimal rate change related to F2 in this population. The reason, of course, is that the fibres are mostly responding to F1, which does not change between these two stimuli.

May et al. (1996) obtained a similar result for discrimination of the F2 frequency using a different method in which a model was fitted to the rate responses to vowels, and d' was computed from the model. The model is based on data like those that are discussed later in Figure 2.10. They also measured the behavioural jnd for F2 frequency in cats (Hienz et al. 1996; May et al. 1996), and showed that the jnd predicted for one optimally chosen AN fibre is very close to the behavioural jnd. 'Optimally chosen' means a fibre with a BF at the peak of the rate difference plot. At low sound levels and in the quiet, high-SR fibres provide the best information (such stimuli may be below threshold for low- and medium-SR fibres) whereas at high levels and in background noise, low- and medium-SR fibres provide the best information, because high-SR fibres' rate responses saturate.

2.8 Responses to consonant–vowel (CV) syllables

The stimuli considered so far are not typical of real-world speech in that they are isolated vowels, with spectra that are constant in time. The study of more realistic stimuli began almost as early as work on vowels, with studies that focussed on stop-consonant–vowel stimuli (Miller & Sachs 1983; Sinex & Geisler 1983; Carney & Geisler 1986). Stops at the beginning of syllables are characterized by a release burst of broadband noisy sound, followed after some delay by the onset of voicing; the voicing produces a vowel-like formant structure in which the formants move rapidly (over a few tens of ms) from frequencies determined by the consonant to frequencies appropriate for the following vowel. The /p/ in Figure 2.1 is an example.

Studies of the neural representation of stops indicated that the principles described above for vowels could be extended to consonants using both rate- and phase-locked representations that varied in time. In these analyses, the spectrum of the stimulus was considered to be a sequence of slightly different spectra in successive time windows. The rate- or phase-locked representation was computed separately in each time window, with results similar to those in Figures 2.4 and 2.6. In particular, the responses were dominated by the formants, either through peaks of discharge rate at BFs that tracked the formants or through phase-locking that was dominated by the formant frequencies. Generally, the rate representation was found to be better for stop consonants than for vowels, probably because the consonants have lower sound levels. Responses to other consonants, including nasals and fricatives, have also been analysed, with similar results (Delgutte & Kiang 1984b; Deng & Geisler 1987b).

The approach of using discriminability as a measure of the quality of a neural representation can provide additional insight into the representation of consonant–vowel syllables. Figure 2.8a shows differences between the rates of AN fibres in response to the synthetic utterances /bab ba/ and /dad da/ (Bandyopadhyay & Young 2004). The data are from a population of fibres, arranged along the ordinate according to BF. The abscissa shows time during the stimulus, and the components of the stimulus are marked above the plot. The first three formant frequencies of the two stimuli are shown by the black lines. The formants are identical during the vowels and diverge during the consonants. /d/ has higher F2 and F3 frequencies than /b/ during the transitions and the F1 frequencies are the same. The colour scale shows the rate differences, with warm colours indicating higher rates to /b/. The neural representation shown here is relatively simple (see also Miller & Sachs (1983); Sinex & Geisler (1983)). During the vowels and silences, the rates do not differ. During the formant transitions (near 0.05, 0.18, and 0.3 s), rate differences are positive (warm colours) at BFs along the /b/ transitions in F2 and negative (cold colours) at BFs along the /d/ transitions in F2, as expected from the spectra. Rate differences are negative at all BFs near F3 because the /d/ has higher overall energy in this frequency range. Particularly strong differences are seen in response to the release bursts of the consonants, which occur just after the beginning of the formant transitions at the syllable onsets and last about 0.01 s. The time delay is the latency of the neural response. The burst is more high-pass in the /d/ than the /b/ (Blumstein & Stevens 1979) so the rate differences tend to be negative at high BFs and positive at low BFs. The rate differences at the highest BFs (above 3 kHz) are mostly due to the burst.

The discriminability of these CV syllables was quantified by calculating the dissimilarity of the spike trains produced by the two stimuli; the dissimilarity measure is a more general form of the d' measure used in Figure 2.7. If the data have Gaussian distributions, the two measures are proportional (dissimilarity = $d'^2/2.\ln 2$). Thus, the interpretation of the dissimilarity measure is the same as for d' (Johnson et al., 2001); the reasons for using the dissimilarity measure and the methods for computing it are beyond the scope of this chapter and are explained by Johnson et al. (2001). The dissimilarity was large during the formant transitions and zero during the vowels and silences.

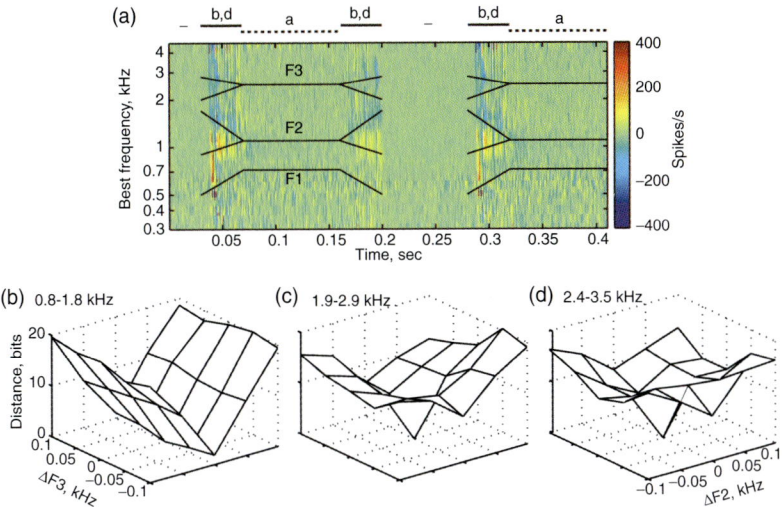

Figure 2.8 (a) Rate differences between responses to /bab ba/ and /dad da/ in a population of 137 AN fibres. Fibre BF is plotted on the ordinate and the abscissa is time during the stimulus. The segments of the stimulus are marked at top by the solid (consonant) and dashed (vowel) lines; hyphens mark the silences. The first three formants of both stimuli are shown as black lines. Fibres were gathered in bins by BF (overlapping 0.25 oct bins spaced at 0.0625 oct) and plots of average rate versus time (in 1 ms bins) were constructed for the fibres in each BF bin. The differences in these rates between the two stimuli are plotted on the colour scale identified at right. Positive rate differences mean a higher rate to /bab ba/. The sound level was 70 dB SPL during the vowels. (b) Information measure of the dissimilarity of the spike trains of a sub-population of model neurones with BFs in the F2 range (0.8–1.8 kHz). Each point is the dissimilarity between the responses to one of 25 stimuli and the centre stimulus (the distance is zero for the centre stimulus). Dissimilarity is plotted against the frequency differences of F2 (abscissa) and F3 (ordinate) in the stimulus relative to the central reference, measured at 0.06 s. These are labelled at the far left and right only, but are the same in b, c, and d. Only spikes during the interval 0.05–0.07 s were used in the calculation. (c), (d) Same as (b) for fibres in higher BF ranges, containing F3 (c, 1.9–2.9 kHz) and above F3 (d, 2.4–3.5 kHz). Contrast the valley shape in (b) with the bowl shapes in (c) and (d).
(Redrawn with permission from Bandyopadhyay & Young (2004)).

The phase-locking analysis (Figure 2.4) suggests that fibres should encode information about formants nearest their BFs. The dissimilarity measure can be used to test this idea. For the analysis, a stimulus set was synthesized that had F2 and F3 frequency transitions that varied independently over the full range from the transitions of /b/ to the transitions of /d/. Five different values of the frequencies at which the F2 and F3 transitions begin were used, and 25 different stimuli with all possible combinations of F2 and F3 transitions were synthesized. The natural /b/ and /d/ lie at two opposite corners of this stimulus set. Responses to the stimulus set were computed using an AN model that has been tuned to produce accurate responses to speech (Bruce et al. 2003). Model data were used because it is difficult to obtain the needed amount of data from AN fibres. Differences between the responses to the 24 outlying stimuli and the central stimulus (i.e. the one with the median F2 and F3 frequencies) were computed using the dissimilarity method discussed above, for three BF ranges.

Figure 2.8b shows the result for model fibres with BFs in the range occupied by the F2 transitions. The plot shows dissimilarity between spike trains as a function of the starting frequency of the F2 (abscissa) and F3 (ordinate) transitions. Distance grows as F2 changes, but not as F3 changes, giving the V-shaped valley shown in the figure. Thus fibres with BFs near F2 code for F2 only and provide little information about F3. In contrast, fibres with higher BFs give bowl-shaped dissimilarity functions (Figure 2.8c and 8d) showing that these fibres convey information about both F2 and F3. This result seems to be inconsistent with the phase-locking analysis of vowels (Figure 2.4), which suggests that fibres with BFs near F3 should respond mainly to F3. It probably reflects the fact that the F2 frequency affects the responses among F3 neurones both through suppression of F3 by F2 and because the stimulus level at F3 changes with the F2 frequency because of the finite bandwidth of the F2 resonance.

2.9 Further comments on the representation in the AN

The analyses in Figures 2.6 and 2.7 suggest that neural codes based on discharge rate are sufficient to represent the first two formants of speech. However, this work considered the relatively unchallenging situation of speech presented in quiet. More difficult situations, such as high sound levels, noisy backgrounds, or when more than one person is talking, may require more information than is encoded by rate (Sachs *et al.* 1983; Geisler & Gamble 1989; Palmer 1990; Silkes & Geisler 1991; Keilson *et al.* 1997). Indeed, measures of the discriminability of vowel F2 frequency based on rate show substantial increases in the predicted 'jnd' a signal-to-noise ratio of 3 dB, because the rate change produced by a formant frequency change decreases in the presence of background noise (May *et al.* 1996; May *et al.* 1998).

In all of these studies, information about the vowel remained in the phase-locked responses of fibres, even at signal-to-noise ratios where there was little or no information in rate. Alternative mechanisms for encoding information about speech and other sounds, have been suggested that take advantage of the information in phase-locking (Shamma 1985; Deng & Geisler 1987a; Carney 1990; Carney *et al.* 2002; Colburn *et al.* 2003). The problem posed by phase-locking is finding a plausible neural mechanism to extract information encoded in this way. One example is a model that detects coincidences between spike trains in AN fibres of different BFs; such a model can produce sharpening of the neural representation over those shown in Figure 2.6 and might also improve performance in noise. Coincidence mechanisms are plausible as a neural mechanism for stimulus representation because they are used in the medial superior olive to extract information about inter-aural time differences (Goldberg & Brown 1969; Yin & Chan 1990).

The assumption implicit in the work described above, that the neural representation of dynamic speech is a rapid temporal sequence of independent responses to different spectra, is not correct. In fact, there are substantial interactions between the responses to successive segments of speech. Delgutte and Kiang (Delgutte 1980; Delgutte & Kiang 1984a) evaluated the effects of a speech sound on the AN response to the following sound. For example, stop consonants like /ba/ have a sudden onset of sound (i.e. increase in sound intensity) where voicing begins; these sound onsets produce brief bursts of high-rate spikes in AN fibres at the time of the onset. If the same physical sound is preceded by a vocal 'murmur' typical of a nasal consonant, then the overall result sounds like /ma/;

in the presence of the murmur, the burst of spikes is significantly reduced or even disappears, even though the vocal portion of the stimulus is unchanged. The burst disappears because of short-term adaptation of the AN fibres by their responses to the /m/ (Smith 1977, 1979). Similar effects were observed when the segment /da/ was preceded by brief sounds that changed the perception of the /da/ to another speech sound (/ada/, /na/, /sha/, or /sa/). Again, the major effect was the loss of the burst of spikes produced by the onset of the vowel. Such an onset burst can also signal the rise-time of the stimulus, as for the difference between 'shoo' (slow rise) and 'chew' (rapid rise). Note that these response bursts are vulnerable to noise masking (Delgutte 1980), so may not provide a robust cue in real-life situations.

2.10 Changes in the neural representation of spectral shape in the cochlear nucleus

The fibres of the AN terminate in the cochlear nucleus, the first auditory centre in the brain. Between the cochlear nucleus and the thalamus, there is a series of complex neural structures making up the brainstem and midbrain auditory circuits (reviewed by Rouiller (1997)). The transformations of the neural representation of sound in these systems vary and can be substantial. In the next few sections, some results on the representation of speech sounds in central auditory nuclei are discussed, focussing on the differences between the representation in the AN and in the central structure and on the potential importance of those differences. Work on the central neural representation of speech is fragmentary at present, so the results discussed below are examples and are far from a comprehensive view of the subject.

The cochlear nucleus contains between five and 10 independent neural subsystems operating in parallel (Rhode & Greenberg 1992; Romand & Avan 1997; Young & Oertel 2003). Each of these receives synaptic inputs from AN fibres of all BFs; in principle, each subsystem constitutes a full neural representation of the sound at the ear. These subsystems differ in terms of synaptic organization, the processing properties of the neurones, and their connections to the rest of the auditory system. Responses to speech have been studied in only a few of these subsystems, mainly the so-called primary-like and chopper neurones. Primary-like neurones relay information important for sound localization to the superior olivary nuclei; for this purpose, it is important to preserve information encoded at high time resolution in the AN (Yin 2002), so these neurones often give responses that are little changed from those of AN fibres. Their responses to speech are generally similar to those described above for AN fibres (Blackburn & Sachs 1990; Winter & Palmer 1990; May *et al.* 1998; Recio & Rhode 2000).

Chopper neurones provide a representation of the spectrum of speech sounds that is improved over that in the AN by being more robust. There are two aspects of the robustness: first, the responses of chopper neurones are less affected by sound level and background noise than are the responses of AN fibres (Blackburn & Sachs 1990; May *et al.* 1998; Recio & Rhode 2000); second, the chopper neurones have a higher sensitivity, or 'gain', for responses to spectral features like the formants (May *et al.* 1996; May *et al.* 1998).

Figure 2.9 shows a comparison of rate profiles in response to the vowel /ɛ/ (Figure 2.9a) from a population of chopper neurones in the cochlear nucleus (heavy solid lines; Blackburn & Sachs 1990) and from AN fibres (dashed and dotted lines; Young & Sachs, 1981).

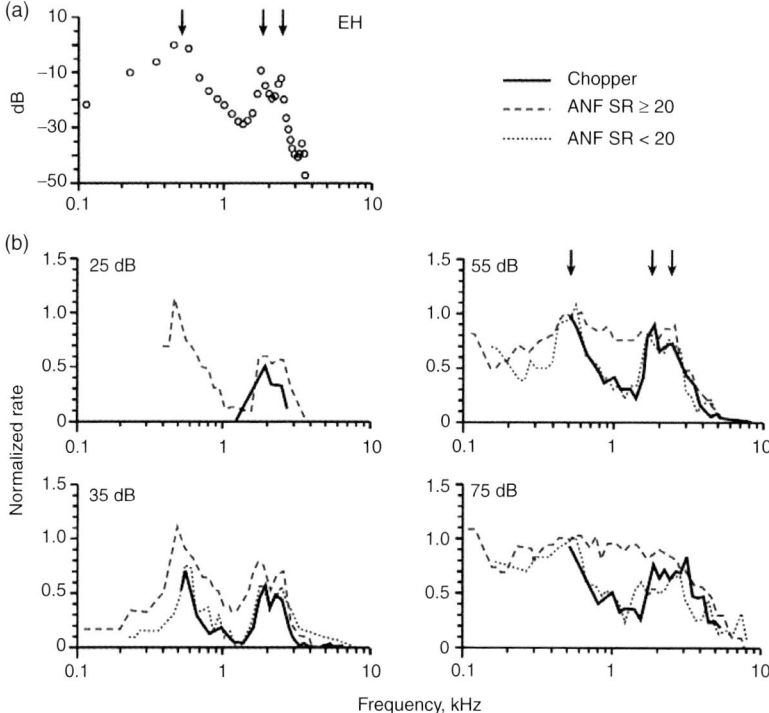

Figure 2.9 Rate profiles for responses of chopper neurones in the cochlear nucleus to the vowel /ɛ/. The data are actually from the subpopulation of chop-T neurones, but are typical of chopper responses in general. A definition of choppers and a description of their characteristics is provided by Blackburn and Sachs (1989). (a) Spectrum of the stimulus, a steady periodic synthetic approximation to /ɛ/ with formants at 0.512, 1.792, and 2.432 kHz (arrows) and fundamental frequency 0.112 kHz. (b) Rate profiles for a population of chop-T neurones and comparison populations of AN fibres, identified in the legend at upper right. Rate is plotted as normalized rate, as in Figure 2.6 and only the log-triangular moving-window averages are shown. AN fibres are divided into low + medium SR fibres (SR ≤ 20 spikes/s) and high SR fibres (SR > 20 spikes/s). Profiles are shown at four sound levels, as marked on the plots in dB SPL. Profiles are not plotted at BFs where few neurones were sampled. Low + medium SR AN fibres mostly did not respond at 25 dB SPL, so their responses are not plotted.
(AN data from Young & Sachs (1981); CN data from Blackburn & Sachs (1990), redrawn with permission).

The profiles are shown at four sound levels, indicated on the plots. As sound level increases, the profiles of the AN fibres change in that rates increase at all BFs; for the high-SR neurones (SR ≥ 20 spikes/s), the rates saturate at the fibres' maximum discharge rate (a normalized rate near 1) at 55–75 dB. In low- and medium-SR neurones (SR < 20 spikes/s), the rates increase but do not saturate at the sound levels shown. The chopper neurones' rates also do not saturate and they retain a rate representation that is as good as that of the low-SR AN fibres. This result shows that chopper neurones have a mechanism for regulating their dynamic range, for keeping their responses within the dynamic portion of their input/output characteristic.

Except at the lowest sound level (25 dB SPL), which is below threshold for most of the low- and medium-SR fibres, the results could be explained if choppers receive synaptic inputs only from low- and medium-SR fibres. However, chopper neurones have low thresholds like high-SR (HSR) AN fibres and respond to the vowel at low sound levels where low- and medium-SR fibres do not. Moreover, individual chopper neurones are known to receive synaptic inputs from all SR groups on the basis of cross-correlation evidence (Young and Sachs, 2008). These characteristics suggest that choppers are able to respond to either high- or low + medium-SR (LMSR) fibre populations and that they switch their responses from the former to the latter as sound level increases – called the selective listening hypothesis (Lai *et al.* 1994).

The changes in the representation in the cochlear nucleus are shown in another way in Figure 2.10 (May *et al.* 1996, 1998), where the goal is to analyse the dynamic range of the neurones' responses to the vowel. For this approach, the spectrum of the vowel /ɛ/ was shifted, by changing the sampling rate of the D/A converter used to present the stimulus, to align different features of the vowel with the BF of the neurone under study. Changing the sampling rate shifts the stimulus spectrum along a logarithmic frequency axis without changing its shape. The features were the first three formants (F1, F2, and F3) and four troughs or minima in the spectrum (T0, T1, T3, and T3), defined in Figure 2.10a. Examples of the resulting spectra for a neurone with BF 2.1 kHz are shown in the inset of Figure 2.10a, for F1, T1, and F2 aligned with BF. The discharge rates for a sample AN fibre are plotted in Figure 2.10a by the solid lines and open circles; this plot shows discharge rate versus the frequency of the feature that is shifted to BF, aligned with the spectrum at the nominal sampling rate (i.e. the one used in previous figures).

Figure 2.10b shows the average driven discharge rates (rate minus spontaneous rate) of two populations of AN fibres (separated by SR) plotted against the sound level of the feature that is aligned with BF, i.e. of the formant or trough harmonic. In each case, data are shown for different values of the overall sound level of the stimulus, plotted with different symbols. The data points taken at one overall sound level were fitted by a straight line, whose slope is a measure of the 'gain' of the neurone, in spikes/(s.dB), at the corresponding sound level. The gain is a measure of the expected quality of a rate profile constructed with these neurones; the larger the gain, the larger the peak height at the F2 frequency.

For the HSR AN fibres, the rates resemble a rate-versus-level function for a BF tone. In particular, the slope of the rate function (or of the lines fitting the points) decreases at high sound levels as the fibre's rate response saturates. In contrast, there is little saturation at levels between 43 dB SPL (filled squares) and 93 dB SPL (down-pointing triangles) for the LMSR fibres. Moreover, there is substantial dynamic range adjustment as the overall level of the stimulus changes, seen by comparing the rates in response to vowels at two different overall sound levels at a fixed feature level. For example, for the LMSR fibres, the rate in response to a feature level of, say, 60 dB SPL is considerably smaller for the 93 dB SPL overall stimulus level (where it is a response with T1 at BF) than for the 63 dB SPL stimulus level (with F1 at BF). This behaviour is also observed for the HSR AN fibres, but it is a smaller effect.

Figure 2.10c shows the same analysis for four populations of cochlear nucleus neurones. The primary-like neurones (top row) show behaviour similar to the AN fibres in Figure 2.10b. The chopper neurones (bottom row) behave like the LMSR fibres; again, the slopes of the lines do not decrease much as the overall stimulus level increases and the dynamic range adjustment, as defined above, is clear.

30   Eric D. Young

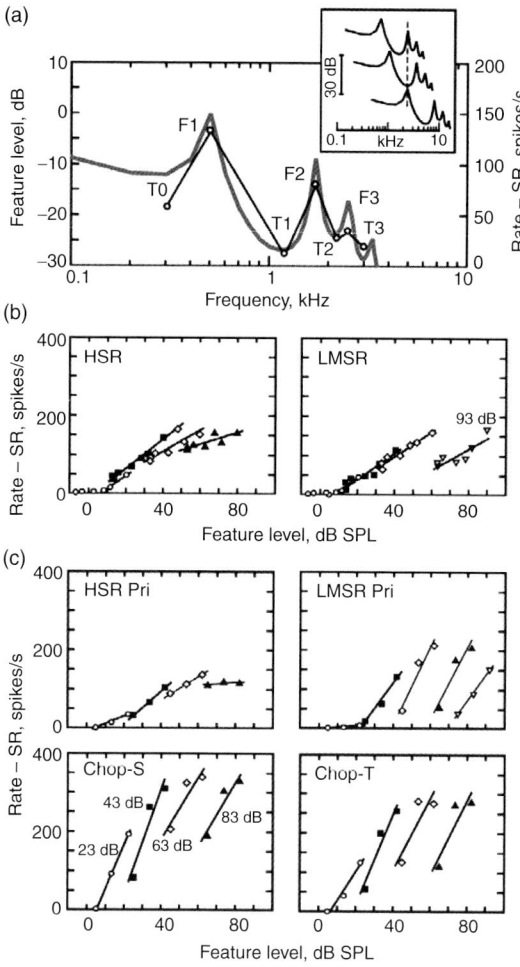

Figure 2.10 (a) Spectrum of the synthetic /ɛ/ (grey solid line, left ordinate) and driven discharge rates (black lines with open symbols, right ordinate) of an AN fibre when the vowel spectrum was shifted along the log-frequency axis by changing the sampling rate so as to align various spectral features with BF. The vertical alignment of the left and right ordinates is arbitrary. The inset shows the spectra of three such shifted vowels that align F2, T1, and F1 (in order top to bottom) with the 2.1 kHz BF of the fibre, at the dashed line. (b) Driven discharge rate (rate–SR) plotted versus the sound level of the stimulus feature aligned on BF; data are averages for high (HSR, left) and low + medium (LMSR, right) SR populations of AN fibres. (c) The same plots for four groups of cochlear nucleus neurones, primary-like (top row) and chopper (bottom row). These neurone types are defined elsewhere (Blackburn & Sachs, 1989) but the differences are not important for this discussion. In both (b) and (c), the shifted vowels were presented at three or four overall sound levels, shown with different symbols. The sound levels are defined in the lower-left plot, as dB SPL, except for the 93 dB SPL symbol at right in (b). The feature level on the abscissa is the sum of the overall sound level of the vowel and the relative level of the feature and is the SPL of the stimulus harmonic that is located at BF. For the AN data, the vowel was shifted to seven positions relative to BF, as in (a); for the cochlear nucleus neurones, only three positions were used (F1, T1, and F2). (redrawn from May et al. (1997)).

Dynamic range adjustment means that the discharge rate with a particular feature at BF depends on whether the feature is located at a peak of the stimulus spectrum or a minimum. The lowest rates are for troughs aligned with BF and the highest rates are for spectral peaks aligned with BF. Presumably, this dynamic range adjustment is a consequence of suppression or inhibition of the response to the weaker components of the stimulus by the stronger components.

There is an important systematic difference between the chopper and LMSR primary-like neurones versus the AN fibres in that the slopes of the lines are significantly higher for the cochlear nucleus neurones, by up to a factor of 2 (May et al. 1998). The increased gain combines with the stability of dynamic range adjustment in the chopper and LMSR primary-like neurones to provide an improved spectral representation. As expected from these data, the jnd for F2 frequency calculated for the chopper neurones is less than or equal to that for the AN and is stable across sound levels (May et al. 1996). The difference extends to the representation in the presence of background noise. The slopes of lines like those in Figure 2.10b decrease in the presence of background noise, but more so for AN fibres than for chopper neurones; the slope decrease between a vowel in quiet and a vowel in background noise at a signal-to-noise ratio of 3 dB is about twice as large in AN fibres as in chopper neurones.

2.11 The temporal envelope of speech in the AN and inferior colliculus

The speech signal is strongly modulated in time. Examples are shown in Figures 2.1 and 2.11A (Delgutte et al. 1998). The latter shows the spectrogram (top) and an oscillogram (bottom) of the sentence 'Wood is best for making toys and blocks'. Associated with the syllables of this sentence is a strong modulation of its envelope, seen as changes in the amplitude of the oscillogram at frequencies of a few Hz. Generally, the vowels have the most energy, shown by the largest amplitudes in the oscillogram and the corresponding dark parts of the spectrogram. Silences associated with the stop consonants are seen as near-zero amplitude parts of the oscillogram and light parts of the spectrogram.

The emphasis of this chapter has been on the neural representation of the frequency content of speech; however, important information is encoded in the temporal envelope as well (Van Tasell et al. 1987; Rosen 1992; Shannon et al. 1995). Examples have been mentioned earlier, such as the stimulus rise-time differences between the consonants 'ch' and 'sh' and the representation of rise-time in terms of transient or onset bursts of spikes in AN fibres (Delgutte & Kiang, 1984a).

Figure 2.11b–d shows data from Delgutte et al. (1998) on the neural representation of the temporal envelope of a sentence, without reference to a particular speech sound. These histograms show, respectively, the average responses of populations of AN fibres, neurones in the cochlear nucleus, and neurones in the inferior colliculus to the sentence in Figure 2.11a. The neurones were grouped by BF into 0.5 octave bins, and the PST histograms of the neurones falling into each bin were averaged. The histograms shown are the averages, identified at left by the centre frequency of the analysis bin.

The AN responses (Figure 2.11b) correspond generally to the envelope of the stimulus, in that rates are high during segments of the stimulus with significant energy. Of course, the rates are modified by the frequency spectrum of the different parts of the stimulus. The analysis was not designed to show the representation of the stimulus spectrum, but

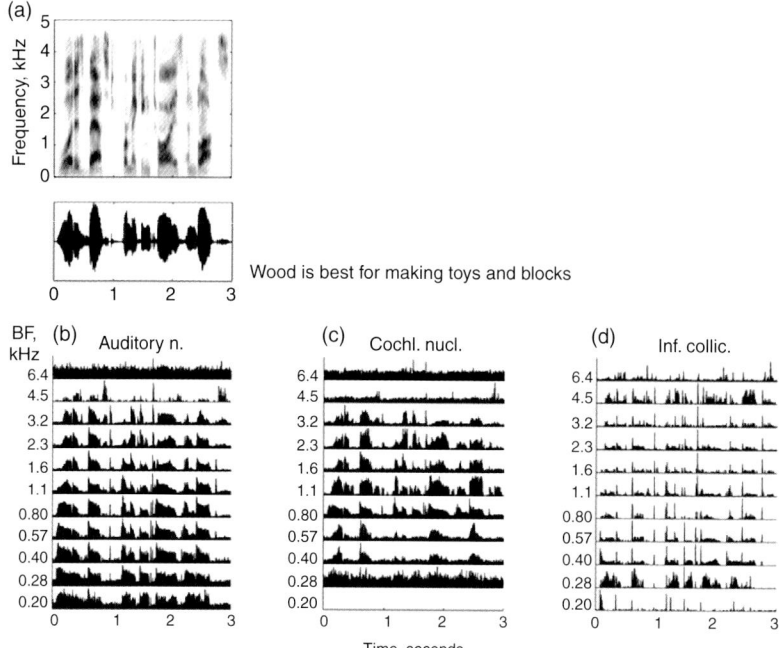

Figure 2.11 (a) Spectrogram (top) and oscillogram (bottom) of the sentence "Wood is best for making toys and blocks". (b), (c), and (d) PST histograms of responses of populations of AN fibres, neurones from cochlear nucleus, and neurones from inferior colliculus to this sentence. The responses of the neurones were recorded in anaesthetized cats. The PST histograms are the averages of responses of populations of neurones gathered into 0.5-octave bands, with centre frequencies indicated at left on each plot.
(Redrawn from Delgutte et al. (1998)).

the effects of the spectrum can be seen, e.g. at the beginning of the stimulus, where the /w/ has mainly low-frequency energy and the rate responses are higher in low-BF fibres. The responses to transient parts of the stimulus, especially onsets of syllables or bursts associated with stop consonants, are evident as sharp peaks in rate that extend broadly across frequency. For example, the responses to the /t/s are clear just before 1 s and at about 1.7 s. Even so, there are substantial AN responses, in the form of maintained rate elevations, to the vowels and the steady consonants (/w/, /s/, /m/, etc.). The responses in the cochlear nucleus (Figure 2.11c) are qualitatively similar, although they appear to be more differentiated across frequency bands.

The data in Figure 2.11d show the responses of neurones in the inferior colliculus. The colliculus is a structure unique to the auditory system, in that other sensory systems do not have an equivalent (see the review by Winer & Schreiner 2005). Its neurones collect axons from a number of sources in the brainstem auditory nuclei, including the cochlear nucleus and the superior olivary complex, as well as descending axons from the thalamus and cortex (Oliver & Huerta 1992; Rouiller 1997). Its outputs travel to the auditory part of the thalamus – the medial geniculate – which projects in turn to the auditory cortex. Although the functional properties of its neurones have been studied extensively (reviewed by Irvine 1992; Ehret 1997; Winer & Schreiner 2005), an understanding of the mechanisms for processing complex stimuli in the colliculus has not been achieved.

The responses in the inferior colliculus (Figure 2.11d) are more transient than those at lower levels. The strongest responses are to the onsets and bursts in the speech and steady responses to the vowels have relatively low discharge rates. These data suggest that the central representation of speech may emphasize transients over steady-state responses, which should have the effect of amplifying the representation of consonants relative to vowels.

Delgutte *et al.* (1998) used a simplified neural model to represent the transformations shown in Figure 2.11. The model consisted of a bandpass filter to model peripheral tuning curves, followed by a rectifier to model compression and to extract the envelope from the responses, followed by a second filter to model the neurone's modulation sensitivity. Without going into details, the modulation sensitivity behaved like a system that receives an excitatory input followed at short latency by an inhibitory input, making it sensitive to changes in the amplitude of the stimulus, but not to a steady stimulus. In the AN and cochlear nucleus, the inhibitory component was either not seen or was weak. Although this model did not predict all the details of the responses, it was qualitatively consistent with the data shown in Figure 2.11d in that it produced phasic responses to speech.

Some information in speech is carried specifically by the time of occurrence of events. An example is the voice onset time (VOT) which is the delay of the onset of voicing from the release time of a stop consonant. For the voiced stops (/b/, /d/, and /g/), voicing begins as soon as the closure of the vocal tract is released. The stops analysed in Figure 2.8 are of this type where the VOT is essentially zero. For the unvoiced stops (/p/, /t/, and /k/) there may be a 40–80 ms delay before voicing begins. This pause is signalled in the acoustic signal as a lack of low-frequency energy in the signal prior to the onset of voicing; thus, there is little energy at F1 prior to the onset of voicing, while F2 and F3 contain energy produced by the noise associated with the release of the stop.

The representation of VOT has been analysed by Sinex and colleagues (Sinex & McDonald 1988; Sinex *et al.* 1991; Sinex 1993; Sinex & Narayan 1994; Chen *et al.* 1996). As expected from the description of the stimulus above, it is mainly low-BF neurones (<1 kHz) that behave differently between voiced and voiceless stops, and the difference is that low-BF neurones show a pause in discharge during the voiceless period and an onset of spiking when voicing begins. The duration of the pause in discharge corresponds well to the VOT (Sinex *et al.* 1991) and serves as a neural correlate of the VOT that could be used by the brain to discriminate voiced and voiceless stops.

The representation of VOT seems to change little between the AN and the inferior colliculus. At both levels, there is a pause in spiking whose duration corresponds to the VOT. Conversion of the duration of the pause to some other form of neural code, like a discharge rate proportional to the pause, apparently does not occur (Sinex & Chen 2000) even though there are duration-tuned neurones in the inferior colliculus in bats (Pinheiro *et al.* 1991; Ehrlich *et al.* 1997; Fuzessery & Hall 1999) and other mammals (Chen 1998; Brand *et al.* 2000).

2.12 Cortical representation of temporal and spectral parameters

Stimulus representations in the auditory cortex have many aspects that differentiate them from the peripheral representations summarized above (e.g. Kaas *et al.* 1999; Schreiner *et al.* 2000; Eggermont 2001; Nelken 2004; Metherate *et al.* 2005). The primary auditory cortex is organized tonotopically, so that there is an orderly arrangement of neurones by BF, with rows of neurones tuned to particular frequencies arranged side by side in order from low to high frequency. Perpendicular to the tonotopic axis (within a row), the neurones are all tuned roughly to the same frequency, but can differ in a number of other response parameters; these include binaural sensitivity, width of tuning, strength and arrangement of inhibitory inputs, and dynamic range properties. Thus the representation of stimuli in the auditory cortex is guaranteed to be more diverse and multifaceted than the straightforward representation in peripheral neurones.

Even for tones, the organization suggested by the tonotopic map can be misleading. The tonotopic map is defined by the BFs of neurones at low sound levels, within a few dB of threshold. At higher sound levels, inhibitory inputs can reshape the responses so that the maximum response to a tone of a particular frequency can lie outside its own tonotopic location, and neurones at the tonotopic location can be inhibited (e g. Schreiner, 1998; Figure 2.3).

There are a number of aspects of cortical physiology that make addressing the question of the cortical 'representation' of speech difficult, or at least substantially different from the more peripheral representations. Three such aspects are discussed below.

First, cortical neurones are often specifically selective for sounds that are behaviourally important for the animal. In marmosets, for example, the responses of cortical neurones to the species' vocalizations are strongly different if the sounds are reversed in time (Wang & Kadia 2001); that this result is not a consequence of general auditory processing follows from the lack of an effect of time reversal when the same sounds are presented to neurones in the cat auditory cortex. The same lack of sensitivity to call reversal was reported in ferret cortex (Schnupp *et al.* 2006); in this case, it was also shown that training the ferrets to discriminate the forward from reverse calls did not change the average strength of responses in cortex, even though there is temporally encoded information that is sufficient to discriminate the calls.

Other well-studied examples of specificity for species-specific sounds are the songbirds' auditory cortex equivalent (field L), where neurones are selective for song relative to similar artificial sounds (Grace *et al.* 2003; Cousillas *et al.* 2005), and the songbirds' auditory–motor song nuclei, where neurones are highly selective amongst songs according to singer (Margoliash 1986; Doupe & Konishi 1991). These examples make the point that cortical structures can be organized to respond specifically to behaviourally meaningful categories of sounds rather than to the general spectrotemporal properties of sounds, i.e. to respond to sounds as representations of auditory sources, not just acoustic signals (Nelken *et al.* 2003).

Second, although neurones in the primary auditory cortex have BFs and receptive fields that can be defined with methods like tuning curves or reverse correlation, the receptive fields are not fixed properties of the neurones (Weinberger 2007). For example, in ferrets trained to perform a behavioural task for a food reward, the BF of a neurone can shift over a significant frequency range until it is centred on the stimulus frequencies important for the task (Fritz *et al.* 2003); the shift may be temporary, and may reverse when the behavioural task is removed, or it may be long-lasting. Another example is provided by the phenomenon of oddball responses, in which the response of a neurone to a particular stimulus depends on its probability of occurring: common sounds evoke weaker responses than rare sounds (Ulanovsky *et al.* 2003). Thus it is misleading to think of the characteristics of cortical neurones as fixed; in fact, they may be adjusted to immediate behavioural contingencies and ongoing sound backgrounds in order to optimize the processing of sound in some way.

Third, neurones in the auditory cortex may respond to sounds in ways that cannot be accurately modelled by straightforward concepts of linear receptive fields. An example is the poor performance of the spectrotemporal receptive field (STRF) model. Such models are fitted to the responses of neurones to a suitable stimulus set, typically broadband noise or some other broadband stimulus (Aertsen & Johannesma 1981; Eggermont 1993; Klein et al. 2000; Theunissen et al. 2001). The method is similar to reverse correlation discussed in connection with Figure 2.2b, except that a measure of the power spectrum of the stimulus preceding spikes is averaged. STRFs provide a basic definition of the neurones' receptive field, but do not capture the nonlinear properties of neurones. When STRFs are tested by using them to predict the responses of the neurone to another stimulus set, they typically account for less than half the variance in the responses of cortical neurones (Yeshurun et al. 1989; Versnel & Shamma 1998; Sen et al. 2001; Machens et al. 2004).

A more subtle indication of the inadequacy of STRF models is that a neurone presented with a complex acoustic signal may not respond to the most intense frequency components near the neurone's BF. For example, when presented with a bird chirp that is accompanied by echoes and background noise, neurones in cat cortex may respond to the whole stimulus differently than to the chirp alone, even when the chirp is at a frequency near BF and has a sound level well above that of the other components (Bar Yosef et al. 2002). This result is consistent with the conclusion drawn above that the responses of cortical neurones are a step removed from the immediate spectrotemporal aspects of the stimulus (Nelken 2004).

An example of the responses of a population of neurones in the auditory cortex of the cat to the syllables /be/ and /pe/ is shown in Figure 2.12 (Schreiner 1998; Wong & Schreiner 2003). The stimuli differ in VOT; their spectra are shown in Figure 2.12a. The formants, marked at right, are reasonably steady in time. The VOT is about 70 ms for the /p/ and much shorter for the /b/. In the spectrogram, the release burst of the /p/ is just visible as a broadband signal near time zero. A similar burst is present for the /b/, where it is accompanied by the large low-frequency (<1 kHz) energy in the voicing. Figure 2.12b shows neurograms of the responses of a population of cortical neurones to the sounds. Neurones are displayed along the ordinate according to their BFs; each horizontal line shows the firing rate of neurones in a particular BF bin plotted on the same time axis as the spectrograms in Figure 2.12a. In this anaesthetized preparation, the neurones respond mainly to the onset of the sound at the release of the /p/ or /b/ (just after time zero) and to the onset of the voicing (at 70 ms) in the case of /pe/. For /be/, there is no change in the voicing to produce a second response. Looking vertically down the ordinate in Figure 2.12b shows the tonotopic distribution of responses. Although there are some changes in latency (time of response) with BF, there are not obvious rate peaks in these data that correspond to the formants of the signals.

The responses to stop-consonant syllables shown in Figure 2.12b are similar to those observed in other studies, performed using anaesthetized cats (Eggermont 1995) and awake monkeys (Steinschneider et al. 1995, 2003). With sufficiently long VOTs, two peaks of response are seen, as for /pe/ in Figure 2.12b – one corresponding to the release of the stop consonant and the second to the onset of voicing. As the VOT is made shorter, the second peak disappears. Although there are other components of the cortical responses

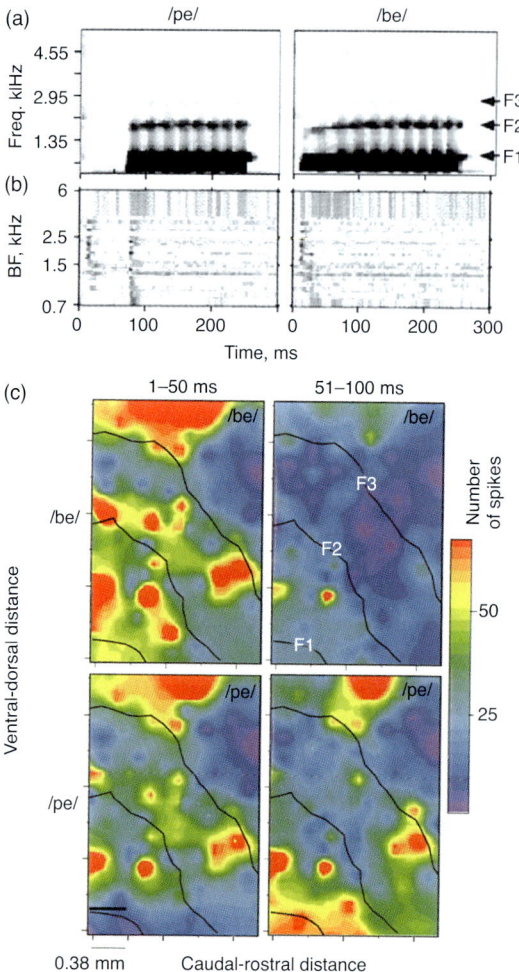

Figure 2.12 Population responses of neurones in anaesthetized cat cortex to two CV syllables. (a) Spectrograms of the syllables /pe/ and /be/ used as stimuli. The formants are roughly constant (0.6, 1.7, and 2.5 kHz) and are marked at right. The VOTs are 70 ms for /pe/ and near-zero for /be/. (b) Neurograms showing discharge rate as a function of time (on the abscissa) for neurones of different BFs (on the ordinate). The darkness of a pixel corresponds to firing rate on a normalized scale from 0 (white) to half the rate at the sound level where the BF-tone rate versus level function begins to saturate (black). There are few neurones for BFs above 3.6 kHz, so the last few BF bins are combined. (c) Plots of discharge rate (colour scale) versus the location of the recording electrode across the surface of auditory cortex. Data are interpolated from responses at 88 recording locations within the map. Maps are shown for /be/ (top row) and /pe/ (bottom row) over two time intervals, indicated at top. The tonotopic map across this piece of cortex is indicated by showing the approximate isofrequency lines for the first three formants, labelled in white in the upper right map. The firing rate is given as the number of spikes in the analysis interval during 20 repetitions of the stimulus. b and c show data from different brains, but the results are qualitatively the same; in both cases the stimulus level was 55 dB SPL.
(Reproduced with permission from Schreiner (1998); Wong & Schreiner (2003)).

to these syllables, especially in awake monkeys, the existing evidence again points towards the pause in discharge between two onset-response peaks as the representation of VOT.

Figure 2.12c shows the distribution of activity across the region of cortex from which recordings were made by Wong and Schreiner (2003). The four plots in Figure 2.12c show maps of the response strength of neurones as a function of location. The estimates of response are based on single- and multi-unit recordings from 88 recording sites distributed uniformly across the map. The colour scale shows the number of spikes during two time windows. The first window (1–50 ms, left column) contains the burst of response to the consonant release and the second window (51–100 ms, right column) contains the onset of voicing. The tonotopic map is marked on the cortical surface by the black lines, which show the approximate locations of neurones tuned to the first three formants of the stimuli.

Consistent with Figure 2.12b, there is no response to /be/ in the second time interval (the upper right map). Looking at the other three maps, one sees patchy distributions of responses which do not necessarily peak at BFs corresponding to the formant frequencies. Although the responses to the two stimuli are clearly different, there is no simple correspondence between the stimulus spectrum and the tonotopic map. It is particularly convincing to look along an isofrequency contour corresponding to one of the formants. Substantial differences in response are observed in all cases, except for F1 in the lower right map. The differences in responses within an isofrequency sheet are as large as the differences across the tonotopic map.

This attempt to reconstruct the response of the neural population across the cortical surface has not been repeated, although several analyses of the responses of populations of cortical neurones to communication sounds have been done. For example, in the marmoset (Wang *et al.* 1995; Wang 2000), neurones give strong burst responses to successive syllables of the twitter call, analogous to the two-peak responses to the VOT stimuli. There is a rough tonotopic representation of call spectrum in this case which varies according to the degree of selectivity of the neurones for the calls. In a similar result, responses to leading consonants in consonant–vowel–consonant (CVC) syllables have been studied in rat cortex (Engineer *et al.* 2008). The analysis focussed on the first burst of spikes in response to the stimulus. These population responses were sufficient to support the performance observed in behavioural discrimination experiments. However, analysis of the representation of syllable identity was not done.

The experiments summarized above illustrate the problems of studying the speech representation in cortex. In thinking about the problem, it is worthwhile to consider the following hypothesis about the organization of cortical responses beyond the tonotopic gradient. Chi *et al.* (2005) have proposed a multiparameter representation of speech modelled on the properties of cortical STRFs, which measure both frequency tuning and temporal responses of neurones. In the frequency domain, the width of frequency tuning and the arrangement of inhibitory versus excitatory regions change the selectivity of the neurone for broad versus narrow resonances in the stimulus spectrum. This aspect is called 'scale'. Each neurone is a bandpass filter for stimulus scale, and is most sensitive to a particular scale; the tuning for scale can be derived from the magnitude of the Fourier transform of the STRF along the frequency axis. Low-scale neurones respond best to speech sounds with broad, widely spaced formants and high-scale neurones prefer narrow, closely spaced formants. A second parameter is 'rate' which measures the selectivity of the neurone for modulation along the time axis or other temporal changes like frequency sweeps. In the model, neurones are sensitive to a range of sweep rates, both upwards and downwards. For example, rate is important in representing formant transitions. A speech sound can be completely represented in terms of the scale and the rate as a function of frequency, meaning that the stimulus can be reconstructed accurately from the description in terms of scale and rate.

Figure 2.13 shows examples of the representation of 10 speech sounds in this system (Mesgarani *et al.* 2008). The sounds were presented in a large number of sentences and responses were recorded in the cortex of awake ferrets. The speech spectrograms (as in Figure 2.1) of four vowels and six consonants are shown in the top row of the figure. Responses of cortical neurones to the phonemes are shown on the same time axis in the second row of the figure. The ordinates of the response plots show the best frequencies of the neurones; thus these plots are like Figure 2.8a or 2.12b–d except that the colour scale shows the average rates of the neurones across several repetitions of the phoneme in different contexts. Consistent with Figure 2.12, there is not a close resemblance of the spectrograms of the stimuli in the top row and the tonotopic representations in the second row; examples of poor resemblance are provided by the responses to /ɛ/, /i/, /f/, and /v/

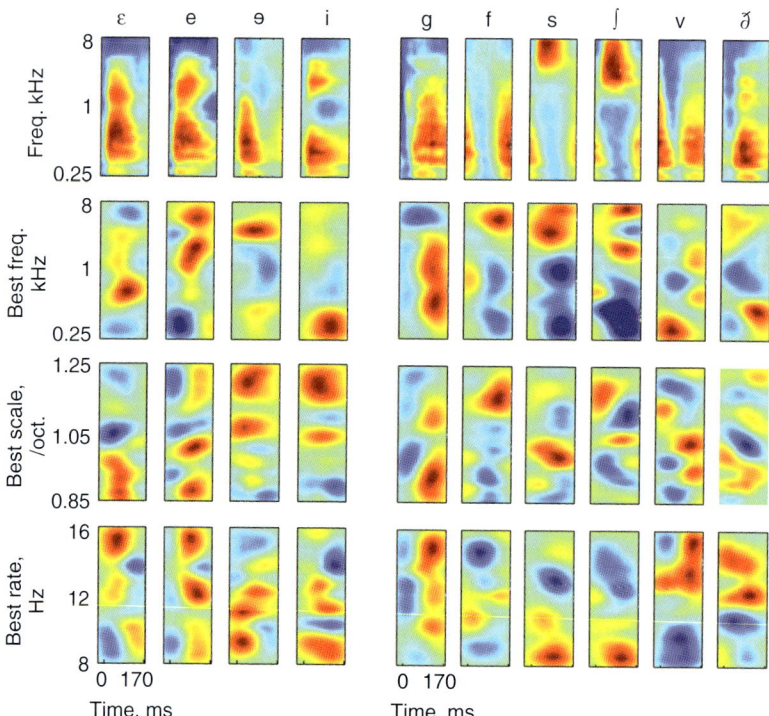

Figure 2.13 Responses of a population of 90 cortical neurones to 10 phonemes presented in sentence context. The average spectrograms of the phonemes are shown on the top row. The population rate responses are shown in the next three rows on axes of time (abscissa) and BF (ordinate, 2nd row), best scale (ordinate, 3rd row), and best rate (ordinate, 4th row). BF, best scale, and best rate are the values of each parameter to which the neurone gives the largest response in a STRF. (redrawn from Mesgarani *et al.* (2008) with permission)

where the regions of high discharge rate do not correspond well to the regions of high energy in the stimulus. The third and fourth rows show the same neural data, except that now the neurones are grouped according to the scale (third row) or rate (fourth row) to which each neurone gives the largest response. In other words, the order of placement of neurones along the ordinate of these plots is by 'best scale' or 'best rate' instead of 'best frequency'. There was no relationship between BF and best scale or rate, so there is a complete reordering of neurons among the second, third, and fourth rows.

The existence of orderly representations by scale and rate is clear. For example, the vowels /ɛ/ and /e/ stimulate low-scale neurones because these vowels have broadband frequency content; the remaining two vowels have narrower bandwidth and thus evoke responses at high scale. The orderly nature of the rate and scale responses suggests that these are useful ways to characterize cortical neurones responding to speech. They, at least, raise the possibility that the cortical speech map involves such measures in addition to frequency.

Now, consider what the tonotopic representation of a speech sound would be in a cortex with rate and scale sensitivity like those shown in Figure 2.13. Along an isofrequency sheet, there would be neurones sensitive to different rate and scale. A sound with a formant at a particular frequency might or might not activate a neurone with a BF at the frequency of the formant, depending on the relative scales and rates of the stimulus and the neurone. Such a representation would be patchy as in Figure 2.12C since there is no orderly relationship between BF and best scale or best rate (Mesgarani *et al.* 2008). This analysis serves as an example of a representation that cannot be easily analysed from data like Figure 2.12c, without some foreknowledge of the principles upon which it is based.

2.13 How is speech represented by neurones?

The nature of the neural representation of speech is clearest in the case of AN fibres, where the spectrotemporal features of the stimulus are isomorphically represented by neural activity. The frequency content of sounds is represented by the distribution of activity across BFs in the population of AN fibres, as displayed in a rate profile. The temporal envelope of connected speech and specific temporal features of the speech signal like the VOT are represented directly by temporal modulation of the neural response.

Less is known about the nature of the representation of speech in central neurones. Because of the large number of different groups of auditory neurones in the brain, only fragmentary results on a few of the potentially important transformations are available. For studies in a non-human animal, questions about the representation of speech have to be translated into general questions about the neural representation of complex auditory stimuli, i.e. there may be specializations for processing speech in the human brain that are not present in animal models and those specializations could be present at levels below the cortex. In the bat, for example, there are specializations for biosonar processing in the inferior colliculus and higher centres (O'Neill *et al.* 1989; Mittmann & Wenstrup 1995).

The changes in the neural representation of speech described for central auditory neurones in this chapter are mainly improvements in the robustness of the representation, as for rate profiles in the cochlear nucleus, and emphasis of some aspects of the response over others, as in the apparent amplification of responses to stimulus transients in the inferior colliculus. These are aspects of generic processing of auditory spectrotemporal representations which might be appropriate for almost any natural sound. Further investigation of the computational mechanisms by which these changes are achieved will be useful, both for the general problem of how the brain represents natural auditory scenes and for the specific problem of how lower parts of the auditory system process speech.

The issue raised at the end of the discussion of the auditory cortex is a particularly important challenge. The assumption in most work on the representation of complex stimuli in the auditory system is that the representation is tonotopic and that there is some additional analysis like the scale and rate dimensions discussed above. However, there is no widely accepted theory of what those dimensions should be, much less how they should be represented in neurones. This remains the central problem of research on the neurophysiology of speech and natural stimuli.

2.14 Appendix: the nature of the data

Studies of the neural representation of speech are based on microelectrode recordings of the activity of single neurones or small groups of neurones in anaesthetized animals. A microelectrode is placed into the neural structure of interest and the activity of the neurones near the electrode tip is recorded as speech-like stimuli are presented in the ear. The experiments are done successively on as many neurones as possible, presenting the same stimulus to each. Thus the experimenter builds up an estimate of the responses to the speech sound in a population of neurones (Pfeiffer & Kim 1975; Sachs & Young 1979). Usually, the electrode is placed so as to isolate the activity of one neurone at a time. Most of the results described here were obtained in this way. In some cases, however, the activity of small groups of unseparated neurones is recorded – so-called multiunit recording – represented here by the data in Figure 2.12.

Many aspects of the activity of neurones can be recorded experimentally. It is beyond the scope of this chapter to describe the physiology of neurones, which can be found in texts devoted to that subject (e.g. Kandel *et al.* 2000; Shepherd 2004). An introduction to the specialized neurophysiology of the auditory system can be found in Pickles (2008). Studies of the neural representation of speech are based on recordings of spike trains, meaning the trains of action potentials produced by the neurone under study. Conventionally, it is assumed that the information in spike trains is encoded entirely in the times of occurrence of action potentials. Thus, it is sufficient, in studying neural representations, just to record and analyse spike times (Rieke *et al.* 1997). Neural responses are the changes in the number or the temporal arrangement of spikes produced by the stimulus, as discussed in Figure 2.3.

Before analysing the responses of a neurone, it is necessary to characterize the neurone in terms of known properties that may be expected to influence its responses to the stimulus. In the simplest case of AN fibres, the neurones differ in terms of the frequencies to which they are sensitive – their frequency tuning (Figure 2.2; Kiang *et al.* 1965; Liberman 1978) – and in terms of the range of sound intensities to which they respond – their thresholds and dynamic ranges (Sachs & Abbas 1974; Yates *et al.* 1990). The importance of these characterizations is evident from the results presented in Figures 2.4–2.7.

In the central nervous system, the analysis problem is more complicated because there may be several different groups of neurones operating in parallel. Each group is made up of cells with different anatomical organization, different electrophysiological properties, and different connections to the rest of the auditory system. In principle, each group of neurones forms an independent representation, with information about various aspects of the stimulus represented differentially in different groups. Characterization of neurones in this case must be more extensive, extending beyond frequency tuning and dynamic range to include an estimate of the type of neurone from which a recording is being made and perhaps other properties. The matter of neural typing is best understood for the cochlear nucleus and is reviewed elsewhere (Rhode & Greenberg 1992; Romand & Avan 1997; Young & Oertel 2003).

An additional complication, for neurones in the central nervous system, is the effect of anaesthesia and the state of the animal in general. Nearly all the studies described here were done in anaesthetized animals, but the effects of anaesthesia on response properties of central neurones are significant (e.g. Kuwada *et al.* 1989; Zurita *et al.* 1994; Ramachandran *et al.* 1999; Gaese & Ostwald 2001; Anderson & Young 2004). The major effects of anaesthetics are a lowering of spontaneous discharge rate, an increase in threshold, and a loss of inhibitory responses. The loss of inhibition is likely to have a particularly important effect on responses to complex stimuli, where inhibitory interactions can predominate. Thus the presence of anaesthesia is a source of uncertainty in interpreting studies of the neural representation of speech in central auditory nuclei (but see May *et al.* (1998) for a counterexample in the cochlear nucleus).

Acknowledgements

Preparation of this chapter was supported by NIH grant DC00109. Thanks are due to Ian Bruce, Michael Heinz, Sharba Bandyopadhyay, Brian Moore, and two anonymous reviewers for comments on a previous version.

References

Aertsen, A. M. H. J. & Johannesma, P. I. M. 1981 The spectro-temporal receptive field. A functional characteristic of auditory neurons. *Biol. Cybern.* **42**, 133–143.
Anderson, M. J.& Young, E. D. 2004 Isoflurane/N2O anesthesia suppresses narrowband but not wideband inhibition in dorsal cochlear nucleus. *J. Speech Hear. Res.* **188**, 29–41.
Bandyopadhyay, S. & Young, E. D. 2004 Discrimination of voiced stop consonants based on auditory-nerve discharges. *J. Neurosci.* **24**, 531–541.
Bar-Yosef, O., Rotman, Y. & Nelken, I. 2002 Responses of neurons in cat primary auditory cortex to bird chirps: effects of temporal and spectral context. *J. Neurosci.* **22**, 8619–8632.
Blackburn, C. C. & Sachs, M. B. 1989 Classification of unit types in the anteroventral cochlear nucleus: PST histograms and regularity analysis. *J. Neurophysiol.* **62**, 1303–1329.
Blackburn, C. C. & Sachs, M. B. 1990 The representations of the steady-state vowel sound /ɛ/ in the discharge patterns of cat anteroventral cochlear nucleus neurons. *J. Neurophysiol.* **63**, 1191–1212.
Blumstein, S. E. & Stevens, K. N. 1979 Acoustic invariance in speech production: evidence from measurements of the spectral characteristics of stop consonants. *J. Acoust. Soc. Am.* **66**, 1001–1017.
Brand, A., Urban, A. & Grothe, B. 2000 Duration tuning in the mouse auditory midbrain. *J. Neurophysiol.* **84**, 1790–1799.
Bruce, I. C., Sachs, M. B. & Young, E. D. 2003 An auditory-periphery model of the effects of acoustic trauma on auditory nerve responses. *J. Acoust. Soc. Am.* **113**, 369–388.
Carney, L. H. 1990 Sensitivities of cells in anteroventral cochlear nucleus of cat to spatiotemporal discharge patterns across primary afferents. *J. Neurophysiol.* **64**, 437–456.

Carney, L. H. 1993 A model for the responses of low-frequency auditory-nerve fibers in cat. *J. Acoust. Soc. Am.* **93**, 401–417.

Carney, L. H. & Geisler, C. D. 1986 A temporal analysis of auditory-nerve fiber responses to spoken stop consonant–vowel syllables. *J. Acoust. Soc. Am.* **79**, 1896–1914.

Carney, L. H. & Yin, T. C. T. 1988 Temporal coding of resonances by low-frequency auditory nerve fibers: single-fiber responses and a population model. *J. Neurophysiol.* **60**, 1653–1677.

Carney, L. H., Heinz, M. G., Evilsizer, M. E., Gilkey, R. H. & Colburn, H. S. 2002 Auditory phase opponency: a temporal model for masked detection at low frequencies. *Acustica-Acta Acustica* **88**, 334–347.

Chen, G.-D. 1998 Effects of stimulus duration on responses of neurons in chinchilla inferior colliculus. *J. Speech Hear. Res.* **122**, 142–150.

Chen, G. D., Nuding, S. C., Narayan, S. S. & Sinex, D. G. 1996 Responses of single neurons in the chinchilla inferior colliculus to consonant–vowel syllables differing in voice onset time. *Aud Neurosci.* 3, 179–198.

Chi, T., Ru, P. & Shamma, S. A. 2005 Multiresolution spectrotemporal analysis of complex sounds. *J. Acoust. Soc. Am.* **118**, 887–906.

Colburn, H.S., Carney, L.H. & Heinz, M.G. 2003 Quantifying the information in auditory-nerve responses for level discrimination. *J. Assoc. Res. Otolaryngol.* **4**, 294–311.

Conley, R. A. & Keilson, S. E. 1995 Rate representation and discriminability of second formant frequencies for /ɛ/-like steady-state vowels in cat auditory nerve. *J. Acoust. Soc. Am.* **98**, 3223–3234.

Cousillas, H., Leppelsack, H. J., Leppelsack, E., Richard, J. P., Mathelier, M. & Hausberger, M. 2005 Functional organization of the forebrain auditory centres of the European starling: a study based on natural sounds. *J. Speech Hear. Res.* **207**, 10–21.

de Boer, E. & de Jongh, H. R. 1978 On cochlear encoding: potentialities and limitations of the reverse-correlation technique. *J. Acoust. Soc. Am.* **63**, 115–135.

Delgutte, B. 1980 Representation of speech-like sounds in the discharge patterns of auditory-nerve fibers. *J. Acoust. Soc. Am.* **68**, 843–857.

Delgutte, B. 1984 Speech coding in the auditory nerve: II. Processing schemes for vowel-like sounds. *J. Acoust. Soc. Am.* **75**, 879–886.

Delgutte, B. 1990 Two-tone rate suppression in auditory-nerve fibers: dependence on suppressor frequency and level. *J. Speech Hear. Res.* **49**, 225–246.

Delgutte, B. & Kiang, N. Y. S. 1984a Speech coding in the auditory nerve: IV. Sounds with consonant-like dynamic characteristics. *J. Acoust. Soc. Am.* **75**, 897–907.

Delgutte, B. & Kiang, N. Y. S. 1984b Speech coding in the auditory nerve: III. Voiceless fricative consonants. *J. Acoust. Soc. Am.* **75**, 887–896.

Delgutte, B. & Kiang, N. Y. S. 1984c Speech coding in the auditory nerve: I. Vowel-like sounds. *J. Acoust. Soc. Am.* **75**, 866–878.

Delgutte, B., Hammond, B. M. & Cariani, P. A. 1998 Neural coding of the temporal envelope of speech: relation to modulation transfer functions. In *Psychophysical and Physiological Advances in Hearing* (eds A. R. Palmer, A. Rees, A. Q. Summerfield & R. Meddis), pp. 595–603. London: Whurr Publ.

Deng, L. & Geisler, C. D. 1987a A composite auditory model for processing speech sounds. *J. Acoust. Soc. Am.* **82**, 2001–2012.

Deng, L. & Geisler, C. D. 1987b Responses of auditory-nerve fibers to nasal consonant-vowel syllables. *J. Acoust. Soc. Am.* **82**, 1977–1988.

Doupe, A. J. & Konishi, M. 1991 Song-selective auditory circuits in the vocal control system of the zebra finch. *PNAS* **88**, 11339–11343.

Eggermont, J. J. 1977 Compound action potential tuning curves in normal and pathological human ears. *J. Acoust. Soc. Am.* **62**, 1247–1251.

Eggermont, J. J. 1993 Wiener and Volterra analyses applied to the auditory system. *J. Speech Hear. Res.* **66**, 177–201.

Eggermont, J. J. 1995 Representation of a voice onset time continuum in primary auditory cortex of the cat. *J. Acoust. Soc. Am.* **98**, 911–920.

Eggermont, J. J. 2001 Between sound and perception: reviewing the search for a neural code. *J. Speech Hear. Res.* **157**, 1–42.
Ehret, G. 1997 The auditory midbrain, a 'shunting yard' of acoustical information processing. In *The Central Auditory System* (eds G. Ehret & R. Romand), pp. 259–316. New York: Oxford University Press.
Ehrlich, D., Casseday, J. H. & Covey, E. 1997 Neural tuning to sound duration in the inferior colliculus of the big brown bat, Eptesicus fuscus. *J. Neurophysiol.* **77**, 2360–2372.
Engineer, C.T., Perez, C.A., Chen, Y.H., Carraway, R.S., Reed, A.C., Shetake, J.A., Jakkamsetti, V., Chang, K.Q. & Kilgard, M.P. 2008 Cortical activity patterns predict speech discrimination ability. *Nat. Neurosci.* **11**, 603–608.
Evans, E. F. 1992 Comparisons of physiological and behavioural properties: Auditory frequency selectivity. In *Auditory Physiology and Perception* (eds N. P. Cooper, Y. Cazals & K. Horner), pp. 159–162. Oxford: Pergamon.
Fant, G. 1970 *Acoustic Theory of Speech Production*. The Hague: Mouton.
Fay, R. R. 1988 *Hearing in Vertebrates: A Psychophysics Databook*. Winnetka, IL: Hill-Fay Assoc.
Fritz, J., Shamma, S., Elhilali, M. & Klein, D. 2003 Rapid task-related plasticity of spectrotemporal receptive fields in primary auditory cortex. *Nat. Neurosci.* **6**, 1216–1223.
Fuzessery, Z. M. & Hall, J. C. 1999 Sound duration selectivity in the pallid bat inferior colliculus. *J. Speech Hear. Res.* **137**, 137–154.
Gaese, B. H. & Ostwald, J. 2001 Anesthesia changes frequency tuning of neurons in the rat primary auditory cortex. *J. Neurophysiol.* **86**, 1062–1066.
Geisler, C. D. 1989 The responses of models of 'high-spontaneous' auditory-nerve fibers in a damaged cochlea to speech syllables in noise. *J. Acoust. Soc. Am.* **86**, 2192–2205.
Geisler, C. D. & Gamble, T. 1989 Responses of 'high-spontaneous' auditory-nerve fibers to consonant-vowel syllables in noise. *J. Acoust. Soc. Am.* **85**, 1639–1652.
Goldberg, J. M. & Brown, P. B. 1969 Response of binaural neurons of dog superior olivary complex to dichotic tonal stimuli: some physiological mechanisms of sound localization. *J. Neurophysiol.* **32**, 613–636.
Grace, J. A., Amin, N., Singh, N. C. & Theunissen, F. E. 2003 Selectivity for conspecific song in the zebra finch auditory forebrain. *J. Neurophysiol.* **89**, 472–487.
Green, D. M. & Swets, J. A. 1966 *Signal Detection Theory and Psychophysics*. New York: John Wiley and Sons, Inc.
Greenwood, D. D. 1961 Critical bandwidth and the frequency coordinates of the basilar membrane. *J. Acoust. Soc. Am.* **33**, 1344–1356.
Greenwood, D. D. 1990 A cochlear frequency-position function for several species—29 years later. *J. Acoust. Soc. Am.* **87**, 2592–2605.
Harrison, R. V., Aran, J. M. & Erre, J.-P. 1981 AP tuning curves from normal and pathological human and guinea pig cochleas. *J. Acoust. Soc. Am.* **69**, 1374–1385.
Hienz, R. D., Aleszczyk, C. M. & May, B. J. 1996 Vowel discrimination in cats: thresholds for the detection of second formant changes in the vowel /ɛ/. *J. Acoust. Soc. Am.* **100**, 1052–1058.
Holmes, S. D., Sumner, C. J., O'Mard, L. P. & Meddis, R. 2004 The temporal representation of speech in a nonlinear model of the guinea pig cochlea. *J. Acoust. Soc. Am.* **116**, 3534–3545.
Irvine, D. R. F. 1992 Physiology of the auditory brainstem. In *The Mammalian Auditory Pathway: Neurophysiology* (eds A. N. Popper & R. R. Fay), pp. 153–231. New York: Springer-Verlag.
Javel, E. 1981 Suppression of auditory nerve responses I: temporal analysis, intensity effects and suppression contours. *J. Acoust. Soc. Am.* **69**, 1735–1745.
Javel, E., Geisler, C. D. & Ravindran, A. 1978 Two-tone suppression in auditory nerve of the cat: rate-intensity and temporal analyses. *J. Acoust. Soc. Am.* **63**, 1093–1104.
Johnson, D. H. 1980a Applicability of white-noise nonlinear system analysis to the peripheral auditory system. *J. Acoust. Soc. Am.* **68**, 876–884.
Johnson, D. H. 1980b The relationship between spike rate and synchrony in responses of auditory-nerve fibers to single tones. *J. Acoust. Soc. Am.* **68**, 1115–1122.
Johnson, D. H., Gruner, C. M., Baggerly, K. & Seshagiri, C. 2001 Information-theoretic analysis of the neural code. *J. Comput. Neurosci.* **10**, 47–69.

Joris, P. X., Ramirez, C.L., McLaughlin, M. & van der Heijden, M. 2006 Spectral and temporal properties of the auditory nerve in old-world monkeys. *Abst Assoc for Res in Otolarygol* **29**, 302.

Kaas, J. H., Hackett, T. A. & Tramo, M. J. 1999 Auditory processing in primate cerebral cortex. *Curr. Opin. Neurobiol.* **9**, 164–170.

Kandel, E. R., Schwartz, J. H. & Jessell, T. M. 2000 *Principles of Neural Science*. New York: McGraw-Hill.

Keilson, S. E., Richards, V. M., Wyman, B. T. & Young, E. D. 1997 The representation of concurrent vowels in the cat anesthetized ventral cochlear nucleus: evidence for a periodicity-tagged spectral representation. *J. Acoust. Soc. Am.* **102**, 1056–1071.

Kiang, N. Y. S., Watanabe, T., Thomas, E. C. & Clark, L. F. 1965 *Discharge Patterns of Single Fibers in the Cat's Auditory Nerve*. Cambridge: MIT Press.

Kiefte, M. & Kluender, K. R. 2001 Synthetic speech stimuli spectrally normalized for nonhuman cochlear dimensions. *ARLO* **3**, 41–46.

Kim, D. O., Siegel, J. H. & Molnar, C. E. 1979 Cochlear nonlinear phenomena in two-tone responses. *Scand. Audiol. Suppl.* **9**, 63–81.

Klein, D. J., Depireux, D. A., Simon, J. Z. & Shamma, S. A. 2000 Robust spectrotemporal reverse correlation for the auditory system: optimizing stimulus design. *J. Comput. Neurosci.* **9**, 85–111.

Kuwada, S., Batra, R. & Stanford, T. R. 1989 Monaural and binaural response properties of neurons in the inferior colliculus of the rabbit: effects of sodium pentobarbital. *J. Neurophysiol.* **61**, 269–282.

Lai, Y. C., Winslow, R. L. & Sachs, M. B. 1994 The functional role of excitatory and inhibitory interactions in chopper cells of the anteroventral cochlear nucleus. *Neural Computation* **6**, 1127–1140.

Lehiste, I. 1967 *Readings in Acoustic Phonetics*. Cambridge, MA: MIT Press.

Lewis, E. R. & Henry, K. R. 1994 Dynamic changes in tuning in the gerbil cochlea. *J. Speech Hear. Res.* **79**, 183–189.

Liberman, M. C. 1978 Auditory-nerve response from cats raised in a low-noise chamber. *J. Acoust. Soc. Am.* **63**, 442–455.

Liberman, M. C. 1980 Morphological differences among radial afferent fibers in the cat cochlea: an electron-microscopic study of serial sections. *J. Speech Hear. Res.* **3**, 45–63.

Liberman, M. C. & Beil, D. G. 1979 Hair cell condition and auditory nerve response in normal and noise-damaged cochleas. *Acta Oto-laryngologica* **88**, 161–176.

Liberman, M. C. & Dodds, L. W. (1984). Single-neuron labeling and chronic cochlear pathology. III. Stereocilia damage and alterations of threshold tuning curves. *J. Speech Hear. Res.* **16**, 55–74.

Liu, C. & Kewley-Port, D. 2004 Formant discrimination in noise for isolated vowels. *J. Acoust. Soc. Am.* **116**, 3119–3129.

Machens, C. K., Wehr, M. S. & Zador, A. M. 2004 Linearity of cortical receptive fields measured with natural sounds. *J. Neurosci.* **24**, 1089–1100.

Margoliash, D. 1986 Preference for autogenous song by auditory neurons in a song system nucleus of the white-crowned sparrow. *J. Neurosci.* **6**, 1643–1661.

May, B. J., LePrell, G. S. & Sachs, M. B. 1998 Vowel representations in the ventral cochlear nucleus of the cat: effects of level, background noise, and behavioral state. *J. Neurophysiol.* **79**, 1755–1767.

May, B. J., Huang, A., Le Prell, G. & Hienz, R. D. 1996 Vowel formant frequency discrimination in cats: comparison of auditory nerve representations and psychophysical thresholds. *Aud Neurosci.* **3**, 135–162.

May, B. J., LePrell, G. S., Hienz, R. D. & Sachs, M. B. 1997 Speech representation in the auditory nerve and ventral cochlear nucleus. In *Acoustical Signal Processing in the Central Auditory System* (ed. J. Syka), pp. 413–429. New York: Plenum Press.

Mesgarani, N., David, S.V., Fritz, J.B. & Shamma, S. 2008 Phoneme representation and classification in primary auditory cortex. *J. Acoust. Soc. Am.* **123**, 899–909.

Metherate, R., Kaur, S., Kawai, H., Lazar, R., Liang, K. & Rose, H. J. 2005 Spectral integration in auditory cortex: mechanisms and modulation. *J. Speech Hear. Res.* **206**, 146–158.

Miller, M. I. & Sachs, M. B. 1983 Representation of stop consonants in the discharge patterns of auditory-nerve fibers. *J. Acoust. Soc. Am.* **74**, 502–517.

Miller, R. L., Calhoun, B. M. & Young, E. D. 1999a Discriminability of vowel representations in cat auditory-nerve fibers after acoustic trauma. *J. Acoust. Soc. Am.* **105**, 311–325.

Miller, R. L., Calhoun, B. M. & Young, E. D. 1999b Contrast enhancement improves the representation of /ɛ/-like vowels in the hearing-impaired auditory nerve. *J. Acoust. Soc. Am.* **106**, 2693–2708.

Miller, R. L., Schilling, J. R., Franck, K. R. & Young, E. D. 1997 Effects of acoustic trauma on the representation of the vowel /ɛ/ in cat auditory nerve fibers. *J. Acoust. Soc. Am.* **101**, 3602–3616.

Mittmann, D. H. & Wenstrup, J. J. 1995 Combination-sensitive neurons in the inferior colliculus. *J. Speech Hear. Res.* **90**, 185–191.

Møller, A. R. 1977 Frequency selectivity of single auditory-nerve fibers in response to broadband noise stimuli. *J. Acoust. Soc. Am.* **62**, 135–142.

Moore, B. C. J. 1995 *Perceptual Consequences of Cochlear Damage*. Oxford: Oxford University Press.

Moore, B. C. J. 2003 *An Introduction to the Psychology of Hearing*. Amsterdam: Elsevier.

Nelken, I. 2004 Processing of complex stimuli and natural scenes in the auditory cortex. *Curr. Opin. Neurobiol.* **14**, 474–480.

Nelken, I., Fishbach, A., Las, L., Ulanovsky, N. & Farkas, D. 2003 Primary auditory cortex of cats: feature detection or something else? *Biol. Cybern.* **89**, 397–406.

O'Neill, W. E., Frisina, R. D. & Gooler, D. M. 1989 Functional organization of mustached bat inferior colliculus: I. Representation of FM frequency bands important for target ranging revealed by 14C-2-deoxyglucose autoradiography and single unit mapping. *J. Comp. Neurol.* **284**, 60–84.

Oliver, D. L. & Huerta, M. F. 1992 Inferior and superior colliculi. In *The Mammalian Auditory Pathway: Neuroanatomy* (eds D. B. Webster, A. N. Popper & R. R. Fay), pp. 168–221. New York: Springer-Verlag.

Oxenham, A. J. & Shera, C. A. 2003 Estimates of human cochlear tuning at low levels using forward and simultaneous masking. *J. Assoc. Res. Otolaryngol.* **4**, 541–554.

Palmer, A. R. 1990 The representation of the spectra and fundamental frequencies of steady-state single- and double-vowel sounds in the temporal discharge patterns of guinea pig cochlear-nerve fibers. *J. Acoust. Soc. Am.* **88**, 1412–1426.

Palmer, A. R. & Russell, I. J. 1986 Phase-locking in the cochlear nerve of the guinea-pig and its relation to the receptor potential of inner hair-cells. *J. Speech Hear. Res.* **24**, 1–15.

Palmer, A. R. & Moorjani, P. A. 1993 Responses to speech signals in the normal and pathological peripheral auditory system. *Prog. Brain Res.* **97**, 107–115.

Palmer, A. R., Winter, I. M. & Darwin, C. J. 1986 The representation of steady-state vowel sounds in the temporal discharge patterns of the guinea pig cochlear nerve and primary-like cochlear nucleus neurons. *J. Acoust. Soc. Am.* **79**, 100–113.

Patterson, R.D., Robinson, K., Holdsworth, J., McKeown, D., Zhang, C. & Allerhand, M. 1992 Complex sounds and auditory images. In: *Auditory Physiology and Perception* (eds Cazals, Y., Horner, K. & Demany, L.), pp. 429–446. Oxford: Pergamon Press.

Pfeiffer, R. R. & Kim, D. O. 1975 Cochlear nerve fiber responses: distribution along the cochlear partition. *J. Acoust. Soc. Am.* **58**, 867–869.

Pickles, J. O. 1979 Psychophysical frequency resolution in the cat as determined by simultaneous masking and its relation to auditory-nerve resolution. *J. Acoust. Soc. Am.* **66**, 1725–1732.

Pickles, J. O. 2008 *An Introduction to the Physiology of Hearing (3rd Ed.)*. London: Academic Press.

Pinheiro, A. D., Wu, M. & Jen, P. H. 1991 Encoding repetition rate and duration in the inferior colliculus of the big brown bat, Eptesicus fuscus. *J. Comp. Physiol.* **169**, 69–85.

Ramachandran, R., Davis, K. A. & May, B. J. 1999 Single-unit responses in the inferior colliculus of decerebrate cats I. Classification based on frequency response maps. *J. Neurophysiol.* **82**, 152–163.

Reale, R. A. & Geisler, C. D. 1980 Auditory-nerve fiber encoding of two-tone approximations to steady-state vowels. *J. Acoust. Soc. Am.* **67**, 891–902.

Recio, A. & Rhode, W. S. 2000 Representation of vowel stimuli in the ventral cochlear nucleus of the chinchilla. *J. Speech Hear. Res.* **146**, 167–184.

Recio, A., Rhode, W. S., Kiefte, M. & Kluender, K. R. 2002 Responses to cochlear normalized speech stimuli in the auditory nerve of cat. *J. Acoust. Soc. Am.* **111**, 2213–2218.

Recio-Spinoso, A., Temchin, A. N., van Dijk, P., Fan, Y. H. & Ruggero, M. A. 2005 Wiener-kernel analysis of responses to noise of chinchilla auditory-nerve fibers. *J. Neurophysiol.* **93**, 3615–3634.

Rhode, W. S. & Greenberg, S. 1992 Physiology of the cochlear nucleus. In *The Mammalian Auditory Pathway: Neurophysiology* (eds A. N. Popper & R. R. Fay), pp. 94–152. Berlin: Springer-Verlag.

Rieke, F., Warland, D., de Ruyter van Steveninck, R. & Bialek, W. 1997 *Spikes, Exploring the Neural Code*. Cambridge, MA: MIT Press.

Robles, L. & Ruggero, M. A. 2001 Mechanics of the mammalian cochlea. *Physiol. Rev.* **81**, 1305–1352.

Romand, R. & Avan, P. 1997 Anatomical and functional aspects of the cochlear nucleus. In *The Central Auditory System* (eds G. Ehret & R. Romand), pp. 97–191. New York: Oxford University Press.

Rose, J. E., Brugge, J. F., Anderson, D. J. & Hind, J. E. 1967 Phase-locked response to low frequency tones in single auditory nerve fibers of the squirrel monkey. *J. Neurophysiol.* **30**, 769–793.

Rosen, S. 1992 Temporal information in speech: acoustic, auditory and linguistic aspects. *Phil Trans R Soc Lond B* **336**, 367–373.

Rouiller, E. M. 1997 Functional organization of the auditory pathways. In *The Central Auditory System* (eds G. Ehret & R. Romand), pp. 3–96. New York: Oxford University Press.

Ruggero, M. A. & Temchin, A. N. 2005 Unexceptional sharpness of frequency tuning in the human cochlea. *PNAS* **102**, 18614–18619.

Ruggero, M. A., Rich, N. C., Recio, A. & Narayan, S. S. 1997 Basilar-membrane responses to tones at the base of the chinchilla cochlea. *J. Acoust. Soc. Am.* **101**, 2151–2163.

Sachs, M. B. 1969 Stimulus-response relation for auditory-noise fibers: two-tone stimuli. *J. Acoust. Soc. Am.* **45**, 1025–1036.

Sachs, M. B. & Abbas, P. J. 1974 Rate versus level functions for auditory-nerve fibers in cats: tone-burst stimuli. *J. Acoust. Soc. Am.* **56**, 1835–1847.

Sachs, M. B. & Young, E. D. 1979 Encoding of steady-state vowels in the auditory nerve: Representation in terms of discharge rate. *J. Acoust. Soc. Am.* **66**, 470–479.

Sachs, M.B., Voigt, H.F. & Young, E.D. 1983 Auditory nerve representation of vowels in background noise. *J. Neurophysiol.* **50**, 27–45.

Salvi, R., Perry, J., Hamernik, R. P. & Henderson, D. 1982 Relationships between cochlear pathologies and auditory nerve and behavioral responses following acoustic trauma. In *New Perspectives on Noise-Induced Hearing Loss* (eds R. P. Hamernik, D. Henderson & R. Salvi), pp. 165–188. New York: Raven.

Schilling, J. R., Miller, R. L., Sachs, M. B. & Young, E. D. 1998 Frequency shaped amplification changes the neural representation of speech with noise-induced hearing loss. *J. Speech Hear. Res.* **117**, 57–70.

Schmiedt, R. A., Zwislocki, J. J. & Hamernik, R. P. 1980 Effects of hair cell lesions on responses of cochlear nerve fibers. I. Lesions, tuning curves, two-tone inhibition, and responses to trapezoidal-wave patterns. *J. Neurophysiol.* **43**, 1367–1389.

Schnupp, J.W.H., Hall, T.M., Kokelaar, R.F. & Ahmed, B. 2006 Plasticity of temporal pattern codes for vocalization stimuli in primary auditory cortex. *J. Neurosci.* **26**, 4785–4795.

Schreiner, C. E. 1998 Spatial distribution of responses to simple and complex sounds in the primary auditory cortex. *Audiol. Neuro-otol.* **3**, 104–122.

Schreiner, C. E., Read, H. L. & Sutter, M. L. 2000 Modular organization of frequency integration in primary auditory cortex. *Ann Rev Neurosci.* **23**, 501–529.

Sen, K., Theunissen, F. E. & Doupe, A. J. 2001 Feature analysis of natural sounds in the songbird auditory forebrain. *J. Neurophysiol.* **86**, 1445–1458.

Shamma, S. A. 1985 Speech processing in the auditory system. II: lateral inhibition and the central processing of speech evoked activity in the auditory nerve. *J. Acoust. Soc. Am.* **78**, 1622–1632.

Shannon, R. V., Zeng, F. G., Kamath, V., Wygonski, J. & Ekelid, M. 1995 Speech recognition with primarily temporal cues. *Science* **270**, 303–304.

Shepherd, G. M. 2004 *The Synaptic Organization of the Brain*. Oxford: Oxford University Press.

Shera, C.A. & Guinan, J.J. 2003 Stimulus-frequency-emission group delay: a test of coherent reflection filtering and a window on cochlear tuning *J. Acoust. Soc. Am.* **113**, 2762–2772.

Shera, C. A., Guinan, J. J. & Oxenham, A. J. 2002 Revised estimates of human cochlear tuning from otoacoustic and behavioral measurements. *PNAS* **99**, 3318–3323.

Siegel, J. H., Cerka, A. J., Recio-Spinoso, A., Temchin, A. N., van Dijk, P. & Ruggero, M. A. 2005 Delays of stimulus-frequency otoacoustic emissions and cochlear vibrations contradict the theory of coherent reflection filtering. *J. Acoust. Soc. Am.* **118**, 2434–2443.

Silkes, S. M. & Geisler, C. D. 1991 Responses of 'lower-spontaneous-rate' auditory-nerve fibers to speech syllables presented in noise. I: general characteristics. *J. Acoust. Soc. Am.* **90**, 3122–3139.

Sinex, D. G. 1993 Auditory nerve fiber representation of cues to voicing in syllable-final stop consonants. *J. Acoust. Soc. Am.* **94**, 1351–1362.

Sinex, D. G. & Geisler, C. D. 1983 Responses of auditory-nerve fibers to consonant-vowel syllables. *J. Acoust. Soc. Am.* **73**, 602–615.

Sinex, D. G. & Geisler, C. D. 1984 Comparison of the responses of auditory nerve fibers to consonant-vowel syllables with predictions from linear models. *J. Acoust. Soc. Am.* **76**, 116–121.

Sinex, D. G. & McDonald, L. P. 1988 Average discharge rate representation of voice onset time in the chinchilla auditory nerve. *J. Acoust. Soc. Am.* **83**, 1817–1827.

Sinex, D. G. & Narayan, S. S. 1994 Auditory-nerve fiber representation of temporal cues to voicing in word-medial stop consonants. *J. Acoust. Soc. Am.* **95**, 897–903.

Sinex, D. G. & Chen, G.-D. 2000 Neural responses to the onset of voicing are unrelated to other measures of temporal resolution. *J. Acoust. Soc. Am.* **107**, 486–495.

Sinex, D. G., McDonald, L. P. & Mott, J. B. 1991 Neural correlates of nonmonotonic temporal acuity for voice onset time. *J. Acoust. Soc. Am.* **90**, 2441–2449.

Slaney, M. 1993 An efficient implementation of the Patterson–Holdsworth auditory filter bank. Apple Computer Technical Report #35.

Smith, R. L. 1977 Short-term adaptation in single auditory nerve fibers:some poststimulatory effects. *J. Neurophysiol.* **40**, 1098–1112.

Smith, R. L. 1979 Adaptation, saturation and physiological masking in single auditory-nerve fibers. *J. Acoust. Soc. Am.* **65**, 166–178.

Steinschneider, M., Fishman, Y. I. & Arezzo, J. C. 2003 Representation of the voice onset time (VOT) speech parameter in population responses within primary auditory cortex of the awake monkey. *J. Acoust. Soc. Am.* **114**, 307–321.

Steinschneider, M., Schroeder, C. E., Arezzo, J. C. & Vaughan Jr., H. G. 1995 Physiologic correlates of the voice onset time boundary in primary auditory cortex (A1) of the awake monkey: temporal response patterns. *Brain and Lang.* **48**, 326–340.

Stevens, K. N. & House, A. S. 1961 An acoustical theory of vowel production. *J. Speech Hear. Res.* **4**, 303–320.

Theunissen, F. E., David, S. V., Singh, N. C., Hsu, A., Vinje, W. E. & Gallant, J. L. 2001 Estimating spatio-temporal receptive fields of auditory and visual neurons from their responses to natural stimuli. *Network: Comput Neural Syst.* **12**, 289–316.

Ulanovsky, N., Las, L. & Nelken, I. 2003 Processing of low-probability sounds by cortical neurons. *Nat. Neurosci.* **6**, 391–398.

Van Tasell, D. J., Soli, S. D., Kirby, V. M. & Widin, G. P. 1987 Speech waveform envelope cues for consonant recognition. *J. Acoust. Soc. Am.* **82**, 1152–1161.

Versnel, H. & Shamma, S. A. 1998 Spectral-ripple representation of steady-state vowels in primary auditory cortex. *J. Acoust. Soc. Am.* **103**, 2502–2514.
Wang, X. 2000 On cortical coding of vocal communication sounds in primates. *PNAS* **97**, 11843–11849.
Wang, X. & Kadia, S. C. 2001 Differential representation of species-specific primate vocalizations in the auditory cortices of marmoset and cat. *J. Neurophysiol.* **86**, 2616–2620.
Wang, X., Merzenich, M. M., Beitel, R. & Schreiner, C. E. 1995. Representation of a species-specific vocalization in the primary auditory cortex of the common marmoset:temporal and spectral characteristics. *J. Neurophysiol.* **74**, 2685–2706.
Weinberger, N. M. 2007 Associative representational plasticity in the auditory cortex: a synthesis of two disciplines. *Learning and Memory* **14**, 1–16.
Winer, J. A. & Schreiner, C. E. 2005 *The Inferior Colliculus* New York: Springer.
Winter, I. M. & Palmer, A. R. 1990 Temporal responses of primarylike anteroventral cochlear nucleus units to the steady-state vowel /i/. *J. Acoust. Soc. Am.* **88**, 1437–1441.
Winter, I. M., Robertson, D. & Yates, G. K. 1990 Diversity of characteristic frequency rate-intensity functions in guinea pig auditory nerve fibres. *J. Speech Hear. Res.* **45**, 191–202.
Wong, J. C., Miller, R. L., Calhoun, B. M., Sachs, M. B. & Young, E. D. 1998 Effects of high sound levels on responses to the vowel /ɛ/ in cat auditory nerve. *J. Speech Hear. Res.* **123**, 61–77.
Wong, S. W. & Schreiner, C. E. 2003 Representation of CV-sounds in cat primary auditory cortex: intensity dependence. *Speech Comm.* **41**, 93–106.
Yates, G. K., Winter, I. M. & Robertson, D. 1990 Basilar membrane nonlinearity determines auditory nerve rate-intensity functions and cochlear dynamic range. *J. Speech Hear. Res.* **45**, 203–220.
Yeshurun, Y., Wollberg, Z. & Dyn, N. 1989 Prediction of linear and non-linear responses of MGB neurons by system identification methods. *Bulletin of Mathematical Biology* **51**, 337–346.
Yin, T. C. T. 2002 Neural mechanisms of encoding binaural localization cues in the auditory brainstem. In *Integrative Functions in the Mammalian Auditory Pathway* (eds D. Oertel, A. N. Popper & R. R. Fay), pp. 99–159. New York: Springer.
Yin, T. C. T. & Chan, J. C. K. 1990 Interaural time sensitivity in medial superior olive of cat. *J. Neurophysiol.* **64**, 465–488.
Young, E. D. & Sachs, M. B. 1979 Representation of steady-state vowels in the temporal aspects of the discharge patterns of populations of auditory-nerve fibers. *J. Acoust. Soc. Am.* **66**, 1381–1403.
Young, E. D. & Sachs, M. B. 1981 Processing of speech in the peripheral auditory system. In *The Cognitive Representation of Speech* (eds T. Myers, J. Laver & J. Anderson), pp. 75–92. Amsterdam: North-Holland.
Young, E. D. & Oertel, D. 2003 The cochlear nucleus. In *Synaptic Organization of the Brain* (ed. G. M. Shepherd), pp. 125–163. New York: Oxford University Press.
Young, E.D. & Sachs, M.B. 2008 Auditory nerve inputs to cochlear nucleus neurons studied with cross-correlation. *Neuroscience* **154**, 127–138.
Zurita, P., Villa, A. E., de Ribaupierre, Y., de Ribaupierre, F. & Rouiller, E. M. 1994 Changes of single unit activity in the cat's auditory thalamus and cortex associated to different anesthetic conditions. *J. Neurosci. Res.* **19**, 303–316.

3

Basic auditory processes involved in the analysis of speech sounds

Brian C. J. Moore

> This chapter reviews basic aspects of auditory processing that play a role in the perception of speech. The frequency selectivity of the auditory system, as measured using masking experiments, is described and used to derive the internal representation of the spectrum (the excitation pattern) of speech sounds. The perception of timbre, and distinctions in quality between vowels, are related to both static and dynamic aspects of the spectra of sounds. The perception of pitch and its role in speech perception is described. Measures of the temporal resolution of the auditory system are described, and a model of temporal resolution based on a sliding temporal integrator is outlined. The combined effects of frequency and temporal resolution can be modelled by calculation of the spectrotemporal excitation pattern, which gives good insight into the internal representation of speech sounds. For speech presented in quiet, the resolution of the auditory system in frequency and time usually markedly exceeds the resolution necessary for the identification or discrimination of speech sounds, which partly accounts for the robust nature of speech perception. However, for people with impaired hearing, speech perception is often much less robust.
>
> **Keywords:** Frequency selectivity, Hearing, Pitch, Temporal resolution, Timbre

3.1 Introduction

This chapter reviews selected aspects of auditory processing, chosen because they play a role in the perception of speech. The review is concerned with relatively basic processes, many of which are strongly influenced by the operation of the peripheral auditory system and which can be characterized using simple stimuli, such as pure tones and bands of noise. More central processes of auditory pattern analysis are described elsewhere in this volume. The resolution of the auditory system in frequency and time is characterized and its role in determining the internal representation of speech sounds is described. It turns out that the resolution of the auditory system in frequency and time, as measured in psychoacoustic experiments, usually markedly exceeds the resolution necessary for the identification or discrimination of speech sounds. This partly accounts for the fact that speech perception is robust, and resistant to distortion of the speech and to background noise. However, hearing impairment usually leads to a reduced ability to analyse and discriminate sounds, and background noise then has much more disruptive effects.

3.2 Frequency selectivity

(a) The concept of the auditory filter

Frequency selectivity refers to the ability to resolve the sinusoidal components in a complex sound, and it plays a role in many aspects of auditory perception, including the perception of loudness, pitch, and timbre. Fletcher (1940), following Helmholtz (1863), suggested that frequency selectivity can be modelled by considering the peripheral auditory system as a bank of bandpass filters, with overlapping passbands. These filters are called the 'auditory filters'. Fletcher thought that the basilar membrane within the cochlea (see chapter by Young, this volume) provided the basis for the auditory filters. Each location on the basilar membrane responds to a limited range of frequencies, so each different point corresponds to a filter with a different centre frequency.

Frequency selectivity can be most readily quantified by studying masking, which is the process by which the threshold of audibility for one sound is raised by the presence of another (masking) sound. The following assumptions are made about a listener trying to detect a sinusoidal signal in a broadband noise background:

(1) The listener makes use of an auditory filter with a centre frequency close to that of the signal. This filter passes the signal, but removes a great deal of the noise;

(2) Only the components in the noise which pass through the filter have any effect in masking the signal;

(3) The threshold for detecting the signal is determined by the amount of noise passing through the auditory filter; specifically, threshold is assumed to correspond to a certain signal-to-noise ratio at the output of the filter.

This set of assumptions is known as the 'power-spectrum model' of masking (Patterson & Moore 1986), since the stimuli are represented by their long-term power spectra, i.e. the short-term fluctuations in the masker are ignored. Although the assumptions of the model are not always valid (Moore 2003a), stimuli can be found for which the assumptions are not strongly violated.

The question considered next is 'What is the shape of the auditory filter?' In other words, how does its relative response change as a function of the input frequency? Most methods for estimating the shape of the auditory filter at a given centre frequency are based on the assumptions of the power-spectrum model of masking. The threshold of a signal whose frequency is fixed is measured in the presence of a masker whose spectral content is varied. It is assumed, as a first approximation, that the signal is detected using the single auditory filter which is centred on the frequency of the signal, and that threshold corresponds to a constant signal-to-masker ratio at the output of that filter. Both the methods described below measure the shape of the filter using this technique.

(b) Psychophysical tuning curves

One method of measuring the shape of the auditory filter involves a procedure which is analogous in many ways to the determination of a neural tuning curve (see the chapter by Young, this volume), and the resulting function is called a psychophysical tuning curve (PTC). To determine a PTC, the signal is fixed in level, usually at a very low level, say, 10 dB above absolute threshold (called 10 dB Sensation Level, SL). The masker can be either a sinusoid or a narrow band of noise.

For each of several masker frequencies, the level of the masker needed just to mask the signal is determined. Because the signal is at a low level, it is assumed that it produces activity primarily at the output of a single auditory filter. It is assumed further that, at threshold, the masker produces a constant output power from that filter, in order to mask the fixed signal. Thus, the PTC indicates the masker level required to produce a fixed output power from the auditory filter as a function of frequency. Normally, a filter characteristic is determined by plotting the output from the filter for an input varying in frequency and fixed in level. However, if the filter is linear, the two methods give the same result. Thus, assuming linearity, the shape of the auditory filter can be obtained simply by inverting the PTC. Examples of some PTCs are given in Figure 3.1.

One problem in interpreting PTCs is that, in practice, the listener may use information from more than one auditory filter. When the masker frequency is above the signal frequency, the listener might do better to use information from a filter centred just below the signal frequency. If the filter has a relatively flat top, and sloping edges, this will considerably attenuate the masker at the filter output, while only slightly attenuating the signal. By using this filter the listener can improve performance. This is known as 'off-frequency listening' or 'off-place listening', and there is good evidence that humans do indeed listen 'off-frequency' when it is advantageous to do so (Johnson-Davies & Patterson 1979; O'Loughlin & Moore 1981b). The result of off-frequency listening is that the PTC has a sharper tip than would be obtained if only one auditory filter were involved (O'Loughlin & Moore 1981a).

Another problem with PTCs is that they can be influenced by the detection of beats, which are amplitude fluctuations caused by the interaction of the signal and the masker. The rate of the beats is equal to the difference in frequency of the signal and masker. Beats of a low rate are more easily detected than beats with a rate above about 120 Hz (Kohlrausch et al. 2000), and slow beats provide a detection cue which results in an increase in the masker level required for threshold for masker frequencies adjacent to the signal frequency. This results in a PTC which has a sharper tip than the underlying auditory filter

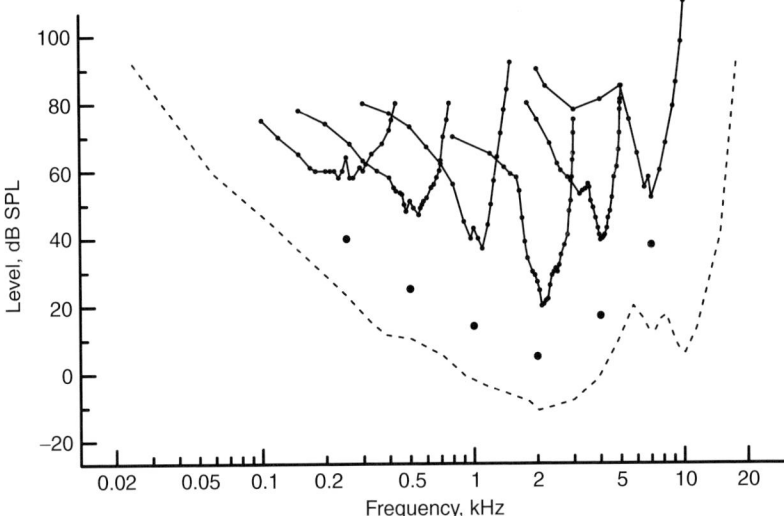

Figure 3.1 Psychophysical tuning curves (PTCs) determined in simultaneous masking, using sinusoidal signals at 10 dB SL. For each curve, the solid circle below it indicates the frequency and level of the signal. The masker was a sinusoid which had a fixed starting phase relationship to the 50-ms signal. The masker level required for threshold is plotted as a function of masker frequency on a logarithmic scale. The dashed line shows the absolute threshold for the signal. Data from Vogten (1978).

(Kluk & Moore 2004). This sharpening effect is greatest when a sinusoidal masker is used, but it occurs even when the masker is a narrowband noise (Kluk & Moore 2004).

(c) The notched-noise method

Patterson (1976) described a method of determining auditory filter shape which limits off-frequency listening and appears not to be influenced by beat detection. The method is illustrated in Figure 3.2. The signal (indicated by the bold, vertical line) is fixed in frequency, and the masker is a noise with a spectral notch centred at the signal frequency. The deviation of each edge of the notch from the centre frequency is denoted by Δf. The width of the notch is varied, and the threshold of the signal is determined as a function of notch width. Since the notch is symmetrically placed around the signal frequency, the method cannot reveal asymmetries in the auditory filter, and the analysis assumes that the filter is symmetric on a linear frequency scale. This assumption appears not unreasonable, at least for the top part of the filter and at moderate sound levels, since PTCs are quite symmetric around the tips. For a signal symmetrically placed in a notched noise, the optimum signal-to-masker ratio at the output of the auditory filter is achieved with a filter centred at the signal frequency, as illustrated in Figure 3.2.

As the width of the spectral notch is increased, less and less noise passes through the auditory filter. Thus the threshold of the signal drops. The amount of noise passing through the auditory filter is proportional to the area under the filter in the frequency range covered by the noise. This is shown as the shaded areas in Figure 3.2. Assuming that threshold corresponds to a constant signal-to-masker ratio at the output of the filter, the change in signal threshold with notch width indicates how the area under the filter varies with Δf. The area under a function between certain limits is obtained by integrating the value of the function over those limits. Hence, by differentiating the function relating threshold to Δf, the relative response of the filter at that value of Δf is obtained. In other words, the relative response of the filter for a given deviation, Δf, from the centre frequency is equal to the slope of the function relating signal threshold to notch width, at that value of Δf.

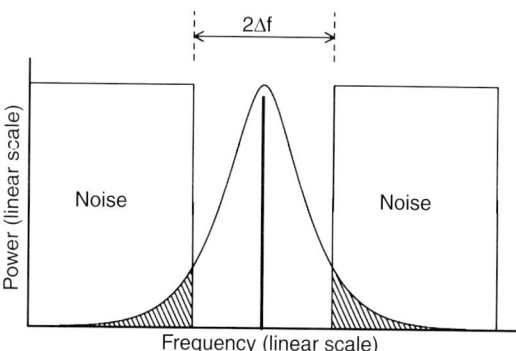

Figure 3.2 Schematic illustration of the technique used by Patterson (1976) to determine the shape of the auditory filter. The threshold of the sinusoidal signal (indicated by the bold vertical line) is measured as a function of the width of a spectral notch in the noise masker. The amount of noise passing through the auditory filter centred at the signal frequency is proportional to the shaded areas.

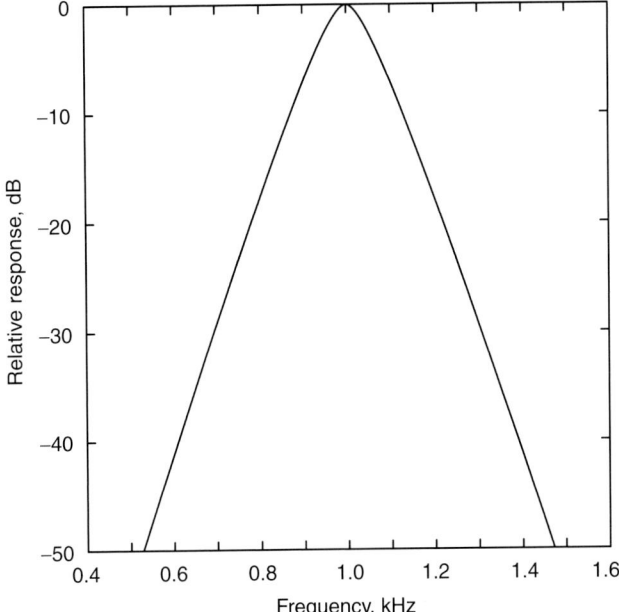

Figure 3.3 A typical auditory filter shape determined using the notched-noise method. The filter is centred at 1 kHz. The relative response of the filter (in dB) is plotted as a function of frequency.

A typical auditory filter derived using this method is shown in Figure 3.3. It has a rounded top and quite steep skirts. The sharpness of the filter is often specified as the bandwidth of the filter at which the response has fallen by a factor of 2 in power, i.e. by 3 dB. The 3-dB bandwidths of the auditory filters derived using the notched-noise method are typically between 10% and 15% of the centre frequency. An alternative measure is the equivalent rectangular bandwidth (ERB), which is the bandwidth of a rectangular filter which has the same peak transmission as the filter of interest and which passes the same total power for a white noise input. The ERB of the auditory filter is a little larger than the 3-dB bandwidth. In what follows, the mean ERB of the auditory filter determined using young listeners with normal hearing and using a moderate noise level is denoted ERB_N (where the subscript N denotes normal hearing). An equation describing the value of ERB_N as a function of centre frequency, F (in Hz), is (Glasberg & Moore 1990):

$$ERB_N = 24.7(0.00437F + 1) \qquad (3.1)$$

Sometimes, it is useful to plot psychoacoustical data on a frequency scale related to ERB_N, called the ERB_N-number scale. For example, the value of ERB_N for a centre frequency of 1 kHz is about 132 Hz, so an increase in frequency from 934 to 1066 Hz represents a step of one ERB_N-number. A formula relating ERB_N-number to frequency is (Glasberg & Moore 1990):

$$ERB_N\text{-number} = 21.4 \log_{10}(0.00437F + 1), \qquad (3.2)$$

where F is frequency in Hz. Each one-ERB_N step on the ERB_N-number scale corresponds, approximately, to a constant distance (0.9 mm) along the basilar membrane (Moore

1986). The ERB_N-number scale is conceptually similar to the Bark scale (Zwicker & Terhardt 1980), which has been widely used by speech researchers, although it differs somewhat in numerical values.

The notched-noise method has been extended to include conditions where the spectral notch in the noise is placed asymmetrically about the signal frequency. This allows the measurement of any asymmetry in the auditory filter, but the analysis of the results is more difficult, and has to take off-frequency listening into account (Patterson & Nimmo-Smith 1980). It is beyond the scope of this chapter to give details of the method of analysis; the interested reader is referred to Patterson and Moore (1986), Moore and Glasberg (1987), Glasberg and Moore (1990; 2000), and Rosen et al. (1998). The results show that the auditory filter is reasonably symmetric at moderate sound levels, but becomes increasingly asymmetric at high levels, the low-frequency side becoming shallower than the high-frequency side. The filter shapes derived using the notched-noise method are quite similar to inverted PTCs (Glasberg et al. 1984), except that PTCs are slightly sharper around their tips, probably as a result of off-frequency listening and beat detection.

(d) Masking patterns and excitation patterns

In the masking experiments described so far, the frequency of the signal was held constant, while the masker was varied. These experiments are most appropriate for estimating the shape of the auditory filter at a given centre frequency. However, in many experiments the masker was held constant in both level and frequency and the signal threshold was measured as a function of the signal frequency. The resulting functions are called masking patterns or masked audiograms.

Masking patterns show steep slopes on the low-frequency side (when the signal frequency is below that of the masker), of between 55 and 240 dB/octave. The slopes on the high-frequency side are less steep and depend on the level of the masker. Figure 3.4 shows a typical set of results, obtained using a narrowband noise masker centred at 410 Hz, with the overall masker level varying from 20 to 80 dB SPL in 10-dB steps (data from Egan & Hake (1950)). Notice that the curve is shallower at the highest level on the high-frequency side. Around the tip of the masking pattern, the growth of masking is approximately linear; a 10-dB increase in masker level leads to, roughly, a 10-dB increase in the signal threshold. However, for signal frequencies well above the masker frequency, in the range from about 1300 to 2000 Hz, when the level of the masker is increased by 10 dB (e.g. from 70 to 80 dB SPL), the masked threshold increases by more than 10 dB; the amount of masking grows non-linearly on the high-frequency side. This has been called the 'upward spread of masking'.

The masking patterns do not reflect the use of a single auditory filter. Rather, for each signal frequency the listener uses a filter centred close to the signal frequency. Thus, the auditory filter is shifted as the signal frequency is altered. One way of interpreting the masking pattern is as a crude indicator of the excitation pattern of the masker (Zwicker & Fastl 1999). The excitation pattern is a representation of the effective amount of excitation produced by a stimulus as a function of characteristic frequency (CF) on the basilar membrane (see the chapter by Young, this volume), and is plotted as effective level (in dB) against CF. In the case of a masking sound, the excitation pattern can be thought of as representing the relative amount of vibration produced by the masker at

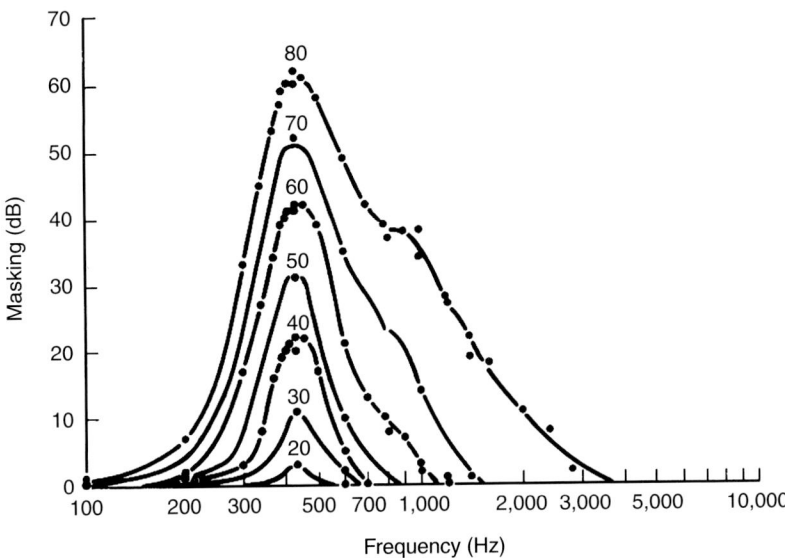

Figure 3.4 Masking patterns for a narrow-band noise masker centred at 410 Hz. Each curve shows the elevation in threshold of a pure-tone signal as a function of signal frequency. The overall noise level in dB SPL for each curve is indicated in the figure.
Data from Egan and Hake (1950).

different places along the basilar membrane. The signal is detected when the excitation it produces is some constant proportion of the excitation produced by the masker at places with CFs close to the signal frequency. Thus, the threshold of the signal as a function of frequency is proportional to the masker excitation level. The masking pattern should be parallel to the excitation pattern of the masker, but shifted vertically by a small amount. In practice, the situation is not so straightforward, since the shape of the masking pattern is influenced by factors such as off-frequency listening, the detection of beats and combination tones (Moore et al. 1998; Alcántara et al. 2000) and by the physiological process of suppression (Delgutte 1990) (see also, the chapter by Young, this volume).

Moore and Glasberg (1983b) have described a way of deriving the shapes of excitation patterns using the concept of the auditory filter. They suggested that the excitation pattern of a given sound can be thought of as the output of the auditory filters plotted as a function of their centre frequency. To calculate the excitation pattern of a sound, it is necessary to calculate the output of each auditory filter in response to that sound, and to plot the output as a function of the centre frequency of the filter. The characteristics of the auditory filters are determined using the notched-noise method described earlier. Figure 3.5 shows excitation patterns calculated in this way for 1000-Hz sinusoids with various levels. The patterns are similar in form to the masking patterns shown in Figure 3.4. Software for calculating excitation patterns can be downloaded from http://hearing.psychol.cam.ac.uk/Demos/demos.html.

It should be noted that excitation patterns calculated as described above do not take into account the physiological process of suppression, whereby the response to a given frequency component can be suppressed or reduced by a strong, neighbouring frequency component (Sachs & Kiang 1968; see also the chapter by Young, this volume). For speech

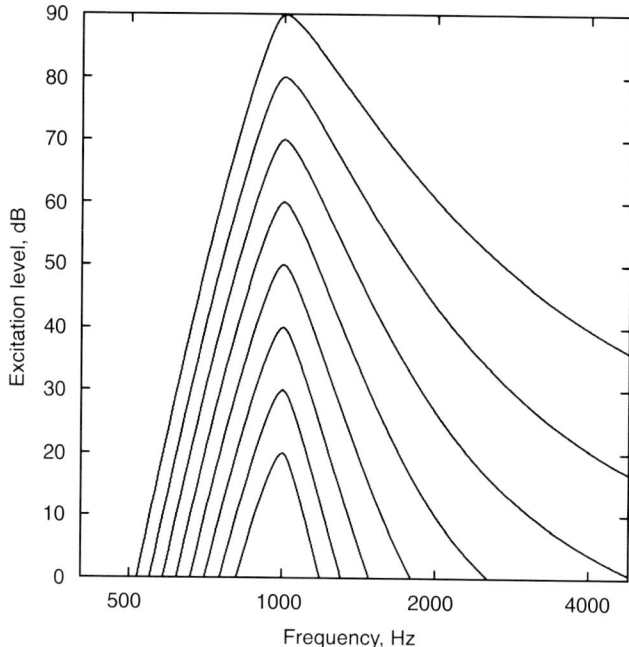

Figure 3.5 Excitation patterns for a 1000-Hz sinusoid at levels ranging from 20 to 90 dB SPL in 10-dB steps.

sounds having spectra with strong peaks and valleys, such as vowels, suppression may have the effect of increasing the peak-to-valley ratio of the excitation pattern (Moore & Glasberg 1983a). Also, the calculated excitation patterns are based on the power-spectrum model of masking, and do not take into account the effects of the relative phases of the components in complex sounds. However, it seems likely that excitation patterns provide a reasonable estimate of the extent to which the spectral features of complex sounds are represented in the auditory system.

(e) Excitation pattern of a vowel sound

The top panel of Figure 3.6 shows the spectrum of a synthetic vowel, /I/ as in 'bit', plotted on a linear frequency scale; this is the way that vowel spectra are often plotted. Each point represents the level of one harmonic in the complex sound (the fundamental frequency was 125 Hz). The middle panel shows the same spectrum plotted on an ERB_N-number scale; this gets somewhat closer to an auditory representation. The bottom panel shows the excitation pattern for the vowel, plotted on an ERB_N-number scale; this is closer still to an auditory representation. Several aspects of the excitation pattern are noteworthy. Firstly, the lowest few peaks in the excitation pattern do not correspond to formant frequencies, but rather to individual lower harmonics; these harmonics are resolved in the peripheral auditory system, and can be heard out as separate tones under certain conditions (Plomp 1964a; Moore & Ohgushi 1993). Hence the centre frequency of the first formant is not directly represented in the excitation pattern; if the frequency of the first formant is relevant for vowel identification (see the

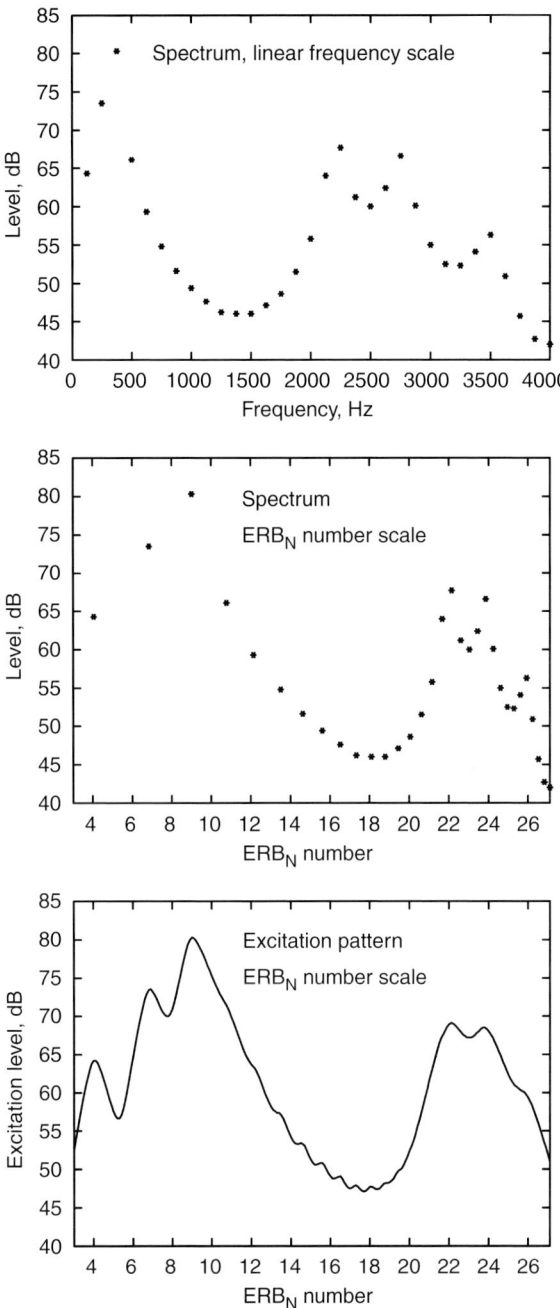

Figure 3.6 Top: the spectrum of a synthetic vowel /I/ plotted on a linear frequency scale. Middle: the same spectrum plotted on an ERB_N-number scale. Bottom: the excitation pattern for the vowel plotted on an ERB_N-number scale.

chapter by Diehl, this volume), then it must be inferred from the relative levels of the peaks corresponding to the individual lower harmonics.

A second noteworthy aspect of the excitation pattern is that, for this specific vowel, the second, third, and fourth formants, which are clearly separately visible in the original spectrum, are not well resolved. Rather, they form a single prominence in the excitation pattern, with only minor ripples corresponding to the individual formants. Assuming that the excitation pattern does give a reasonable indication of the internal representation of the vowel, the perception of this vowel probably depends more on the overall prominence than on the frequencies of the individual formants. For other vowels, the higher formants often lead to separate peaks in the excitation pattern (see Figure. 3.8).

(f) Frequency selectivity in cases of impaired hearing

In the developed countries, the most common cause of hearing loss is damage to the cochlea. This is usually associated with reduced frequency selectivity; the auditory filters are broader than normal (Moore 1998). As a result, the excitation patterns of complex sounds, such as vowels, are 'blurred' relative to those for normally hearing listeners. This makes it more difficult to distinguish the timbres of different vowel sounds. It also leads to increased susceptibility to masking by background sounds. For example, when trying to listen to a target talker in the presence of an interfering talker, a hearing-impaired person will be less able than a normal-hearing person to take advantage of differences in the short-term spectra of the two talkers, as described by Darwin (this volume).

3.3 Across-channel processes in masking

The discrimination and identification of complex sounds, including speech, requires comparison of the outputs of different auditory filters. This section reviews data on across-channel processes in auditory masking, and their relevance for speech perception.

(a) Co-modulation masking release

Hall *et al.* (1984) were among the first to demonstrate that across-filter comparisons could enhance the detection of a sinusoidal signal in a fluctuating noise masker. The crucial feature for achieving this enhancement was that the fluctuations should be correlated across different frequency bands. One of their experiments was similar to a classic experiment of Fletcher (1940). The threshold for detecting a 1000-Hz, 400-ms sinusoidal signal was measured as a function of the bandwidth of a noise masker, keeping the spectrum level constant. The masker was centred at 1000 Hz. They used two types of masker. One was a random noise; this has irregular fluctuations in amplitude, and the fluctuations are independent in different frequency regions. The other was a random noise which was modulated in amplitude at an irregular, low rate; a noise lowpass filtered at 50 Hz was used as a modulator. The modulation resulted in fluctuations in the amplitude of the noise which were the same in different frequency regions. This across-frequency correlation was called 'co-modulation' by Hall *et al.* (1984). Figure 3.7 shows the results of this experiment.

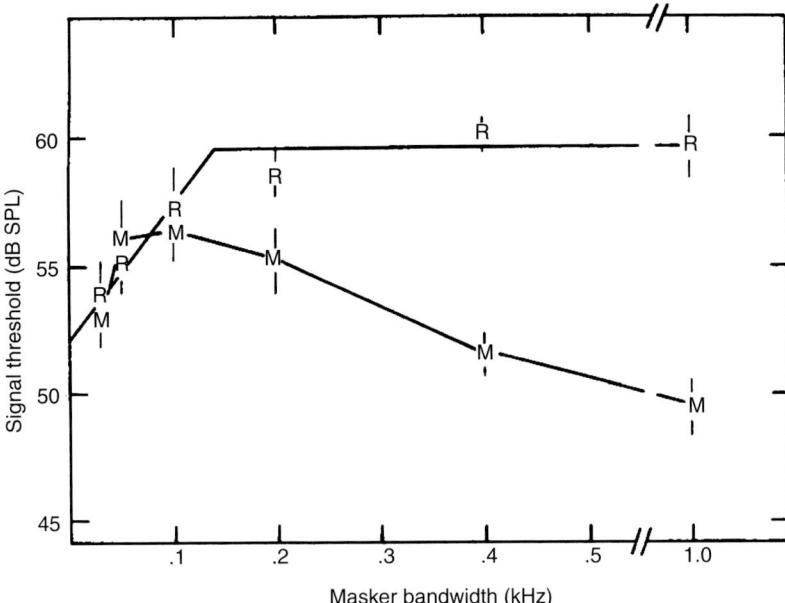

Figure 3.7 The points labelled 'R' are thresholds for detecting a 1-kHz signal centred in a band of random noise, plotted as a function of the bandwidth of the noise. The points labelled 'M' are the thresholds obtained when the noise was amplitude modulated at an irregular, low rate.
From Hall et al. (1984), by permission of the authors and the Journal of the Acoustical Society of America.

For the random noise (denoted by R), the signal threshold increases as the masker bandwidth increases up to about 100–200 Hz, and then remains constant – a result similar to that of Fletcher (1940). The value of ERB_N at this centre frequency is about 130 Hz. Hence, for noise bandwidths up to 130 Hz, increasing the bandwidth results in more noise passing through the filter. However, increasing the bandwidth beyond 130 Hz does not substantially increase the noise power passing through the filter, so threshold does not increase. The pattern for the modulated noise (denoted by M) is quite different. For noise bandwidths greater than 100 Hz, the signal threshold decreases as the bandwidth increases. This suggests that subjects can compare the outputs of different auditory filters to enhance signal detection (See, however, Verhey et al. (1999)). The fact that the decrease in threshold with increasing bandwidth only occurs with the modulated noise indicates that fluctuations in the masker are critical and that the fluctuations need to be correlated across frequency bands. Hence, this phenomenon has been called 'co-modulation-masking release' (CMR).

It seems likely that across-filter comparisons of temporal envelopes are a general feature of auditory pattern analysis, which may play an important role in extracting signals from noisy backgrounds, or separating competing sources of sound (see the chapter by Darwin, this volume). As pointed out by Hall et al. (1984): 'Many real-life auditory stimuli have intensity peaks and valleys as a function of time in which intensity trajectories are highly correlated across frequency. This is true of speech, of interfering noise such as 'cafeteria' noise, and of many other kinds of environmental stimuli'. However, the importance of CMR for speech perception remains controversial.

Some studies have suggested that it plays only a very minor role in the detection and identification of speech sounds in modulated background noise (Grose & Hall 1992; Festen 1993), although common modulation of target speech and background speech can lead to reduced intelligibility (Stone & Moore 2004). For synthetic speech in which the cues are impoverished compared to normal speech (sinewave speech; see Remez *et al.* (1981)), co-modulation of the speech (amplitude modulation by a sinusoid) can markedly improve the intelligibility of the speech, both in quiet (Carrell & Opie 1992) and in background noise (Carrell 1993). The amplitude modulation may help because it leads to perceptual fusion of the components of the sinewave speech, so as to form an auditory object (see the chapter by Darwin, this volume).

(b) Profile analysis

Green and his colleagues (Green 1988) have carried out a series of experiments demonstrating that, even for stimuli without distinct envelope fluctuations, subjects are able to compare the outputs of different auditory filters to enhance the detection of a signal. They investigated the ability to detect an increment in the level of one component in a complex sound relative to the level of the other components; the other components are called the 'background'. Usually, the complex sound has been composed of a series of equal-amplitude sinusoidal components, uniformly spaced on a logarithmic frequency scale. To prevent subjects from performing the task by monitoring the magnitude of the output of the single auditory filter centred at the frequency of the incremented component, the overall level of the whole stimulus was varied randomly from one stimulus to the next, over a relatively large range (typically about 40 dB). This makes the magnitude of the output of any single filter an unreliable cue to the presence of the signal.

Subjects were able to detect changes in the relative level of the signal of only 1–2 dB. Such small thresholds could not be obtained by monitoring the magnitude of the output of a single auditory filter. Green and his colleagues have argued that subjects performed the task by detecting a change in the shape or profile of the spectrum of the sound; hence the name 'profile analysis'. In other words, subjects can compare the outputs of different auditory filters, and can detect when the output of one changes relative to that of others, even when the overall level is varied. This is equivalent to detecting changes in the shape of the excitation pattern.

Speech researchers will not find the phenomenon of profile analysis surprising. It has been known for many years that spectral shape is one of the main factors determining the timbre or quality of a sound (see Section 4). Our everyday experience tells us that we can recognize and distinguish familiar sounds, such as the different vowels, regardless of the levels of those sounds. When we do this, we are distinguishing different spectral shapes in the face of variations in overall level. This is functionally the same as profile analysis. The experiments on profile analysis can be regarded as a way of quantifying the limits of our ability to distinguish changes in spectral shape. In this context, it is noteworthy that the differences in spectral shape between different vowels result in differences in the excitation patterns evoked by those sounds which are generally far larger than the smallest detectable changes as measured in profile-analysis experiments. This is illustrated in Figure 3.8, which shows excitation patterns for three vowels – /i/, /a/, and /u/ – plotted on an ERB_N-number scale. Each vowel had an overall level of about 58 dB SPL. It can be seen that the differences in the shapes of the excitation patterns are considerable.

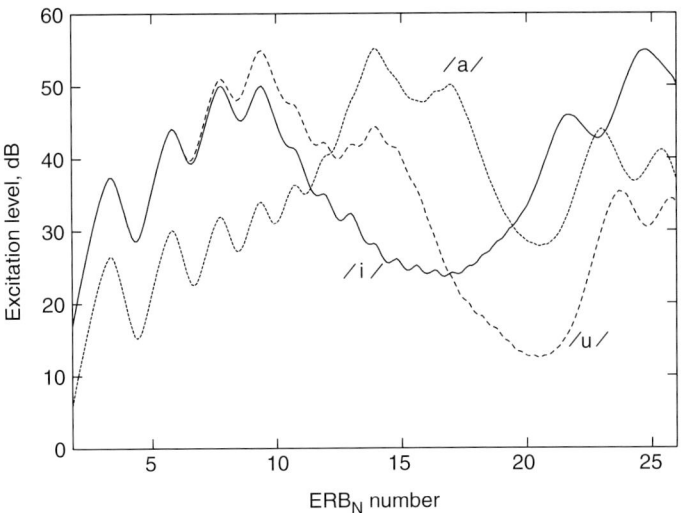

Figure 3.8 Excitation patterns for three vowels, /i/, /a/, and /u/, plotted on an ERB$_N$-number scale.

(c) Modulation-discrimination interference

In some situations, the detection or discrimination of a signal is impaired by the presence of frequency components remote from the signal frequency. Usually, this happens when the task is either to detect modulation of the signal, or to detect a change in depth of modulation of the signal. Yost and Sheft (1989) showed that the threshold for detecting sinusoidal amplitude modulation (AM) of a sinusoidal carrier was increased in the presence of another carrier – amplitude modulated at the same rate – even when the second carrier was remote in frequency from the first. They called this modulation-detection interference (MDI). They showed that MDI did not occur if the second carrier was unmodulated.

Moore *et al.* (1991) determined how thresholds for detecting an increase in modulation depth (sinusoidal AM or frequency modulation) of a 1000-Hz carrier frequency (the target) were affected by modulation of carriers (interferers) with frequencies of 230 and 3300 Hz. They found that modulation-increment thresholds were increased (worsened) when the remote carriers were modulated. This MDI effect was greatest when the target and interferers were modulated at similar rates, but the effect was broadly tuned for modulation rate. When the target and interfering sounds were both modulated at 10 Hz, there was no significant effect of the relative phase of modulation of the target and interfering sounds. A lack of effect of relative phase has also been found by other researchers (Moore 1992; Hall *et al.* 1995).

The explanation for MDI remains unclear. Yost and Sheft (1989) suggested that MDI might be a consequence of perceptual grouping; the common AM of the target and interfering sounds might make them fuse perceptually, making it difficult to 'hear out' the modulation of the target sound (see the chapter by Darwin, this volume). However, certain aspects of the results on MDI are difficult to reconcile with an explanation in terms of perceptual grouping (Moore & Shailer 1992). One would expect that widely spaced frequency components would only be grouped perceptually if their modulation pattern was very similar. Grouping would not be expected, e.g. if the components were modulated out of phase or at different rates, but, in fact, it is possible to obtain large amounts of MDI under these conditions.

An alternative explanation for MDI is that it reflects the operation of 'channels' specialized for detecting and analysing modulation (Kay & Mathews 1972; Dau *et al.* 1997a, 1997b). Yost *et al.* (1989) suggested that MDI might arise in the following way. The stimulus is first processed by an array of auditory filters. The envelope at the output of each filter is extracted. When modulation is present, channels tuned for modulation rate are excited. All filters responding with the same modulation rate excite the same channel, regardless of the centre frequency of the filter. Thus, modulation at one centre frequency can adversely affect the detection and discrimination of modulation at other centre frequencies.

The purpose of the hypothetical modulation channels remains unclear. Since physiological evidence suggests that such channels exist in animals (Schreiner & Urbas 1986; Langner & Schreiner 1988), we can assume that they did not evolve for the purpose of speech perception. Nevertheless, it is possible, even likely, that speech analysis makes use of the modulation channels. There is evidence that amplitude modulation patterns in speech are important for speech recognition (Steeneken & Houtgast 1980; Drullman *et al.* 1994a; Shannon *et al.* 1995). Thus, anything that adversely affects the detection and discrimination of the modulation patterns would be expected to impair intelligibility. One way of describing MDI is: modulation in one frequency region may make it more difficult to detect and discriminate modulation in another frequency region. Thus, it may be the case that MDI makes speech recognition more difficult in situations where there is a background sound that is modulated, such as one or more people talking (Brungart *et al.* 2005).

3.4 Timbre perception

Timbre is usually defined as 'that attribute of auditory sensation in terms of which a listener can judge that two sounds similarly presented and having the same loudness and pitch are dissimilar' (ANSI, 1994). The distribution of energy over frequency is one of the major determinants of timbre. However, timbre depends upon more than just the frequency spectrum of the sound; fluctuations over time can play an important role, as discussed below.

Timbre is multidimensional; there is no single scale along which the timbres of different sounds can be compared or ordered. Thus, a way is needed of describing the spectrum of a sound which takes into account this multidimensional aspect, and which can be related to the subjective timbre. For steady sounds, a crude first approach is to look at the overall distribution of spectral energy. The 'brightness' or 'sharpness' (von Bismarck 1974) of sounds seems to be related to the spectral centroid. However, a much more quantitative approach has been described by Plomp and his colleagues (Plomp 1970, 1976). They showed that the perceptual differences between different steady sounds, such as vowels, were closely related to the differences in the spectra of the sounds, when the spectra were specified as the levels in 18 1/3-octave frequency bands. A bandwidth of 1/3 octave is slightly greater than the ERB_N of the auditory filter over most of the audible frequency range. Thus, timbre is related to the relative level produced at the output of each auditory filter. Put another way, the timbre of a steady sound is related to the excitation pattern of that sound.

It is likely that the number of dimensions required to characterize the timbre of steady sounds is limited by the number of ERB_Ns required to cover the audible frequency range. This would give a maximum of about 37 dimensions. For a restricted class of sounds, such as vowels, a much smaller number of dimensions may be involved. It appears to be generally true – both for speech and non-speech sounds – that the timbres of steady tones are determined primarily by their magnitude spectra, although the relative phases of the components may also play a small role (Plomp & Steeneken 1969; Patterson 1987).

Differences in spectral shape are not always sufficient to allow the absolute identification of an 'auditory object', such as a musical instrument or a speech sound. One reason for this is that the magnitude and phase spectrum of the sound may be markedly altered by the transmission path and room reflections (Watkins, 1991). In practice, the recognition of a particular timbre, and hence of an 'auditory object', may depend upon several other factors. Schouten (1968) has suggested that these include: (1) whether the sound is periodic, having a tonal quality for repetition rates between about 20 and 20 000 periods per second, or irregular, and having a noise-like character; (2) whether the waveform envelope is constant, or fluctuates as a function of time, and, in the latter case, what the fluctuations are like; (3) whether any other aspect of the sound (e.g. spectrum or periodicity) is changing as a function of time; and (4) what the preceding and following sounds are like.

A powerful demonstration of the last factor may be obtained by listening to a stimulus with a particular spectral structure and then switching rapidly to a stimulus with a flat spectrum, such as white noise. A white noise heard in isolation may be described as 'colourless'; it has no pitch and has a neutral timbre. However, when a white noise follows immediately after a stimulus with spectral structure, the noise sounds 'coloured'. The coloration corresponds to the inverse of the spectrum of the preceding sound. For example, if the preceding sound is a noise with a spectral notch, the white noise has a pitch-like quality, with a pitch value corresponding to the centre frequency of the notch (Zwicker 1964). It sounds like a noise with a small spectral peak. A harmonic complex tone with a flat spectrum may be heard as having a vowel-like quality if it is preceded by a harmonic complex having a spectrum which is the inverse of that of a vowel (Summerfield et al. 1987).

The cause of this effect is not clear. Three types of explanation have been advanced, based on adaptation in the auditory periphery (see the chapter by Young, this volume), perceptual grouping (see the chapter by Darwin, this volume) and comparison of spectral shapes of the preceding and test sounds (Summerfield & Assmann 1987). All may play a role to some extent, depending on the exact properties of the stimuli. Whatever the underlying mechanism, it appears that the auditory system is especially sensitive to 'changes' in spectral patterns over time (Kluender et al. 2003). This may be of value for communication in situations where the spectral shapes of sounds are (statically) altered by room reverberation or by a transmission channel with a non-flat frequency response.

Perceptual compensation for the effects of a non-flat frequency response has been studied extensively by Watkins and co-workers (Watkins 1991; Watkins & Makin 1996a; 1996b). In one series of experiments (Watkins & Makin 1996b), they investigated how the identification of vowel test sounds was affected by filtering of preceding and following sounds. All sounds were edited and processed from natural speech spoken with a British accent. Listeners identified words from continua between /ɪtʃ/ and /ɛtʃ/ (itch and etch), /æpt/ and /ɒpt/ (apt and opt), or /sləʊ/ and /fləʊ/ (slow and flow). The parts of the stimuli other than the vowels (e.g., the /tʃ/ or the /pt/) were filtered with complex frequency responses corresponding to the difference of spectral envelopes from the end-point test sounds (the vowels). An example of a 'difference filter' is shown in the bottom panel of

Figure 3.9. The shift in the phoneme boundary of the vowels was used to measure perceptual compensation for the effects of the spectral distortion of the consonants. When the words were presented without a precursor phrase, the results indicated perceptual compensation. Thus, information from the consonants modified the perception of the preceding or following vowels. When a precursor phrase 'the next word is' was used, and was filtered in the same way as the consonant, the effects were larger. However, the large effects were somewhat reduced when the precursor phrase was filtered but the following consonant was not. This clearly indicates a role for sounds following a vowel. The effects of following sounds found in these experiments clearly indicate that factors other than adaptation play a role. Presumably, these effects reflect relatively central perceptual compensation mechanisms.

Overall, the results described in this section indicate that the perceived timbre of brief segments of sounds can be strongly influenced by sounds that precede and follow those segments. In some cases, the observed effects appear to reflect relatively central perceptual compensation processes.

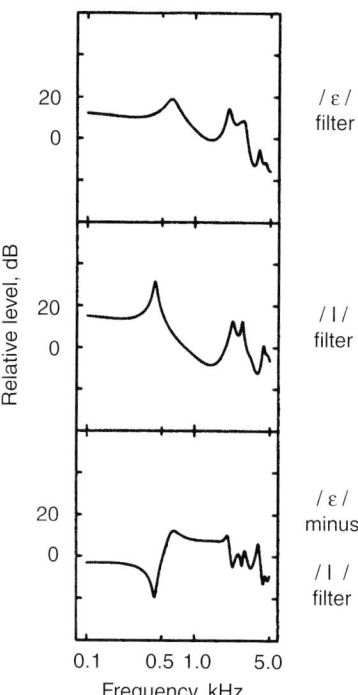

Figure 3.9 Illustration of the filters used by Watkins and Makin (1996a). The top and middle panels show 'filters' corresponding to the spectral envelopes of the vowels /ɛ/ and /ɪ/. The bottom panel shows the filter corresponding to the difference between the spectral envelopes of the /ɛ/ and /ɪ/.

3.5 The perception of pitch

Pitch is usually defined as 'that attribute of auditory sensation in terms of which sounds can be ordered on a scale extending from low to high' (ANSI 1994). In other words, variations in pitch give rise to a sense of melody. For speech sounds, variations in voice pitch over time convey intonation information, indicating whether an utterance is a question or a statement, and helping to identify stressed words. Voice pitch can also convey information about the sex, age, and emotional state of the speaker (Rosen & Fourcin 1986). In some languages ('tone' languages), pitch and variations in pitch distinguish different lexical items.

Pitch is related to the repetition rate of the waveform of a sound; for a pure tone this corresponds to the frequency and for a periodic complex tone to the fundamental frequency, $F0$. There are, however, exceptions to this simple rule. Since voiced speech sounds are complex tones, this section will focus on the perception of pitch for complex tones.

(a) The phenomenon of the missing fundamental

Although the pitch of a complex tone usually corresponds to its $F0$, the component with frequency equal to $F0$ does not have to be present for the pitch to be heard. Consider, as an example, a sound consisting of short impulses (clicks) occurring 200 times per second. This sound has a low pitch, which is very close to the pitch of a 200-Hz sinusoid, and a sharp timbre. It contains harmonics with frequencies 200, 400, 600, 800 ... etc., Hz. However, if the sound is filtered so as to remove the 200-Hz component, the pitch does not alter; the only result is a slight change in the timbre of the note. Indeed, all except a small group of mid-frequency harmonics can be eliminated, and the low pitch still remains, although the timbre becomes markedly different.

Schouten (1970) called the low pitch associated with a group of high harmonics the 'residue'. He pointed out that the residue is distinguishable, subjectively, from a fundamental component which is physically presented or from a fundamental which may be generated (at high sound-pressure levels) by nonlinear distortion in the ear. Thus the perception of a residue pitch does not require activity at the point on the basilar membrane which would respond maximally to a pure tone of similar pitch. Several other names have been used to describe residue pitch, including 'periodicity pitch', 'virtual pitch', and 'low pitch'. This chapter will use the term low pitch. Even when the fundamental component of a complex tone is present, the low pitch of the tone is usually determined by harmonics other than the fundamental. Thus, the perception of a low pitch should not be regarded as unusual. Rather, low pitches are normally heard when listening to complex tones, including speech. For example, when listening over the telephone, the fundamental component for male speakers is usually inaudible, but the pitch of the voice can still be easily heard.

(b) The principle of dominance

Ritsma (1967) carried out an experiment to determine which components in a complex sound are most important in determining its pitch. He presented complex tones in which the frequencies of a small group of harmonics were multiples of an $F0$ which was slightly higher or lower than the $F0$ of the remainder. The subject's pitch judgements were used to determine whether the pitch of the complex as a whole was affected by the shift in the group of harmonics. Ritsma found that:

> For fundamental frequencies in the range 100 Hz to 400 Hz, and for sensation levels up to at least 50 dB above threshold of the entire signal, the frequency band consisting of the third, fourth and fifth harmonics tends to dominate the pitch sensation as long as its amplitude exceeds a minimum absolute level of about 10 dB above threshold.

This finding has been broadly confirmed in other ways (Plomp 1967), although the data of Moore *et al.* (1984, 1985) show that there are large individual differences in which harmonics are dominant, and for some subjects the first two harmonics play an important role. Other data also show that the dominant region is not fixed in terms of harmonic number, but depends somewhat on absolute frequency (Plomp 1967; Patterson & Wightman 1976). For high $F0$s (above about 1000 Hz), the fundamental is usually the dominant component, while for very low $F0$s, around 50 Hz, harmonics above the fifth may be dominant (Moore & Glasberg 1988; Moore & Peters 1992). Finally, the dominant region shifts somewhat towards higher harmonics with decreasing duration (Gockel *et al.* 2005). For speech sounds, the dominant harmonics usually lie around the frequency of the first formant.

(c) *Discrimination of the pitch of complex tones*

When the $F0$ of a periodic complex tone changes, all of the components change in frequency by the same ratio, and a change in low pitch is heard. The ability to detect such changes is better than the ability to detect changes in a sinusoid at $F0$ (Flanagan & Saslow 1958), and can be better than the ability to detect changes in the frequency of any of the sinusoidal components in the complex tone (Moore *et al.* 1984). This indicates that information from the different harmonics is combined or integrated in the determination of low pitch. This can lead to very fine discrimination; changes in $F0$ of about 0.2% can often be detected for $F0$s in the range 100–400 Hz.

The discrimination of $F0$ is usually best when low harmonics are present (Hoekstra & Ritsma 1977; Moore & Glasberg 1988; Shackleton & Carlyon 1994). Somewhat less good discrimination (typically 1–4%) is possible when only high harmonics are present (Houtsma & Smurzynski 1990). $F0$ discrimination can be impaired (typically by about a factor of two) when the two sounds to be discriminated also differ in timbre (Moore & Glasberg 1990); this can be the situation with speech sounds, where changes in $F0$ are usually accompanied by changes in timbre.

In speech, intonation is typically conveyed by differences in the pattern of $F0$ change over time. When the stimuli are dynamically varying, the ability to detect $F0$ changes is markedly poorer than when the stimuli are steady. Klatt (1973) measured thresholds for detecting differences in $F0$ for an unchanging vowel (i.e. one with static formant frequencies) with a flat $F0$ contour, and also for a series of linear glides in $F0$ around an $F0$ of 120 Hz. For the flat contour, the threshold was about 0.3 Hz. When both contours were falling at the same rate (30 Hz over the 250-ms duration of the stimulus), the threshold increased markedly to 2 Hz. When the steady vowel was replaced by the sound /ya/, whose formants change over time, thresholds increased further by 25–65%.

Generally, the $F0$ changes that are linguistically relevant for conveying stress and intonation are much larger than the limits of $F0$ discrimination measured psychophysically using steady stimuli. This is another reflection of the fact that information in speech is conveyed using robust cues that do not severely tax the discrimination abilities of the auditory system. Again, however, this may not be true for people with impaired hearing, for whom *F0* discrimination is often much worse than normal (Moore & Carlyon 2005).

It should be noted that in natural speech the period (corresponding to the time between successive closures of the vocal folds) varies randomly from one period to the next (Fourcin & Abberton 1977). This jitter conveys information about the emotional state of the talker, and is required for a natural voice quality to be perceived. Human listeners can detect jitter of 1–2% (Pollack 1968; Kortekaas & Kohlrausch 1999). Large amounts of jitter are associated with voice pathologies, such as hoarseness (Yumoto *et al.* 1982).

(d) Perception of pitch in speech

Data on the perception of *F0* contours in a relatively natural speech context were presented by Pierrehumbert (1979). She started with a natural nonsense utterance 'ma-MA-ma-ma-MA-ma', in which the prosodic pattern was based on the sentence 'The baker made bagels'. The stressed syllables (MA) were associated with peaks in the *F0* contour. She then modified the *F0* of the second peak, over a range varying from below to above the *F0* of the first peak. Subjects listened to the modified utterances, and were required to indicate whether the first or second peak was higher in pitch. The results reflected what she called 'normalization for expected declination'; when the two stressed syllables sounded equal in pitch, the second was actually lower in *F0*. For first peak values of 121 and 151 Hz, the second peak had to be shifted over a range of about 20 Hz to change judgements from 75% 'second peak lower' to 75% 'second peak higher'. This indicates markedly poorer discriminability than found for steady stimuli. Similarly, 't Hart (1981) found that about a 19% difference was necessary for successive pitch movements in the same direction to be reliably heard as different in extent.

Hermes and van Gestel (1991) studied the perception of the excursion size of prominence-lending *F0* movements in utterances resynthesized in different *F0* registers. The task of the subjects was to adjust the excursion size in a comparison stimulus in such a way that it lent equal prominence to the corresponding syllable in a fixed test stimulus. The comparison stimulus and the test stimulus had *F0*s running parallel on either a logarithmic frequency scale, an ERB_N-number scale, or a linear frequency scale. They found that stimuli were matched in such a way that the average excursion sizes in different registers were equal when the ERB_N-number scale was used. Put another way, the perceived prominence of *F0* movements is related to the size of those movements expressed on an ERB_N-number scale.

3.6 Temporal analysis

Time is a very important dimension in hearing, since almost all sounds change over time. For speech, much of the information appears to be carried in the changes themselves, rather than in the parts of the sounds which are relatively stable (Kluender *et al.* 2003). In characterizing temporal analysis, it is essential to take account of the filtering that takes place in the peripheral auditory system. Temporal analysis can be considered as resulting from two main processes: analysis of the time pattern occurring within each frequency channel; and comparison of the time patterns across channels. This chapter focusses on the first of these.

A major difficulty in measuring the temporal resolution of the auditory system is that changes in the time pattern of a sound are generally associated with changes in its magnitude spectrum – the distribution of energy over frequency. Thus, the detection of a change in time pattern can sometimes depend not on temporal resolution *per se*, but on the detection of the spectral change. At times, the detection of spectral changes can lead to what appears to be extraordinarily fine temporal resolution. For example, a single click can be distinguished from a pair of clicks when the gap between the two clicks in a pair is only a few tens of microseconds – an ability that depends upon spectral changes at very high frequencies (Leshowitz 1971). Although spectrally based detection of temporal changes can occur for speech sounds, this chapter focusses on experimental situations which avoid the confounding effects of spectral cues.

There have been two general approaches to avoiding the use of cues based on spectral changes. One is to use signals whose magnitude spectrum is not changed when the time pattern is altered. For example, the magnitude spectrum of white noise remains flat if a gap is introduced into the noise. The second approach uses stimuli whose spectra are altered by the change in time pattern, but extra background sounds are added to mask the spectral changes. Both of these approaches will be considered.

(a) Within-channel temporal analysis using broadband sounds

The experiments described next all use broadband sounds whose long-term magnitude spectrum is unaltered by the temporal manipulation being performed. For example, interruption or amplitude modulation of a white noise does not change its long-term magnitude spectrum, and time-reversal of any sound also does not change its long-term magnitude spectrum.

The threshold for detecting a gap in a broadband noise provides a simple and convenient measure of temporal resolution. The gap threshold is typically 2–3 ms (Plomp 1964b). The threshold increases at very low sound levels, when the level of the noise approaches the absolute threshold, but is relatively invariant with level for moderate to high levels.

Ronken (1970) used as stimuli pairs of clicks differing in amplitude. One click, labelled A, had an amplitude greater than that of the other click, labelled B. Typically the amplitude of A was twice that of B. Subjects were required to distinguish click pairs differing in the order of A and B: either AB or BA. The ability to do this was measured as a function of the time interval or gap between A and B. Ronken found that subjects could distinguish the click pairs for gaps down to 2–3 ms. Thus the limit to temporal resolution found in this task is similar to that found for the detection of a gap in broadband noise. It should be noted that, in this task, subjects do not hear the individual clicks within a click pair. Rather, each click pair is heard as a single sound with its own characteristic quality. For example, the two click pairs AB and BA might sound like 'tick' and 'tock'.

The experiments described above each give a single value to describe temporal resolution. A more general approach is to measure the threshold for detecting changes in the amplitude of a sound as a function of the rapidity of the changes. In the simplest case, white noise is sinusoidally amplitude modulated, and the threshold for detecting the modulation is determined as a function of modulation rate. The function relating threshold to modulation rate is known as a temporal modulation transfer function (TMTF) (Viemeister 1979). An example of a TMTF is shown in Figure 3.10 (Bacon &

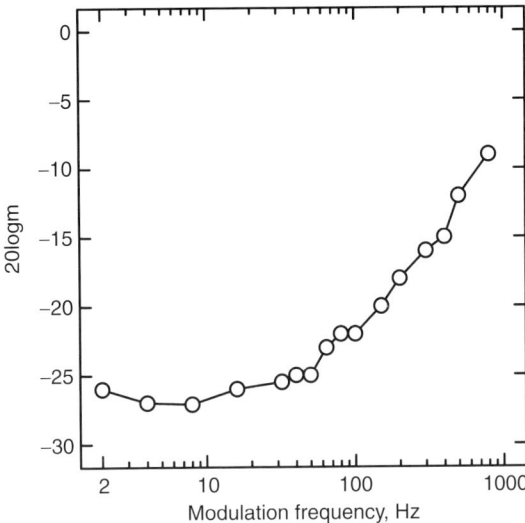

Figure 3.10 A temporal modulation transfer function (TMTF). A broadband white noise was sinusoidally amplitude modulated, and the threshold amount of modulation required for detection is plotted as a function of modulation rate. The amount of modulation is specified as 20log$m$, where m is the modulation index. The higher the sensitivity to modulation, the more negative is this quantity. Data from Bacon and Viemeister (1985).

Viemeister 1985). The thresholds are expressed as 20log$m$, where m is the modulation index ($m = 0$ corresponds to no modulation and $m = 1$ corresponds to 100% modulation). For low modulation rates, performance is limited by the amplitude resolution of the ear, rather than by temporal resolution. Thus, the threshold is independent of modulation rate for rates up to about 50 Hz. As the rate increases beyond 50 Hz, temporal resolution starts to have an effect; performance worsens, and for rates above about 1000 Hz the modulation is hard to detect at all. Thus, sensitivity to modulation becomes progressively less as the rate of modulation increases. The shapes of TMTFs do not vary much with overall sound level, but the ability to detect the modulation does worsen at low sound levels. Over the range of modulation rates important for speech perception, below about 50 Hz (Steeneken & Houtgast 1980; Drullman *et al.* 1994a, 1994b), the sensitivity to modulation is rather good.

(b) Within-channel temporal analysis using narrowband sounds

Experiments using broadband sounds provide no information regarding the question of whether the temporal resolution of the auditory system varies with centre frequency. This issue can be examined by using narrowband stimuli that excite only one, or a small number, of auditory channels.

Green (1973) used stimuli where each stimulus consisted of a brief pulse of a sinusoid in which the level of the first-half of the pulse was 10 dB different from that of the second-half. Subjects were required to distinguish two signals, differing in whether the half with the high level was first or second. Green measured performance as a function of the total duration of the stimuli. The threshold was similar for centre frequencies of

2 and 4 kHz, and was between 1 and 2 ms. However, the threshold was slightly higher for a centre frequency of 1 kHz, being between 2 and 4 ms.

Performance in this task was actually a non-monotonic function of duration. Performance was good for durations in the range 2–6 ms, worsened for durations around 16 ms, and then improved again as the duration was increased beyond 16 ms. For the very short durations, subjects listened for a difference in quality between the two sounds – rather like the 'tick' and 'tock' described earlier for Ronken's stimuli. At durations around 16 ms, the tonal quality of the bursts became more prominent, and the quality differences were harder to hear. At much longer durations the soft and loud segments could be separately heard, in a distinct order. It appears, therefore, that performance in this task was determined by two separate mechanisms, one based on timbre differences associated with the difference in time pattern, and the other based on the perception of a distinct succession of auditory events.

Several researchers have measured thresholds for detecting gaps in narrowband sounds, either noises (Fitzgibbons 1983; Shailer & Moore 1983; Buus & Florentine 1985; Eddins *et al.* 1992) or sinusoids (Shailer & Moore 1987; Moore *et al.* 1993). When a temporal gap is introduced into a narrowband sound, the spectrum of the sound is altered. Energy 'splatter' occurs outside the nominal frequency range of the sound. To prevent the splatter being detected, the sounds are presented in a background sound, usually a noise, designed to mask the splatter.

Gap thresholds for noise bands decrease with increasing bandwidth, but show little effect of centre frequency when the bandwidth is held constant. For noises of moderate bandwidth (a few hundred Hz), the gap threshold is typically about 10 ms. Gap thresholds for narrowband noises tend to decrease with increasing sound level for levels up to about 30 dB above absolute threshold, but remain roughly constant after that.

Shailer and Moore (1987) showed that the detectability of a gap in a sinewave was strongly affected by the phase at which the sinusoid was turned off and on to produce the gap (Shailer & Moore 1987). Only the simplest case is considered here, called 'preserved phase' by Shailer and Moore (1987). In this case, the sinusoid was turned off at a positive-going zero crossing (i.e. as the waveform was about to change from negative to positive values) and it started (at the end of the gap) at the phase it would have had if it had continued without interruption. Thus, for the preserved-phase condition it was as if the gap had been 'cut out' from a continuous sinusoid. For this condition, the detectability of the gap increased monotonically with increasing gap duration.

Shailer and Moore (1987) found that the threshold for detecting a gap in a sinewave was roughly constant at about 5 ms for centre frequencies of 400, 1000, and 2000 Hz. Moore *et al.* (1993) found that gap thresholds were almost constant at 6–8 ms over the frequency range 400–2000 Hz, but increased somewhat at 200 Hz, and increased markedly, to about 18 ms, at 100 Hz. Individual variability also increased markedly at 100 Hz.

Overall, the results of experiments using narrowband stimuli indicate that temporal resolution does not vary markedly with centre frequency, except perhaps for a worsening at very low frequencies (200 Hz and below). Gap thresholds for narrowband stimuli are typically higher than those for broadband noise. However, for moderate noise bandwidths gap thresholds are typically around 10 ms or less. The smallest detectable gap is usually markedly larger than temporal gaps that are relevant for speech perception (e.g. 'sa' and 'sta' may be distinguished by a temporal gap lasting several tens of milliseconds).

(c) Modelling temporal resolution

Most models of temporal resolution are based on the idea that there is a process at levels of the auditory system higher than the auditory nerve which is 'sluggish' in some way, thereby limiting temporal resolution. The models assume that the internal representation of stimuli is 'smoothed' over time, so that rapid temporal changes are reduced in magnitude but slower ones are preserved. Although this smoothing process almost certainly operates on neural activity, the most widely used models are based on smoothing a simple transformation of the stimulus, rather than its neural representation.

Most models include an initial stage of bandpass filtering, reflecting the action of the auditory filters. Each filter is followed by a nonlinear device. This nonlinear device is meant to reflect the operation of several processes that occur in the peripheral auditory system such as amplitude compression on the basilar membrane and neural transduction, whose effects resemble half-wave rectification (see the chapter by Young, this volume). The output of the nonlinear device is fed to a 'smoothing' device, which can be implemented either as a lowpass filter (Viemeister 1979) or (equivalently) as a sliding temporal integrator (Moore *et al.* 1988; Plack & Moore 1990). The device determines a kind of weighted average of the output of the compressive nonlinearity over a certain time interval or 'window'. This weighting function is sometimes called the 'shape' of the temporal window. The window is assumed to slide in time, so that the output of the temporal integrator is a weighted running average of the input. This has the effect of smoothing rapid fluctuations while preserving slower ones. When a sound is turned on abruptly, the output of the temporal integrator takes some time to build up. Similarly, when a sound is turned off, the output of the integrator takes some time to decay. The shape of the window is assumed to be asymmetric in time, such that the build up of its output in response to the onset of a sound is more rapid than the decay of its output in response to the cessation of a sound. The output of the sliding temporal integrator is fed to a decision device. The decision device may use different 'rules' depending on the task required. For example, if the task is to detect a brief temporal gap in a signal, the decision device might look for a 'dip' in the output of the temporal integrator. If the task is to detect amplitude modulation of a sound, the device might assess the amount of modulation at the output of the sliding temporal integrator (Viemeister 1979).

3.7 Calculation of the internal representation of sounds

I describe next a method of calculating the internal representation of sounds, including speech, based on processes that are known to occur in the auditory system and taking into account the frequency and temporal resolution of the auditory system. The spectrogram is often regarded as a crude representation of the spectrotemporal analysis that takes place in the auditory system, although this representation is inaccurate in several ways (Moore 2003a). Presented below is the outline of a model that probably gives a better representation, although it is still oversimplified in several respects. The model is based on the assumption that there are certain 'fixed' processes in the peripheral auditory system, which can be modelled as a series of stages including:

(1) Fixed filters representing transfer of sound through the outer ear (Shaw 1974) and middle ear (Aibara *et al.* 2001). The overall transfer function through the outer and

middle ear for a frontally incident sound in free field has been estimated by Glasberg & Moore (2002) and is shown in their Figure 1. The effect of this transfer function is that low frequencies (below 500 Hz) and high frequencies (above 5000 Hz) are attenuated relative to middle frequencies.
(2) An array of bandpass filters (the auditory filters).
(3) Each auditory filter is followed by nonlinear processes reflecting the compression that occurs on the basilar membrane (Ruggero *et al.* 1997; Oxenham & Moore 1994; see also Young, this volume). The compression is weak for very low sound levels (below about 30 dB SPL), and perhaps for very high levels (above 90 dB SPL), but it has a strong influence for mid-range sound levels. Half-wave or full-wave rectification may also be introduced, to mimic the transformation from basilar-membrane vibration to neural activity effected by the inner hair cells (Viemeister 1979).
(4) An array of devices (sliding temporal integrators) that 'smooth' the output of each nonlinearity. As described earlier, the smoothing is assumed to reflect a relatively central process, occurring after the auditory nerve.

In some models, the filtering and the compressive nonlinearity are combined in a single non-linear filter bank (Irino & Patterson 2001; Lopez-Poveda & Meddis 2001; Zhang *et al.* 2001). Also, the transformation from basilar membrane vibration to neural activity can be simulated more accurately using a hair-cell model (Sumner *et al.* 2002). However, the basic features of the internal representation can be represented reasonably well using models of the type defined by stages (1)–(4) above. The internal representation of a given stimulus can be thought of as a three-dimensional array, with centre frequency as one axis (corresponding to the array of auditory filters with different centre frequencies) and time and magnitude as the other axes (corresponding to the output of each temporal integrator plotted as a function of time). The resulting pattern can be called a spectrotemporal excitation pattern (STEP) (Moore 1996). An example is shown in Figure 3.11, adapted from Moore (2003c). The figure shows the calculated STEP of the word 'tips'. In this figure, the frequency scale has been transformed to an ERB_N-number scale, as described earlier. The corresponding frequency is also shown. It should be noted that the STEP does not represent information that is potentially available in the temporal 'fine structure' at the output of each auditory filter. The role of this fine structure in speech perception is uncertain, and some have suggested that it plays little role (Shannon *et al.* 1995). However, temporal fine structure may play a role in the perception of pitch (Moore *et al.* 2006), in the separation of simultaneous sounds (Moore 2003a; Hopkins *et al.* 2008) and in distinguishing voiced from voiceless sounds.

3.8 Concluding remarks

This chapter has reviewed several aspects of auditory perception that are relevant to the perception of speech. These aspects include frequency selectivity, timbre perception, the perception of pitch, and temporal analysis. A recurring theme has been the finding that the basic discrimination abilities of the auditory system – measured using simple non-speech stimuli – are very good when considered relative to the acoustic differences that distinguish speech sounds. This partially accounts for the robust nature of speech perception. Indeed, it is remarkable that speech remains reasonably intelligible even under conditions of extreme distortion, such as infinite peak clipping (Licklider & Pollack 1948), time reversal of segments of speech (Saberi & Perrott 1999), representation of speech by three or four sinewaves tracking the formant frequencies (Remez *et al.* 1981),

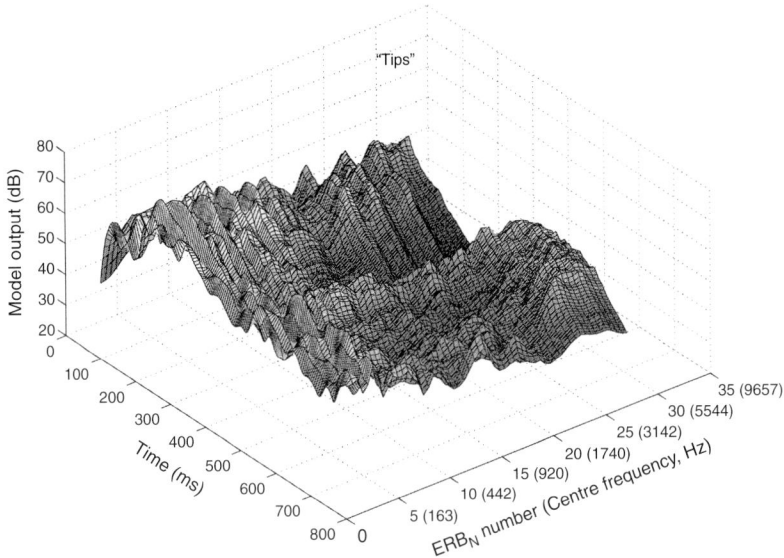

Figure 3.11 Spectro-temporal excitation pattern (STEP) of the word "tips". The figure was produced by Prof. C.J. Plack.
(Adapted from Moore (2003c)).

or representation of speech by a few amplitude-modulated noise bands (Shannon *et al.* 1995). However, this robustness applies to speech presented in quiet. Speech is often heard under much less ideal conditions. For example, reverberation, background noise, and competing talkers may be present. Under these conditions, many of the cues in speech become less discriminable, and some cues may be completely inaudible. Speech perception then becomes much less robust, especially if the functioning of the auditory system is impaired (Moore 2003b, 2007). The perception of speech when competing sounds are present is reviewed in the chapter by Darwin, this volume.

Acknowledgements

I thank Tom Baer and Chris Darwin for helpful comments on an earlier version of this chapter.

References

Aibara, R., Welsh, J. T., Puria, S. & Goode, R. L. 2001 Human middle-ear sound transfer function and cochlear input impedance. *J. Speech Lang. Hear. Res.* **152**, 100–109.
Alcántara, J. I., Moore, B. C. J. & Vickers, D. A. 2000 The relative role of beats and combination tones in determining the shapes of masking patterns at 2 kHz: I. Normal-hearing listeners. *J. Speech Lang. Hear. Res.* **148**, 63–73.
ANSI 1994 *ANSI S1.1-1994. American National Standard Acoustical Terminology*. New York: American National Standards Institute.
Bacon, S. P. & Viemeister, N. F. 1985 Temporal modulation transfer functions in normal-hearing and hearing-impaired subjects. *Audiolog.* **24**, 117–134.

Brungart, D. S., Simpson, B. D., Darwin, C. J., Arbogast, T. L. & Kidd, G., Jr. 2005 Across-ear interference from parametrically degraded synthetic speech signals in a dichotic cocktail-party listening task. *J. Acoust. Soc. Am.* **117**, 292–304.

Buus, S. & Florentine, M. 1985 Gap detection in normal and impaired listeners: the effect of level and frequency. In *Time Resolution in Auditory Systems* (ed. A. Michelsen), pp. 159–179. New York: Springer-Verlag.

Carrell, T. 1993 The effect of amplitude comodulation on extracting sentences from noise: Evidence from a variety of contexts. *J. Acoust. Soc. Am.* **93**, 2327.

Carrell, T. D. & Opie, J. M. 1992 The effect of amplitude comodulation on auditory object formation in sentence perception. *Percept. Psychophys.* **52**, 437–445.

Dau, T., Kollmeier, B. & Kohlrausch, A. 1997a Modeling auditory processing of amplitude modulation. I. Detection and masking with narrowband carriers. *J. Acoust. Soc. Am.* **102**, 2892–2905.

Dau, T., Kollmeier, B. & Kohlrausch, A. 1997b Modeling auditory processing of amplitude modulation. II. Spectral and temporal integration. *J. Acoust. Soc. Am.* **102**, 2906–2919.

Delgutte, B. 1990 Physiological mechanisms of psychophysical masking: Observations from auditory-nerve fibers. *J. Acoust. Soc. Am.* **87**, 791–809.

Drullman, R., Festen, J. M. & Plomp, R. 1994a Effect of reducing slow temporal modulations on speech reception. *J. Acoust. Soc. Am.* **95**, 2670–2680.

Drullman, R., Festen, J. M. & Plomp, R. 1994b Effect of temporal envelope smearing on speech reception. *J. Acoust. Soc. Am.* **95**, 1053–1064.

Eddins, D. A., Hall, J. W. & Grose, J. H. 1992 Detection of temporal gaps as a function of frequency region and absolute noise bandwidth. *J. Acoust. Soc. Am.* **91**, 1069–1077.

Egan, J. P. & Hake, H. W. 1950 On the masking pattern of a simple auditory stimulus. *J. Acoust. Soc. Am.* **22**, 622–630.

Festen, J. M. 1993 Contributions of comodulation masking release and temporal resolution to the speech-reception threshold masked by an interfering voice. *J. Acoust. Soc. Am.* **94**, 1295–1300.

Fitzgibbons, P. J. 1983 Temporal gap detection in noise as a function of frequency, bandwidth and level. *J. Acoust. Soc. Am.* **74**, 67–72.

Flanagan, J. L. & Saslow, M. G. 1958 Pitch discrimination for synthetic vowels. *J. Acoust. Soc. Am.* **30**, 435–442.

Fletcher, H. 1940 Auditory patterns. *Rev. Mod. Phys.* **12**, 47–65.

Fourcin, A. J. & Abberton, E. 1977 Laryngograph studies of vocal-fold vibration. *Phonetica*, **34**, 313–315.

Glasberg, B. R. & Moore, B. C. J. 1990 Derivation of auditory filter shapes from notched-noise data. *J. Speech Lang. Hear. Res.* **47**, 103–138.

Glasberg, B. R. & Moore, B. C. J. 2000 Frequency selectivity as a function of level and frequency measured with uniformly exciting notched noise. *J. Acoust. Soc. Am.* **108**, 2318–2328.

Glasberg, B. R. & Moore, B. C. J. 2002 A model of loudness applicable to time-varying sounds. *J. Audio Eng. Soc.* **50**, 331–342.

Glasberg, B. R., Moore, B. C. J., Patterson, R. D. & Nimmo-Smith, I. 1984 Dynamic range and asymmetry of the auditory filter. *J. Acoust. Soc. Am.* **76**, 419–427.

Gockel, H., Carlyon, R. P. & Plack, C. J. 2005 Dominance region for pitch: effects of duration and dichotic presentation. *J. Acoust. Soc. Am.* **117**, 1326–1336.

Green, D. M. 1973 Temporal acuity as a function of frequency. *J. Acoust. Soc. Am.* **54**, 373–379.

Green, D. M. 1988 *Profile Analysis*. Oxford: Oxford University Press.

Grose, J. H. and Hall, J. W. 1992 Comodulation masking release for speech stimuli. *J. Acoust. Soc. Am.* **91**, 1042–1050.

Hall, J. W., Grose, J. H. & Mendoza, L. 1995 Across-channel processes in masking. In *Hearing* (ed. B. C. J. Moore), pp. 243–266. San Diego: Academic Press.

Hall, J. W., Haggard, M. P. & Fernandes, M. A. 1984 Detection in noise by spectro-temporal pattern analysis. *J. Acoust. Soc. Am.* **76**, 50–56.

Helmholtz, H. L. F. 1863 *Die Lehre von den Tonempfindungen als physiologische Grundlage für die Theorie der Musik (On the Sensations of Tone as a Physiological Basis for the Theory of Music)*. Braunschweig: F. Vieweg.

Hermes, D. J. & van Gestel, J. C. 1991 The frequency scale of speech intonation. *J. Acoust. Soc. Am.* **90**, 97–102.
Hoekstra, A. & Ritsma, R. J. 1977 Perceptive hearing loss and frequency selectivity. In *Psychophysics and Physiology of Hearing* (eds. E. F. Evans & J. P. Wilson), pp. 263–271. London, England: Academic.
Hopkins, K., Moore, B. C. J. & Stone, M. A. 2008 Effects of moderate cochlear hearing loss on the ability to benefit from temporal fine structure information in speech. *J. Acoust. Soc. Am.* **123**, 1140–1153.
Houtsma, A. J. M. & Smurzynski, J. 1990 Pitch identification and discrimination for complex tones with many harmonics. *J. Acoust. Soc. Am.* **87**, 304–310.
Irino, T. & Patterson, R. D. 2001 A compressive gammachirp auditory filter for both physiological and psychophysical data. *J. Acoust. Soc. Am.* **109**, 2008–2022.
Johnson-Davies, D. & Patterson, R. D. 1979 Psychophysical tuning curves: restricting the listening band to the signal region. *J. Acoust. Soc. Am.* **65**, 765–770.
Kay, R. H. & Mathews, D. R. 1972 On the existence in human auditory pathways of channels selectively tuned to the modulation present in frequency-modulated tones. *J. Physiol.* **225**, 657–677.
Klatt, D. H. 1973 Discrimination of fundamental frequency contours in speech: implications for models of pitch perception. *J. Acoust. Soc. Am.* **53**, 8–16.
Kluender, K. R., Coady, J. A. & Kiefte, M. 2003 Sensitivity to change in perception of speech. *Speech Comm.* **41**, 59–69.
Kluk, K. & Moore, B. C. J. 2004 Factors affecting psychophysical tuning curves for normally hearing subjects. *J. Speech Lang. Hear. Res.* **194**, 118–134.
Kohlrausch, A., Fassel, R. & Dau, T. 2000 The influence of carrier level and frequency on modulation and beat-detection thresholds for sinusoidal carriers. *J. Acoust. Soc. Am.* **108**, 723–734.
Kortekaas, R. W. & Kohlrausch, A. 1999 Psychoacoustical evaluation of PSOLA. II. Double-formant stimuli and the role of vocal perturbation. *J. Acoust. Soc. Am.* **105**, 522–535.
Langner, G. & Schreiner, C. E. 1988 Periodicity coding in the inferior colliculus of the cat. I. Neuronal mechanisms. *J. Neurophysiol.* **60**, 1799–1822.
Leshowitz, B. 1971 Measurement of the two-click threshold. *J. Acoust. Soc. Am.* **49**, 426–466.
Licklider, J. C. R. & Pollack, I. 1948 Effects of differentiation, integration and infinite peak clipping upon the intelligibility of speech. *J. Acoust. Soc. Am.* **20**, 42–52.
Lopez-Poveda, E. A. & Meddis, R. 2001 A human nonlinear cochlear filterbank. *J. Acoust. Soc. Am.* **110**, 3107–3118.
Moore, B. C. J. 1986 Parallels between frequency selectivity measured psychophysically and in cochlear mechanics. *Scand. Audiol.* **25**, 139–152.
Moore, B. C. J. 1992 Across-channel processes in auditory masking. *J. Acoust. Soc. Jpn.(E).* **13**, 25–37.
Moore, B. C. J. 1996 Masking in the human auditory system. In *Collected Papers on Digital Audio Bit-Rate Reduction* (eds. N. Gilchrist & C. Grewin), pp. 9–19. New York: Audio Engineering Society.
Moore, B. C. J. 1998 *Cochlear Hearing Loss*. London: Whurr.
Moore, B. C. J. 2003a *An Introduction to the Psychology of Hearin.* 5th Edn. San Diego: Academic Press.
Moore, B. C. J. 2003b Speech processing for the hearing-impaired: Successes, failures, and implications for speech mechanisms. *Speech Comm.* **41**, 81–91.
Moore, B. C. J. 2003c Temporal integration and context effects in hearing. *J. Phonetics* **31**, 563–574.
Moore, B. C. J. 2007 *Cochlear Hearing Loss: Physiological, Psychological and Technical Issues,* 2nd Edn. Chichester: Wiley.
Moore, B. C. J., Alcántara, J. I. & Dau, T. 1998 Masking patterns for sinusoidal and narrowband noise maskers. *J. Acoust. Soc. Am.* **104**, 1023–1038.
Moore, B. C. J. & Carlyon, R. P. 2005 Perception of pitch by people with cochlear hearing loss and by cochlear implant users. In *Pitch Perception* (eds C. J. Plack, A. J. Oxenham, R. R. Fay & A. N. Popper), pp. 234–277. New York: Springer.

Moore, B. C. J. & Glasberg, B. R. 1983a Masking patterns of synthetic vowels in simultaneous and forward masking. *J. Acoust. Soc. Am.* **73**, 906–917.

Moore, B. C. J. & Glasberg, B. R. 1983b Suggested formulae for calculating auditory-filter bandwidths and excitation patterns. *J. Acoust. Soc. Am.* **74**, 750–753.

Moore, B. C. J. & Glasberg, B. R. 1987 Formulae describing frequency selectivity as a function of frequency and level and their use in calculating excitation patterns. *J. Speech Lang. Hear. Res.* **28**, 209–225.

Moore, B. C. J. & Glasberg, B. R. 1988 Effects of the relative phase of the components on the pitch discrimination of complex tones by subjects with unilateral cochlear impairments. In *Basic Issues in Hearing* (eds H. Duifhuis, H. Wit & J. Horst), pp. 421–430. London: Academic Press.

Moore, B. C. J. & Glasberg, B. R. 1990 Frequency discrimination of complex tones with overlapping and non-overlapping harmonics. *J. Acoust. Soc. Am.* **87**, 2163–2177.

Moore, B. C. J., Glasberg, B. R., Flanagan, H. J. & Adams, J. 2006 Frequency discrimination of complex tones: assessing the role of component resolvability and temporal fine structure. *J. Acoust. Soc. Am.* **119**, 480–490.

Moore, B. C. J., Glasberg, B. R., Gaunt, T. & Child, T. 1991 Across-channel masking of changes in modulation depth for amplitude- and frequency-modulated signals. *Q. J. Exp. Psychol.* **43A**, 327–347.

Moore, B. C. J., Glasberg, B. R. & Peters, R. W. 1985 Relative dominance of individual partials in determining the pitch of complex tones. *J. Acoust. Soc. Am.* **77**, 1853–1860.

Moore, B. C. J., Glasberg, B. R., Plack, C. J. & Biswas, A. K. 1988 The shape of the ear's temporal window. *J. Acoust. Soc. Am.* **83**, 1102–1116.

Moore, B. C. J., Glasberg, B. R. & Shailer, M. J. 1984 Frequency and intensity difference limens for harmonics within complex tones. *J. Acoust. Soc. Am.* **75**, 550–561.

Moore, B. C. J. & Ohgushi, K. 1993 Audibility of partials in inharmonic complex tones. *J. Acoust. Soc. Am.* **93**, 452–461.

Moore, B. C. J. & Peters, R. W. 1992 Pitch discrimination and phase sensitivity in young and elderly subjects and its relationship to frequency selectivity. *J. Acoust. Soc. Am.* **91**, 2881–2893.

Moore, B. C. J., Peters, R. W. & Glasberg, B. R. 1993 Detection of temporal gaps in sinusoids: effects of frequency and level. *J. Acoust. Soc. Am.* **93**, 1563–1570.

Moore, B. C. J. & Shailer, M. J. 1992 Modulation discrimination interference and auditory grouping. *Phil. Trans. R. Soc. Lond. B* **336**, 339–346.

O'Loughlin, B. J. & Moore, B. C. J. 1981a Improving psychoacoustical tuning curves. *J. Speech Lang. Hear. Res.* **5**, 343–346.

O'Loughlin, B. J. & Moore, B. C. J. 1981b Off-frequency listening: effects on psychoacoustical tuning curves obtained in simultaneous and forward masking. *J. Acoust. Soc. Am.* **69**, 1119–1125.

Oxenham, A. J. & Moore, B. C. J. 1994 Modeling the additivity of nonsimultaneous masking. *J. Speech Lang. Hear. Res.* **80**, 105–118.

Patterson, R. D. 1976 Auditory filter shapes derived with noise stimuli. *J. Acoust. Soc. Am.* **59**, 640–654.

Patterson, R. D. 1987 A pulse ribbon model of monaural phase perception. *J. Acoust. Soc. Am.* **82**, 1560–1586.

Patterson, R. D. & Moore, B. C. J. 1986 Auditory filters and excitation patterns as representations of frequency resolution. In *Frequency Selectivity in Hearing* (ed. B. C. J. Moore), pp. 123–177. London: Academic.

Patterson, R. D. & Nimmo-Smith, I. 1980 Off-frequency listening and auditory filter asymmetry. *J. Acoust. Soc. Am.* **67**, 229–245.

Patterson, R. D. & Wightman, F. L. 1976 Residue pitch as a function of component spacing. *J. Acoust. Soc. Am.* **59**, 1450–1459.

Pierrehumbert, J. 1979 The perception of fundamental frequency declination. *J. Acoust. Soc. Am.* **66**, 363–368.

Plack, C. J. & Moore, B. C. J. 1990 Temporal window shape as a function of frequency and level. *J. Acoust. Soc. Am.* **87**, 2178–2187.
Plomp, R. 1964a The ear as a frequency analyzer. *J. Acoust. Soc. Am.* **36**, 1628–1636.
Plomp, R. 1964b The rate of decay of auditory sensation. *J. Acoust. Soc. Am.* **36**, 277–282.
Plomp, R. 1967 Pitch of complex tones. *J. Acoust. Soc. Am.* **41**, 1526–1533.
Plomp, R. 1970 Timbre as a multidimensional attribute of complex tones. In *Frequency Analysis and Periodicity Detection in Hearing* (eds R. Plomp & G. F. Smoorenburg), pp. 397–414. Leiden: Sijthoff.
Plomp, R. 1976. *Aspects of Tone Sensation.* London: Academic Press.
Plomp, R. & Steeneken, H. J. M. 1969 Effect of phase on the timbre of complex tones. *J. Acoust. Soc. Am.* **46**, 409–421.
Pollack, I. 1968 Periodicity discrimination for auditory pulse trains. *J. Acoust. Soc. Am.* **43**, 1113–1119.
Remez, R. E., Rubin, P. E., Pisoni, D. B. & Carrell, T. D. 1981 Speech perception without traditional speech cues. *Science* **212**, 947–950.
Ritsma, R. J. 1967 Frequencies dominant in the perception of the pitch of complex sounds. *J. Acoust. Soc. Am.* **42**, 191–198.
Ronken, D. 1970 Monaural detection of a phase difference between clicks. *J. Acoust. Soc. Am.* **47**, 1091–1099.
Rosen, S., Baker, R. J. & Darling, A. 1998 Auditory filter nonlinearity at 2 kHz in normal hearing listeners. *J. Acoust. Soc. Am.* **103**, 2539–2550.
Rosen, S. & Fourcin, A. 1986 Frequency selectivity and the perception of speech. In *Frequency Selectivity in Hearing* (ed. B. C. J. Moore), pp. 373–487. London: Academic.
Ruggero, M. A., Rich, N. C., Recio, A., Narayan, S. S. & Robles, L. 1997 Basilar-membrane responses to tones at the base of the chinchilla cochlea. *J. Acoust. Soc. Am.* **101**, 2151–2163.
Saberi, K. & Perrott, D. R. 1999 Cognitive restoration of reversed speech. *Natur.* **398**, 760.
Sachs, M. B. & Kiang, N. Y. S. 1968 Two-tone inhibition in auditory nerve fibers. *J. Acoust. Soc. Am.* **43**, 1120–1128.
Schouten, J. F. 1968 The perception of timbre. *6th International Conference on Acoustic.* **1**, GP-6-2.
Schouten, J. F. 1970 The residue revisited. In *Frequency Analysis and Periodicity Detection in Hearing* (eds R. Plomp & G. F. Smoorenburg), pp. 41–54. Leiden, The Netherlands: Sijthoff.
Schreiner, C. E. & Urbas, J. V. 1986 Representation of amplitude modulation in the auditory cortex of the cat. I. The anterior auditory field (AAF). *J. Speech Lang. Hear. Res.* **21**, 227–241.
Shackleton, T. M. & Carlyon, R. P. 1994 The role of resolved and unresolved harmonics in pitch perception and frequency modulation discrimination. *J. Acoust. Soc. Am.* **95**, 3529–3540.
Shailer, M. J. & Moore, B. C. J. 1983 Gap detection as a function of frequency, bandwidth and level. *J. Acoust. Soc. Am.* **74**, 467–473.
Shailer, M. J. & Moore, B. C. J. 1987 Gap detection and the auditory filter: phase effects using sinusoidal stimuli. *J. Acoust. Soc. Am.* **81**, 1110–1117.
Shannon, R. V., Zeng, F.-G., Kamath, V., Wygonski, J. & Ekelid, M. 1995 Speech recognition with primarily temporal cues. *Scienc.* **270**, 303–304.
Shaw, E. A. G. 1974 Transformation of sound pressure level from the free field to the eardrum in the horizontal plane. *J. Acoust. Soc. Am.* **56**, 1848–1861.
Steeneken, H. J. M. & Houtgast, T. 1980 A physical method for measuring speech-transmission quality. *J. Acoust. Soc. Am.* **69**, 318–326.
Stone, M. A. & Moore, B. C. J. 2004 Side effects of fast-acting dynamic range compression that affect intelligibility in a competing speech task. *J. Acoust. Soc. Am.* **116**, 2311–2323.
Summerfield, A. Q. & Assmann, P. 1987 Auditory enhancement in speech perception. In *The Psychophysics of Speech Perception* (ed. M. E. H. Schouten), pp. 140–150. Dordrecht: Martinus Nijhoff.
Summerfield, A. Q., Sidwell, A. S. & Nelson, T. 1987 Auditory enhancement of changes in spectral amplitude. *J. Acoust. Soc. Am.* **81**, 700–708.

Sumner, C. J., Lopez-Poveda, E. A., O'Mard, L. P. & Meddis, R. 2002 A revised model of the inner-hair cell and auditory-nerve complex. *J. Acoust. Soc. Am.* **111**, 2178–2188.

't Hart, J. 1981 Differential sensitivity to pitch distance, particularly in speech. *J. Acoust. Soc. Am.* **69**, 811–821.

Verhey, J. L., Dau, T. & Kollmeier, B. 1999 Within-channel cues in comodulation masking release (CMR): experiments and model predictions using a modulation-filterbank model. *J. Acoust. Soc. Am.* **106**, 2733–2745.

Viemeister, N. F. 1979 Temporal modulation transfer functions based on modulation thresholds. *J. Acoust. Soc. Am.* **66**, 1364–1380.

Vogten, L. L. 1978 Low-level pure-tone masking: a comparison of 'tuning curves' obtained with simultaneous and forward masking. *J. Acoust. Soc. Am.* **63**, 1520–1527.

von Bismarck, G. 1974 Sharpness as an attribute of the timbre of steady sounds. *Acustica*, **30**, 159–172.

Watkins, A. J. 1991 Central, auditory mechanisms of perceptual compensation for spectral-envelope distortion. *J. Acoust. Soc. Am.* **90**, 2942–2955.

Watkins, A. J. & Makin, S. J. 1996a Effects of spectral contrast on perceptual compensation for spectral-envelope distortion. *J. Acoust. Soc. Am.* **99**, 3749–3757.

Watkins, A. J. & Makin, S. J. 1996b Some effects of filtered contexts on the perception of vowels and fricatives. *J. Acoust. Soc. Am.* **99**, 588–594.

Yost, W. A. & Sheft, S. 1989 Across-critical-band processing of amplitude-modulated tones. *J. Acoust. Soc. Am.* **85**, 848–857.

Yost, W. A., Sheft, S. & Opie, J. 1989 Modulation interference in detection and discrimination of amplitude modulation. *J. Acoust. Soc. Am.* **86**, 2138–2147.

Yumoto, E., Gould, W. J. & Baer, T. 1982 Harmonics-to-noise ratio as an index of the degree of hoarseness. *J. Acoust. Soc. Am.* **71**, 1544–1549.

Zhang, X., Heinz, M. G., Bruce, I. C. & Carney, L. H. 2001 A phenomenological model for the responses of auditory-nerve fibers: I. Nonlinear tuning with compression and suppression. *J. Acoust. Soc. Am.* **109**, 648–670.

Zwicker, E. 1964 'Negative afterimage' in hearing. *J. Acoust. Soc. Am.* **36**, 2413–2415.

Zwicker, E. & Fastl, H. 1999 *Psychoacoustics – Facts and Models, Second Edition*. Berlin, Springer-Verlag.

Zwicker, E. & Terhardt, E. 1980 Analytical expressions for critical band rate and critical bandwidth as a function of frequency. *J. Acoust. Soc. Am.* **68**, 1523–1525.

4

Acoustic and auditory phonetics: the adaptive design of speech sound systems

Randy L. Diehl

> Speech perception is remarkably robust. This chapter examines how acoustic and auditory properties of vowels and consonants help to ensure intelligibility. First, the source–filter theory of speech production is briefly described, and the relationship between vocal-tract properties and formant patterns is demonstrated for some commonly occurring vowels. Next, two accounts of the structure of preferred sound inventories, quantal theory and dispersion theory, are described and some of their limitations are noted. Finally, it is suggested that certain aspects of quantal and dispersion theories can be unified in a principled way so as to achieve reasonable predictive accuracy.
>
> **Keywords:** acoustic phonetics; auditory phonetics; speech sounds

4.1 Introduction

Speech sounds tend to be accurately perceived even in unfavourable listening conditions. Moore, this volume discusses several aspects of basic auditory processing that contribute to this perceptual robustness. The present paper considers how acoustic and auditory properties of commonly occurring speech sounds also help to ensure high levels of intelligibility. First, the source–filter theory of speech production is outlined, and the relationship between vocal-tract (VT) cavity size and shape and formant patterns is illustrated. Second, two theories intended to account for cross-language preferences in sound inventories, quantal theory and dispersion theory, are described and evaluated. Finally, it is suggested that a version of dispersion theory that incorporates certain aspects of quantal theory may have greater predictive success than either theory in its original form.

4.2 Source–filter theory of speech production

The mapping between VT properties and acoustic signals has been investigated over many decades (Chiba & Kajiyama 1941; Stevens & House 1955, 1961; Fant 1960; Flanagan 1972; Stevens 1998) and, as documented in the last of these cited works, is now reasonably well understood for the major classes of speech sounds. At the core of this understanding lies the assumption that speech outputs can be analysed as the response of a set of VT filters to one or more sources of sound energy. A further assumption, that holds to a first approximation in most cases, is that the source and filter properties of the vocal tract are independent.

A source in the vocal tract is any modulation of the airflow that creates audible energy. Such sound-producing modulations occur in the vicinity of constrictions either at the glottis (i.e. the space between the vocal folds of the larynx) or in the supralaryngeal regions of the vocal tract. Several types of source may be distinguished. One is (quasi-) periodic and consists of cycles of varying airflow attributable to vocal-fold vibration or voicing. Sounds produced with a voiced source have a fundamental frequency (F0) equal to the repetition rate of vocal-fold vibration. They include vowels (e.g. /a/ and /u/), nasal consonants (e.g. /m/), liquids (e.g. /r/ and /l/) and glides (e.g. /w/). Other sources are aperiodic and include (i) turbulence noise generated as air flows rapidly through an open, non-vibrating glottis (referred to as 'aspiration'), (ii) turbulence noise generated as air flows rapidly through a narrow supralaryngeal constriction (referred to as 'frication'),[1] and (iii) a brief pulse of excitation caused by a rapid change in oral air pressure (referred to as a 'transient source'). Examples of the use of these aperiodic sources are, respectively, the aspirated /h/, the fricatives /f/ and /s/ and the stop consonants /p/ and /t/ (both of which, in stressed-syllable-initial position, tend to be associated with a rapid reduction in oral air pressure at the moment of VT opening). Some speech sounds have multiple sources operating simultaneously or in succession. For example, the fricative /z/ is produced with a voiced source and a simultaneous turbulence noise (i.e. frication) source, while the stop consonant /t/ may be produced with, in quick succession, a transient source, a frication source and an aspiration source, as the mouth opens (Fant 1973).

All of these sources—both periodic and aperiodic—are well suited for evoking responses from the VT filters. Under normal conditions, each source has an energy level sufficient to generate highly audible speech sounds. Moreover, each source has an amplitude spectrum that is fairly broadband, ensuring that even VT filters in the higher-frequency range (1–5 kHz) will tend to be excited.

How, then, does the vocal tract act to filter sound energy generated by the sources? Any fully or partially enclosed volume of air has certain natural frequencies of vibration, or resonance frequencies, that are determined mainly by the size and shape of the volume, and by the extent and character of the enclosing surfaces. When the volume of air is exposed to a broadband energy source, it will respond strongly to source frequencies at or near its resonance frequencies and weakly to other source frequencies. This relative response as a function of frequency defines the filter, or transfer, function of the vocal tract in a given configuration.[2]

Figure 4.1 illustrates the source–filter theory for four different vowel sounds. Figure 4.1a(i) displays an idealized spectrum of the glottal airflow waveform corresponding to a voiced source. The value of F0 is 100 Hz, and the slope of this spectrum is –12 dB per octave, values typical of an adult male voice. Since efficiency of sound transmission from the mouth (known as the 'radiation characteristic') increases at frequencies above 300–500 Hz at a rate of 6 dB per octave, the effective glottal spectrum slope with respect to the listener is –6 dB per octave (see dotted curve). Figure 4.1b shows filter functions for the vowels /ə/, or schwa, (as in the first syllable of 'about'), /u/ (as in 'boot'), /i/ (as in 'beet') and /a/ (as in American English 'hot'), with each function including three resonance peaks within the 0–3 kHz range. Figure 4.1c represents the acoustic output spectra of the four vowels. On the assumption of source-filter independence, the spectrum of the output sound is considered to be the product of three terms: the source spectrum; the VT filter function; and the radiation characteristic (Stevens & House 1961).

An important consequence of the near independence of VT sources and filters is that the speech signal can transmit linguistic information at higher rates than would otherwise be possible. For example, in most languages, F0 variations are used to convey lexical, grammatical and paralinguistic (e.g. attitudinal or emotional) information in parallel

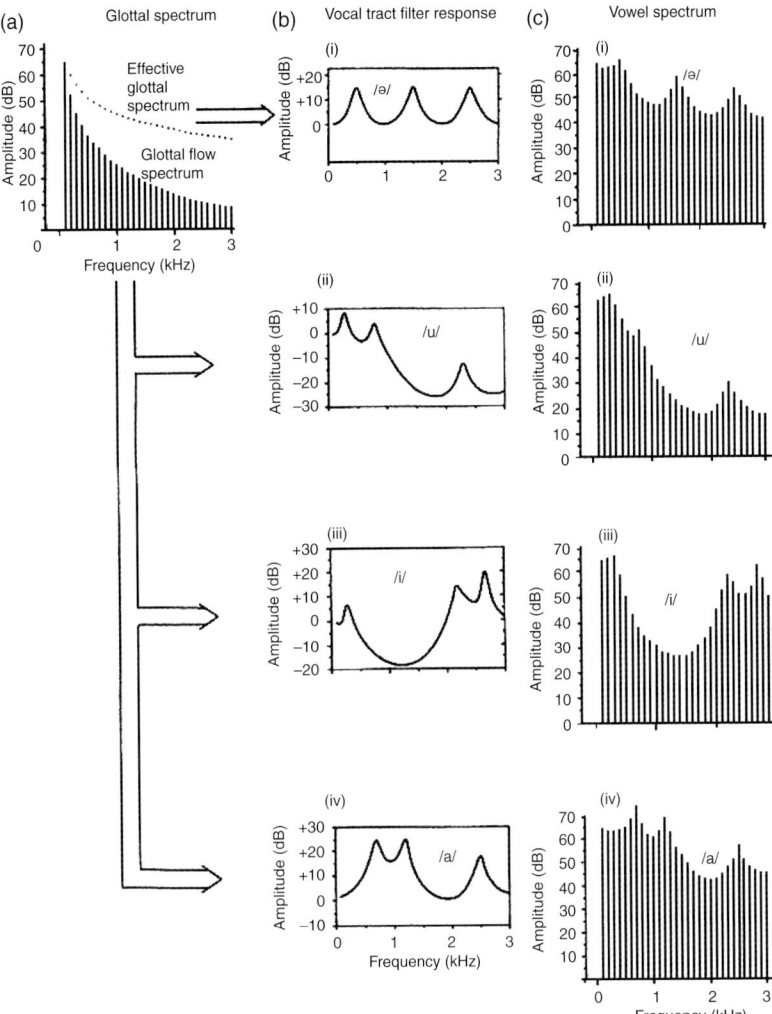

Figure 4.1 Source–filter theory of speech production illustrated for the vowels /ə/, /u/, /i/ and /a/. (a) An idealized spectrum of the glottal airflow waveform, with a slope of −12 dB per octave, is displayed. The effective glottal spectrum slope (dotted curve) is −6 dB per octave owing to more efficient sound transmission from the mouth at higher frequencies. (b) Filter functions for the four vowels. (c) Product of the glottal source spectrum and the filter functions yields the acoustic output spectra. (Adapted with permission from Pickett (1999), Allyn & Bacon; adapted from Fant (1960) and Stevens & House (1961)).

with that provided by the sequencing of vowels and consonants. Moreover, consonant sounds in most languages are distinguished on the basis of both source and filter properties of the vocal tract (Maddieson 1984). A significant degree of independence also characterizes the relationship between different VT sources (e.g. voicing and frication) and between different VT filter properties (e.g. place of articulation and nasality, see later), further increasing the information content of speech. This principle of independence is

4.3 Vocal-tract cavity properties and formant frequencies

A key concept in acoustic phonetics is the 'formant'. It refers to the acoustic realization of an underlying resonance peak in the VT filter function and is illustrated by the envelope peaks in the output spectra of each of the vowels represented in Figure 4.1. A formant is characterized by a centre frequency, a relative amplitude and a bandwidth. For the acoustic description of vowel sounds, the most important parameters are the centre frequencies of the lowest three or four formants, referred to as the 'formant pattern' collectively. Perceived vowel identity (e.g. whether a vowel token is heard as an instance of /i/ or /u/) is strongly influenced by the formant pattern but only modestly affected (across a sizable range of values) by the relative amplitudes or bandwidths of the formants (Klatt 1982). Given any formant pattern and a glottal source spectrum, the acoustic theory of vowel production (Fant, 1960, 1973; Stevens & House 1961) makes reasonably accurate predictions about formant bandwidths and relative amplitudes and, hence, the overall shape of the spectral envelope.

To understand the relationship between the size and shape of the vocal tract and the formant pattern, consider first the vowel /ə/ (the top-most vowel represented in Figure 4.1). During production of this vowel, the cross-sectional area of the vocal tract is approximately uniform from the region just above the glottis all the way to the lips. This uniform tube is open at the lips and effectively closed at the glottal end because the average size of the glottis during vocal-fold vibration is very small relative to the cross-sectional area of the supralaryngeal vocal tract. When a pressure wave is generated by airflow through the vibrating vocal folds, it travels to the lip opening where it is almost completely reflected owing to the very small impedance outside the mouth. There is a boundary condition of essentially zero acoustic pressure at the lips. For each frequency component in the source, the corresponding forward-going wave from the glottis and the reflected wave from the lips combine to form a standing wave with a wavelength inversely related to frequency. The amplitude of the standing wave varies sinusoidally along the length of the vocal tract, with nodes (i.e. zero points corresponding to steady atmospheric pressure) located at the lips and at every half-wavelength back to the glottis and antinodes (points of maximum positive and negative deviations from atmospheric pressure) located at odd multiples of the quarter-wavelength distance from the lips. Each standing pressure wave has a corresponding standing volume–velocity (airflow) wave with nodes and antinodes located, respectively, at the antinodes and nodes of the standing pressure wave. Resonance occurs at just those frequencies for which a pressure antinode (a volume–velocity node) occurs at the glottal end of the vocal tract.[3]

Figure 4.2 shows the standing pressure waves for the three lowest resonance frequencies (500, 1500 and 2500 Hz) of /ə/, given a VT length, l, of 17.5 cm, a typical adult male value. Notice that in each case the boundary conditions described in the previous paragraph are met. The lowest frequency resonance, corresponding to the first formant, is represented at the bottom. It may be observed that the standing pressure wave extending from the glottis to the lips amounts to one-quarter of a sinusoidal cycle; thus, the

Acoustic and auditory phonetics: the adaptive design of speech sound systems 83

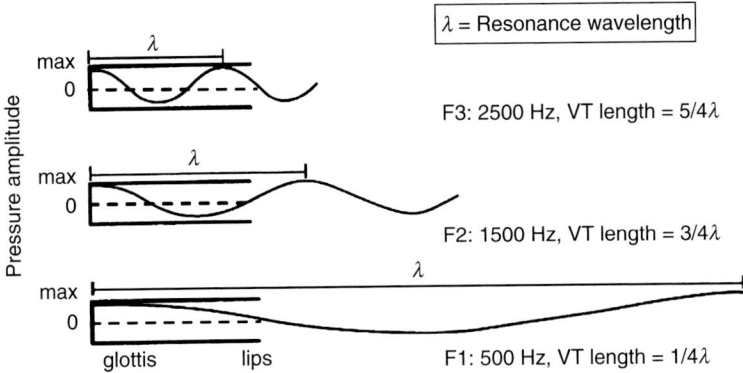

Figure 4.2 The standing pressure waves for the three lowest resonance frequencies (500, 1500, 2500 Hz) of the vowel /ə/, produced with a vocal-tract (VT) length of 17.5 cm. Each standing wave satisfies the boundary conditions that an antinode exists at the closed (glottal) end of the vocal tract and a node exists at the open (lip) end. F1, F2 and F3 refer to the first three formants, corresponding to the first three resonances of the vocal tract. (Adapted with permission from Johnson (1997), Blackwell Publishers).

wavelength, λ, equals $4l$ or 70 cm. Accordingly, tubes closed at one end and open at the other end are called 'quarter-wave resonators'. The corresponding resonance or formant frequency, F1, is calculated from the formula $f = c/\lambda$ (where c is equal to the speed of sound, approx. 35 000 cm s^{-1}), yielding a value of 500 Hz. Analogous calculations for the second and third formant frequencies (F2 and F3) give values of 1500 and 2500 Hz. These frequencies are consistent with the resonance and formant peaks shown for /ə/ in Figure 4.1.

Models of VT configurations for vowels other than /ə/ require either tubes of non-uniform cross-sectional area or else a series of two or more uniform tubes with different cross-sectional areas. In the case of the vowel /a/, the jaw and tongue body are lowered, creating a large oral (front) cavity, and the tongue is also retracted, creating a narrow pharyngeal (back) cavity. This configuration can be modelled as a series of two quarter-wave resonators, that is, a series of two uniform tubes each effectively closed at the input end and open at the output end. The back cavity is treated as open at its output end, while the front cavity is treated as closed at its input end owing to the relatively large difference in cross-sectional area between the two cavities. The size of this difference also implies that the two tubes can be considered acoustically independent, at least to a first approximation. This means that the filter function for the entire VT configuration can be estimated by combining the separate resonance frequencies of the front and back cavities. Each of the two quarter-wave resonators used in the production of /a/ is, of course, shorter than the single one used to produce /ə/, and thus their lowest resonance frequencies are higher in value. In addition, the two resonators are comparable in length, yielding F1 and F2 values that are relatively close together. These acoustic properties of /a/ are shown in Figure 4.1.

In the case of the vowel /i/, the jaw and tongue body are raised, creating a narrow front cavity, and the tongue body is moved forward, enlarging the back cavity. This configuration can be modelled as a series of two uniform tubes with very different cross-sectional areas.[4] The wide back cavity is effectively closed at both the glottal end and the forward end that communicates with the narrow front cavity, while the front cavity is open at both ends. For a uniform tube closed at both ends, resonance occurs only at frequencies for which there are antinodes in the standing pressure wave at the closed ends of the tube and a node at the very middle of the tube. If a tube is open at both ends, resonance occurs only at frequencies for which there is a node at each end and an antinode at the middle. In both cases, the lowest frequency standing wave extending across the length of the tube is one-half of a sinusoidal wavelength, and the tubes are thus referred to as 'half-wave resonators'. Other things being equal, half-wave resonators have a lowest natural frequency that is double that of quarter-wave resonators. In Figure 4.1, the relatively high F2 and F3 of /i/ correspond to the lowest frequency resonances of the front and back tubes. The low F1 of this vowel is attributable to a 'Helmholtz resonator' comprising both the wide back and the narrow front cavities. The natural frequency of a Helmholtz resonator increases with the square root of the cross-sectional area of the front cavity, and decreases with the square root of the length of the front cavity and the volume of the back cavity. Given the dimensions of VT cavities, such a resonator has a low natural frequency.

For the vowel /u/, the tongue body is raised and retracted, producing a wide front cavity and a narrow constriction between the front and back cavities, the tongue root (the lower back portion of the tongue that forms the front wall of the pharynx) is moved forward, creating a wide back cavity, and the lips are rounded, creating a narrow opening. This configuration can be modelled both as a series of four uniform tubes (wide–narrow–wide–narrow), all of which are half-wave resonators, and as a series of two Helmholtz resonators. As shown in Figure 4.1, /u/ has a low F1 and F2, which are produced by the two Helmholtz resonators, and a high F3 which is produced by the longest of the half-wave resonators.

More realistic non-uniform tube models based on the accurate measurements of VT dimensions are, of course, possible. Nevertheless, simplified models consisting of uniform tubes and Helmholtz resonators suffice to illustrate some of the main principles underlying the relationship between VT properties and formant patterns.

4.4 Adaptive design of speech sound inventories

(a) The restricted character of speech sound systems

Among the vowels and consonants that have been observed in the world's languages, some occur commonly, whereas most are relatively rare (Crothers 1978; Maddieson 1984). What factors might explain such cross-language preferences for certain sounds over others? One possible factor, long discussed by linguists (Passy 1890; Jakobson 1941; Martinet 1955), is the requirement that speech sounds be audible and distinctive (i.e. not confusable with other speech sounds). A second possible factor is a general tendency towards efficiency in human behaviour (Zipf 1949) such that goals—in this case, successful speech communication—are achieved with minimum effort. These two factors will be referred to, respectively, as 'listener-oriented' and 'talker-oriented' constraints on sound selection.

During the last several decades, two theories, quantal theory (Stevens 1972, 1989, 1998) and dispersion theory (Liljencrants & Lindblom 1972; Lindblom 1986; Diehl *et al.* 2003; Diehl & Lindblom 2004), have been developed to account for cross-language preferences

(b) Quantal theory: vowels

Quantal theory is based on the observation that certain nonlinearities exist in the mappings between articulatory (i.e. VT) configurations of talkers and acoustic outputs and also between the speech signals and the auditory responses of listeners. Such a nonlinearity in the articulatory-to-acoustic mapping is represented in Figure 4.3. In regions I and III, perturbations in the articulatory parameter result in small changes in the acoustic output, whereas in region II, comparably sized perturbations yield large acoustic changes. Given these alternating regions of acoustic stability and instability, an adaptive strategy for a language community is to select sound categories that occupy the stable regions and that are separated by the unstable region. Locating sound categories in stable regions allows talkers to achieve an acceptable acoustic output with less articulatory precision than would otherwise be necessary, thus helping to satisfy the talker-oriented goal of minimum effort. In addition, separating two sound categories by a region of acoustic instability ensures that they are acoustically very different, and thus helps to satisfy the listener-oriented requirement of sufficient auditory distinctiveness. Another advantage for listeners is that vowels produced in acoustically stable regions should be relatively invariant. According to quantal theory, this convergence of talker- and listener-oriented selection criteria leads to a preference for certain 'quantal' vowels and consonants.

Consider, for example, the VT model in Figure 4.4a consisting of a series of two quarter-wave resonators with lengths l_1 (back cavity) and l_2 (front cavity) and cross-sectional areas A_1 and A_2. (See the earlier discussion of the two-tube model for the vowel /a/.) Figure 4.4b is a nomogram representing the effects on the first four resonance frequencies

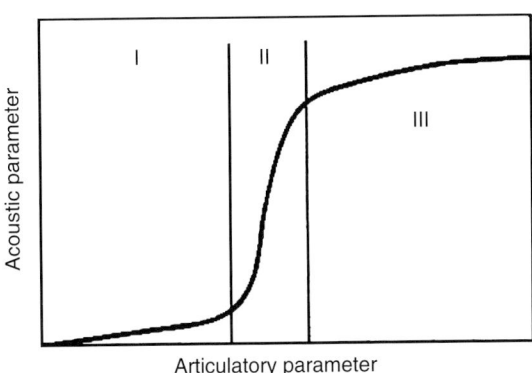

Figure 4.3 A schematic of a quantal nonlinearity in the mapping between an articulatory parameter of the vocal tract and the acoustic output. Regions I and III are acoustically quite stable with respect to perturbations in the articulatory parameter, whereas region II is acoustically unstable. Speech sound categories are assumed to be located in regions I and III. (Adapted with permission from Stevens (1989), Academic Press).

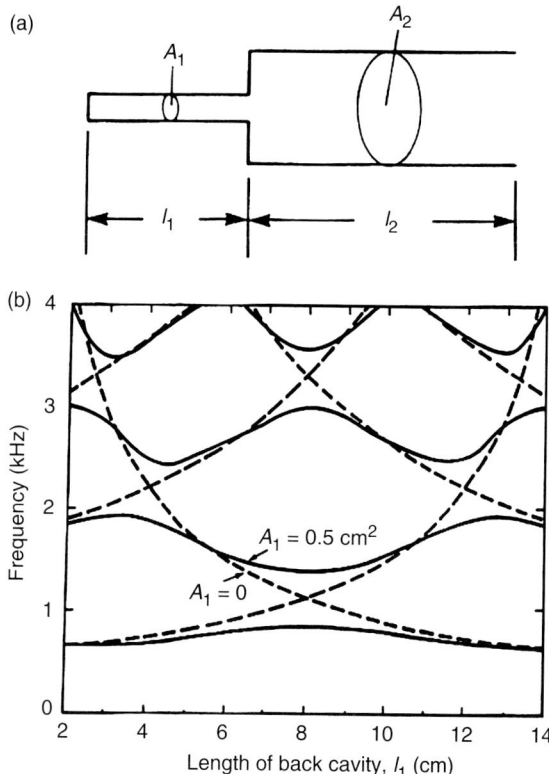

Figure 4.4 (a) A two-tube model of the vocal tract, with each tube effectively closed at the input end and open at the output end. The lengths of the left (back) and right (front) tubes are l_1 and l_2, and the cross-sectional areas are A_1 and A_2, respectively, (b) A nomogram representing the first four resonance frequencies for the two-tube model in (a) as the length l_1 of the back cavity is varied, with overall length $l_1 + l_2$ held constant at 16 cm and $A_2 = 3$ cm². The dashed curves represent the case where $A_1 \ll A_2$; the solid curves represent the case where $A_1 = 0.5$ cm². (Adapted with permission from Stevens (1989), Academic Press).

of varying l_1 while the total length of the model, $l_1 + l_2$, is held constant at 16 cm and $A_2 = 3$ cm². If the ratio of A_1 to A_2 is very small, the resonance frequencies of one tube are roughly independent of those of the other. The idealized case of complete independence is represented by the dashed curves in the nomogram. The resonance frequency curves with an upward trend as a function of l_1 are associated with the front cavity; those with a downward trend are associated with the back cavity. Note that the correspondence between cavities and numbered formants in the acoustic output changes at the crossover points in the curves.

The solid curves in Figure 4.4b represent resonance frequencies of the two-tube model when A_1 is increased to 0.5 cm², a value large enough to yield the non-turbulent airflow characteristic of vowels. The larger ratio of A_1 to A_2 in this case causes a modest degree of acoustic coupling between the two cavities such that the resonance frequencies of each cavity are influenced by the other. A main result is that the curves no longer intersect but rather approach each other and then diverge. In the region of convergence, perturbations in l_1 have small effects on formant frequencies, whereas in adjacent regions such perturbations have larger effects. It turns out that at an l_1 value of approximately 8 cm, with maximum convergence between F1 and F2 and also between F3 and F4, the two-tube model corresponds rather closely to the VT configuration for the vowel /a/.

Stevens (1972, 1989) used this example to illustrate the notion of a quantal vowel and presented similar arguments with respect to the vowels /i/ and /u/. Recall that, as in the case of /a/, /i/ is produced with a large difference in cross-sectional area between the front and back cavities (but in the opposite direction), whereas /u/ is produced with a narrow constriction separating the back and front cavities. For all three vowels, therefore, the front and back cavities are only weakly coupled acoustically; however, the coupling is sufficient to yield acoustically stable quantal regions that alternate with acoustically unstable regions, where stability is defined with respect to variation in cavity length parameters. As may be inferred from these examples, weak yet non-negligible coupling between different VT resonators is a necessary (though plainly not a sufficient) condition for quantal vowel status. Zero coupling would produce no regions of acoustic stability (see the dashed curves in Figure 4.4b), while a high degree of coupling would flatten the peaks and troughs of the resonance curves, creating large stable regions but without the adjacent regions of acoustic instability that confer auditory distinctiveness (Diehl 1989).

Stevens (1972, 1989) emphasized another property of quantal vowels that should be advantageous for listeners. When two resonances are in close proximity, they reinforce each other creating a relatively intense spectral region. This may be confirmed by examining the filter functions and output spectra for the non-schwa vowels in Figure 4.1. For /u/ and /a/ there are prominences in the low- and mid-frequency regions, respectively, owing to the convergence of F1 and F2, whereas for /i/ there is a similar prominence in the high-frequency region due to the convergence of F2 and F3.

To summarize, quantal vowels appear to satisfy listener-oriented selection criteria in at least three ways: (i) being produced in acoustically stable regions of phonetic space, they are relatively invariant, (ii) being separated from nearby vowels by regions of acoustic instability, their formant patterns are auditorily distinctive, and (iii) being characterized by relatively intense spectral prominences, they are likely to be resistant to masking by background noise (Darwin, this volume; Moore, this volume).

A common misinterpretation of quantal theory is that it predicts greater stability for quantal vowels such as /i/ or /u/ than for non-quantal vowels like /ʌ/ (as in American English 'cup'), which is produced without a major VT constriction (Ladefoged *et al.* 1977; Pisoni 1980; Syrdal & Gopal 1986). Several results inconsistent with this putative prediction have been reported. For example, Pisoni (1980) found that when talkers repeatedly mimicked synthetic vowel sounds, within-talker variances in F2 were smaller for /ʌ/ than for /i/ or /u/.

However, as Diehl (1989) noted, quantal theory does not, in fact, predict that a quantal vowel is more stable than *any* non-quantal vowel. Rather, as shown in Figure 4.4b, it predicts that a quantal vowel will be more stable than any non-quantal vowel that is produced with different cavity length parameters but with the same cross-sectional area parameters. A vowel like /ʌ/ is actually predicted to be among the most stable of vowels by quantal theory (with respect to perturbations in l_1) because the back and front cavities differ little in cross-sectional area and are, therefore, highly coupled acoustically. Recall that greater acoustic coupling results in a flattening of the peaks and troughs of the resonance frequency curves in the nomogram. (When A_1 and A_2 are equal, creating a uniform tube corresponding to schwa, variation in 'l_1' obviously does not alter the configuration at all, and the frequency curves become perfectly horizontal.) Accordingly, the vowel /ʌ/ lacks quantal status not because it occupies an unstable region of phonetic space, but because it is not bounded by regions of acoustic instability that confer auditory distinctiveness vis-à-vis other nearby vowels. In other words, /ʌ/ satisfies the talker-oriented, but not the listener-oriented, selection criteria of quantal theory.

Although quantal theory is not falsified by the evidence that non-quantal vowels like /ʌ/ are more stable than quantal vowels such as /i/ or /u/, the theory nevertheless faces a major difficulty regarding the claim that the quantal vowels are relatively stable. Recall that acoustic stability is defined in quantal theory with respect to variation in some cavity length parameter such as l_1 in Figure 4.4a. However, it is reasonable to ask how stable quantal vowels are with respect to variations in other VT parameters such as A_1. In Figure 4.4b, it may be seen that the largest effects on resonance frequencies of varying the A_1 parameter occur in just those regions that are most stable with respect to perturbations in l_1. Thus, quantal vowels are actually the *least* stable vowels with respect to changes in cross-sectional area parameters. This would perhaps not be a serious problem for quantal theory if the acoustic effects of varying cavity width were relatively small. However, as noted by Diehl (1989), perturbations in the width of a VT cavity tend to yield changes in resonance frequencies that are at least equal to—and often greater than— those caused by comparable perturbations in cavity length. The relative stability of quantal vowels is, therefore, questionable.

This argument in no way undermines quantal theory's claims that quantal vowels are favoured by the listener-oriented criteria of auditory distinctiveness and audibility in noise. It is possible that these criteria alone may be sufficient to generate accurate predictions about cross-language preferences in the structure of vowel inventories. In the UCLA Phonological Segment Inventory Database (UPSID, Maddieson 1984) of 317 diverse languages, the most commonly occurring vowels are /i/, /a/ and /u/, which appear in 92, 88 and 84% of the languages, respectively. As was discussed earlier, each of these vowels clearly meets the listener-oriented criteria for quantal status and, accordingly, their high frequency of occurrence is consistent with quantal theory.

However, /i/, /a/ and /u/ are not the only quantal vowels. When a tongue configuration appropriate for /i/ is combined with lip rounding, the resulting vowel is /y/ (as in the first syllable of the German word 'über'). Relative to /i/, both F2 and F3 are shifted downward for this vowel, but otherwise the nomogram is very similar (Stevens 1989). Both vowels have closely spaced values of F2 and F3 and each is produced in a stable region of the l_1 dimension that is bounded by unstable regions. (For /y/ the stable region occurs at a somewhat higher value of l_1.) If /i/ is a quantal vowel, so too is /y/. Presumably, then, quantal theory would predict a frequency of occurrence for /y/ that is comparable to that for /i/. However, /y/ occurs in only approximately 8% of the languages of the UPSID sample (Maddieson 1984) and is virtually absent in languages with small vowel inventories. The large discrepancy in frequency of occurrence between /i/ and /y/ is difficult to explain within the framework of quantal theory.

After /i/, /a/ and /u/, the most commonly occurring vowels in the UPSID sample (Maddieson 1984) are the mid-front vowels /e/ (as in Spanish 'tres') or /ɛ/ (as in English 'bet') and the mid-back vowels /o/ (as in Spanish 'dos') or /ɔ/ (as in American English 'bought'). Given the degree of VT constriction during their production and the proximity of their F1 and F2 values, /o/ and /ɔ/ appear to be quantal vowels, and their high frequency of occurrence is thus predicted by quantal theory. However, the same is not true for vowels /e/ and /ɛ/. During the production of these vowels, the vocal tract is relatively unconstricted (Fant 1960; Perkell 1979) and, as in the case of /ʌ/, such vowels cannot be considered quantal because they lack surrounding regions of acoustic instability that yield auditory distinctiveness.

Quantal theory thus has only mixed success as an account of preferred vowel inventories. Although it correctly predicts the frequent occurrence across languages of /i/, /a/, /u/ and /o/ (or /ɔ/), it fails to predict the high frequency of /e/ (or /ɛ/) and the low frequency of /y/.

(c) Quantal theory: consonants

Most of Stevens's work on quantal theory has focused on vowel sounds; however, the quantal notion is intended to apply also to certain consonant sounds and to the distinction between vowels and consonants. Thus, for example, there are quantal contrasts between stop consonants (with the airflow completely blocked across the region of articulatory closure), fricative consonants (with the articulatory constriction sufficient to produce turbulence noise) and vowel-like sounds (i.e. liquids, glides and true vowels, all of which have sufficient articulatory opening to produce mainly laminar, or non-turbulent airflow; Stevens 1972). The quantal character of these contrasts is reflected in the nonlinear mapping between articulatory settings (e.g. cross-sectional area of the constriction) and the acoustic signal.

Contrasts based on varying degrees of VT constriction, such as those described in the previous paragraph, are often referred to as distinctions in *manner of articulation*. Other important consonant distinctions include: *oral* versus *nasal* (e.g. /b/ versus /m/, or /d/ versus /n/), reflecting whether the soft palate, or velum, is raised or lowered, the latter case resulting in acoustic coupling of the nasal and oral cavities; *place of articulation* (e.g. /b/ versus /d/ versus /g/, as in 'go'), corresponding to the location in the vocal tract where the most prominent articulatory closure or constriction occurs; and *voiced* versus *voiceless* (e.g. /b/ versus /p/, or /d/ versus /t/), indicating whether or not vocal-fold vibration occurs in the temporal vicinity of the consonant constriction and/or release.

The oral/nasal consonant distinction is quantal in a way analogous to the distinction between stop consonants and continuant consonants, such as fricatives and glides. In both cases, there is either complete occlusion of an airway or some degree of opening, and the difference between these two states can be modelled as a region of acoustic instability. Moreover, in both cases, the occluded state is acoustically stable with respect to a range of muscular forces, whereas the non-occluded state is relatively stable with respect to a range of constriction sizes. The occurrence of oral/nasal consonant contrasts in approximately 97% of the languages in the UPSID sample (Maddieson 1984) is, therefore, consistent with quantal theory.

Just as there are quantal regions for vowels along the back-cavity length (l_1) dimension, Stevens (1989) noted that there are several quantal regions for consonants along the place-of-articulation dimension. (As in the case of vowels, these regions correspond to intersections between resonance frequency curves such that two formants are in close proximity and are relatively stable in frequency with respect to perturbations in place of articulation). These quantal regions occur at the velar place of articulation (i.e. the oral occlusion is between the tongue body and velum), where F2 and F3 are close together, and at the retroflex place of articulation (i.e. the tongue tip is raised and retracted to occlude the vocal tract at the hard palate), where F3 and F4 are close together. Velar stop consonants (/g/ or /k/) occur in more than 99% of languages in the UPSID sample (Maddieson 1984), which counts as a successful prediction of quantal theory. However, retroflex stops (e.g. /ɖ/ in Hindi) occur in only approximately 12% of languages in the UPSID sample. It is unclear how quantal theory can account for the differing frequencies of velar and retroflex stops without appealing to principles outside the theory.

Even more problematic for quantal theory is the high cross-language frequency (more than 99% of the languages in the UPSID sample, Maddieson 1984) of stops having a labial place of articulation (/b/ or /p/) and those having a dental/alveolar place of articulation (/d/ or /t/). ('Labial' refers to occlusion at the lips; 'dental' refers to occlusion between the tongue tip and the rear surfaces of the upper teeth; and 'alveolar' refers to occlusion between the tongue tip or blade and the upper gum or alveolar ridge.) As noted by Diehl (1989), neither labials nor dentals/alveolars satisfy Stevens's criteria for quantal status. This is because the front-cavity resonances for these place values are too high in frequency to intersect with the back-cavity resonances within a perceptually significant frequency range, thus removing the possibility of acoustically stable regions with formants close together. (See Diehl (1989) for similar arguments with respect to fricative consonants).

In all of the cases of possible quantal effects discussed so far, the putative nonlinearities occur in the mapping between articulatory configurations and acoustic outputs. Recall that Stevens (1989) noted that nonlinear relations between acoustic signals and auditory responses might also yield preferences for certain sound categories or sound category contrasts. There is evidence, for example, that an auditory nonlinearity (not specifically cited by Stevens) may help to account for the widespread use of consonant voicing contrasts among languages (61% of the UPSID sample, Maddieson 1984).

In a cross-language study of syllable-initial stops, Lisker & Abramson (1964) identified an important phonetic correlate of voicing contrasts, namely, *voice onset time* (VOT), the interval between the release of the articulators (e.g. opening of the lips) and the start of vocal-fold vibration. Across languages, stops in initial position tend to occur within three VOT sub-ranges: long negative VOTs (voicing onset precedes the articulatory release by more than 45 ms); short positive VOTs (voicing onset follows the release by no more than 20 ms); and long positive VOTs (voicing onset follows the release by more than 35 ms). From these three VOT sub-ranges, languages typically select two adjacent ones to implement their voicing contrasts. For example, Spanish and Dutch use long negative VOT values for their voiced category and short positive VOT values for their voiceless category, whereas English and Cantonese use the short positive VOT sub-range for their voiced category and the long positive VOT sub-range for their voiceless category. Thai is a rare example of a language that exploits all three VOT sub-ranges to implement a three-way voicing contrast.

Initial stops with negative VOT values have a low-frequency 'voice bar' during the closure interval that is followed, starting at the moment of articulatory release, by a broad-band of mainly periodic energy concentrated in the formant regions. For stops with positive VOT values, there is no voice bar during the closure, and the VOT interval is characterized by a strongly attenuated first formant and by higher formants that are excited aperiodically.

Lisker & Abramson (1970; Abramson & Lisker 1970) examined perception of synthetic VOT syllables by native-speaking listeners of English, Spanish and Thai. They found that for each of the three language groups VOT perception is 'categorical' in the sense that listeners show (i) sharp identification boundaries between the categories relevant for their language, (ii) relatively good discrimination of stimulus pairs that straddle category boundaries and (iii) relatively poor discrimination of stimulus pairs drawn from the same voicing category. (This pattern of perceptual performance has been demonstrated for a variety of consonant contrasts. For reviews, see Repp (1984) and Diehl *et al.* (2004).)

Considered in isolation, categorical perception of speech sounds by adult human listeners does not provide convincing evidence for the existence of quantal effects based on auditory nonlinearities. Enhanced discriminability at category boundaries might simply reflect language-specific experience in categorizing speech sounds. Several lines of evidence, however, support the conclusion that discrimination peaks near voicing category boundaries are at least in part of a general auditory character. First, infants from a Spanish-speaking environment showed enhanced discrimination of VOT differences that straddle either the Spanish or the English voicing boundaries (Lasky *et al.* 1975), and a similar pattern of results was found for infants from an English-speaking environment (Aslin *et al.* 1981). Second, adult listeners' discrimination functions for non-speech analogues of VOT stimuli exhibit similar regions of peak performance (Miller *et al.* 1976; Pisoni 1977). Third, non-human animals (chinchillas) that were first trained to respond differently to two endpoint stimuli from a synthetic alveolar VOT series (/da/, 0 ms VOT; and /ta/, 80 ms VOT), and then tested on the full series, showed identification functions very similar to those of English-speaking humans (Kuhl & Miller 1978).

A neural correlate of heightened discriminability near the English voicing category boundary was reported by Sinex *et al.* (1991). They recorded auditory nerve responses in chinchilla to stimuli from a VOT series (/da/–/ta/). For stimuli that were well within either the voiced or the voiceless category, there was high response variability across neurones with different best frequencies. However, for the 30 ms VOT and 40 ms VOT stimuli, located near, but on opposite sides of, the English /d/–/t/ boundary, the response to the onset of voicing was highly synchronized across the same sample of neurones. This pattern of neural responses is shown in Figure 4.5.

Consistent with quantal theory, the above findings suggest that, in order to achieve enhanced distinctiveness, languages may exploit certain auditory nonlinearities in selecting sound categories.[5]

(d) Dispersion theory: vowels

As remarked earlier, the idea that speech sound inventories are structured to maintain perceptual distinctiveness has a long history in linguistics. However, the first investigators to express this idea quantitatively were Stevens (1972), in his first paper on quantal theory, and Liljencrants & Lindblom (1972), who took a quite different approach to the problem. Whereas in quantal theory distinctiveness characterizes the relationship between sound categories in localized regions of phonetic space (*viz.* where the categories are separated by acoustically unstable zones), Liljencrants & Lindblom viewed distinctiveness as a global property of an entire inventory of sound categories. A vowel or consonant inventory was said to be maximally distinctive if the sounds were maximally dispersed (i.e. separated from each other) in the available phonetic space.[6]

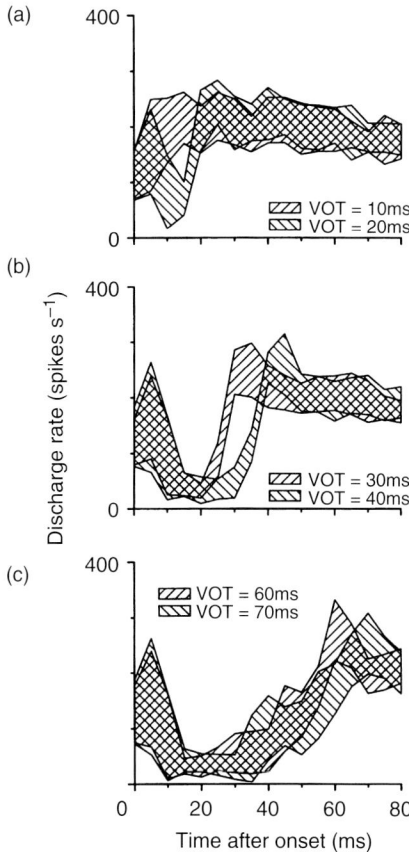

Figure 4.5 Auditory nerve responses in chinchilla to pairs of alveolar VOT stimuli in which the VOT difference was 10 ms. Each cross-hatched area encloses the mean ±1 s.d. of the average discharge rates of neurones. (Adapted with permission of the first author from Sinex *et al.* (1991)).

Dispersion theory was primarily applied to vowel sounds, and predicted vowel inventories will be the main focus of discussion. Liljencrants & Lindblom (1972) began by specifying the available phonetic space from which particular vowel sounds might be selected. This space comprises the set of possible acoustic outputs of a computational model of the vocal tract (Lindblom & Sundberg 1971) that reflects natural articulatory degrees of freedom for the jaw, lips, tongue and larynx. The model outputs were restricted to simulations of non-nasal vowel-like sounds (i.e. sounds produced without coupling between the oral and nasal cavities and without a narrow constriction in the pharyngeal or oral portions of the vocal tract). The outputs were stationary in frequency and were represented as points in a Mel-scaled F1 × F2′ space where F2′ corresponds to an effective F2 value corrected for the influence of F3. The F1 × F2′ space was densely sampled at equal-Mel intervals to create a large set of candidate vowel sounds. (The Mel scale is a measure of subjective frequency similar to the Bark scale,[7] and also related to the ERB_N-number scale described by Moore (this volume)).

To simulate preferred vowel inventories, Liljencrants & Lindblom (1972) next applied the following selection criterion: for any given size of vowel inventory, choose those candidate vowels that maximize pairwise Euclidean distances within the F1 × F2′ space. The results of these simulations are shown in Figure 4.6 for inventory sizes between 3 and 12. Solid curves represent the range of possible outputs from the articulatory model of Lindblom & Sundberg (1971), and filled circles correspond to the vowels selected according to the maximum distance criterion. (Note that the formant frequency axes are now scaled in kilohertz, though distances were calculated in Mel units.) Figure 4.6d pools all vowels selected across the 10 inventory sizes.

It is evident that a large majority of selected vowels are located on the periphery of the vowel space; central vowels appear only in the larger inventories. This pattern is consistent with the vowel inventory structure of most natural languages (Crothers 1978; Maddieson 1984). To evaluate in more detail how well dispersion theory predicts favoured vowel inventories, consider the simulations for the 3-, 5- and 7-vowel systems. (The vowels represented by the filled circles in these three cases may be identified by referring to Figure 4.7). For the 3-vowel inventory in Figure 4.6, the predicted system includes /i/, /u/ and a slightly fronted version of /a/. Recall that these vowel categories are the most

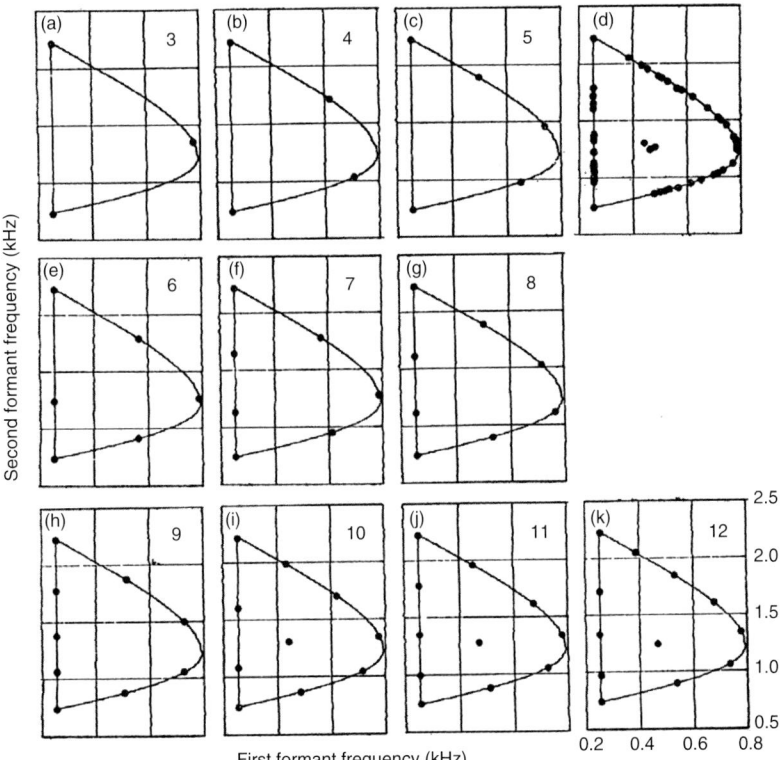

Figure 4.6 Results of simulations by Liljencrants & Lindblom (1972) of preferred vowel systems ranging in size from (a–c, e–k) 3 to 12. Solid curves represent the range of possible outputs from the articulatory model of Lindblom & Sundberg (1971), and filled circles correspond to the vowels selected according to a maximum distance criterion. (d) Pools all vowels selected across the 10 inventory sizes. (Adapted with permission of the second author from Liljencrants & Lindblom (1972). Linguistic Society of America).

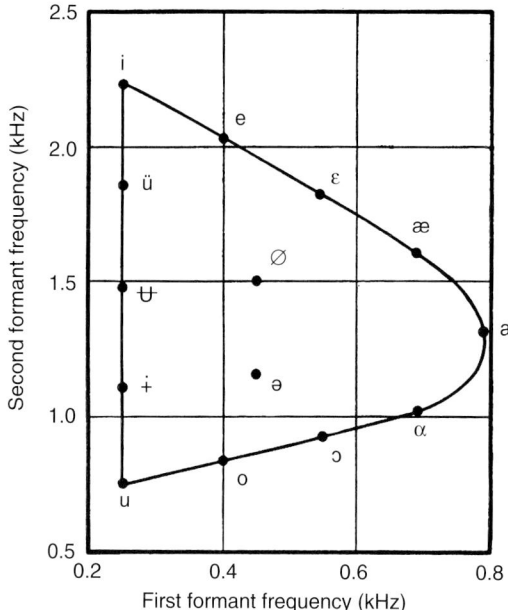

Figure 4.7 Approximate locations of major vowel categories within the space of outputs from the articulatory model of Lindblom & Sundberg (1971). (Adapted with permission of the second author from Liljencrants & Lindblom (1972), Linguistic Society of America).

frequently occurring among the world's languages, and they all appear in the most common 3-vowel inventories (Crothers 1978; Maddieson 1984). Given their locations at the extreme points of the vowel space, it is not at all surprising that these vowels would be selected by a criterion of maximum dispersion.

For the 5-vowel inventory, the predicted system in Figure 4.6 again includes /i/ and /u/ and an even more fronted version of /a/ (approaching /æ/, as in American English 'bat'). In addition, this system includes a mid-front vowel between /e/ and /ɛ/ and mid-to-low back vowel between /ɔ/ and /ɑ/ (a retracted version of /a/, as in 'father'). The most frequently occurring 5-vowel inventory (e.g. Spanish) is /i e (or ɛ) a o (or ɔ) u/ (Crothers 1978; Maddieson 1984). The predicted system deviates somewhat from the commonly observed system, especially with respect to the position of the mid-back vowel, but the overall fit is still reasonably good.

The predicted 7-vowel system in Figure 4.6 resembles the predicted 5-vowel system, except for the addition of two new high vowels between /i/ and /u/. These additional high vowels are the back, non-lip-rounded /ɨ/ and the front lip-rounded /ü/ (equivalent to /y/). However, the most commonly observed 7-vowel system (e.g. Italian) is /i e ɛ a ɔ o u/, which includes pairs of mid-front and mid-back vowels but no high vowels between /i/ and /u/.

Thus, the dispersion theory of Liljencrants & Lindblom (1972) performs as well as quantal theory in predicting the frequent occurrence of /i/, /a/, /u/ and /o/ (or /ɔ/), and it outperforms quantal theory in correctly predicting that /e/ (or /ɛ/) will occur frequently and that /y/ will be less common than /i/. The most important failure of the Liljencrants &

Lindblom simulations is the prediction of too many high vowels for inventories of seven or more vowels.

Similar to quantal theory, dispersion theory includes both listener- and talker-oriented selection criteria. In the simulations of Liljencrants & Lindblom (1972), the talker-oriented selection criteria were implemented by restricting the available phonetic space (i.e. the output of the vowel production model of Lindblom & Sundberg (1971)) to certain 'basic' articulatory types (e.g. non-nasalized vowels). Later versions of dispersion theory (Lindblom 1986, 1990) attempted to account not only for the structure of preferred vowel inventories and but also for variation in speech clarity in running speech. In these versions of the theory, the goal of talkers is to produce speech that is intelligible to listeners but to do so with as little effort as necessary. In other words, talkers try to achieve sufficient, rather than maximal, distinctiveness, and they thus tend to vary their utterances from reduced ('hypo-speech') forms to clear ('hyper-speech') forms depending on the communication conditions that apply. In general, as information content increases, clarity of speech production also increases (for a review, see Hay *et al.* (2006)). Since the focus of the present paper is on the structure of preferred sound inventories rather than on phonetic variation in running speech, the role of talker-oriented selection factors will not be further discussed.

(e) Auditory enhancement hypothesis

It is useful to consider in more detail how talkers implement a listener-oriented strategy of vowel dispersion. One simple approach is to select relatively extreme tongue body and jaw positions since acoustic distinctiveness tends to be correlated with articulatory distinctiveness. As was discussed earlier, the vowels /i/, /a/ and /u/ are each characterized by articulatory extremes in this sense. However, the acoustic dispersion of these vowels is only partly explained by the positioning of the tongue body and jaw. A fuller account is provided by the auditory enhancement hypothesis (Diehl & Kluender 1989*a*, *b*; Diehl *et al.* 1990; Kingston & Diehl 1994), which attempts to explain common patterns of phonetic covariation on listener-oriented grounds. The hypothesis states that phonetic properties of sound categories covary as they do largely because language communities tend to select properties that have mutually enhancing auditory effects. In the case of vowels, auditory enhancement is most typically achieved by combining articulatory properties that have similar—and hence reinforcing—acoustic consequences. (The auditory enhancement hypothesis is closely related to the theory of redundant features independently developed by Stevens and his colleagues (Stevens *et al.* 1986; Stevens & Keyser 1989)).

The high back vowel /u/ offers a good example of how auditory enhancement works. In Figure 4.7 it may be seen that /u/ is distinguished from lower vowels in having a low F1 and from more anterior vowels in having a low F2. Articulatory properties that produce a lowering of F1 and F2 thus contribute to the distinctiveness of /u/. From acoustic theory (e.g. Fant 1960), it is known that for a tube-like configuration such as the vocal tract, there are several ways to lower a resonance frequency. These are: (i) to lengthen the tube at either end, (ii) to constrict the tube at any node in the standing pressure wave corresponding to the resonance (see earlier discussion), and (iii) to dilate the tube at any antinode in the same standing pressure wave. It happens that in carefully articulated versions of /u/, every one of these options is exploited.

For the purpose of this analysis, the vocal tract is treated in its initial configuration as a quarter-wave resonator, which is then subjected to such perturbations as (i)–(iii) to achieve a target configuration (*viz.* a sequence of two Helmholtz resonators) corresponding to /u/. VT lengthening can be achieved by protruding the lips, a typical component of the lip-rounding gesture that accompanies the production of /u/. Lengthening is also achieved by larynx lowering, and this too has been observed during /u/ production (MacNeilage 1969; Riordan 1977). Both F1 and F2 are lowered as a result of these gestures. The two major VT constrictions characteristic of /u/ occur at the lips (contraction of the lip orifice is another component of the rounding gesture) and in the velar region near the junction between the pharyngeal and oral cavities. As the lip orifice is located at nodes in the standing pressure waves corresponding to the first and second resonances, the effect of lip constriction is to lower F1 and F2. The constriction in the velar region, which is located near another node in the standing pressure wave pattern for the second resonance, yields additional lowering of F2. Finally, VT dilations near the mid-palate and in the lower pharynx (both corresponding to second resonance antinodes) contribute to yet more F2 lowering. The dilation near the mid-palate may be viewed as a by-product of tongue-body retraction and raising, but the pharyngeal dilation is largely the result of tongue-root advancement, which appears to be, anatomically speaking, partly independent of tongue height (Lindau 1979). In summary, the VT configuration for /u/ is optimally tailored to yield an acoustically distinctive vowel. Analogous arguments may be made with respect to other commonly occurring vowels such as /i/ and /a/.

(f) Attempts to unify dispersion theory and (aspects of) quantal theory

Following up on the work of Lindblom and his colleagues (Liljencrants & Lindblom 1972; Lindblom 1986) and Stevens (1972, 1989), Schwartz *et al.* (1997) proposed the dispersion–focalization theory of vowel systems. The theory attempts to predict favoured vowel inventories by summing the effects of two perceptual components: global dispersion and local focalization. The first component is based on distances among vowels in a formant frequency space similar to that used by Liljencrants & Lindblom. The second component is based on spectral salience of individual vowels (i.e. the presence or absence of relatively intense spectral regions) and is related to proximity between formants, thus giving weight to an important property of quantal vowels. Simulations of preferred vowel systems are controlled by two free parameters: one sets the relative contribution of F1 versus higher formant frequencies in determining inter-vowel distances and the other sets the relative contributions of the dispersion and the focalization components. Schwartz *et al.* identified a range of values of these parameters that provided reasonably good fits to the structure of the most common vowel inventories. In particular, by giving F1 more weight than higher formant frequencies, the tendency to predict the occurrence of too many high vowels was eliminated (cf. Liljencrants & Lindblom 1972). This outcome is expected because F2 and F3 are primarily related to lip configuration and the front-back position of the tongue, whereas F1 is primarily related to tongue and jaw height. Accordingly, reducing the weight of higher formants effectively compresses the auditory extent of the front-back dimension relative to the height dimension, allowing less room for high vowels such as /i/ and /y/.

Is it possible to approximate the success of the dispersion–focalization theory without using free parameters to obtain an acceptable fit to the data? One approach is to try to improve on formant-based measures of inter-vowel distance by taking into account aspects of the auditory representation of speech signals. For example, Lindblom (1986) adopted a measure of auditory distance based on auditory representations of whole spectra rather than formant frequencies. The representations are derived from a model incorporating auditory filtering (see Moore this volume) as well as pitch and loudness scaling. Auditory distance is defined as the Euclidean distance between two vowels (with the same F0) in an n-dimensional space, where n is the number of auditory filters, and the value for a given vowel on any dimension is equal to the output of the corresponding filter (scaled in Sones/Bark).[7] When vowel system simulations were carried out using this measure of auditory distance (along with the same inter-vowel distance maximization criterion applied in earlier simulations), the results were disappointing. Although there was some small improvement in predictive accuracy relative to the simulations of Liljencrants & Lindblom (1972), the problem of too many high vowels remained.

The auditory filter outputs used in the simulations by Lindblom (1986) are whole-spectrum representations intended to model (albeit very roughly) the average firing rate of auditory neurones as a function of their best frequency (see Young this volume). Such representations are incomplete in one important respect, namely, temporal information about stimulus frequency (phase locking) is not included. This omission may be significant because such temporal information tends to be more resistant to noise degradation than information contained in average firing rates alone (Sachs *et al.* 1982; Greenberg 1988; Young this volume). In particular, spectral peaks (e.g. formants) are temporally coded by neurones not only with best frequencies closest to the peaks but also with somewhat different best frequencies (Delgutte & Kiang 1984). Thus, temporal coding yields a redundant and fairly noise resistant representation of prominent regions in the stimulus spectrum. Since normal speech communication takes place in the presence of background noise, a measure of auditory distance that ignores temporal coding may yield inaccurate predictions about preferred speech sound inventories.

With these considerations in mind, Diehl *et al.* (2003) conducted a new series of vowel system simulations with an auditory model that incorporates an analogue of average firing rate as a function of best frequency as well as a dominant frequency representation based on temporal coding. For any two vowels with the same F0, these two forms of representation are multiplied to form a single spatio-temporal measure of auditory distance.[8] The vowel systems predicted by these simulations show a reasonably good fit to the most commonly occurring systems. For example, the predicted 7-vowel system includes /i/, /a/ and /u/ as well as two mid-front vowels and two mid-back vowels, similar to Italian and most other 7-vowel systems. In other words, the problem of too many high vowels (Liljencrants & Lindblom 1972; Lindblom 1986) is eliminated.

Why does inclusion of temporal coding information improve the accuracy of vowel system simulations? Owing to redundant specification of relatively intense frequency components, formant peaks contribute disproportionately to auditory representations and hence to measures of auditory distance. The first formant plays an especially large role owing to its greater intensity relative to higher formants. This produces a perceptual warping of the vowel space such that the vowel height dimension (corresponding to F1) can perceptually accommodate more vowel contrasts than the front–back dimension (corresponding to F2 and F3), and this in turn reduces the likelihood that high vowels between /i/ and /u/ will be selected by the maximum distance criterion. Temporal coding also boosts the contrastive value of quantal vowels, since their closely spaced formants give rise to spectrally salient regions that are redundantly specified in the auditory representation.

4.5 Concluding remarks

The modelling approach of Diehl *et al.* (2003) yields results that are generally similar to those of the dispersion-focalization theory (Schwartz *et al.* 1997). However, Diehl *et al.* do not treat global dispersion and local focalization (spectral salience) as separate perceptual components. Instead, spectral salience directly enhances global dispersion as an automatic consequence of spatio-temporal coding of frequency. In this way, key elements of dispersion and quantal theories are unified within a single explanatory framework.

Acknowledgements

I thank Björn Lindblom for many years of productive discussion of these issues. I am also grateful to Andrew Lotto, Jessica Hay, Sarah Sullivan, Brian Moore, Thomas Baer and Christopher Darwin for their very helpful comments on an earlier draft of this chapter and to Sarah Sullivan for her help in preparing the figures. Preparation of this chapter was supported by NIH grant R01 DC000427-15.

Endnotes

1 Catford (1977) distinguishes between two types of frication source: 'channel turbulence', which is produced simply by airflow through a channel, and 'wake turbulence', which is created downstream from the edge of an obstacle (e.g. teeth or upper lip) oriented perpendicular to the airflow.
2 In addition to resonances, a VT filter function may be characterized by antiresonances, which have the opposite effect on the spectrum. At or near the frequency of an antiresonance, energy from a source is absorbed and hence greatly attenuated in the output spectrum. Antiresonances are introduced into the filter function if (i) the vocal tract has a side branch or bifurcated airways, as in the production of nasal consonants or nasalized vowels or (ii) there is an occlusion or narrow constriction of the vocal tract, as in the production of stop or fricative consonants (Kent & Read 1992).
3 Tom Baer's helpful suggestions about the wording of this paragraph are gratefully acknowledged.
4 A more realistic model for /i/ would include a third tube at the front of the vocal tract larger in diameter than the second tube. This extended model incorporates the effects of lip spreading (Stevens 1998).
5 Consonants tend to be briefer in duration and less intense than vowels, but their perception is generally robust. As discussed in §4c, quantal properties of certain consonants help to explain this robustness. However, another important factor is the dynamic character of consonant production, which gives rise to a rich set of time-distributed perceptual cues. For example, the identity of a word-medial stop is signalled by properties of the vowel–consonant transitions, the occlusion interval (e.g. its duration), the transient burst of energy at the articulatory release, the fricative and/or aspiration following the burst and the consonant-vowel transitions (Pickett 1999). For further discussion of cue redundancy in consonant perception, see Kingston & Diehl (1994).

6 This notion of maximal dispersion is apparently what earlier linguists such as Passy (1890), Jakobson (1941) and Martinet (1955) had in mind when they discussed the role of perceptual contrast in the structure of speech sound systems.
7 Sones are units of subjective loudness; Barks are units of subjective frequency, with one Bark corresponding to a step in frequency equal to one critical band (Zwicker & Terhardt 1980; also see Moore this volume).
8 To model temporal coding, an inverse FFT is performed on the spectral output of each auditory filter and the resulting time-domain signal is input to a dominant frequency detector, which specifies dominant frequency in terms of zero crossings (Carlson & Granström 1982). The output of these detectors (one per auditory filter) is the dominant frequency representation for a given vowel. To calculate auditory distance between two vowels, the product of the average firing rate representation and the dominant frequency representation is computed for each filter, and the Euclidean distances are then calculated as in Lindblom (1986).

References

Abramson, A. S. & Lisker, L. 1970 Discriminability along the voicing continuum: cross-language tests. In *Proc. 6th Int. Cong, of Phonetic Sciences, Prague, 1967*, pp. 569–573. Prague, Czech Republic: Academia.
Aslin, R. N., Pisoni, D. B., Hennessy, B. L. & Perey, A. J. 1981 Discrimination of voice onset time by human infants: new findings and implications for the effects of early experience. *Child Dev.* **52**, 1135–1145. (doi:10.2307/1129499)
Carlson, R. & Granström, B. 1982 Towards an auditory spectrograph. In *The representation of speech in the peripheral auditory system* (eds R. Carlson & B. Granström), pp. 109–114. Amsterdam, The Netherlands: Elsevier Biomedical.
Catford, J. C. 1977 *Fundamental problems in phonetics*. Bloomington, IN: Indiana University Press.
Chiba, T. & Kajiyama, M. 1941 *The vowel: its nature and structure*. Tokyo, Japan: Tokyo-Kaisekan. (Reprinted by the Phonetic Society of Japan 1958.)
Chomsky, N. & Halle, M. 1968 *The sound pattern of English*. New York, NY: Harper & Row.
Crothers, J. 1978 Typology and universals of vowel systems. In *Universals of human language*, vol. 2 (eds J. H. Greenberg, C. A. Ferguson & E. A. Moravcsik), pp. 99–152. Stanford, CA: Stanford University Press.
Delgutte, B. & Kiang, N. Y.-S. 1984 Speech coding in the auditory nerve I: vowel-like sounds. *J. Acoust. Soc. Am.* **75**, 866–878. (doi:10.1121/1.390596)
Diehl, R. L. 1989 Remarks on Stevens' quantal theory of speech. *J. Phonet.* **17**, 71–78.
Diehl, R. L. & Kluender, K. R. 1989a On the objects of speech perception. *Ecol. Psychol.* **1**, 121–144. (doi:10.1207/sl5326969eco0102_2)
Diehl, R. L. & Kluender, K. R. 1989b Reply to commentators. *Ecol. Psychol.* **1**, 195–225. (doi:10.1207/sl5326969 eco0102_6)
Diehl, R. L. & Lindblom, B. 2004 Explaining the structure of feature and phoneme inventories: the role of auditory distinctiveness. In *Speech processing in the auditory system* (eds S. Greenberg, W. A. Ainsworth, A. N. Popper & R. R. Fay), pp. 101–162. New York, NY: Springer.
Diehl, R. L., Kluender, K. R. & Walsh, M. A. 1990 Some auditory bases of speech perception and production. In *Advances in speech, hearing and language processing*, vol. 1 (ed. W. A. Ainsworth), pp. 243–268. London, UK: JAI Press.
Diehl, R. L., Lindblom, B. & Creeger, C. P. 2003 Increasing realism of auditory representations yields further insights into vowel phonetics. In *Proc. 15th Int. Cong, of Phonetic Sciences*, vol. 2, pp. 1381–1384. Adelaide, Australia: Causal Publications.
Diehl, R. L., Lotto, A. J. & Holt, L. L. 2004 Speech perception. *Annu. Rev. Psychol.* **55**, 149–179. (doi:10.1146/annurev.psych.55.090902.142028)
Fant, G. 1960 *Acoustic theory of speech production*. The Hague, The Netherlands: Mouton. Fant, G. 1973 *Speech sounds and features*. Cambridge, MA: MIT Press.

Flanagan, J. L. 1972 *Speech analysis synthesis and perception*. Berlin, Germany: Springer.
Greenberg, S. 1988 Acoustic transduction in the auditory periphery. *J. Phonet.* **16**, 3–17.
Hay, J. F., Sato, M., Coren, A. E., Moran, C. L. & Diehl, R. L. 2006 Enhanced contrast for vowels in utterance focus: a cross-language study. *J. Acoust. Soc. Am.* **119**, 3022–3033. (doi:10.1121/1.2184226)
Jakobson, R. 1941 *Kindersprache, Aphasie und allgemeine Lautgesetze*, pp. 1–83. Uppsala, Sweden: Uppsala Universitets Arsskrift.
Jakobson, R., Fant, G. & Halle, M. 1963 *Preliminaries to speech analysis*. Cambridge, MA: MIT Press.
Johnson, K. 1997 *Acoustic and auditory phonetics*. Cambridge, MA: Blackwell.
Kent, R. D. & Read, C. 1992 *The acoustic analysis of speech*. San Diego, CA: Singular Publishing.
Kingston, J. & Diehl, R. L. 1994 Phonetic knowledge. *Language* **70**, 419–54. (doi:10.2307/416481)
Klatt, D. H. 1982 Prediction of perceived phonetic distance from critical-band spectra: a first step. In *Proc. IEEE Int. Conf. Speech, Acoustic Signal Process*, vol. **82**, pp. 1278–1281.
Kuhl, P. K. & Miller, J. D. 1978 Speech perception by the chinchilla: identification functions for synthetic VOT stimuli. *J. Acoust. Soc. Am.* **63**, 905–917. (doi:10.1121/ 1.381770)
Ladefoged, P., Harshman, R., Goldstein, L. & Rice, L. 1977 Vowel articulation and formant frequencies. *UCLA Working Papers Phonet.* **38**, 16–40.
Lasky, R. E., Syrdal-Lasky, A. & Klein, R. E. 1975 VOT discrimination by four to six and a half month old infants from Spanish environments. *J. Exp. Chid Psychol.* **20**, 215–225. (doi:10.1016/0022–0965(75)90099–5)
Liljencrants, J. & Lindblom, B. 1972 Numerical simulation of vowel quality systems: the role of perceptual contrast. *Language* **48**, 839–862. (doi:10.2307/411991)
Lindau, M. 1979 The feature expanded. *J. Phonet.* **7**, 163–176.
Lindblom, B. 1986 Phonetic universals in vowel systems. In *Experimental phonology* (eds J. J. Ohala & J. J. Jaeger), pp. 13–44. Orlando, FL: Academic Press.
Lindblom, B. 1990 Explaining phonetic variation: a sketch of the H & H theory. In *Speech production and speech modeling* (eds W. J. Hardcastle & A. Marchal), pp. 403–39. Dordrecht, The Netherlands: Kluwer.
Lindblom, B. & Sundberg, J. 1971 Acoustical consequences of lip, tongue, jaw and larynx movement. *J. Acoust. Soc. Am.* **50**, 1166–1179. (doi:10.1121/1.1912750)
Lisker, L. & Abramson, A. S. 1964 A cross-language study of voicing in initial stops: acoustical measurements. *Word* **20**, 384–422.
Lisker, L. & Abramson, A. S. 1970 The voicing dimension: some experiments in comparative phonetics. In *Proc. 6th Int. Cong, of Phonetic Sciences, Prague 1967*, pp. 563–567'. Prague, Czech Republic: Academia.
MacNeilage, P. M. 1969 A note on the relation between tongue elevation and glottal elevation. Monthly Internal Memorandum, University of California, Berkeley, January 1969, pp. 9–26.
Maddieson, I. 1984 *Patterns of sound*. Cambridge, UK: Cambridge University Press.
Martinet, A. 1955 *Économie des Changements Phonétiques*. Berne, Switzerland: Francke.
Miller, J. D., Wier, C. C., Pastore, R. E., Kelly, W. J. & Dooling, R. J. 1976 Discrimination and labeling of noise-buzz sequences with varying noise-lead times: an example of categorical perception. *J. Acoust. Soc. Am.* **60**, 410–417. (doi:10.1121/0.381097)
Passy, P. 1890 *Études sur les Changements Phonétiques et Leurs Caractéres Généraux*. Paris, France: Librairie Firmin-Didot.
Perkell, J. S. 1979 On the nature of distinctive features: implications of a preliminary vowel production study. In *Frontiers of speech communication research* (eds B. Lindblom & S. Öhman), pp. 365–380. New York, NY: Academic Press.
Pickett, J. M. 1999 *The acoustics of speech communication*. Needham Heights, MA: Allyn & Bacon.
Pisoni, D. B. 1977 Identification and discrimination of the relative onset time of two component tones: implications for voicing perception in stops. *J. Acoust. Soc. Am.* **61**, 1352–1361. (doi:10.1121/1.381409)
Pisoni, D. B. 1980 Variability of vowel formant frequencies and the quantal theory of speech: a first report. *Phonetica* **37**, 285–305.

Repp, B. H. 1984 Categorical perception: issues, methods, findings. In *Speech and language: advances in basic research and practice*, vol. **10** (ed. N. J. Lass), pp. 243–335. New York, NY: Academic Press.

Riordan, C. J. 1977 Control of vocal-tract length in speech. *J. Acoust. Soc. Am.* **62**, 998–1002. (doi:10.1121/1.381595)

Sachs, M., Young, E. & Miller, M. 1982 Encoding of speech features in the auditory nerve. In *The representation of speech in the peripheral auditory system* (eds R. Carlson & B. Granström), pp. 115–130. Amsterdam, The Netherlands: Elsevier Biomedical.

Schwartz, J.-L., Boë, L.-J., Vallée, N. & Abry, C. 1997 The dispersion–focalization theory of vowel systems. *J. Phonet.* 25, 255–286. (doi:10.1006/jpho.1997.0043)

Sinex, D. G., McDonald, L. P. & Mott, J. B. 1991 Neural correlates of nonmonotonic temporal acuity for voice onset time. *J. Acoust. Soc. Am.* **90**, 2441–2449. (doi:10. 1121/1.402048)

Stevens, K. N. 1972 The quantal nature of speech: evidence from articulatory-acoustic data. In *Human communication: a unified view* (eds E. E. David & P. B. Denes), pp. 51–66. New York, NY: McGraw-Hill.

Stevens, K. N. 1989 On the quantal nature of speech. *J. Phonet.* **17**, 3–45.

Stevens, K. N. 1998 *Acoustic phonetics*. Cambridge, MA: MIT Press.

Stevens, K. N. & House, A. S. 1955 Development of a quantitative description of vowel articulation. *J. Acoust. Soc. Am.* **27**, 484–493. (doi:10.1121/1.1907943)

Stevens, K. N. & House, A. S. 1961 An acoustical theory of vowel production and some of its implications. *J. Speech Hear. Res.* **4**, 303–320.

Stevens, K. N. &Keyser, S. J. 1989 Primary features and their enhancement in consonants. *Language* **65**, 81–106. (doi:10.2307/414843)

Stevens, K. N., Keyser, S. J. & Kawasaki, H. 1986 Toward a phonetic and phonological theory of redundant features. In *Invariance and variability in speech processes* (eds J. S. Perkell & D. H. Klatt), pp. 426–449. Hillsdale, NJ: Erlbaum.

Syrdal, A. K. & Gopal, H. S. 1986 A perceptual model of vowel recognition based on the auditory representation of American English vowels. *J. Acoust. Soc. Am.* **79**, 1086–1100. (doi:10.1121/1.393381)

Zipf, G. K. 1949 *Human behavior and the principle of least effort: an introduction to human ecology*. Cambridge, MA: Addison-Wesley.

Zwicker, E. & Terhardt, E. 1980 Analytical expressions for critical band rate and critical bandwidth as a function of frequency. *J. Acoust. Soc. Am.* **68**, 1523–1525. (doi:10. 1121/1.385079)

5

Early language acquisition: phonetic and word learning, neural substrates, and a theoretical model

Patricia K. Kuhl

> Infants learn language(s) with apparent ease, and behavioral and brain studies are providing valuable information about the mechanisms that underlie this capacity. Noninvasive, safe brain technologies have now been proven feasible for use with children starting at birth. The past decade has produced an explosion in neuroscience research examining young children's processing of language at the phonetic, word, and sentence levels, and studies have begun to explore how children develop bilingual language skills. At all levels of language, the neural signatures of learning can be documented at remarkably early points in development. Individual continuity in linguistic development is seen in data showing that infants' responses to phonemes in the first year of life predicts those same children's language abilities in the second and third year of life, a finding with theoretical and clinical implications. Developmental neuroscience studies using language are beginning to answer questions about the origins of humans' language faculty.

5.1 Introduction

Infants begin life with the capacity to detect phonetic distinctions across all languages, and develop a language-specific phonetic capacity and acquire early words before the end of the first year (Jusczyk 1997; Werker & Curtin 2005; Kuhl *et al.* 2008). The tools of modern developmental neuroscience are bringing us closer to understanding how the interaction between biology and culture produces the human capacity for language. Neuroscientific studies will also provide valuable information that may allow us to diagnose developmental disabilities at a stage in development when interventions are more likely to improve children's lives.

Remarkable progress has been made in the last decade in scientists' abilities to examine the young infant brain while its owner processes language, reacts to social stimuli such as faces, listens to music, or hears their mother's voice. This review focusses on the new techniques and what they are teaching us about the earliest phases of language acquisition.

Neuroscientific studies on infants and young children now extend from phonemes to words to sentences. These studies fuel the hope that an understanding of development in typically developing children and in children with developmental disabilities will be achieved. Studies show that exposure to language in the first year of life begins to set the neural architecture in a way that vaults the infant forward in the acquisition of language. The goal of this chapter is to explore what we have learned about the neural mechanisms that underlie language in typically developing children, and how they differ in children with developmental disabilities that involve language, such as autism.

5.2 Neuroscience techniques measure language processing in the young brain

Rapid advances have been made in the development of noninvasive techniques to examine language processing in infants and young children (Figure 5.1). These methods include electroencephalography (EEG)/event related potentials (ERPs), magnetoencephalography (MEG), functional magnetic resonance imaging (fMRI), and near-infrared spectroscopy (NIRS).

ERPs have been widely used to study speech and language processing in infants and young children (for reviews, see Kuhl 2004; Friederici 2005; Conboy *et al*. 2008b; and Kuhl & Rivera-Gaxiola 2008). ERPs, a part of the EEG, reflect electrical activity that is time-locked to the presentation of a specific sensory stimulus (e.g. syllables or words) or

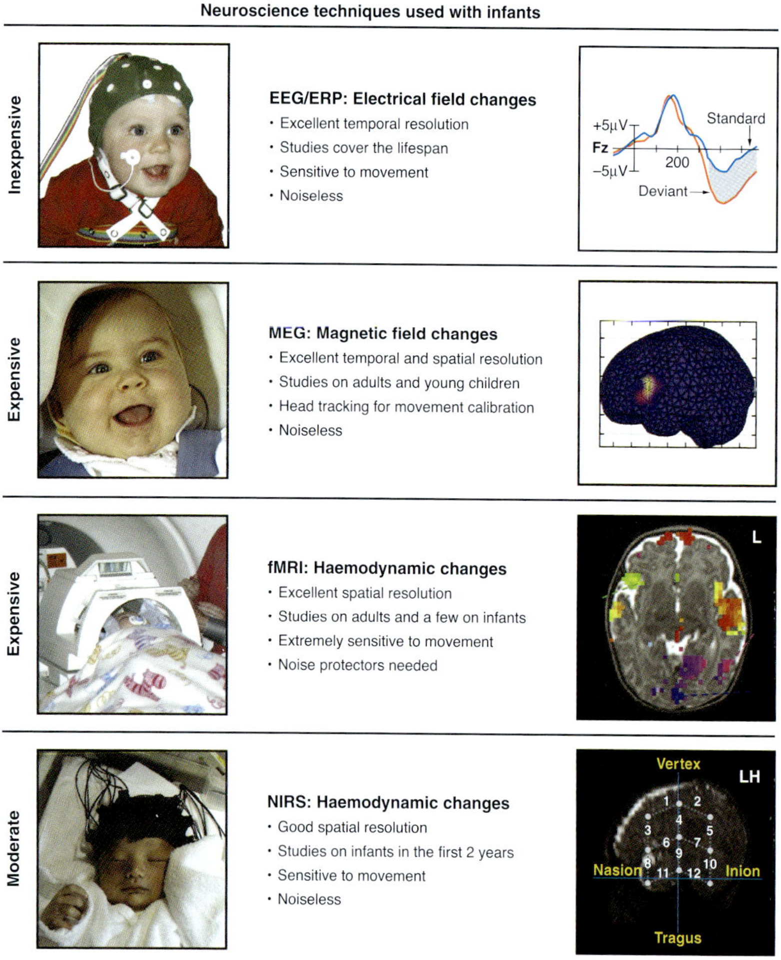

Figure 5.1 Four neuroscience techniques now used with infants and young children to examine the brain's responses to linguistic signals.
From Kuhl & Rivera-Gaxiola (2008).

a cognitive process (recognition of a semantic violation within a sentence or phrase). By placing sensors on a child's scalp, the activity of neural networks firing in a coordinated and synchronous fashion in open-field configurations can be measured, and voltage changes occurring as a function of cortical neural activity can be detected. ERPs provide precise time resolution (milliseconds), making them well suited for studying the high-speed and temporally ordered structure of human speech. ERP experiments can also be carried out in populations who, because of age or cognitive impairment, cannot provide overt responses. Spatial resolution of the source of brain activation is, however, limited.

MEG is another brain imaging technique that tracks activity in the brain with exquisite temporal resolution. MEG (as well as EEG) techniques are safe and noiseless, allowing data collection while infants listen to language in a quiet environment. The SQUID (superconducting quantum interference device) sensors located within the MEG helmet measure the minute magnetic fields associated with electrical currents that are produced by the brain when it is performing sensory, motor, or cognitive tasks. MEG allows precise localization of the neural currents responsible for the sources of the magnetic fields, and has been used to test phonetic discrimination in adults (Kujala *et al.* 2004).

Recently, a genuine advance has been documented by the first MEG studies testing awake infants in the first year of life (Cheour *et al.* 2004; Imada *et al.* 2006; Bosseler *et al.* 2008; Imada *et al.* 2008). In these studies, the use of sophisticated head-tracking software and hardware allows correction for infants' head movements, so infants are free to move comfortably during the tests. MEG studies allow whole-brain imaging during speech discrimination, providing data on the location and timing of brain activation in critical regions (Broca's and Wenicke's) involved in language acquisition (see Imada *et al.* 2006; Bosseler *et al.* 2008; Imada *et al.* 2008).

MEG and/or EEG can be combined with MRI, a technique that provides static structural/anatomical pictures of the brain. Using mathematical modelling methods, the specific brain regions that produce the magnetic or electrical signals can be identified in the human brain with high spatial resolution (millimeter). Structural MRIs allow measurement of anatomical changes in white and grey matter in specific brain regions across the lifespan. MRIs can be superimposed on the physiological activity detected by MEG or EEG to refine the spatial localization of brain activities for individual participants.

fMRI is now considered a standard method of neuroimaging in adults because it provides high spatial-resolution maps of neural activity across the entire brain (e.g. Gernsbacher & Kaschak 2003). However, unlike EEG and MEG, fMRI does not directly detect neural activity, but rather the changes in blood-oxygenation that occur in response to neural activation/firing. Neural events happen in milliseconds, while the blood-oxygenation changes that they induce are spread out over several seconds, thereby severely limiting fMRI's temporal resolution. Adult studies are employing new fMRI data-analysis methods for speech stimuli, and correlating the fMRI data to behavioural data. For example, Raizada and colleagues (2009), using a multivariate pattern classifier, showed that English – but not Japanese – speakers exhibited distinct neural activity patterns for /ra/ and /la/ in the primary auditory cortex. Subjects who behaviourally distinguished the sounds most accurately also had the most distinct neural activity patterns.

fMRI techniques would be very valuable with infants, but few studies have attempted fMRI with infants (Dehaene-Lambertz *et al.* 2002; 2006). The technique requires subjects to be perfectly still, and the MRI device produces loud sounds making it necessary to shield infants' ears while delivering language stimuli.

NIRS also measures cerebral haemodynamic responses in relation to neural activity, but employs the absorption of light, which is sensitive to the concentration of haemoglobin, to measure activation (Aslin & Mehler 2005). NIRS utilizes near-infrared light to measure changes in blood oxy- and deoxy-haemoglobin concentrations in the brain as well as total blood-volume changes in various regions of the cerebral cortex. The NIRS system can determine where and how active the specific regions of the brain are by continuously monitoring blood haemoglobin levels, and reports have begun to appear on infants in the first 2 years of life (Peña *et al.* 2003; Homae *et al.* 2006; Bortfeld *et al.* 2007; Taga & Asakawa 2007). Homae *et al.*, e.g. provided data using NIRS that suggest that sleeping 3-month-old infants process the prosodic information in sentences in the right temporoparietal region. As with other techniques relying on haemodynamic changes such as fMRI, NIRS does not provide good temporal resolution. One of the most important aspects of this technique is that co-registration with other testing techniques such as EEG and MEG may be possible.

The use of these techniques with infants and young children has produced an explosion of neuroscience studies using stimuli that tap all levels of language – phoneme, word, and sentence. In the next sections, examples of recent findings will be described to give a sense of the promise of neuroscience for the study of language acquisition in children.

5.3 Neural signatures of phonetic learning in typically developing children

Perception of the basic units of speech – the vowels and consonants that make up words – is one of the most widely studied behaviours in infancy and adulthood, and studies using ERPs have advanced our knowledge of development and learning.

Behavioural studies demonstrated that, at birth, young infants exhibit a universal capacity to detect differences between phonetic contrasts used in the world's languages (Eimas *et al.* 1971). We have referred to this as Phase 1 in development (Kuhl *et al.* 2008). This universal capacity is dramatically altered by language experience starting as early as 6 months for vowels and by 10 months for consonants: over time, native language phonetic abilities significantly increase (Kuhl *et al.* 1992; Cheour *et al.* 1998; Rivera-Gaxiola *et al.* 2005b; Kuhl *et al.* 2006; Sundara *et al.* 2006), while the ability to discriminate phonetic contrasts that are not relevant to the language of the culture declines (Werker & Tees 1984; Cheour *et al.* 1998; Best & McRoberts 2003; Rivera-Gaxiola *et al.* 2005b; Kuhl *et al.* 2006).

By the end of the first year, the infant brain is no longer universally prepared for all languages, but primed to acquire the specific one(s) to which they have been exposed. We refer to this as Phase 2 in infant phonetic development (Kuhl *et al.* 2008). The explanation of this transition from Phase 1 to Phase 2 has become the focus of intense study because it illustrates the interaction between biology and culture – between infants' initial state and infants' abilities to learn. Speech offers the opportunity to study the brain's ability to be shaped implicitly by experience.

Kuhl *et al.* (2008) examined whether the transition in phonetic perception from a language general ability to a language-specific one – from Phase 1 to Phase 2 – can be linked to the growth of language. The work provided a critical test stemming from the native language neural commitment (NLNC) hypothesis (Kuhl 2004). According to NLNC, initial native language learning involves 'neural commitment' to the patterned regularities contained in ambient speech, with bidirectional effects: neural coding facilitates the detection of more complex language units (words) that build on initial learning, while simultaneously reducing attention to alternative patterns, such as those of a foreign language.

This formulation suggests that infants with excellent phonetic learning skills should advance more quickly towards language. In contrast, foreign language phonetic perception reflects the degree to which the infant brain remains uncommitted to native language patterns – still in Phase 1 as it were – at a more universal and immature phase of development. Infants in Phase 1 remain 'open' to non-native speech patterns. As an open system reflects uncommitted circuitry, infants who remain highly skilled at discriminating foreign language phonetic units would be expected to show a slower progression towards language.

New ERP studies of infants support the NLNC assertion. Kuhl *et al.* (2008) measured infants' ERPs at 7.5 months of age in response to changes in native (/p-t/), non-native (Mandarin /ɕ-tɕʰ/, and Spanish /t-d/) phonemes. The Mismatch Negativity (MMN), which has been shown in adults to be a neural correlate of phonetic discrimination (Näätänen *et al.* 1997), was calculated for both the native and non-native phonemes for each infant. Individual variation was observed for both native and non-native discrimination, representing either 'noise' or meaningful differences among infants.

The results of our analysis supported the idea that the differences among infants were meaningful. MMN measurements taken at 7.5 months – for both the native and the non-native phonetic contrasts – predicted later language abilities. However, and in accord with the NLNC hypothesis, the native and non-native contrasts predicted language growth in opposing directions (Kuhl *et al.* 2008).

The MMN component was elicited in individual infants (Figure 5.2a). Native and non-native contrasts were measured in counterbalanced order, and the MMN was observed between 250 and 400 ms (Figure 5.2b). For the infant shown in Figure 5.2a, greater negativity of the MMN – indicating better neural discrimination – was shown for the native than for the non-native phonetic contrast; other infants showed equal discrimination for the two contrasts, or better discrimination of the non-native contrast. Infants' language abilities were measured at four later points in time – 14, 18, 24, and 30 months of age – using the MacArthur–Bates Communicative Development Inventories (CDI), a reliable and valid measure assessing language and communication development from 8 to 30 months of age (Fenson *et al.* 1993).

The MMN measures taken at 7.5 months of age were related to the language measures taken between 14 and 30 months of age. For the native contrast, the strength of the MMN (better discrimination) predicted accelerated word production at 24 months, greater sentence complexity at 24 months, and longer mean length of utterance at 30 months of age. In contrast, for the non-native stimulus pair, the strength of the MMN at the same age in the same infants predicted slower language development at the same future points in time. Behavioural (Kuhl *et al.* 2005b) and brain measures (Kuhl *et al.* 2008), collected on the same infants, were significantly correlated.

This pattern, showing differential effects of good discrimination for the native and non-native contrasts, can be readily seen in the growth of vocabulary from 14 to 30 months (Figure 5.2c). Hierarchical linear growth curve modelling (Raudenbush *et al.* 2005) shows that both native and non-native discrimination at 7.5 months significantly predict vocabulary growth, but the effects of good phonetic discrimination are reversed for the native

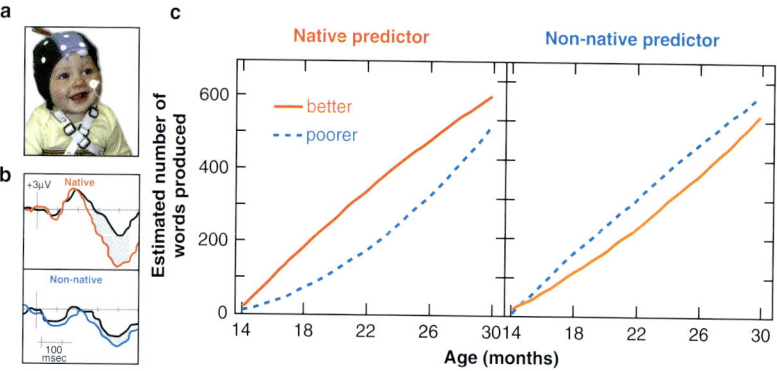

Figure 5.2 (a) A 7.5-month-old infant wearing an ERP electrocap. (b) Infant ERP waveforms at one sensor location (CZ) for one infant are shown in response to a native (English) and non-native (Mandarin) phonetic contrast at 7.5 months. The mismatch negativity (MMN) is obtained by subtracting the standard waveform (*black*) from the deviant waveform (*colour*). This infant's response suggests that native language learning has begun because the MMN negativity in response to the native English contrast is considerably stronger (more negative) than that to the non-native contrast. (c) Hierarchical linear growth modelling of vocabulary growth between 14 and 30 months is shown for two groups of children, those whose MMN values at 7.5 months indicated better discrimination (−1 SD) and those whose MMN values indicated poorer discrimination (+1 SD). Vocabulary growth was significantly faster for infants with better MMN phonetic discrimination for the native contrast at 7.5 months of age (*c, left*). In contrast, for the non-native contrasts, infants with better discrimination (−1 SD), as indicated by MMN at 7.5 months, showed slower vocabulary growth (*c, right*). Both contrasts predict vocabulary growth but the effects of better discrimination are reversed for the native and non-native contrasts.
From Kuhl & Rivera-Gaxiola (2008).

and non-native predictors. Better native phonetic discrimination predicts accelerated vocabulary growth, whereas better non-native phonetic discrimination predicts slower vocabulary growth (Kuhl *et al*. 2008). These results support the NLNC hypothesis.

Rivera-Gaxiola and colleagues (Rivera-Gaxiola *et al*. 2005a) demonstrated a similar pattern of prediction using a different non-native contrast. They recorded auditory ERP complexes in 7- and 11-month-old American infants in response to both Spanish and English voicing contrasts. Two patterns of ERP response were observed, an early positive-going wave (P150–250), and a later negative-going wave (N250–550) (Rivera-Gaxiola *et al*. 2005b). Further work examined the patterns of the same auditory ERP positive-negative complexes in a larger sample of 11-month-old monolingual American infants using the same contrasts as for the developmental study, and found that infants' response to the non-native contrast predicted the number of words produced at 18, 22, 25, 27, and 30 months of age (Rivera-Gaxiola *et al*. 2005a). Infants showing an N250–550 to the foreign contrast at 11 months of age (indexing better neural discrimination) produced significantly fewer words at all ages than infants showing a less negative response. Scalp distribution analyses on 7-, 11-, 15-, and 20-month-old infants revealed that the P150–250 and the N250–550 components differ in distribution (Rivera-Gaxiola *et al*. 2007). Thus, in both Kuhl *et al*. (2008) and Rivera-Gaxiola *et al*. (2005a), an enhanced negativity in response to the non-native contrast is associated with slower language development.

The continuity in language development documented in these studies using infants' early phonetic skills to predict concurrent language (Conboy *et al.* 2005), and later language (Tsao *et al.* 2004; Kuhl *et al.* 2005b; Rivera-Gaxiola *et al.* 2005a; Kuhl *et al.* 2008), is also seen in studies that use infants' early pattern-detection skills for speech to predict later language (Newman *et al.* 2006), and in studies that use infants' early processing efficiency for words to predict later language (Fernald *et al.* 2006). Taken as a whole, these studies form bridges between the early precursors to language in infancy and measures of language competencies in early childhood – bridges that are important to theory building as well as to clinical populations with developmental disabilities that involve language.

ERP studies at the phonetic level suggest that the young brain's response to the elementary building blocks of language matters, and that initial native language phonetic learning is a pathway to language (Kuhl 2008). The data also suggest that discriminating non-native phonetic contrasts for a longer period of time in early development – reflecting infants' initial, more immature state – can be linked to slower language development. In infants exposed to a single language, the ability to attend to changes in the phonetic contrasts that are relevant to the culture's language, while at the same time reducing attention to phonetic contrasts from other languages that are discriminable but irrelevant to the language of their culture, appears to be an important first step towards the acquisition of language. Behavioural studies by Conboy *et al.*, (2008c) indicate that non-native phonetic perception is significantly correlated with cognitive control abilities – especially those that tap inhibitory control – but that native phonetic abilities are not similarly linked. Moreover, Conboy *et al.* (2008c) show that native phonetic abilities are strongly linked to concurrent vocabulary skills, whereas non-native phonetic abilities are not. Taken together, these results suggest that infants' abilities to attend to and process native-language phonetic categories, while at the same time disregarding discriminable non-native categories, predict more rapid advancements in language. What the tools of modern neuroscience may allow us to do in the future is more fully understand this interaction between language learning and cognitive development, and its relation to the 'critical period' for language development (Kuhl *et al.* 2005b).

5.4 Learning from exposure to a second language

Recent studies in my laboratory have shown that young infants are capable of phonetic learning at 9 months of age from exposure to a new language, but only when exposure occurs during live human presentation; television or audio-only exposure did not produce learning (Kuhl *et al.* 2003). Social interaction appears to be a critical component for language learning. This finding ties early communicative learning in speech to examples of communicative learning in neurobiology more generally, as shown by the importance of social factors in song learning in birds (e.g. Brainard & Knudsen 1998). These second language exposure studies have been used to argue that the social brain may 'gate' the computational mechanisms underlying language learning during the earliest stages of human language acquisition (Kuhl 2007).

The social gating hypothesis was tested in a set of studies using ERP as a measure of learning from foreign language exposure to Spanish (Conboy & Kuhl 2007; Conboy *et al.* 2008a). In the study, American monolingual infants were exposed to Spanish at 9 months of age by native Spanish speakers. Infants were tested after exposure to see if they learned phonemes and words from this foreign language experience. The study tested the social hypothesis by examining whether the infants' tendency to interact in socially sophisticated ways during the exposure sessions would predict the degree to which individual infants learned phonemes and words from the new language.

Infants' ERPs in response to English and Spanish phonemes, as well as their ERP responses to Spanish words, were measured before and after exposure to Spanish. As in the Mandarin study, exposure consisted of live interaction with foreign language 'tutors' during 12 sessions, each of which lasted 25 min. All sessions were videotaped using a four-camera system, and detailed measures of shared visual attention between the infants and their tutors were taken by an independent observer.

The ERP results demonstrated that the MMN response to the Spanish contrast was not present before exposure, but that following exposure to Spanish, the MMN was robust (Conboy & Kuhl 2007), replicating the phonetic learning results that were measured behaviourally in the Mandarin study (Kuhl *et al.* 2003). The new results provide convincing evidence of infants' ability to learn phonetically from exposure to a foreign language at 9 months of age. Extending these previous findings beyond phoneme learning, Conboy and Kuhl also showed that infants learned Spanish words that were presented during the exposure sessions. When compared to Spanish words that had not been presented, infants' ERPs to the Spanish words revealed the classic components related to known words (Conboy & Kuhl 2007).

The social gating hypothesis was also strongly supported. Infants' degree of social engagement – e.g. the degree to which infants alternated their visual attention between a newly presented toy and the tutor's eyes, as opposed to simply focussing on the toy or on the tutor – predicted the degree of learning both for phonemes and for words (Conboy *et al.* 2008a). The fact that an individual infant's social interest during the 12 language sessions predicted the degree of learning supports the argument that the social factors may 'gate' language learning (Kuhl 2007). Gaze following has previously been shown to predict word learning in infants (Brooks & Meltzoff 2008). The present results show that the relationship between social interaction and language learning can be demonstrated experimentally for new learning of language material at 9 months of age.

Finally, the results of the study suggest the possibility that exposure to a new language provides cognitive enhancement. Pre- and post-exposure measures of 'cognitive control,' the ability to attend selectively and inhibit pre-potent responses, and one previously shown to be enhanced in bilingual adults (Bialystok 1999) and children (Carlson & Meltzoff 2008), were also obtained from the children involved in the language exposure experiments. These measures indicated that cognitive control skills are enhanced after, but not prior to, Spanish exposure, linking bilingual learning to the enhancement of particular cognitive skills (Conboy *et al.* 2008c).

In sum, ERPs provide a highly sensitive measure of learning for both phonemes and words in a variety of experiments. ERP responses to speech not only predict the growth of language over the first 30 months (Rivera-Gaxiola *et al.* 2005a; Kuhl *et al.* 2008), but are also sufficiently sensitive to reflect the effects of differences in subtle abilities that contribute to infant learning, such as infants' social eye gaze following (Conboy *et al.* 2008a). Complex natural language learning may demand social interaction, because language evolved in a social setting. The neurobiological mechanisms underlying language likely utilized interactional cues made available only in a social setting. In the future, whole-brain measures, such as those provided by MEG, will allow us to observe brain activation during live presentations of language versus those that are merely televised to explore hypotheses about why human interaction is essential to language learning (Kuhl *et al.* 2003). Moreover, using 'social' robots, we are now conducting studies that will define what constitutes a social agent for a young child (Virnes *et al.* 2008).

5.5 Neural signatures of word learning

A sudden increase in vocabulary typically occurs between 18 and 24 months of age – a 'vocabulary explosion' (Ganger & Brent 2004; Fernald *et al.* 2006), but word learning starts much earlier. Infants show recognition of their own name at 4.5 months (Mandel *et al.* 1995). At 6 months, infants use their own names or the word 'Mommy' in an utterance to identify word boundaries (Bortfeld *et al.* 2005), and look appropriately to pictures of their mother or father when hearing 'Mommy' or 'Daddy' (Tincoff & Jusczyk 1999). By 7 months, infants listen longer to passages containing words they have previously heard than to passages containing words they have not heard (Jusczyk & Hohne 1997), and by 11 months infants prefer to listen to words that are highly frequent in language input over infrequent words (Halle & de Boysson-Bardies 1994).

Behavioural studies indicate that infants learn words using both 'statistical learning' strategies in which the transitional probabilities between syllables are exploited to identify likely words (Saffran 2003; Saffran *et al.* 1996; Newport & Aslin 2004), and pattern-detection strategies in which infants use the typical pattern of metric stress that characterizes ambient language to segment running speech into likely words (Cutler & Norris 1988; Johnson & Jusczyk 2001; Nazzi *et al.* 2006; Hohle *et al.* 2009).

How is word recognition evidenced in the brain? ERPs in response to words index word familiarity as early as 9 months of age and word meaning by 13–17 months of age: ERP studies have shown differences in amplitude and scalp distributions for components that are related to words that are known versus unknown to the child (Molfese 1990; Molfese *et al.* 1990, 1993; Mills *et al.* 1993, 1997, 2005; Thierry *et al.* 2003).

As early as 9 months of age, ERPs indicate word familiarity, and by 13–17 months of age, studies show ERP components that reliably signal the brain's coding of words that are known versus unknown by the child (Mills *et al.* 1993, 1997, 2005; Thierry *et al.* 2003). Toddlers with larger vocabularies tend to have a more focalized and larger N200 for known words – they show an enhanced negativity to known versus unknown words only at left temporal and parietal electrode sites – whereas children with smaller vocabularies show more broadly distributed effects (Mills *et al.* 1993), features that also distinguish typically developing preschool children from preschool children with autism (Coffey-Corina *et al.* 2007).

Processing efficiency for phonemes and words can be seen as well in the relative focalization and duration of brain activation in adult MEG studies (Zhang *et al.* 2005), indicating that these features index language experience and proficiency not only in children (Friederici 2005; Conboy *et al.* 2008b), but over the lifespan. Individual differences in the response latency to a familiar word at the age of 2 are related to both lexical and grammatical measures collected between 15 and 25 months, providing more evidence that processing speed is associated with greater language facility (Fernald *et al.* 2006).

Mills *et al.* (2005) used ERPs in 20-month-old toddlers to examine new word learning. The children listened to known and unknown words, and to non-words that were phonotactically legal in English. ERPs were recorded as the children were presented with novel objects paired with the non-words. After the learning period, ERPs to the non-words that had been paired with novel objects were shown to be similar to those of previously known words, suggesting that new words may be encoded in the same neural regions as previously learned words.

ERP studies on German infants reveal the development of word-segmentation strategies based on the typical stress patterns of German words. When presented with bi-syllabic strings with either a trochaic (typical in German) or iambic pattern, infants who heard a trochaic pattern embedded in an iambic string showed the N200 ERP component – similar to that elicited in response to a known word, whereas infants presented with the iambic bi-syllable embedded in the trochaic pattern showed no response (Weber *et al.* 2004). The data suggest that German infants at this age are applying a metric segmentation strategy, consistent with the behavioural data of Hohle *et al.* (2009).

5.6 Infants' early lexicons

There is evidence suggesting that young children's word representations are phonetically underspecified. Children's growing lexicons must code words in a way that distinguishes words from one another. Given that by the end of the first year infants' phonetic skills are language specific (Werker & Tees 1984; Best & McRoberts 2003; Kuhl *et al.* 2006), it was assumed that children's early word representations were phonetically detailed. However, studies suggest that learning new words taxes young children's capacities, and that as a result, new word representations are not phonetically complete.

Reactions to mispronunciations – the age at which children no longer accept *tup* for *cup* or *bog* for *dog* – provide information about phonological specificity. Studies across languages suggest that by 1 year of age mispronunciations of common words (Jusczyk & Aslin 1995; Fennel & Werker 2003), words in stressed syllables (Vihman *et al.* 2004), or monosyllabic words (Swingley 2005), are not accepted as target words, indicating well-specified representations. Other studies using visual fixation of two targets (e.g. apple and ball) while one is named ('Where's the ball?') show that, between 14 and 25 months, children's tendencies to fixate the target item when it is mispronounced diminish over time (Swingley & Aslin 2000, 2002; Bailey & Plunkett 2002; Ballem & Plunkett 2005).

However, behavioural and neural evidence suggest that learning new words can tax children's phonological skills. Stager and Werker (1997) demonstrated that 14-month-old infants fail to learn new words when similar-sounding phonetic units are used to distinguish those words ('bih' and 'dih'), but do learn if the two new words are distinct phonologically ('leef' and 'neem'). By 17 months of age, infants can learn to associate similar-sounding nonsense words to novel objects (Bailey & Plunkett 2002; Werker *et al.* 2002). Infants with larger vocabularies succeeded on this task even at the younger age, suggesting the possibility that infants with greater phonetic learning skills acquire new words more rapidly, consistent with studies showing that better native phonetic learning skills are associated with advanced word learning skills (Tsao *et al.* 2004; Kuhl *et al.* 2005b; Rivera-Gaxiola *et al.* 2005a; Kuhl *et al.* 2008).

Mills *et al.* (2004) used ERPs to corroborate these results. They compared ERP responses to familiar words that were either correctly pronounced or mispronounced, as well as non-words. At the earliest age tested, 14 months, a negative ERP component (N200–400) distinguished known versus dissimilar nonsense words (*bear* vs. *kobe*) but not known versus phonetically similar nonsense words (*bear* vs *gare*). By 20 months, this same ERP component distinguished correct pronunciations, mispronunciations, and non-words, supporting the idea that between 14 and 20 months, children's phonological representations of early words become increasingly detailed. Other evidence of early processing limitations stems from infants' failure to learn a novel word when its auditory label closely resembles a word they already know (*gall* which closely resembles *ball*), suggesting lexical competition effects (Swingley & Aslin 2007).

How phonetic and word learning interact – and whether the progression is from phonemes to words, words to phonemes, or bidirectional – is a topic of strong interest that will be aided by the use of neuroscientific methods. Recent theoretical models of early language acquisition such as NLM-e (Kuhl *et al.* 2008) and PRIMER (Werker & Curtin 2005) suggest that phonological and word learning may bidirectionally influence one another. Infants with better phonetic learning skills advance more quickly towards language because phonetic skills assist the detection of phonotactic patterns, the detection of transitional probabilities in adjacent syllables, and the ability to phonologically distinguish minimally contrastive words (Kuhl *et al.* 2005b). On the other hand, the more words children learn, the more crowded lexical space becomes, putting pressure on children to attend to the phonetic units that distinguish them (Swingley & Aslin 2007). Further studies examining both phoneme and word learning in the same children, as in the studies using exposure to a foreign language and ERP measures as assessments of learning, will help address this issue (Conboy & Kuhl 2007).

ERP research shows that the young brain has difficulty representing phonetic detail when focussed on the task of assigning a new auditory label to a novel object. ERP results with toddlers also show that brain signatures distinguish words that are known from ones that are unfamiliar. ERPs recorded to words in the first 2 years suggest that experience with words results in the formation of neural representations of those words that are increasingly well specified towards the end of the second year of life.

5.7 Neural signatures of early sentence processing

To understand sentences, the child must have exquisite phonological abilities that allow segmentation of the speech signal into words, and the ability to extract word meaning. In addition, the relationship among words composing the sentence – between a subject, its verb, and its accompanying object – must be deciphered to arrive at a full understanding of the sentence. Human language is based on the ability to process hierarchically structured sequences (Friederici *et al.* 2006).

Electrophysiological components have been recorded in children, and contribute to our knowledge of when and how the young brain decodes syntactic and semantic information in sentences. In adults, specific neural systems process semantic versus syntactic information within sentences, and the ERP components elicited in response to syntactic and semantic anomalies are well established (Figure 5.3). For example, a negative ERP wave occurring between 250 and 500 ms that peaks around 400 ms, referred to as the N400, is elicited to semantically anomalous words in sentences (Kutas 1997). A late-positive wave peaking at about 600 ms and largest at parietal sites, known as the P600, is elicited in response to syntactically anomalous words in sentences (Friederici 2002).

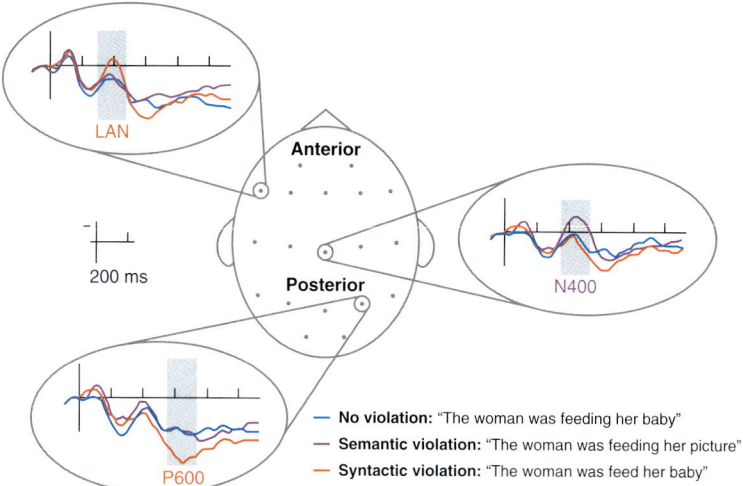

Figure 5.3 ERP responses to normal sentences and sentences with either semantic or syntactic anomalies show distinct distribution and polarity differences in adults.
From Kuhl & Rivera-Gaxiola (2008).

And a negative wave over frontal sites between 300 and 500 ms, known as the 'late anterior negativity' (LAN), is elicited in response to syntactic and morphological violations (Friederici 2002).

ERP data on sentence processing in children suggest that adult-like components in response to semantic and syntactic violations can be elicited starting in the second year of life, but also that there are differences in the latencies and scalp distributions of these components in children and adults (Harris 2001; Friederich & Friederici 2005, 2006; Oberecker et al. 2005; Silva-Pereyra et al. 2005a, 2005b, 2007; Oberecker & Friederici 2006). Holcomb, et al., (1992) reported the N400 in response to the semantic anomaly effect in children from 5 years of age to adolescence; the latency of the effect was shown to decline systematically with age (see also Neville et al. (1993) and Hahne et al. (2004)). Studies also show that syntactically anomalous sentences elicit the P600 in children between 7 and 13 years of age (Hahne et al. 2004).

Recent studies have examined these ERP components in preschool children. Harris (2001) reported an N400-like effect in 36–38-month-old children, which was largest over posterior regions of both hemispheres, unlike the adult scalp distribution. Friederich and Friederici (2006) observed an N400-like wave to semantic anomalies in 19- and 24-month-old German-speaking children.

Silva-Pereyra et al. (2005b) recorded ERPs in children between 36 and 48 months of age in response to semantic and syntactic anomalies. In both cases, the ERP effects in children were more broadly distributed and elicited at later latencies than in adults. In work with even younger infants (30-month-olds), Silva-Pereyra et al. (2005a) used the same stimuli and observed late positivities distributed broadly at posterior electrode sites in response to syntactic anomalies and anterior negativities in response to semantically anomalous sentences. In each case, the 30-month-old children's responses had longer latencies than seen in the older children and in adults (Figure 5.4) – a pattern observed repeatedly and attributed to the immaturities and inefficiencies of the developing processing mechanisms.

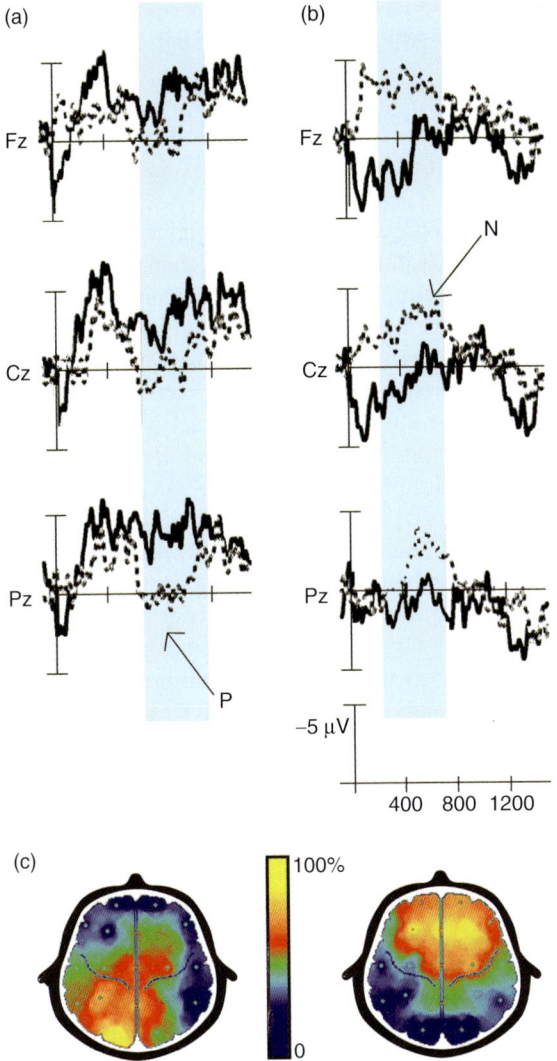

Figure 5.4 ERP waveforms elicited from 30-month-old children in response to sentences with syntactic (a) or semantic (b) violations. Children's ERP responses resemble those of adults (see Figure 5.3), but have longer latencies and are more broadly distributed (c).
From Silva-Pereyra *et al.* (2005a).

Syntactic processing of sentences with semantic content information removed – 'jabberwocky sentences' – has also been tested using ERP measures with children. Silva-Pereyra and colleagues (2007) recorded ERPs to phrase-structure violations in 36-month-old children using sentences in which the content words were replaced with pseudowords while leaving grammatical function words intact. The ERP components elicited to the jabberwocky phrase-structure violations versus the same violations in real sentences differed. Two negative components, one from 750–900 ms and the other from 950–1050 ms, rather than the positivities seen in response to phrase-structure violations in real sentences in the same children, were observed. Jabberwocky studies with adults

(Munte *et al.* 1997; Canseco-Gonzalez 2000; Hahne & Jeschenick 2001) have also reported negative-going waves for jabberwocky sentences, though at much shorter latencies.

5.8 ERP measures of early language processing in children with autism spectrum disorder (ASD)

Scientific discoveries on the progression towards language by typically developing children are now providing new insights into the language deficit shown by children with autism spectrum disorder (ASD). Neural measures of language processing in children with autism – involving both phonemes and words, when coupled with measures of social interest in speech – are revealing a tight coupling between social interaction skills and language acquisition in children with ASD. These measures hold promise as potential diagnostic markers of risk for autism in very young children, and therefore, there is a great deal of excitement surrounding the application of these basic measures of speech processing in very young children with autism.

In typically developing children, ERP responses to simple speech syllables such as 'pa' and 'ta' predict the growth of language to the age of 30 months (Kuhl *et al.* 2008). It is therefore interesting to test whether ERP measures of autism at the phonetic level are sensitive to the degree of severity of autism, and also the degree to which the brain's responses to syllables can be predicted by other factors, such as a social interest in speech.

Kuhl *et al.* (2005a) conducted the first study examining phonetic perception in preschool children with autism using ERP methods. ERPs to a simple change in two speech syllables, as well as a measure of social interest in speech, were taken. In these experiments, a listening choice test allowed children with autism to choose motherese or non-speech signals in which the formant frequencies of speech were matched by pure tones – the resulting signal was a computer warble that exactly followed the frequencies and amplitudes of the 5-s speech samples over time. Slight head turns to one direction versus the other allowed the toddlers to choose their preferred signal on each trial. The goal was to compare performance at the group level between typically developing children and children with ASD, as well as to examine the relationship between brain measures of speech perception and measures of social processing of speech in children with ASD.

Considering first the ERP measures of phonetic perception, the results showed that, as a group, children with ASD exhibited no MMN to the simple change in syllables. However, when children with ASD were sub-grouped on the basis of their preference for infant-directed (ID) speech (often called 'motherese'), very different results were obtained.

The results showed that while typically developing children listened to both signals, children with autism strongly preferred the non-speech analogue signals. Moreover, the degree to which they did so was significantly correlated with both the severity of autism symptoms and individual children's MMN responses to speech syllables. Toddlers with ASD who preferred motherese produced MMN responses that resembled those of typically developing children, whereas those who preferred the non-speech analogue did not show an MMN response to the change in a speech syllable.

These results underscore the importance of social interest in speech early in development, especially an interest in motherese. Research has shown that the phonetic units in motherese are acoustically exaggerated, making them more distinct from one another (Kuhl *et al.* 1997; Burnham *et al.* 2002; Liu *et al.* 2003, 2007; Englund 2005). Infants whose mothers use the exaggerated phonetic patterns to a greater extent when talking to them show significantly better performance in phonetic discrimination tasks (Liu *et al.* 2003). In the absence of a listening preference for motherese, children with autism would miss the benefit these exaggerated phonetic cues provide.

ID speech also produces unique brain responses in typically developing infants. Brain measures of typical infants' response to ID speech, used by Peña *et al.* (2003) in the first study using NIRS, showed more activation in left temporal areas when infants were presented with ID speech as opposed to backward speech or silence. Bortfeld *et al.* (2007) obtained analogous results using NIRS in a sample of 6–9-month-old infants presented with ID speech and visual stimulation. It will be of interest to examine brain activation while children with autism listen to motherese as opposed to acoustically matched non-speech signals. In children with ASD, brain activation to carefully controlled speech versus non-speech signals may provide clues to their aversion to the highly intonated speech signals typical of motherese.

Recent studies extend the findings on children with autism to word processing using ERP measures (Coffey-Corina *et al.* 2007, 2008). In this study, 24 toddlers with ASD between 18 and 31 months of age were separated into high-functioning and low-functioning subgroups defined by the severity of their social symptoms. ERP measures were recorded in response to known words, unknown words, and words played backwards. They were compared to ERPs elicited from a group of 20 typically developing toddlers between the ages of 20 and 31 months.

The results for typically developing toddlers showed a highly localized response to the difference between known and unknown words at a left temporal electrode site (T3) in the 200–500 ms and 500–700 ms windows (Figure 5.5a). These data replicate previous studies of typically developing children published by Mills *et al.* (1993), and indicate that highly focalized responses are a marker of increasing developmental sophistication in the processing of words in typically developing children. It was, therefore, of interest to observe that toddlers with ASD showed a very diffuse response to known and unknown words. Known words elicited a greater negativity than unknown words across all electrode sites, and at a longer latency than age-matched typically developing children (Figure 5.5b). This pattern of more diffuse activation and longer response latency has been observed in younger, typically developing, children (Mills *et al.* 1997).

Replicating the pattern seen in the studies of phonetic perception in children with autism, the word processing results for children with ASD differed markedly depending on the children's social skills. High-functioning toddlers with ASD produced ERP responses that were similar to those of typically developing children – they exhibited a localized left-hemisphere response to known and unknown words. Significant word-type effects were observed only at the left parietal electrode site (P3) in the 200–500 ms time-window (Figure 5.5c). In contrast, ERP waveforms of low-functioning toddlers with ASD exhibited a diffuse response to words. ERPs for known words were significantly more negative than those for unknown words at multiple electrode sites and in all measurement windows (Figure 5.5d).

The idea that ERP measures in response to syllables and words may allow us to predict future language outcomes in young children with ASD is exciting. Towards that end, we note that children with ASD exhibited highly significant correlations between their ERP components at the initial test time and their verbal intelligence quotient (IQ) scores measured one year after ERP data collection (Figure 5.6).

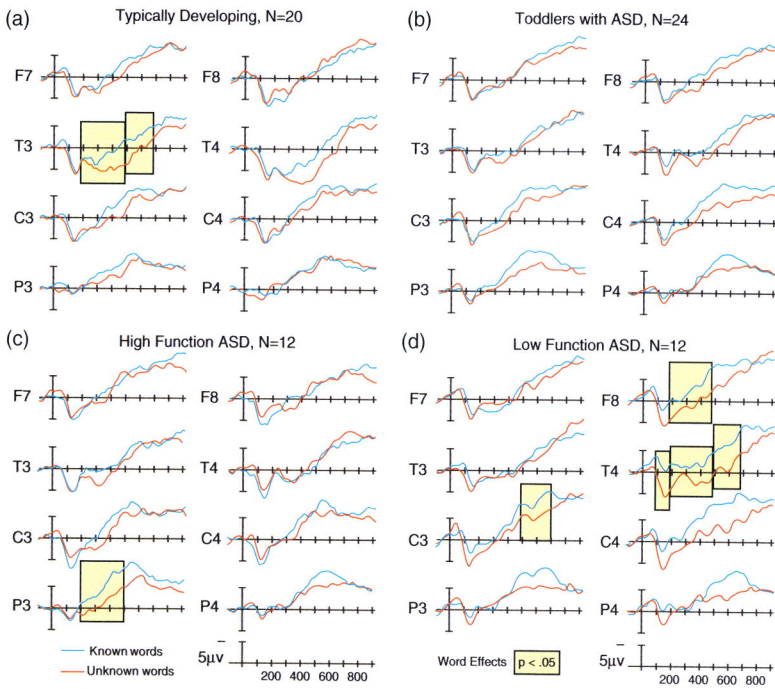

Figure 5.5 Group data showing ERP waveforms for typically developing toddlers (a) and toddlers with autism spectrum disorder (b). TD toddlers exhibit a localized response with significant differences between known and unknown words at the left temporal electrode site (T3). Toddlers with ASD exhibit a diffuse response to known and unknown words but the differences are significant across all electrode sites in the 500–700 ms measurement window. Subgroup analysis shows that ERP waveforms for high-functioning toddlers with ASD are similar to those of typically developing children, exhibiting a localized response with significant differences between known and unknown words at a parietal electrode site in the left hemisphere (P3) (c). Low-functioning toddlers with ASD exhibit a diffuse response to known and unknown words with significant differences in multiple time windows and electrode sites, a significant effect when collapsed across all electrode sites in the 500–700 ms measurement window (d).
From Coffey-Corina *et al.* (2008).

In new studies with the siblings of children with ASD, we are now exploring whether these early brain and behavioural responses to syllables and words, and listening preferences for speech, are diagnostic markers for autism. The interest in these measures is that they can be used reliably in infants as early as 6 months of age, an age at which intervention measures might be more effective in changing the course of development for children at risk for autism.

5.9 Mirror neurones and shared brain systems

Neuroscience studies focussed on shared neural systems for perception and action have a long tradition in speech research (Liberman & Mattingly 1985; Fowler 2006). The discovery of 'mirror neurones' for social cognition (Gallese 2003; Meltzoff & Decety 2003;

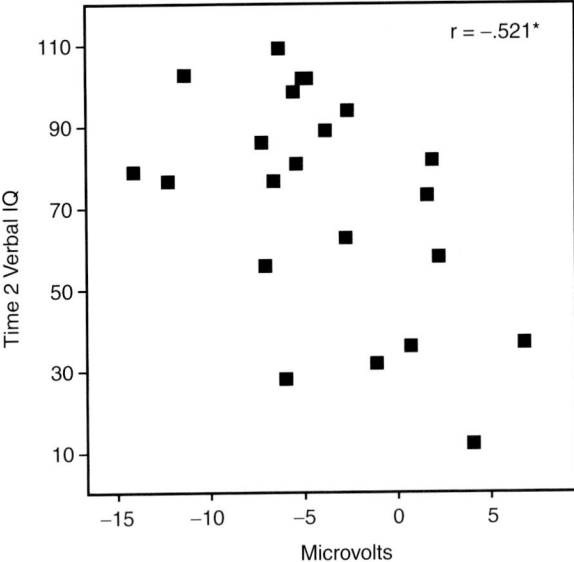

Figure 5.6 Predictive correlations for children with ASD between the mean amplitude of ERPs to known words at the left parietal electrode site (P3) and Verbal IQ measured 1 year later. A more negative response predicted significantly higher verbal IQ ($r = -0.521$, $p = 0.013$).
From Coffey-Corina et al. (2008).

Rizzolatti & Craighero 2004; Pulvermuller 2005; Rizzolatti 2005) has re-invigorated this tradition. Neuroscience studies using speech and whole-brain imaging techniques have the capacity to examine the origins of shared brain systems in infants from birth (Imada et al. 2006; Bosseler et al. 2008).

In speech, the theoretical linkage between perception and action came in the form of the original *Motor Theory* (Liberman et al. 1967) and in a different formulation of the direct perception of gestures, named *Direct Realism* (Fowler 1986). Both posited close interaction between speech perception and speech production. The perception–action link for speech was viewed as innate by the original motor theorists (Liberman & Mattingly 1985). Alternatively, it was viewed as being forged early in development through experience with speech motor movements and their auditory consequences (Kuhl & Meltzoff 1982, 1996). Two new infant studies shed some light on the developmental issue.

Imada et al. (2006) used MEG, studying newborns, 6-month-old infants and 12-month-old infants while they listened to non-speech signals, harmonics, and syllables (Figure 5.7). Dehaene-Lambertz et al. (2006) used fMRI to scan 3-month-old infants while they listened to sentences. Both studies show activation in brain areas responsible for speech production (the inferior frontal, Broca's area, etc.) in response to auditorally presented speech. Imada et al. reported synchronized activation in response to speech in auditory and motor areas at 6 and 12 months, and Dehaene-Lambertz et al. reported activation in motor speech areas in response to sentences in 3-month-olds.

Is activation of Broca's area by speech sounds present at birth? Newborns tested by Imada et al. (2006) showed no activation in motor speech areas for any signals, whereas auditory areas responded robustly to all signals, suggesting the possibility that perception–action linkages for speech develop by 3 months of age as infants produce vowel-like sounds.

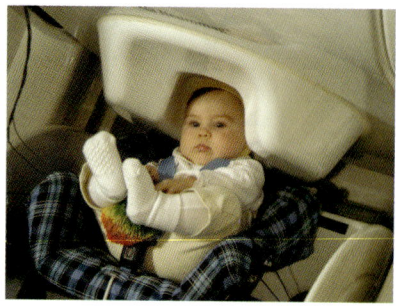

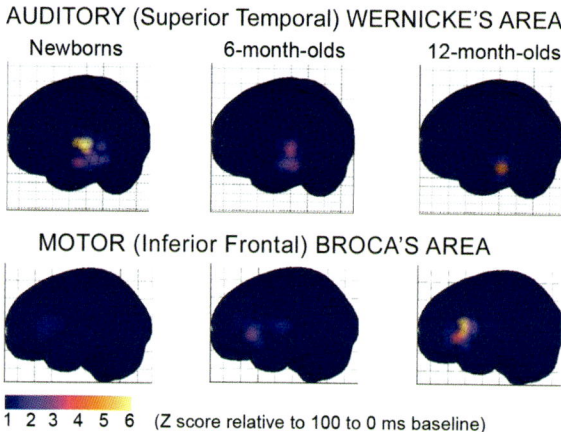

Figure 5.7 Neuromagnetic signals were recorded using MEG in newborns, 6-month-old, and 12-month-old infants while listening to speech (shown) and nonspeech auditory signals. Brain activation recorded in auditory (*top row*) and motor (*bottom row*) brain regions revealed no activation in the motor speech areas in the newborn in response to auditory syllables. However, activation increased in the motor areas in response to speech (but not non-speech) in 6- and 12-month-old infants, and it was temporally synchronized between the auditory and motor brain regions. From Imada *et al.* (2006).

But further work must be done to answer the question. How the binding of perception and action takes place, and whether it requires experience, is one of the exciting questions that can now be addressed with infants from birth using the tools of modern neuroscience.

We now know a great deal about the linkages and the circuitry underlying language processing in adults (Kuhl & Damasio in press). What is unknown, but waiting to be discovered, is the state of this circuitry at birth and how refined connections are forged in early infancy as perception and action are jointly experienced.

5.10 Bilingual infants: two languages, one brain

One of the most interesting questions is how infants map two distinct languages in the brain. From phonemes to words, and then to sentences, how do infants simultaneously bathed in two languages develop the neural networks necessary to respond in a native-like manner to two different codes?

Bilingual language experience could potentially have an impact on development – both because the learning process requires the development of two codes – and because it could take a longer period of time for sufficient data from both languages to be experienced than in the monolingual case. Infants learning two first languages simultaneously might reach the developmental change in perception at a later point in development than infants learning either language monolingually. This could depend on factors such as the number of people in the infants' environment producing the two languages in speech directed towards the child, and the amount of input they provide. These factors could change the rate of development in bilingual infants.

There are very few studies that address this question thus far, and the data that do exist provide somewhat mixed results. Some studies suggest that infants exposed to two languages show later acquisition of phonetic skills in the two languages when compared to monolingual infants (Bosch & Sebastian-Galles 2003a,b). This is especially the case when infants are tested on contrasts that are phonemic in only one of the two languages; this has been shown both for vowels (Bosch & Sebastian-Galles 2003b) and consonants (Bosch & Sebastian-Galles 2003a). However, others studies report no change in the timing of the developmental transition in phonetic skills in the two languages of bilingual infants (Burns *et al.* 2007; Sundara *et al.* 2008). For example, Sundara *et al.*, testing monolingual English and monolingual French as well as bilingual French-English infants, examined discrimination of dental (French) and alveolar (English) consonants. They demonstrated that, at 6–8 months, infants in all three language groups succeeded; at 10–12 months, monolingual English infants and French-English bilingual infants, but not monolingual French infants, distinguished the English contrast. Thus, bilingual infants performed on par with their English monolingual peers and better than their French monolingual peers. Moreover, data from an ERP study of Spanish-English bilingual infants show that, at both 6–9 and 9–12 months of age, bilingual infants show MMN responses to both Spanish and English phonetic contrasts (Rivera-Gaxiola & Romo 2006), distinguishing them from English-learning monolingual infants who fail to respond to the Spanish contrast at the later age (Rivera-Gaxiola *et al.* 2005b).

ERP studies on word development in bilingual children have just begun to appear. Conboy and Mills (2006) recorded ERPs to known and unknown English and Spanish words in bilingual children at 19–22 months. Expressive vocabulary sizes were obtained in both English and Spanish, and were used to determine language dominance for each child. A conceptual vocabulary score was calculated by summing the total number of words in both languages and then subtracting the number of times a pair of conceptually equivalent words (e.g. '*water*' and '*agua*') occurred in the two languages.

ERP differences to known and unknown words in the dominant language occurred as early as 200–400 and 400–600 ms in these 19–22-month-old infants, and were broadly distributed over the left and right hemispheres, resembling patterns observed in younger (13- to 17-month-old) monolingual children (Mills *et al.* 1997). In the nondominant language of the same children, these differences were not apparent until late in the waveform, from 600 to 900 ms. Moreover, children with high versus low conceptual vocabulary scores produced greater responses to known words in the left hemisphere, particularly for the dominant language (Conboy & Mills 2006).

Researchers have just begun to explore the nature of the bilingual brain, and it is one of the areas in which neuroscience techniques will be of strong interest. Using whole-brain imaging, we may be able to understand whether learning a second language at different ages – in infancy as opposed to adulthood – recruits different brain structures. These kinds of data may play a role in our eventual understanding of the 'critical period' for language learning.

5.11 A theoretical model and future research

How do we integrate the body of work showing the effects of early language experience on the brain? From the early brain measures recorded at 7 months in response to phonetic units to those recorded at 30 months in response to sentences that are syntactically correct versus anomalous, infants' brain responses show that the brain is altered by exposure to a language.

A theory developed in my laboratory, NLM-e (*Native Language Magnet-Expanded*), provides a framework for understanding these data and makes specific predictions that will structure future research. The framework described by NLM-e indicates that early exposure to a specific language establishes a neural architecture – specific connections and tissue – that is 'neurally committed' to the acoustic patterns typical of that language.

This *Native Language Neural Commitment* (NLNC) hypothesis captures the idea that learning results in the formation of new neural networks that are specialized for a specific language. Dedicated networks do two very interesting things: first, the neural networks detect patterns (such as those typical of phonetic units in the language) that promote the development of higher order language units (such as words), and second, these neural networks allow infants to begin to inhibit responses to linguistic units that are characteristic of other languages. These two factors promote attentional focus, which produces rapid advancement in language acquisition.

NLM-e is schematically described in Figure 5.8, and it encompasses four phases of development. In Phase 1, early in life, infants discriminate all phonetic units in the world's languages, and factors such as acoustic salience and directional asymmetries explain the degree to which infant performance varies across phonetic contrasts. Infants' initial performance leaves room for substantial improvement, especially for those contrasts that are acoustically fragile. In Phase 1, infants' phonetic abilities are relatively crude, reflecting general auditory constraints (see Kuhl *et al.* (2008) for more details). The critical feature of the initial state stipulated by the model is that infants begin life with a capacity to discriminate the acoustic cues that code differences among phonetic units. Infants' initial ability to discriminate phonetic units, albeit crudely, assists their language development in Phase 2.

Phase 2 represents the core of the NLM-e model. At this stage in development, infants' sensitivity to the distributional patterns (Kuhl *et al.* 1992; Maye *et al.* 2002) and exaggerated cues of ID speech (Liu *et al.* 2003) causes phonetic learning. As depicted, learning occurs earlier for vowels than consonants (e.g. Werker & Tees 1984; Kuhl *et al.* 1992; Polka & Werker 1994; Best & McRoberts 2003). This difference could reflect the availability of exaggerated cues in ID speech: consonants are not as easily exaggerated as vowels, because exaggeration can change the category (e.g. stretching the formant transitions of /b/ produces /w/). Alternatively, there may be differences in the availability and/or prominence of distributional differences for consonants (e.g. consonants like /th/ occur in function words, which are lower in energy and do not capture infant attention, see Sundara *et al.* (2006)). Understanding how these two aspects of environmental input – exaggerated

Early language acquisition: phonetic and word learning, neural substrates, and a theoretical model 123

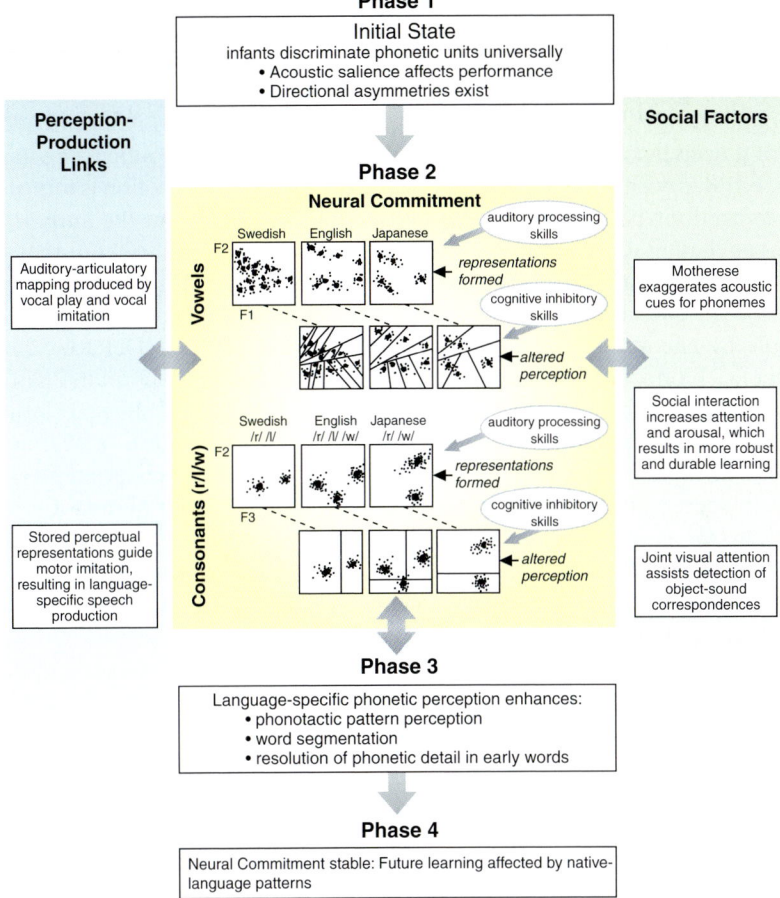

Figure 5.8 The four phases posited by NLM-e. (see text for description). From Kuhl et al. (2008).

acoustics and distributional properties – interact to support infants' perception of categories will be important for future studies.

In Phase 2, NLM-e shows an important new component – social interaction – as playing a role in phonetic and early word learning. The new data reviewed earlier show that social factors strongly influence infants' computational learning in the domain of language (Kuhl et al. 2003; Kuhl 2007). Future studies using MEG will allow us to determine whether the increased attention and arousal that occurs during social interaction, the specific information provided during social interaction (such as joint visual attention to an object), or both factors, are responsible for the facilitative effect social interaction has on language learning. Either a general 'motivational' explanation involving attention or arousal, or a more specific 'informational' explanation could account for the effects of social interaction on learning, and both are likely to play a role (Kuhl et al. 2003).

In complex natural communicative settings, social interaction may serve to 'gate' computational learning (Kuhl 2007). Why might this be the case? One only needs to watch the complexity of the social interaction that takes place as infants are exposed to natural language to make the assumption that the attention demanded by interacting with people, and the ability to glean information from those complex social interactions, would prove to significantly modulate language learning in individual children. The dual deficits in social understanding and language acquisition in children with ASD provide further evidence of a deep linkage between social cognition and linguistic processing in humans.

Finally, NLM-e posits that during Phase 2 a link to speech production is forged. Infants develop connections between their own production of speech and the auditory signals this causes as they practice and play with vocalizations, and imitate those they hear. As speech production improves, imitation of the learned patterns stored in memory leads to language-specific speech production. It has been suggested that speech production itself plays a role by encouraging the use of learned motor patterns (DePaolis 2005), and NLM-e depicts bidirectional effects between perception and production in Phase 2 as the link between them is formed. As the MEG data reviewed earlier showed (Imada *et al.* 2006), 6- and 12-month-old infants exhibit synchronized activation in Wernicke's and Broca's area during the perception of speech. This synchronized activation was shown to be unique to speech, and may indicate the inception of a 'mirror system' for speech in human infants.

By the end of Phase 2, infants' perception is altered and attentional focus shifts towards the native language and away from non-native patterns. Language learning has begun. The detection of native language phonetic cues is enhanced, while detection of non-native phonetic patterns is reduced. At this stage, infant perception has been 'warped' by experience and begins to reflect attunement between infant perception and the language and culture in which infants are being raised.

In Phase 3, enhanced speech perception abilities improve three independent skills that propel infants towards word acquisition: the detection of phonotactic patterns (Friederici & Wessels 1993; Mattys *et al.* 1999), the detection of transitional probabilities between segments and syllables (Goodsitt *et al.* 1993; Saffran *et al.* 1996; Newport & Aslin 2004), and the association between sound patterns and objects (Swingley & Aslin 2002; Werker *et al.* 2002; Ballem & Plunkett 2005). Each of these skills – detection of phonotactic patterns, detection of word-like units, and the resolution of phonetic detail in early words – is likely to predict future language, though empirical studies have just begun to test these relationships (Newman *et al.* 2006). Bidirectional effects are indicated at this stage – native language phonetic learning would assist the detection of word patterns, and the learning of phonetically close words would be expected to sharpen awareness of phonetic distinctions.

By Phase 4, analysis of incoming language has produced relatively stable neural representations – and these representations start to restrict the learning of new languages. In infancy, neural networks are not completely 'set,' and do not constrain learning. Infants are capable of learning from multiple languages, as shown in everyday life, and also as shown by experimental interventions in which children learn from exposure to new language material (Maye *et al.* 2002; Kuhl *et al.* 2003). By adulthood, representations *are* stable, and it is much more difficult to learn by listening to a new language. Thus, exposure to a new language does not automatically create new neural structure as we age. The principle underlying the model is that the degree of 'plasticity' in learning a second language depends on the stability of the underlying perceptual representations, and therefore on the degree of neural commitment.

Future studies will benefit from the tools of modern neuroscience, which will allow us to examine the 'neural commitment' process using both functional MEG measures and structural MRI measures on the developing infant brain. Using functional brain measures, we will also be able to examine how social interaction affects the language areas of the brain, and what defines a 'social agent' for a child. We hypothesize that early language learning in a social context results in a highly robust and very durable form of learning, one that creates patterns of perception and production that will affect us all our lives.

Two populations will be especially interesting as we move towards understanding the neurobiology of language – bilingual children and children with developmental disabilities affecting language. Both will allow us to test the strong assumptions underlying NLM-e and other models of language acquisition.

5.12 Conclusions

Knowledge of infant language acquisition is now beginning to reap benefits from information obtained by experiments that directly examine the human brain's response to linguistic material as a function of experience. EEG, MEG, fMRI, and NIRS technologies – all safe, noninvasive, and proven feasible – are now being used in studies with very young infants, including newborns, as they listen to the phonetic units, words, and sentences of a specific language. Brain measures now document the neural signatures of learning as early as 7 months for native-language phonemes, 9 months for familiar words, and 30 months for semantic and syntactic anomalies in sentences. Theoretical models, such as NLM-e explain these data and suggest new experiments that will further our understanding of the neurobiology of language. Studies show continuity from the earliest phases of language learning in infancy to the complex processing evidenced at the age of 3 when all typically developing children show the ability to carry on a sophisticated conversation. Individual variation in language-specific processing at the phonetic level – at the cusp of the transition from phase 1, in which all phonetic contrasts are discriminated, to phase 2, in which infants focus on the distinctions relevant to their native language – is strongly linked to infants' abilities to process words and sentences 2 years later. This is important theoretically but is also vital to the eventual use of these early precursors to speech to diagnose children with developmental disabilities that involve language. In fact, new studies suggest the possibility that early measures of the brain's responses to speech may provide a diagnostic marker for ASD. The fact that language experience affects brain processing of both the signals being learned (native patterns) and the signals to which the infant is not exposed (non-native patterns) may play a role in our understanding of the brain mechanisms underlying the critical period. At the phonetic level, the data suggest that learning itself, not merely time, may contribute to the critical period phenomenon. Whole-brain imaging now allows us to examine multiple brain areas during speech processing, including both auditory and motor brain regions, revealing the possible existence of a shared brain system (a 'mirror' system) for speech. Researchers have also begun to use these measures to understand how the bilingual brain maps two distinct languages. Answers to the classic questions about the unique human capacity to acquire language will be enriched by studies that utilize the tools of modern neuroscience to peer into the infant brain.

Acknowledgements

The author is supported by the National Science Foundation's Science of Learning Center grant to the University of Washington LIFE Center (SBE 0354453), by grants from the National Institutes of Health (HD 37954; MH066399; HD34565; HD55782), by core grants (P30 HD02274; P30 DC04661), and by a grant from the Cure Autism Now Foundation. This chapter utilizes and updates the material in Kuhl *et al.*, *Philosophical Transactions of the Royal Society B* (2008), Kuhl & Rivera-Gaxiola, *Annual Review of Neuroscience* (2008), Kuhl, *The Cognitive Neurosciences IV* (in press), and Kuhl & Damasio, *Principles of Neuroscience V* (in press).

References

Aslin, R. N. & Mehler, J. 2005 Near-infrared spectroscopy for functional studies of brain activity in human infants: promise, prospects, and challenges. *J. Biomed. Optics*, 1001–1009.
Bailey, T. M. & Plunkett, K. 2002 Phonological specificity in early words. *Cog. Dev.* **17**, 1265–1282.
Ballem, K. D. & Plunkett, K. 2005 Phonological specificity in children at 1;2. *J. Child Lang.* **32**, 159–173.
Best, C. C. & McRoberts, G. W. 2003 Infant perception of non-native consonant contrasts that adults assimilate in different ways. *Lang. Speech* **46**, 183–216.
Bialystok, E. 1999 Cognitive complexity and attentional control in the bilingual mind. *Child Dev.* **70**, 636–644.
Bortfeld, H., Morgan, J. L., Golinkoff, R. M. & Rathbun, K. 2005 Mommy and me: familiar names help launch babies into speech-stream segmentation. *Psychol. Science* **16**, 298–304.
Bortfeld, H., Wruck, E. & Boas, D. A. 2007 Assessing infants' cortical response to speech using near-infrared spectroscopy. *NeuroImage* **34**, 407–415.
Bosch, L. & Sebastian-Galles, N. 2003a Language experience and the perception of a voicing contrast in fricatives: infant and adult data. In *International Congress of Phonetic Sciences* (eds M. J. Sole, D. Recasens & J. Romero), pp. 1987–1990. Barcelona, Spain: UAB.
Bosch, L. & Sebastian-Galles, N. 2003b Simultaneous bilingualism and the perception of a language-specific vowel contrast in the first year of life. *Lang. Speech* **46**, 217–243.
Bosseler, A. N., Imada, T., Pihko, E., Mäkelä, J., Taulu, S., Ahonen, A & Kuhl, P. K. 2008 Neural correlates of speech and non-speech processing: role of language experience in brain activation. *J. Acoust. Soc. Am.* **123**, 3333.
Brainard, M. S. & Knudsen, E. I. 1998 Sensitive periods for visual calibration of the auditory space map in the barn owl optic tectum. *J. Neurosci.* **18**, 3929–3942.
Brooks, R. & Meltzoff, A. N. 2008 Gaze following and pointing predicts accelerated vocabulary growth through two years of age: a longitudinal growth curve modeling study. *J. Child Lang.* **35**, 207–220.
Burnham, D., Kitamura, C. & Vollmer-Conna, U. 2002 What's new, pussycat? On talking to babies and animals. *Science* **296**, 1435.
Burns, T. C., Yoshida, K. A., Hill, K. & Werker, J. F. 2007 The development of phonetic representation in bilingual and monolingual infants. *App. Psycholinguist.* **28**, 455–474.
Canseco-Gonzalez, E. 2000 Using the recording of event-related brain potentials in the study of sentence processing. In *Language and the Brain: Representation and Processing* (eds Y. Grodzinsky, L. Shapiro & D. Swinney), pp. 229–266. San Diego: Academic.
Carlson, S. M. & Meltzoff, A. N. 2008 Bilingual experience and executive functioning in young children. *Dev. Sci.* **11**, 282–298.
Cheour, M., Ceponiene, R., Lehtokoski, A., Luuk, A., Allik, J., Alho, K. & Näätänen, R. 1998 Development of language-specific phoneme representations in the infant brain. *Nat. Neurosci.* **1**, 351–353.

Cheour, M., Imada, T., Taulu, S., Ahonen, A., Salonen, J. & Kuhl, P. K. 2004 Magnetoencephalography is feasible for infant assessment of auditory discrimination. *Exp. Neurol.* **190**, 44–51.

Coffey-Corina, S., Padden, D., Kuhl, P. K. & Dawson, G. 2007 Electrophysiological processing of single words in toddlers and school-age children with Autism Spectrum Disorder. Paper presented at the *Annual Meeting Cognitive Neuroscience Society, New York, NY. May 5–May 8, 2007*.

Coffey-Corina, S., Padden, D., Kuhl, P. K. & Dawson, G. 2008 ERPs to words correlate with behavioral measures in children with Autism Spectrum Disorder. *J. Acoust. Soc. Am.* **123**, 3742.

Conboy, B. T. & Kuhl, P. K. 2007 ERP mismatch negativity effects in 11-month-old infants after exposure to Spanish. Paper presented at the *Biennial Meeting of the Society for Research in Child Development, Boston, March 29–April 1, 2007*.

Conboy, B. T. & Mills, D. L. 2006 Two languages, one developing brain: event-related potentials to words in bilingual toddlers. *Dev. Sci.* **9**, F1–F12.

Conboy, B. T., Rivera-Gaxiola, M., Klarman, L., Aksoylu, E. & Kuhl, P. K. 2005 Associations between native and nonnative speech sound discrimination and language development at the end of the first year. In *Supplement to the Proceedings of the 29th Boston University Conference on Language Development* (eds A. Brugos, M. R. Clark-Cotton, and S. Ha). See http://www.bu.edu/linguistics/APPLIED/BUCLD/

Conboy, B. T., Brooks, R., Taylor, M., Meltzoff, A. N. & Kuhl, P. K. 2008a Joint engagement with language tutors predicts brain and behavioral responses to second-language phonetic stimuli. Paper presented at the *XVIth Biennial International Conference on Infant Studies, Vancouver, BC. March 27–March 29, 2008*.

Conboy, B. T., Rivera-Gaxiola, M., Silva-Pereyra, J. & Kuhl, P. K. 2008b Event-related potential studies of early language processing at the phoneme, word, and sentence levels. In *Early Language Development: Vol. 5. Bridging Brain and Behavior, Trends in Language Acquisition Research* (eds A. D. Friederici and G. Thierry), pp. 23–64. Amsterdam/The Netherlands: John Benjamins.

Conboy, B. T., Sommerville, J. & Kuhl, P. K. 2008c Cognitive control factors in speech perception at 11 months. *Dev. Psychol.* **44**, 1505–1512.

Cutler, A. & Norris, D. 1988 The role of strong syllables in segmentation for lexical access. *J. Exp. Psychol. Hum.: Percep. Perform.* **14**, 113–121.

Dehaene-Lambertz, G., Dehaene, S. & Hertz-Pannier, L. 2002 Functional neuroimaging of speech perception in infants. *Science* **298**, 2013–2015.

Dehaene-Lambertz, G., Hertz-Pannier, L., Dubois, J., Meriaux, S. & Roche, A. 2006 Functional organization of perisylvian activation during presentation of sentences in preverbal infants. *Proc. Natl. Acad. Sci.* **103**, 14240–14245.

DePaolis, R. 2005 The influence of production on perception: output as input. In *Supplement to the Proceedings of the 29th Boston University Conference on Language Development* (eds A. Brugos, M. R. Clark-Cotton & S. Ha). See http://www.bu.edu/linguistics/APPLIED/BUCLD/

Eimas, P. D., Siqueland, E. R., Jusczyk, P. & Vigorito, J. 1971 Speech perception in infants. *Science* **171**, 303–306.

Englund, K. T. 2005 Voice onset time in infant directed speech over the first six months. *First Lang.* **25**, 219–234.

Fennell, C. T. & Werker, J. F. 2003 Early word learners' ability to access phonetic detail in well-known words. *Lang. Speech* **46**, 245–264.

Fenson, L., Dale, P., Reznick, J. S., Thal, D., Bates, E., Hartung, J., Pethick, S. & Reilly, J. S. 1993 *MacArthur Communicative Development Inventories: User's Guide and Technical Manual*. San Diego, CA: Singular Publishing Group.

Fernald, A., Perfors, A. & Marchman, V. A. 2006 Picking up speed in understanding: Speech processing efficiency and vocabulary growth across the 2nd year. *Dev. Psychol.* **42**, 98–116.

Fowler, C. A. 1986 An event approach to the study of speech perception from a direct-realist perspective. *J. Phonet.* **14**, 3–28.

Fowler, C. A. 2006 Compensation for coarticulation reflects gesture perception, not spectral contrast. *Percep. Psychophys.* **68**, 161–177.

Friederici, A. D. 2002 Towards a neural basis of auditory sentence processing. *Trends Cogn. Sci.* **6**, 78–84.

Friederici, A. D. 2005 Neurophysiological markers of early language acquisition: from syllables to sentences. *Trends Cogn. Sci.* **9**, 481–488.

Friederici, A. D. & Wessels, J. M. I. 1993 Phonotactic knowledge of word boundaries and its use in infant speech perception. *Percep. Psychophys.* **54**, 287–295.

Friederici, A. D., Fiebach, C. J., Schlesewsky, M., Bornkessel, I. D. & von Cramon, D. Y. 2006 Processing linguistic complexity and grammaticality in the left frontal cortex. *Cereb. Cortex* **16**, 1709–1717.

Friedrich, M. & Friederici, A. D. 2005 Lexical priming and semantic integration reflected in the event-related potential of 14-month-olds. *NeuroReport* **16**, 653–656.

Friedrich, M. & Friederici, A. D. 2006 Early N400 development and later language acquisition. *Psychophysiol.* **43**, 1–12.

Gallese, V. 2003 The manifold nature of interpersonal relations: the quest for a common mechanism. *Phil. Trans. R. Soc. B* **358**, 517–528.

Ganger, J. & Brent, M. R. 2004 Reexamining the vocabulary spurt. *Dev. Psychol.* **40**, 621–632.

Gernsbacher, M. A. & Kaschak, M. P. 2003 Neuroimaging studies of language production and comprehension. *Ann Rev Psychol* **54**, 91–114.

Goodsitt, J. V., Morgan, J. L. & Kuhl, P. K. 1993 Perceptual strategies in prelingual speech segmentation. *J. Child Lang.* **20**, 229–252.

Hahne, A. & Jescheniak, J. D. 2001 What's left if the Jabberwock gets the semantics? An ERP investigation into semantic and syntactic processes during auditory sentence comprehension. *Cogn. Brain Res.* **11**, 199–212.

Hahne, A., Eckstein, K. & Friederici, A. D. 2004 Brain signatures of syntactic and semantic processes during children's language development. *J. Cogn. Neurosci.* **16**, 1302–1318.

Halle, P. A. & de Boysson-Bardies, B. 1994 Emergence of an early receptive lexicon: infants' recognition of words. *Inf. Behav. Dev.* **17**, 119–129.

Harris, A. M. 2001 Processing semantic and grammatical information in auditory sentences: electrophysiological evidence from children and adults. *Diss. Abstr. Internat.* **61**, 6729B.

Hohle, B., Bijcljac-Babic, R., Herold, B., Weissenborn, J., & Nazzi, T. 2009. Language specific prosodic preferences during the first half year of life: evidence from German and French infants. *Inf. Behav. Dev.*, (doi: 10.1016/j.infbeh.2009.03.004)

Holcomb, P. J., Coffey, S. A. & Neville, H. J. 1992 Visual and auditory sentence processing: a developmental analysis using event-related brain potentials. *Dev. Neuropsychol.* **8**, 203–241.

Homae, F., Watanabe, H., Nakano, T., Asakawa, K. & Taga, G. 2006 The right hemisphere of sleeping infant perceives sentential prosody. *Neurosci. Res.* **54**, 276–280.

Imada, T., Zhang, Y., Cheour, M., Taulu, S., Ahonen, A. & Kuhl, P. K. 2006 Infant speech perception activates Broca's area: a developmental magnetoencephalography study. *NeuroReport* **17**, 957–962.

Imada, T., Bosseler, A. N., Taulu, S., Pihko, E., Mäkelä, J., Ahonen, A., & Kuhl, P. K. 2008 Magnetoencephalography as a tool to study speech perception in awake infants. *J. Acoust. Soc. Am.* **123**, 3742.

Johnson, E. K. & Jusczyk, P. W. 2001 Word segmentation by 8-month-olds: when speech cues count more than statistics. *J. Mem. Lang.* **44**, 548–567.

Jusczyk, P. W. 1997 Finding and remembering words: some beginnings by English-learning infants. *Curr. Direct. Psychol. Sci.* **6**, 170–174.

Jusczyk, P. W. & Aslin, R. N. 1995 Infants' detection of the sound patterns of words in fluent speech. *Cogn. Psychol.* **29**, 1–23.

Jusczyk, P. W. and Hohne, E. A. 1997 Infants' memory for spoken words. *Science*, **277**, 1984–1986.

Kuhl, P. K. 2004 Early language acquisition: cracking the speech code. *Nat. Rev. Neurosci.* **5**, 831–841.

Kuhl, P. K. 2007 Is speech learning 'gated' by the social brain? *Dev. Sci.* **10**, 110–120.

Kuhl, P. K. 2008 Linking infant speech perception to language acquisition: phonetic learning predicts language growth. In *Infant Pathways to Language: Methods, Models, and Research Directions* (eds P. McCardle, J. Colombo & L. Freund), pp. 213–243. New York: Erlbaum.

Kuhl, P. K. In press Early language acquisition: neural substrates and theoretical models. In *The Cognitive Neurosciences IV* (ed. M. S. Gazzaniga). Cambridge, MA: MIT Press.

Kuhl, P. K. & Damasio, A. In press. Language. In *Principles of Neural Science: 5th Edn* (eds E. R. Kandel, J. H. Schwartz, T. M. Jessell, S. Siegelbaum & J. Hudspeth). New York: McGraw Hill.

Kuhl, P. K. & Meltzoff, A. N. 1982 The bimodal perception of speech in infancy. *Science* **218**, 1138–1141.

Kuhl, P. K. & Meltzoff, A. N. 1996 Infant vocalizations in response to speech: vocal imitation and developmental change. *J. Acoust. Soc. Am.* **100**, 2425–2438.

Kuhl, P. K. & Rivera-Gaxiola, M. 2008 Neural substrates of language acquisition. *Ann Rev Neurosci* **31**, 511–534.

Kuhl, P. K., Williams, K. A., Lacerda, F., Stevens, K. N. & Lindblom, B. 1992 Linguistic experience alters phonetic perception in infants by 6 months of age. *Science* **255**, 606–608.

Kuhl, P. K., Andruski, J. E., Chistovich, I. A., Chistovich, L. A., Kozhevnikova, E. V., Ryskina, V. L., Stolyarova, E., Sundberg, U. & Lacerda, F. 1997 Cross-language analysis of phonetic units in language addressed to infants. *Science* **277**, 684–686.

Kuhl, P. K., Tsao, F. -M. & Liu, H. -M. 2003 Foreign-language experience in infancy: effects of short-term exposure and social interaction on phonetic learning. *Proc. Natl. Acad. Sci.* **100**, 9096–9101.

Kuhl, P. K., Coffey-Corina, S., Padden, D. & Dawson, G. 2005a Links between social and linguistic processing of speech in preschool children with autism: behavioral and electrophysiological measures. *Dev. Sci.* **8**, F1–F12.

Kuhl, P. K., Conboy, B. T., Padden, D., Nelson, T. & Pruitt, J. 2005b Early speech perception and later language development: implications for the 'critical period.' *Lang. Learn. Dev.*, **1**, 237–264.

Kuhl, P. K., Stevens, E., Hayashi, A., Deguchi, T., Kiritani, S. & Iverson, P. 2006 Infants show a facilitation effect for native language phonetic perception between 6 and 12 months. *Dev. Sci.* **9**, F13–F21.

Kuhl, P. K., Conboy, B. T., Coffey-Corina, S., Padden, D., Rivera-Gaxiola, M. & Nelson, T. 2008 Phonetic learning as a pathway to language: new data and native language magnet theory expanded (NLM-e). *Phil. Trans. R. Soc. B* **363**, 979–1000.

Kujala, A., Alho, K., Service, E., Ilmoniemi, R. J. & Connolly, J. F. 2004 Activation in the anterior left auditory cortex associated with phonological analysis of speech input: localization of the phonological mismatch negativity response with MEG. *Cogn. Brain Res.* **21**, 106–113.

Kutas, M. 1997 Views on how the electrical activity that the brain generates reflects the functions of different language structures. *Psychophysiology* **34**, 383–398.

Liberman, A. M. & Mattingly, I. G. 1985 The motor theory of speech perception revised. *Cognition*, **21**, 1–36.

Liberman, A. M., Cooper, F. S., Shankweiler, D. P. & Studdert-Kennedy, M. 1967 Perception of the speech code. *Psychol. Rev.* **74**, 431–461.

Liu, H. -M., Kuhl, P. K. & Tsao, F.-M. 2003 An association between mothers' speech clarity and infants' speech discrimination skills. *Dev. Sci.* **6**, F1–F10.

Liu, H. -M., Tsao, F. -M. & Kuhl, P. K. 2007 Acoustic analysis of lexical tone in Mandarin infant-directed speech. *Dev. Psychol.* **43**, 912–917.

Mandel, D. R., Jusczyk, P. W. & Pisoni, D. 1995 Infants' recognition of the sound patterns of their own names. *Psychol. Sci.* **6**, 314–317.

Mattys, S. L., Jusczyk, P. W., Luce, P. A. & Morgan, J. L. 1999 Phonotactic and prosodic effects on word segmentation in infants. *Cog. Psychol.* **38**, 465–494.

Maye, J., Werker, J. F. & Gerken, L. 2002 Infant sensitivity to distributional information can affect phonetic discrimination. *Cognition* **82**, B101–B111.

Meltzoff, A. N. & Decety, J. 2003 What imitation tells us about social cognition: a rapprochement between developmental psychology and cognitive neuroscience. *Phil. Trans. R. Soc. B* **358**, 491–500.

Mills, D. L., Coffey-Corina, S. A. & Neville, H. J. 1993 Language acquisition and cerebral specialization in 20-month-old infants. *J. Cogn. Neurosci.* **5**, 317–334.

Mills, D. L., Coffey-Corina, S. A. & Neville, H. J. 1997 Language comprehension and cerebral specialization from 13–20 months. *Dev. Neuropsychol.* **13**, 233–237.

Mills, D. L., Prat, C., Zangl, R., Stager, C. L., Neville, H. J. & Werker, J. F. 2004 Language experience and the organization of brain activity to phonetically similar words: ERP evidence from 14- and 20-month-olds. *J. Cogn. Neurosci.* **16**, 1452–1464.

Mills, D. L., Plunkett, K., Prat, C. & Schafer, G. 2005 Watching the infant brain learn words: effects of vocabulary size and experience. *Cogn. Dev.* **20**, 19–31.

Molfese, D. L. 1990 Auditory evoked responses recorded from 16-month-old human infants to words they did and did not know. *Brain Lang.* **38**, 345–363.

Molfese, D. L., Morse, P. A. & Peters, C. J. 1990 Auditory evoked responses to names for different objects: cross-modal processing as a basis for infant language acquisition. *Dev. Psychol.* **26**, 780–795.

Molfese, D. L., Wetzel, W. & Gill, L. A. 1993 Known versus unknown word discriminations in 12-month-old human infants: electrophysiological correlates. *Dev. Neuropsychol.* **9**, 241–258.

Munte, T. F., Matzke, M. & Johanes, S. 1997 Brain activity associated with syntactic incongruencies in words and psuedowords. *J. Cogn. Neurosci.* **9**, 318–329.

Näätänen, R., Lehtokoski, A., Lennes, M., Cheour, M., Huotilainen, M., Iivonen, A., Vainio, M. Alku, P., Ilmoniemi, R. J., Luuk, A., Allik, J., Sinkkonen, J. & Alho, K. 1997 Language-specific phoneme representations revealed by electric and magnetic brain responses. *Nature* **385**, 432–434.

Nazzi, T., Iakimova, G., Bertoncini, J., Frédonie, S. & Alcantara, C. 2006 Early segmentation of fluent speech by infants acquiring French: emerging evidence for crosslinguistic differences. *J. Mem. Lang.* **54**, 283–299.

Neville, H. J., Coffey, S. A., Holcomb, P. J. & Tallal, P. 1993 The neurobiology of sensory and language processing in language-impaired children. *J. Cogn. Neurosci.* **5**, 235–253.

Newman, R. N., Ratner, B., Jusczyk, A. M., Jusczyk, P. W. & Dow, K. A. (2006). Infants' early ability to segment the conversational speech signal predicts later language development: a retrospective analysis. *Dev. Psychol.* **42**, 643–655.

Newport, E. L. & Aslin, R. N. 2004 Learning at a distance I. Statistical learning of non-adjacent dependencies. *Cogn. Psychol.* **48**, 127–162.

Oberecker, R. & Friederici, A. D. 2006 Syntactic event-related potential components in 24-month-olds' sentence comprehension. *NeuroReport* **17**, 1017–1021.

Oberecker, R., Friedrich, M. & Friederici, A. D. 2005 Neural correlates of syntactic processing in two-year-olds. *J. Cogn. Neurosci.* **17**, 1667–1678.

Peña, M., Maki, A., Kovacic, D., Dehaene-Lambertz, G., Koizumi, H., Bouquet, F. & Mehler, J. 2003 Sounds and silence: an optical topography study of language recognition at birth. *Proc. Natl. Acad. Sci.* **100**, 11702–11705.

Polka, L. & Werker, J. F. 1994 Developmental changes in perception of nonnative vowel contrasts. *J. Exp. Psychol. Hum. Percep. Perform.* **20**, 421–435.

Pulvermuller, F. 2005 *The Neuroscience of Language: On Brain Circuits of Words and Serial Order*. Cambridge, UK: Cambridge University Press.

Raizada, R. D. S., Tsao, F. -M., Liu, H. -M. & Kuhl, P. K. 2009. Quantifying the adequacy of neural representations for a cross-language phonetic discrimination task: prediction of individual differences. *Celebral Cortex*, (doi:10.1093/cercor/bhp076)

Raudenbush, S. W., Bryk, A. S., Cheong, Y. F. & Congdon, R. 2005 *HLM-6: Hierarchical Linear and Nonlinear Modeling*. Lincolnwood, IL, Scientific Software International.

Rivera-Gaxiola, M. & Romo, H. 2006 Infant head-start learners: brain and behavioral measures and family assessments. Paper presented at *From Synapse to Schoolroom: The Science of Learning. NSF Science of Learning Centers Satellite Symposium, Society for Neuroscience Annual Meeting, Atlanta, GA, October 13, 2006.*

Rivera-Gaxiola, M., Klarman, L., Garcia-Sierra, A. & Kuhl, P. K. 2005a Neural patterns to speech and vocabulary growth in American infants. *NeuroReport* **16**, 495–498.

Rivera-Gaxiola, M., Silva-Pereyra, J. & Kuhl, P. K. 2005b Brain potentials to native and non-native speech contrasts in 7- and 11-month-old American infants. *Dev. Sci.* **8**, 162–172.

Rivera-Gaxiola, M., Silva-Pereyra, J., Klarman, L., Garcia-Sierra, A., Lara-Ayala, L., Cadena-Salazar, C. & Kuhl, P. K. 2007 Principal component analyses and scalp distribution of the auditory P150-250 and N250-550 to speech contrasts in Mexican and American infants. *Dev. Neuropsychol.* **31**, 363–378.

Rizzolatti, G. 2005 The mirror neuron system and its function in humans. *Anat. Embryol.* **210**, 419–421.

Rizzolatti, G. & Craighero, L. 2004 The mirror-neuron system. *Ann Rev Neurosci.* **27**, 169–192.

Saffran, J. R. 2003 Statistical language learning: Mechanisms and constraints. *Curr. Direct. Psychol. Sci.* **12**, 110–114.

Saffran, J. R., Aslin, R. N. & Newport, E. L. 1996 Statistical learning by 8-month-old infants. *Science* **274**, 1926–1928.

Silva-Pereyra, J., Klarman, L., Lin, J. F. & Kuhl, P. K. 2005a Sentence processing in 30-month-old children: an ERP study. *NeuroReport* **16**, 645–648.

Silva-Pereyra, J., Rivera-Gaxiola, M. & Kuhl, P. K. 2005b An event-related brain potential study of sentence comprehension in preschoolers: semantic and morphosyntactic processing. *Cogn. Brain Res.* **23**, 247–258.

Silva-Pereyra, J., Conboy, B. T., Klarman, L. & Kuhl, P. K. 2007 Grammatical processing without semantics? An event-related brain potential study of preschoolers using Jabberwocky sentences. *J. Cogn. Neurosci.* **19**, 1050–1065.

Stager, C. L. & Werker, J. F. 1997 Infants listen for more phonetic detail in speech perception than in word-learning tasks. *Nature* **388**, 381–382.

Sundara, M., Polka, L. & Genesee, F. 2006 Language-experience facilitates discrimination of /d-ð/ in monolingual and bilingual acquisition of English. *Cognition* **100**, 369–388.

Sundara, M., Polka, L. & Molnar, M. 2008 Development of coronal stop perception: bilingual infants keep pace with their monolingual peers. *Cognition* **108**, 232–242.

Swingley, D. 2005 11-month-olds' knowledge of how familiar words sound. *Dev. Sci.* **8**, 432–443.

Swingley, D. & Aslin, R. N. 2000 Spoken word recognition and lexical representation in very young children. *Cognition* **76**, 147–166.

Swingley, D. & Aslin, R. N. 2002 Lexical neighborhoods and the word-form representations of 14-month-olds. *Psychol. Sci.* **13**, 480–484.

Swingley, D. & Aslin, R. N. 2007 Lexical competition in young children's word learning. *Cognitive Psychol.* **54**, 99–132.

Taga, G. & Asakawa, K. 2007 Selectivity and localization of cortical response to auditory and visual stimulation in awake infants aged 2 to 4 months. *NeuroImage* **36**, 1246–1252.

Thierry, G. C. A., Vihman, M. & Roberts, M. 2003 Familiar words capture the attention of 11-month-olds in less than 250 ms. *NeuroReport* **14**, 2307–2310.

Tincoff, R. & Jusczyk, P. W. 1999 Some beginnings of word comprehension in 6-month-olds. *Psychol. Sci.* **10**, 172–175.

Tsao, F.-M., Liu, H.-M. & Kuhl, P. K. 2004 Speech perception in infancy predicts language development in the second year of life: a longitudinal study. *Child Dev.* **75**, 1067–1084.

Vihman, M. M., Nakai, S., DePaolis, R. A. & Halle, P. 2004 The role of accentual pattern in early lexical representation. *J. Mem. Lang.* **50**, 336–353.

Virnes, M., Cardillo, G., Kuhl, P. K. & Movellan, J. R. 2008 LIFE-TDLC NSF Science of Learning Center Collaboration: Social Robots for Learning Language.

Weber, C., Hahne, A., Friedrich, M. & Friederici, A. D. 2004 Discrimination of word stress in early infant perception: electrophysiological evidence. *Cogn. Brain Res.* **18**, 149–161.

Werker, J. F. & Curtin, S. 2005 PRIMIR: a developmental framework of infant speech processing. *Lang. Learn. Dev.* **1**, 197–234.

Werker, J. F. & Tees, R. C. 1984 Cross-language speech perception: evidence for perceptual reorganization during the first year of life. *Inf. Behav. Dev.* **7**, 49–63.

Werker, J. F., Fennell, C. T., Corcoran, K. M. & Stager, C. L. 2002 Infants' ability to learn phonetically similar words: effects of age and vocabulary size. *Infancy* **3**, 1–30.

Zhang, Y., Kuhl, P. K., Imada, T., Kotani, M. & Tohkura, Y. 2005 Effects of language experience: neural commitment to language-specific auditory patterns. *NeuroImage* **26**, 703–720.

6

The processing of audio-visual speech: empirical and neural bases
Ruth Campbell

In this selective review, I outline a number of ways in which seeing the talker affects auditory perception of speech, including, but not confined to, the McGurk effect. To date, studies suggest that all linguistic levels are susceptible to visual influence, and that two main modes of processing can be described: a *complementary* mode, whereby vision provides information more efficiently than hearing for some under-specified parts of the speech stream, and a *correlated* mode, whereby vision partially duplicates information about dynamic articulatory patterning.

Cortical correlates of seen speech suggest that at the neurological as well as the perceptual level, auditory processing of speech is affected by vision, so that 'auditory speech regions' are activated by seen speech. The processing of natural speech, whether it is heard, seen or heard and seen, activates the perisylvian language regions (left > right). It is highly probable that activation occurs in a specific order. First, superior temporal, then inferior parietal and finally inferior frontal regions (left > right) are activated. There is some differentiation of the visual input stream to the core perisylvian language system, suggesting that complementary seen speech information makes special use of the visual ventral processing stream, while for correlated visual speech, the dorsal processing stream, which is sensitive to visual movement, may be relatively more involved.

Keywords: speech reading; audiovisual speech processing; visual speech.

6.1 Introduction

Language most shows a man: speak, that I might see thee! (Ben Jonson 1572–1637; Timber: Or Discoveries, 1640)

That speech has visible as well as auditory consequences, and that watching the talker can be beneficial for speech understanding, has been acknowledged for many years. Fifty years ago, it was established that seeing the talker could give an improvement in the comprehension of auditory speech in noise equivalent to that produced by an increase of up to 15 dB in signal-to-noise ratio (Sumby & Pollack 1954). This was widely interpreted to mean that the effects of vision on audition were only apparent at low signal-to-noise (S–N) ratios. However, a re-examination of Sumby and Pollack's findings (Remez 2005) clearly shows that the benefit of seeing the talker was not limited to adverse acoustic conditions, but was apparent at all S–N ratios. Nevertheless, throughout the 1960s and 1970s, the impression was that in order for vision to affect speech perception, acoustic information needed to be suboptimal.

Things changed in the 1970s and 1980s. First came the demonstration that the perception of certain speech segments could be strongly influenced by vision even when acoustic conditions were good, and indeed that some audio-visual pairings could lead to illusory perceptions. The original discovery of the McGurk effect (McGurk & MacDonald 1976) was accidental. The investigators were researching young children's imitation of auditory speech patterns. They dubbed a number of different video to auditory syllable tokens with the aim of distracting the child from the auditory imitation task. The syllables had varied consonant-vowel forms, including ba, ga, da, ka, ta and pa. When a seen 'ga' was dubbed to a heard 'ba', all participants thought that 'da' had been said—and the technician was reprimanded for not dubbing the 'ga' and 'ba' correctly. Only when he insisted that the tokens had the required form, and participants tested their perceptions by closing their eyes and watching the silent videotape closely, did it become apparent that 'da' was illusory. That is, under these specific conditions, the perceiver heard an event which was not present in either the visual or the auditory stimulus. The illusion also held for the unvoiced synthesis (visual 'ka'; auditory 'pa'… hear 'ta'). It was as marked for children as for adults and was found to be relatively insensitive to knowledge of its bases or to lexical or other expectations. The McGurk illusion thus added a new impetus to studies of audio-visual speech.[1] In another set of studies, Reisberg et al. (1987), using natural rather than dubbed audio-visual speech, and extended passages of speech rather than isolated tokens, reported that, even when hearing conditions are excellent, there is a gain in speech comprehension under audio-visual compared with auditory-alone conditions. This occurred for hard-to-understand but easy-to-hear passages. What is it about audio-visual processing that can deliver such outcomes? How does vision affect the primary modality of audition for understanding speech?

This chapter presents some experimental findings, including behavioural and neurological data, that explore the idea that understanding speech requires that we take into account its visual concomitants. I will consider both speech-reading in the absence of hearing (silent speech-reading) and audio-visual speech perception, and will offer some suggestions concerning multimodal mechanisms.

6.2 The source-filter model of speech: some applications to speech-reading

A model of speech production can be constructed based on the physical characteristics of the system that is used to produce speech (Fant 1960; Diehl, this volume). One source of the initial acoustic event is the vibration of the vocal folds (the rate of vibration determining fundamental voice pitch); the filter function describes the effects on the resultant acoustic waveform of its passage through the rest of the vocal apparatus—the vocal cavities, the hard and soft palates, the tongue and mouth. It may be possible to detect some visual correlates of source function. For example, Munhall et al. (2004) report that in sentential utterances, head movements are quite well temporally aligned with the onset and offset of voicing—and hence there are correspondences between the kinematics of head actions and the dynamic sound pattern over the period of the utterance as seen in the speech spectrum. When we speak, our vocal folds do not function independently of other bodily actions, as you can confirm for yourself when watching a talker from behind. The onset and offset of speech, in particular, are relatively easy to detect from head movements.

As well as source effects, many aspects of the filter function are visible. In women and children, the length of the vocal tract is generally shorter than in men. Gender and age are predominantly identified by sight. As for the configuration of the vocal tract, mouth opening and closure, as well as mouth shape, are all highly visible. Visible configurations of the lips, teeth and tongue allow us to distinguish 'map' from 'nap', 'threat' from 'fret', 'tap' from 'tack' and 'him' from 'ham' by eye. While place of articulation can often be determined visually, manner of articulation can also sometimes be seen: for instance, the late voicing of 'p' in 'park' can be accompanied by a visible lip-puff, which is absent when 'bark' is uttered.

Source–filter models are appropriate for the description of speech production, and can account for some specific visual as well as acoustic properties of speech. But are these just a few local features, or do speech production characteristics have broader applicability to speech-reading? Yehia *et al.* (2002) measured the visual kinematics as well as the spectral (acoustic) properties of some spoken phrases. The kinematics of the talker's face and head were correlated with spectral events in these utterances, to the extent that the visible motion characteristics could be used to estimate and predict (i.e. recover) almost all of the speech acoustic patterns. Movements of the mouth and lips, while contributing to the synthesis, were not themselves as useful as head, face (eyebrows especially) and mouth movements. Such demonstrations suggest that purely visual spatio-temporal speech patterns afford reliable access to representations of phrase-length utterances, and, moreover, that speech-reading may usefully consider actions of the face and head beyond those of the mouth and lips.

Summerfield (1987) suggested that when we perceive speech, we reconstruct the patterns of articulation used by the talker, irrespective of the modality of input. Visual and even haptic processes (Fowler & Dekle 1991) can affect the impression of what was heard. While information from these modalities may be integrated with acoustic information via purely associative mechanisms, it seems probable that the processing system will make use of the correspondences between visual, somaesthetic and acoustic events to inform processing. Since these are all the consequences of the act of speaking, implicit knowledge about articulatory processes is likely to influence multimodal speech processing. Naturally, if we hear well, speech representations will be dominated by our acoustic impressions, but, nevertheless, speech perception cannot be considered to be exclusively auditory. Many explorations of how non-acoustic impressions of the talker can moderate segmental speech perception have been undertaken. Green and colleagues performed some of the most convincing of these. Among other things, Green *et al.* (1991) showed that the visual impression of a talker's gender could shift the perception of a clearly heard but ambiguous auditory consonant from 'sh' to 's'. 's' is produced with the tongue immediately behind the teeth, while for 'sh' the place of articulation is more posterior. The perceiver can gain an impression of the vocal-tract characteristics of the talker by vision alone; in this case, the estimation is of the place of articulation in relation to the probable depth of the mouth cavity.

Whether speech is considered at the 'fine-grain' level of phonetic context for phoneme discrimination, or the 'coarse-grain' level of the spectral characterization of a 2 or 3 s utterance of connected speech, there is good evidence that the talker's seen actions can contribute to the perception of speech.

6.3 Binding: some preliminaries

It is one thing to claim that articulatory events can be perceived amodally (or supramodally, or cross-modally), but quite another to describe exactly how, when one perceives natural speech from a talker who one sees as well as hears, visual and auditory information may combine to allow the speech processor to select the appropriate fit. What is required for an audio-visual speech event to be processed? First, is attention needed or is audio-visual processing automatic and mandatory? It has long been claimed that McGurk effects are automatic. McGurk effects do not require attention to be explicitly directed to them to be experienced (e.g. Soto-Faraco *et al.* 2004). Infants who are not yet able to speak or respond to attentional instruction are sensitive to audio-visual fusions. From the age of six months or so, infants who have habituated to a 'McGurk' audio-visual 'da' (the stimulus comprises an auditory 'ba' dubbed to a visual 'ga') dishabituate when a congruent audio-visual 'ba' is played to them, and fail to respond with a dishabituation response when a 'real' audio-visual 'da', derived from a visual 'da' and an auditory 'da', is played to them (Burnham & Dodd (2004) and see also Rosenblum *et al.* (1997)). Audio-visual speech processing capabilities are in place even before the child has useful speech.[2] Nevertheless, studies with adults suggest that there can be an attentional cost to processing audio-visual fusions of this sort. Tiippana *et al.* (2004) and Alsius *et al.* (2005) independently showed that a visual distractor task could reduce the incidence of illusory McGurk percepts. In Tiippana *et al.*'s study, the distractor element was a moving image of a leaf randomly floating across the face of the talker (but not obscuring the mouth). In the Alsius *et al.* experiment, line drawings of objects were overlaid on the image of the talker. In both cases, it might be claimed that the visual distractor degraded the visual input, so the reduction of the McGurk effect resulted from perceptual rather than attentional processes. However, Alsius *et al.* (2005) showed that *auditory* distraction during the presentation of McGurk stimuli also reduced vulnerability to McGurk effects. Thus, paradoxically, adding an auditory task to the identification of an audio-visual syllable *increased the* probability of an auditory response to an incongruent audio-visual item. This strongly suggests that it is the process of integrating vision and hearing which is vulnerable to additional attentional load, rather than the processing of each input stream prior to their integration. A deficit in attentional control may also account for reports of defective audiovisual binding in clinical groups, including schizophrenia (e.g. Ross *et al.* 2007)

In McGurk-type experiments, participants are presented with well-synchronized visual and auditory tokens, apparently emanating from a single talker; that is, the auditory and visual parts of the McGurk stimuli are spatially and temporally coherent and coextensive. It might be thought that this is a strong cue to their co-processing, to their 'binding'. In fact, audio-visual effects, including McGurk effects and an audio-visual advantage for bimodal compared with purely auditory processing, are reported under relatively large desynchronizations. The effects of vision on audition occur for asynchronies (vision leading) of 250 ms or more, depending on the task (Grant *et al.* 2004). To some extent, the loose temporal fit of vision to aftercoming sound may reflect anticipatory coarticulation—appropriate movements of the mouth often occur prior to vocalization. Certainly, tolerance of vision-led asynchronies is greater than for audition-led asynchronies.

In addition to tolerance of asynchronies, displacement of a voice in space is not accurately perceived. Wherever its actual source, perceivers locate an artificial speech source at the position of a visually perceived apparent talker (the ventriloquism illusion), i.e. vision 'captures' the location of the auditory event (Radeau & Bertelson 1974). When the auditory channel comprises two 'overlaid' voices, heard to be speaking simultaneously, and they issue from a single central loudspeaker, the perceiver not only uses a video display corresponding to one of the utterances to locate and shadow that talker effectively, but, surprisingly, may also be more able to shadow the *unseen* talker, who is now perceived to be spatially separated from apparent location of the visible talker (Driver 1996).

It seems that the speech processor is relatively unconcerned about the fine spectro-temporal and locational details of the match between vision and audition. Any model of the pattern-matching system that binds auditory and visual speech events must take account of this looseness of fit. The looseness may arise from the relative dominance and salience of audition in the perception of speech, so that vision, as the secondary input stream, is not required to match the auditory spectro-temporal markings precisely. A flexible system, using attentional resources where necessary, appears to be at work, allowing any analysis-by-synthesis approach of vision and audition to exercise variable constraints depending on the salience of the input event (cf. Grant et al. 2004). An alternative proposal is that, if articulatory plans are incorporated into the representations to which such perceptual events map, they may specify the acoustic and visual correlates of an utterance less with respect to location and temporal patterning and more in relation to the somaesthetic consequences of a speech act.

6.4 What does vision deliver? The art of 'hearing by eye'

The McGurk effect results from perceptual integration of a visible open mouth syllable (e.g. 'da') with a heard one (e.g. 'ga'), which has spectral similarities to the syllable that is perceived following combination (e.g. 'ba'). The combinatorial rules and processes involved have exercised psychologists for many years (see Bernstein et al. 2004a for review), but this work has focused on processing at the phoneme level. The findings that phonetic context perceived by eye can shift phonemic category boundaries (Green et al. 1991; van Linden & Vroomen, 2007), and that prelinguistic babies are sensitive to McGurk effects (Burnham & Dodd 2004), demonstrate that audio-visual integration can occur 'pre-phonemically'. That said, the phonemic level of linguistic structure offers the most approachable entry point for examining many aspects of the perception of seen speech in the *absence* of hearing—that is, silent speech-reading. Some speech-read segments are relatively unambiguous. For instance, labiodental consonants and English point vowels enjoy a high level of audio-visual mapping consistency ('what you see is what you hear'). However, a seen speech event usually maps onto several (acoustically defined) phonological categories. Visually confusable phonemes ('visemes') can be considered to constitute a phonemically equivalent class (PEC; Auer & Bernstein 1997). The number of PECs will vary from person to person, depending on their speech-reading skill and on the visibility of the talker's speech. Auer & Bernstein (1997) found that 12 PECs were sufficient to identify most English words. This number corresponds well with theoretical studies, suggesting that this number of distinctions should suffice for useful visual speech-reading as an aid to hearing, and contrasts with estimates of approximately 40 phonemes available 'by ear' in spoken English. The reason why a relatively small number of PECs can, in principle, suffice for identifying individual spoken words is that most words in English are relatively unique in their segmental and syllabic structure. That is, lexical space in English is relatively sparsely occupied and well distributed. Heard speech can, on this type of analysis, be considered to be overdetermined, containing a great deal of structural redundancy. This perspective allows us to understand the finding (Reisberg et al. 1987) that seeing as well as hearing the talker improves understanding of conceptually difficult texts, even under excellent perceptual conditions. We can assume that understanding such texts requires allocation of limited cognitive resources. If the audio-visual speech signal is highly redundant, it may require minimal cognitive processing to determine what words were spoken, thus freeing cognitive resources to allow better interpretation of the utterance.

The redundancy of speech can also explain why some people attain good speech comprehension by sight alone—at least under optimal talking and viewing conditions—as demonstrated by 'super-speech-readers' (e.g. Andersson & Lidestam 2005). These are people, often deaf from an early age, whose speech-reading abilities allow them to follow silently spoken conversations with high accuracy. If just 12 PECs are required to identify more than 90% of English words, one can understand how, in principle, such accuracy can be achieved by speech-reading alone—especially given that higher-level constraints (topic, discourse constraints, syntax and meaning) can also be used to aid comprehension. From this perspective, the fact that most hearing people are relatively poor speech-readers may reflect relative (over-) reliance on acoustic parameters of the speech stream.

6.5 The visible speech stream: varieties of information

An interesting feature of speech is that segments that are confusable acoustically (for instance, 'm' and 'n', and 'th' and 'f') are often visually distinctive—and vice versa ('p' and 'b' are acoustically distinct, but visually confusable; see Summerfield (1987) for illustrations). While this may implicate vision in some aspects of the evolution of spoken languages, a more pressing concern is to try to answer the question—what are the visual features of such distinctive speech segments? Summerfield (1979) showed that speech-reading relied mainly on mouth shape, mouth opening and the visible position of the tongue (and sometimes teeth). Is this just because we are disposed to perceive a simple correspondence between the diameter of a probable sound source and its amplitude? This cannot be a critical feature, as the identification of speech in noise is not helped by the perception of an annulus whose diameter is controlled by the amplitude of the acoustic signal (also see Bernstein *et al.* 2004*b*; Ghazanfar *et al.* 2005). This (among other demonstrations) suggests that, for the most efficient speech-reading, mouth opening and closing and the tongue position should be clearly visible. Figure 6.1 indicates how still images of faces producing speech can be reliably classified in terms of their speech characteristics. Vowels and consonants can be easily identified from a closed set, when mouth shape and the configuration of lips, teeth and tongue are visible.

A number of demonstrations, however, suggest that the configuration of the mouth, tongue and teeth, as captured in a still image, may not fully explain the audiovisual advantage or account fully for McGurk effects. McGurk effects can be obtained at viewing distances too great for mouth disposition to be clearly visible (Jordan & Sergeant 2000). Rosenblum and his colleagues have used point-light-illuminated faces to explore the time-varying aspects of seeing speech. Typically, 12–20 points on the face surface, videotaped at normal speed, give information about dynamic deformation of the face surface in the absence of any facial features. Such sparse stimuli induce McGurk effects (albeit at a reduced level compared with full facial images; Rosenblum & Saldaña 1996) and an audio-visual gain for speech perceived in noise (Rosenblum *et al.* 1996). Another manipulation that differentially affects the visibility of specific face features is spatial frequency filtering of the image. As long as the temporal characteristics of the signal are maintained, low-pass spatial frequency filtering (down to 19 cycles per face) blurs the image without markedly compromising audio-visual gain for understanding spoken

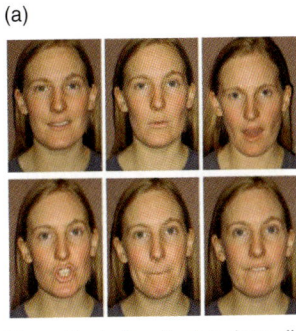

(a) While in the scanner, participants made the decision 'vowel' (top row) or 'consonant' (bottom row), when these images were presented singly.
Baseline condition — detect the lateral movement of a central cross on a monotone field

(b)

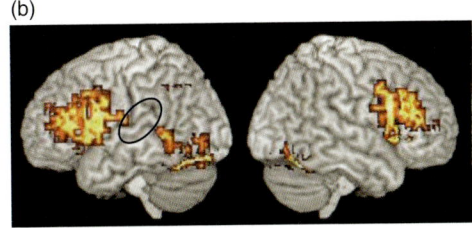

The activated cortical regions did not include the left pSTS, a region consistently activated when watching natural speech (within black circle)

Figure 6.1 (*a*) Speech images used in the fMRI experiment reported by Capek *et al.* (2005). Vowels are shown in the top row and consonants in the bottom row. (*b*) Rendered group activation maps for the task of distinguishing vowel and consonant lip shapes. Images were presented singly for decision. The baseline task was to detect the movement of a cross on a blank background (Capek *et al.* 2005). pSTS (black circle) was not activated. Significant foci of activation (*x*, *y*, *z*, coordinates) included: (i) inferior temporal cortex/fusiform gyrus (–29, –78, –17), (ii) right inferior frontal cortex extending into dlpfc (47, 11, 26), (iii) left inferior frontal cortex extending into dlpfc (–47, 7, 33), (iv) left inferior parietal lobule (–25, –63, 43), and (v) caudal anterior cingulate gyrus (–3, 11, 50).

sentences (Munhall *et al.* 2004). Visible kinematics can deliver critical components of the audio-visual advantage. Also critical to useful visual speech is the temporal sampling rate. When this is below 10 frames s^{-1}, the audio-visual advantage may decrease or disappear. The average rate of opening and closing of the vocal tract in normal speech is approximately 12 Hz. This is likely to be the minimal sampling rate for effective audio-visual speech processing.

6.6 Complementarity and redundancy in the speech stream

To summarize, audio-visual processing is more effective than auditory processing of natural speech for two reasons. First, some segmental contrasts can be seen clearly, thus aiding speech comprehension, especially where those segments are acoustically confusable. Second, many features of an utterance can be perceived by both ear and eye: the audible and the visible patterns are highly correlated, reflecting the underlying dynamics of speech production. The speech processing system makes use of the redundancies offered by the similar pattern of time-varying signal change across the modalities. Thus, there are, in principle, two modes whereby seen speech can affect what is heard: a *complementary* mode, whereby vision provides information about some aspects of the speech event that are hard to hear, and which may depend on the shape and contour of the lower face being clearly visible; and a *correlated* mode, where the crucial feature of the speech stream is its temporo-spectral signature, which will show regions of similar dynamic patterning across both audible and visible channels. This latter mode must require the perception of visible motion, and studies of patients with acquired lesions differentially affecting visual processing of form and motion bear this out (Campbell *et al.* 1997). Different cross-modal binding principles may apply depending on the relative importance of complementary and correlated information in the utterance to be processed.

6.7 Neural mechanisms for audio-visual and visual speech processing

Reviews of the neural bases of visual and audio-visual speech processing can be found elsewhere (e.g. Callan *et al.* 2004; Calvert & Lewis 2004; Capek *et al.* 2004; Miller & D'Esposito 2005). Some well-established findings are itemized here, which will then be examined in relation to the 'two-mode' sketch outlined in §5. They apply to hearing people: the special case of speech-reading in profound prelingual deafness is different again (Sadato *et al.* 2002).

Before summarizing these findings, it is worth pointing out that many aspects of the patterns of brain activation to be described here may also be found in non-human primates, when animals are presented with species-specific calls unimodally or bimodally. Since distinctive activation patterns can be traced in the brains of such non-speaking species—and regions homologous to those in humans appear to be implicated—within-species communication, rather than speech-specific, mechanisms may underlie many of the findings related to multimodal audio-visual speech processing (Ghazanfar *et al.* 2005) One study has suggested that visible mouth movements can affect auditory brain stem responses to speech (Musacchia *et al.* 2006), but, to date, cortical events have proved to be the most reliable indicators of how seen and heard speech interact.

(i) Speech-reading in the absence of any auditory input (silent speech-reading) activates auditory cortex. This may include activation within core regions of primary auditory cortex (A1; Pekkola *et al.* 2005), although the extent and specificity of activation within auditory cortex is problematic. While all investigators agree that parts of the superior temporal plane adjoining the upper part of the superior temporal gyrus are activated by silent speech, there had been disagreement concerning the extent to which primary auditory cortex within Heschl's gyrus might be activated by seen silent speech (Calvert *et al.* 1997; Bernstein *et al.* 2002). Pekkola *et al.*'s findings, using more powerful scanning techniques than earlier studies, have unambiguously demonstrated that primary auditory cortex can be activated by silent speech-reading. Now what needs to be determined is the specificity of activation in A1 to speech-like events.

(ii) Speech-reading tends to generate left-lateralized or bilateral activation (e.g. Calvert & Lewis 2004; Capek *et al.* 2004). This is in contrast to the usual finding for other face actions, such as perception of gaze direction or facial expression, which tend to show more extensive right-lateralized activation.

(iii) The middle and posterior parts of the superior temporal gyrus, including the posterior superior temporal sulcus (pSTS), are reliably and consistently activated by silent speech-reading, and also by audio-visual speech. This region usually constitutes the principal focus of activation in fMRI studies of speech-reading (e.g. Calvert *et al.*

1997, 2000; Ludman *et al.* 2000; MacSweeney *et al.* 2002; Wright *et al.* 2003; Callan *et al.* 2004; Capek *et al.* 2004; Hall *et al.* 2005; Skipper *et al.* 2005).

(iv) (Left) pSTS can (but does not always) show differential activation for congruent and incongruent audio-visual speech. It can show supra-additive activation for (congruent) audio-visual speech compared with unimodal seen or heard speech (e.g. Calvert *et al.* 2000; Wright *et al.* 2003; Miller & D'Esposito 2005). Inhibitory (i.e. sub-additive) activation for audio-visual compared with unimodal input can be observed in other parts of the superior temporal gyrus (e.g. Wright *et al.* 2003), and for incongruent audio-visual pairings within pSTS. While the findings are variable, they are generally consistent with the idea that pSTS is a primary binding site for audiovisual speech processing.

(v) Inferior frontal regions, including Broca's region (BA 44/45), and extending into anterior parts of the insula, are activated by speech-reading. Often, watching speech generates greater activation in this region than observing other actions or listening to speech (e.g. Buccino *et al.* 2001; Campbell *et al.* 2001; Santi *et al.* 2003; Watkins *et al.* 2003; Ojanen *et al.* 2005; Skipper *et al.* 2005).

6.8 Time course of cortical activation

The time course of functional cortical activation has been studied primarily using scalp-recorded event-related potentials (ERP) and magnetoencephalography (MEG), both of which offer online methods of tracking brain events and have good temporal resolution. The results obtained to date suggest the following sequence of events, many of which are illustrated in Figure 6.2.

(i) Following unimodal processing within the relevant primary sensory cortices, superior temporal regions are implicated in binding the different sensory streams in speech events (Miller & D'Esposito 2005). So, for example, a distinctive electrophysiological signature for a McGurk audio-visual stimulus compared with an audio-visual congruent one is likely to reflect a signal source in superior temporal regions—within approximately 150 ms of auditory signal onset (e.g. Sams *et al.* 1991; Colin *et al.* 2002; Möttönen *et al.* 2004). One study using McGurk-type stimuli suggests that the amplitude and latency of the auditory evoked potential related to auditory identification (the N1/P2 complex) can be reduced when vision leads audition slightly (van Wassenhove *et al.* 2005). The better the visual event predicts the following auditory one, the smaller the auditory ERP. This waveform and its sensitivity to information in the visual stream probably arise within pSTS Additionally, a cortical correlate of auditory phonemic speech adaptation (N1), visible over left temporal regions, has been shown to be affected not only by an auditory, but also a silent speech-read adaptor token (Jääskeläinen *et al.* 2004; 2008)

(ii) Activation in pSTS extends posteriorly to the junction with the parietal lobe. Activation for silent speech-reading, like that for audio-visual speech, then extends anteriorly to inferior frontal regions (Nishitani & Hari 2002), via temporo-parieto-frontal junction activation (Figure 6.2).

(iii) Activation can also project from (secondary) multisensory sites, such as pSTS, to (primary) unimodal visual and auditory processing regions (back projection; Calvert *et al.* 2000). Back projection provides the most likely mechanism for activation in auditory cortex produced by silent speech-reading (see Calvert & Lewis 2004).

(iv) Activation in somatosensory cortex has been reported for silent speech-reading (Möttönen *et al.* 2005). This recent finding adds weight to the consideration of speech perception in terms of all of its multimodal properties.

I have suggested that speech-reading may be considered in terms of two different processing modes: one largely dependent on perceiving the configuration of the mouth, lips and tongue, which can provide complementary visual information to that in the auditory speech stream, and the other reflecting correlations between the kinematics of heard and seen speech and carrying useful informational redundancy in the multimodal speech stream. If this conceptualization is valid, then the different modes may have distinctive cortical activation characteristics. Studies exploring the effects of manipulating the display, and the response task, may offer further insight into this.

6.9 Posterior superior temporal sulcus: speech actions, but not all speech images

There are hints that pSTS is especially sensitive to dynamic aspects of seen speech. This might mean that visible speech information that drives activation in this region is related primarily to the dynamic aspects of the heard speech stream (correlated mode), rather than to the visibility of specific facial configurations (complementary mode). Callan *et al.* (2004) used spatial filtering to vary the amount of facial detail visible in spoken sentences, presented audio-visually in noise (speech babble). When fine spatial detail was accessible (natural and middle-pass filtered video), there was more activation in the middle temporal gyrus (MTG) than when the video was low-pass filtered. Under low-pass filtering, which

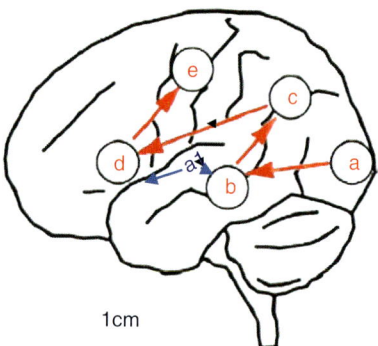

Figure 6.2. Schematic of the left hemisphere showing locations and activation sequence for the processing of visual speech (adapted from Nishitani & Hari (2002)). Participants in this MEG study of silent speech-reading identified vowel forms from videoclips. The following regions were activated, in sequence (a) visual cortex, including visual movement regions, (b) superior temporal gyrus (secondary auditory cortex), (c) pSTS and inferior parietal lobule, (d) inferior frontal and (e) premotor cortex. Auditory inputs (primary auditory cortex, A1) are hypothesized to access this system at (b).

reduced the visibility of facial detail, pSTS was activated, while under normal and middle-pass filtering both pSTS and MTG were activated.

Calvert & Campbell (2003) compared activation in response to natural visible (silent) speech and a visual display comprising sequences of still photo images captured from the natural speech sequence. Spoken VCV disyllables were seen. The still images were captured at the apex of the gesture—so for 'th', the image clearly showed the tongue between the teeth, and for the vowels, the image captured was that which best showed the vowel's identity in terms of mouth shape. The still series thus comprised just three images: vowel; consonant; and vowel again. However, the video sequence was built up so that the natural onset and offset time signatures of the vowel and consonant were preserved (i.e. multiple frames of vowel, then consonant and then vowel again). The overall duration of the still lip series was identical to that for the normal speech sample, and care was taken to avoid illusory movement effects as the vowel and consonant images changed, by using visual pink noise frames interleaved with those of the speech series. The visual impression was of a still image of a vowel (approx. 0.5 s), followed by a consonant (approx. 0.25 s), and again followed by a vowel. Participants in the scanner were asked to detect a consonantal target ('v') among the disyllables seen. Although pSTS was activated in both natural and still conditions, it was activated more strongly by normal movement than by the still image series. In a complementary finding, Santi *et al.* (2003) found that point-light-illuminated speaking faces—which carried no information about visual form—generated activation in pSTS. Finally, a recent study (Capek *et al.* 2005, in preparation) used stilled photo images of lip actions, each presented for 1 s, for participants in the scanner to classify as vowels or consonants. Under these conditions, no activation of pSTS was detectable at the group level—nor in individual scans (Figure 6.1). Images of lips and their possible actions are not always sufficient to generate activation of this region; pSTS activation requires that either visual motion be available in the stimulus or the task requires access to a dynamic representation (of heard or seen speech). This sole negative finding concerning the involvement of pSTS underlines its crucial role for most speech-reading and audio-visual perception. It strongly suggests that pSTS is especially involved in the analysis of natural speech whose visual movement characteristics correspond with auditory spectro-temporal features. However, the functional role of pSTS is not confined to multimodal or unimodal speech processing. It is activated extensively in the integration of biological visible form and motion (Puce *et al.* 2003). pSTS is activated when imitating or observing imitations of the actions of others (for review, see Buccino *et al.* 2001; Brass & Heyes 2005). pSTS is also activated by the presentation of learned, arbitrary audio-visual pairings (for discussion, see Miller & D'Esposito (2005)). This must inform theorizing concerning its role in visual and audio-visual speech perception.

6.10 Photographs of speech: more routes to visible speech processing and audiovisual integration?

pSTS was not activated when participants examined photographs to decide whether a vowel or consonant was being spoken. There must therefore be networks that can support some aspects of seen speech perception that are not directly reliant on pSTS. Where could these be? Stilled lip images generated relatively more activation than moving lips within primary visual areas V1 and V2 (Calvert & Campbell 2003). These primary visual regions project to the ventral visual system, including inferior temporal cortex. Projections radiate from this region to the middle and (anterior) superior temporal cortex. These parts of the temporal lobe support a range of associative processes—they are traditionally regarded as 'secondary association areas' for categorizing and associating inputs from the senses. These inferior and middle temporal regions of the ventral visual system may have been accessed in the case of the motion-blind patient, LM, who could identify speech sounds that were associated with isolated seen speech photographs, but was unable to identify natural visible speech, and showed no effect of vision on audition when presented with naturally moving McGurk stimuli (Campbell et al. 1997). LM had bilateral damage to the lateral occipital cortices, including area V5, which project to pSTS. Here is one means by which complementary information concerning the precise place of articulation may be made available to the cognitive system. It should also be noted that another multimodal region, the left inferior parietal lobule (IPL), a region dorsal to pSTS, which is often activated when people are engaged in segmental speech analysis (Scott 2005), was activated by seen speech in the experiment of Capek et al. (Figure 6.1). IPL may be accessed not only by projections from pSTS, but also from other projections, possibly from inferior frontal regions.

6.11 The broader picture

Current theorizing about auditory speech processing suggests two major processing streams emanating from primary auditory cortex to generate activity in the perisylvian regions (superior temporal and inferior frontal) of the left hemisphere (Scott 2005; Patterson & Johnsrude this volume). One stream runs anteriorly along the upper surface of the temporal lobe. This becomes increasingly sensitive to the semantic characteristics of the utterance (a 'what' stream). The other runs dorsally through the superior temporal gyrus to the temporo-parieto-frontal junction and is especially sensitive to the segmental properties of speech. This may constitute part of a 'how' stream, concerned with the specification of the segmental properties of speech in articulatory and acoustic terms. Both 'how' and 'what' streams project to inferior frontal regions including Broca's area, though through different tracts. One function for the posterior left-lateralized network as a whole, including its projections to frontal regions, may be to 'align' the segmental specifications of speech whether it is planned, produced or perceived (Hickok & Poeppel 2004). That is, the frontal component, related to the planned articulation of speech, and the temporo-parietal component, concerned with the acoustic specification of the speech segment, need to interact to develop representations of segmental speech forms. By contrast, the anterior stream, including its frontal projections, may be differentially specialized for analysing meaning in larger linguistic units (Thompson-Schill 2005).

The picture sketched earlier in this review suggested two modes whereby seen speech may influence the processing of heard speech. One was described as a complementary mode, making use of face information to distinguish speech segments by eye which may be hard to distinguish by ear. The other is a correlated mode, for which information in the visible speech stream that is dynamically similar to that in the auditory stream provides useful redundancy. My contention is that these are reflected not in completely discrete cortical processing systems, but rather in relatively differentiated access to two major streams for the processing of natural language—a 'what' and a 'how' stream. The 'what' stream makes particular use of the inferior occipito-temporal regions and the ventral visual processing stream, which can specify image details effectively. It can therefore serve as a useful route for complementary visual information to be processed. A major projection of this stream is towards association areas in middle and superior temporal cortex. To the extent that seen speech (whether alone or in combination with sound) activates language meanings and associations, it will engage these anterior and middle temporal regions, and possibly bilaterally rather than left-lateralized. This stream could be accessed effectively by still speech images. When the addition of vision to audition generates changes in meaning, especially at the level of the phrasal utterance, functional activation in these cortical regions could be differentially engaged. By and large, the process whereby visual and auditory characteristics are bound together within this route is primarily associative. That is, learned associations of images and sounds of speech are associated with a specific response pattern. Thus, in the study of Capek *et al.* (2005), the requirement to identify images of vowels and distinguish them from consonants may have used a 'general purpose' associative mechanism. The specificity of associative processing in this region is untested. It may be limited to object-based associations (glass shatters, ducks quack). In contrast to this, the 'how' stream for the analysis of auditory speech may be readily accessed by natural visible speech, characterized by dynamic features that correspond with those available acoustically. Processing that requires sequential segmental analysis (e.g. identifying syllables or words individually or in lists) will differentially engage this posterior stream. It is in this stream that the *correlational* structure of seen and heard speech is best reflected. The visual input to these analyses arises primarily in the lateral temporo-occipital regions that track visual movement, which project primarily to pSTS. pSTS has been shown to play a range of roles in intra- and intermodal processing, but the suggestion here is that it may have a crucial role in processing the supramodal dynamic patterns that characterize natural audio-visual speech by abstracting relevant features from both the visual and the auditory stream. One should be cautious, though, in predicting that this is the only network involved in audio-visual speech binding. Among other things, we do not yet know the extent to which the hypothesized posterior audio-visual stream is responsible for cross-modal integration of vision and audition in the perception of speech prosody, or for the interplay of vision and hearing in the perception of spoken discourse (but see Skipper *et al.* 2005).

When we learn to speak, our developing vocalizations tune to those that we hear—we imitate the vocal patterns of our language teachers. Clearly, hearing other talkers—and matching our own utterances to those we hear—is a crucial part of the development of speech. Yet how we do this, and the relative contributions of the perisylvian regions to the development of amodal representations that capture articulatory as well as auditory and other sensory features, remains mysterious. There is little doubt that our experiences of performing actions leave strong traces in the representations we use to perceive those actions. Following the classical studies of Meltzoff and colleagues, which showed that human neonates imitate visually observed mouth movements (Meltzoff & Moore 1983), the ability to imitate the visually observed actions of others has become the focus of studies that suggest that frontal cortical regions may contain mirror-neurone assemblies, specialized to respond to the perception of particular actions, as well as their planning and execution (Rizzolatti & Arbib 1998). Several studies now confirm that Broca's area within the inferior lateral left frontal lobe, classically understood to be involved primarily in the selection of speech acts for production, is especially active in processing seen speech (e.g. Buccino *et al.* 2001; Campbell *et al.* 2001; Watkins *et al.* 2003; Skipper *et al.* 2005), even when no overt speech action is required. The studies do not, however, 'prove' the mirror-neurone hypothesis, which places the primary perception-action link in a specific frontal region, and whose homology to Broca's area is uncertain. Rather, it seems that input from the primary visual sensory regions drives activation in specific temporal regions that are in turn connected to inferior parietal–inferior frontal circuits. Audio-visual and visual speech perceptions thus bring about cortical activation of action plans and sequences, as well as some somaesthetic consequences of speaking. Does the extent of inferior frontal activation then determine or constrain the *perception* of an audio-visual or a visual speech gesture? There are claims that this is so (Skipper *et al.* 2005; van Wassenhove *et al.* 2005), and there is no doubt that activation in Broca's area can be shown to play a distinctive role in speech-reading and audiovisual speech processing (Sams *et al.* 2005; Pekkola *et al.* 2006). However, whether such activation is a necessary component of speech perception is unproven. That said, to have non-auditory sense—by sight, by 'feel' and by articulatory knowledge—of both talk and the talker is a vital (although possibly not a sufficient) component of speech mastery.

Acknowledgements

I am grateful to Cheryl Capek for reading drafts of this paper, and to the Wellcome Trust, the Medical Research Council of Great Britain, the Royal Society of London and the Economic and Social Research Council for financial support related to the work reported here.

Endnotes

1 The McGurk effect is not illusory in the sense that it is a distortion of 'normal perception'. Its illusory nature lies in the specific and unique combination of visual (velar stop consonant ('k', 'g')) and auditory (bilabial stop consonant ('p', 'b')) inputs giving rise to an apparent alveolar stop consonant ('t', 'd'), which did not occur in either input stimulus. Massaro (1987), through many empirical studies, has shown that the McGurk effect can be accounted for within a broader pattern processing perspective. On Massaro's scheme, the effects of vision on audition reflect Bayesian rules for the combination of auditory and visual inputs, working at the level of phoneme identification information. That is, McGurk stimuli, while producing illusory identifications, nevertheless behave systematically with respect to *all* combinations of possible visual and auditory syllables to which the perceiver is exposed in the experiment. Identical principles apply to the combination of, for instance, heard and written syllables.
2 In itself, this finding militates against a completely motoric theory of audio-visual speech perception, for at the age of six months the child has no useful speech production abilities.

References

Alsius, A., Navarra, J., Campbell, R. & Soto-Faraco, S. S. 2005 Audiovisual integration of speech falters under high attention demands. *Curr. Biol.* **15**, 839–843. (doi:10.1016/j.cub.2005.03.046)
Andersson, U. & Lidestam, B. 2005 Bottom-up driven speechreading in a speechreading expert: the case of AA (JK023). *Ear Hear.* **26**, 214–224. (doi:10.1097/00003446-200504000-00008)

Auer Jr, E. T. & Bernstein, L. E. 1997 Speechreading and the structure of the lexicon: computationally modelling the effects of reduced phonetic distinctiveness on lexical uniqueness. *J. Acoust. Soc. Am.* **102**, 3704–3710. (doi:10.1121/1.420402)

Bernstein, L. E., Auer, E. T., Moore, J. K., Ponton, C. W., Don, M. & Singh, M. 2002 Visual speech perception without primary auditory cortex activation. *Neuroreport* **13**, 311–315. (doi:10.1097/00001756-200203040-00013)

Bernstein, L. E., Auer Jr, E. T. & Moore, J. K. 2004a Audiovisual speech binding: convergence or association? In *The handbook of multisensory perception* (eds G. A. Calvert, C. Spence & B. E. Stein), pp. 203–224. Cambridge, MA: MIT Press.

Bernstein, L. E., Auer, E. T. & Takayanagi, S. 2004b Auditory speech detection in noise enhanced by lipreading. *Speech Commun.* **44**, 5–18. (doi:10.1016/j.specom.2004.10.011)

Brass, M. & Heyes, C. 2005 Imitation: is cognitive neuroscience solving the correspondence problem? *Trends Cogn. Sci.* **9**, 489–495. (doi:10.1016/j.tics.2005.08.007)

Buccino, G. *et al.* 2001 Action observation activates premotor and parietal areas in a somatotopic manner: an fMRI study. *Eur. J. Neurosci.* **13**, 400–404. (doi:10.1046/j.1460-9568.2001.01385.x)

Burnham, D. & Dodd, B. 2004 Auditory–visual speech integration by prelinguistic infants: perception of an emergent consonant in the McGurk effect. *Dev. Psychobiol.* **45**, 204–220. (doi:10.1002/dev.20032)

Callan, D. E., Jones, J. A., Munhall, K., Kroos, C., Callan, A. M. & Vatikiotis-Bateson, E. 2004 Multisensory integration sites identified by perception of spatial wavelet filtered visual speech gesture information. *J. Cogn. Neurosci.* **16**, 805–816. (doi:10.1162/089892904970771)

Calvert, G. A. & Campbell, R. 2003 Reading speech from still and moving faces: the neural substrates of seen speech. *J. Cognit. Neurosci.* **15**, 57–70. (doi:10.1162/089892903321107828)

Calvert, G. A. & Lewis, J. W. 2004 Hemodynamic studies of audiovisual interaction. In *The handbook of multisensory perception* (eds G. A. Calvert, C. Spence & B. E. Stein), pp. 483–502. Cambridge, MA: MIT Press.

Calvert, G. A., Bullmore, E., Brammer, M. J., Campbell, R., Woodruff, P., McGuire, P., Williams, S., Iversen, S. D. & David, A. S. 1997 Activation of auditory cortex during silent speechreading. *Science* **276**, 593–596. (doi:10.1126/science.276.5312.593)

Calvert, G. A., Campbell, R. & Brammer, M. 2000 Evidence from functional magnetic resonance imaging of crossmodal binding in the human heteromodal cortex. *Curr. Biol.* **10**, 649–657. (doi:10.1016/S0960-9822(00)00513-3)

Campbell, R., Zihl, J., Massaro, D. W., Munhall, K. & Cohen, M. M. 1997 Speechreading in the akinetopsic patient. *Brain* **121**, 1794–1803.

Campbell, R., MacSweeney, M., Surguladze, S., Calvert, G. A., McGuire, P. K., Brammer, M. J., David, A. S. & Suckling, J. 2001 Cortical substrates for the perception of face actions: an fMRI study of the specificity of activation for seen speech and for meaningless lower-face acts (gurning). *Cognit. Brain Res.* **12**, 233–243. (doi:10.1016/S0926-6410(01)00054-4)

Capek, C. M., Bavelier, D., Corina, D., Newman, A. J., Jezzard, P. & Neville, H. J. 2004 The cortical organization of audio-visual sentence comprehension: an fMRI study at 4 Tesla. *Cognit. Brain Res.* **20**, 111–119. (doi:10.1016/j.cogbrainres.2003.10.014)

Capek, C. M., Campbell, R., MacSweeney, M., Woll, B., Seal, M., Waters, D., Davis, A. S., McGuire, P. K. & Brammer, M. J. 2005 The organization of speechreading as a function of attention. Cognitive Neuroscience Society Annual Meeting, poster presentation, San Francisco, CA: Cognitive Neuroscience Society.

Capek, C. *et al.* In preparation. Cortical correlates of the processing of stilled speech images—effects of attention to task.

Colin, C., Radeau, M., Soquet, A., Demolin, D., Colin, F. & Deltenre, P. 2002 Mismatch negativity evoked by the McGurk–MacDonald effect: a phonetic representation within short-term memory. *Clin. Neurophysiol.* **113**, 495–506. (doi:10.1016/S1388-2457(02)00024-X)

Driver, J. 1996 Enhancement of selective listening by illusory mislocation of speech sounds due to lip-reading. *Nature* **381**, 66–68. (doi:10.1038/381066a0)

Fant, G. *1960 Acoustic theory of speech production.* The Hague, The Netherlands: Mouton.

Fowler, C. A. & Dekle, D. 1991 Listening with eye and hand: crossmodal contributions to speech perception. *J. Exp. Psychol. Hum. Percept. Perform.* **17**, 816–828. (doi:10.1037/0096-1523.17.3.816)

Ghazanfar, A. A., Maier, J. X., Hoffman, K. L. & Logothetis, N. K. 2005 Multisensory integration of dynamic faces and voices in rhesus monkey auditory cortex. *J. Neurosci.* **25**, 5004–5012. (doi:10.1523/JNEUROSCI.0799-05.2005)

Grant, K. W., Greenberg, S., Poeppel, D. & van Wassenhove, V. 2004 Effects of spectro-temporal asynchrony in auditory and auditory–visual speech processing. *Semin. Hear.* **25**, 241–255. (doi:10.1055/s-2004-832858)

Green, K. P., Kuhl, P. K., Meltzoff, A. N. & Stevens, E. B. 1991 Integrating speech information across talkers, gender, and sensory modality: female faces and male voices in the McGurk effect. *Percept. Psychophys.* **50**, 524–536.

Hall, D. A., Fussell, C. & Summerfield, A. Q. 2005 Reading fluent speech from talking faces: typical brain networks and individual differences. *J. Cogn. Neurosci.* **17**, 939–953. (doi:10.1162/0898929054021175)

Hickok, G. & Poeppel, D. 2004 Dorsal and ventral streams: a framework for understanding aspects of the functional anatomy of language. *Cognition* **92**, 67–99. (doi:10.1016/j.cognition.2003.10.011)

Jääskeläinen, I. P. et al (2004) Adaptation of neuromagnetic N1 responses to phonetic stimuli by visual speech in humans. Neuroreport **15**, 2741–2744.

Jääskeläinen, I. P., Kauramäki, J., Tujunen, J. & Sams, M. (2008) Formant-transition specific adaptation by lipreading of left auditory cortex N1m. Neuroreport **19**, 93–97.

Jordan, T. R. & Sergeant, P. C. 2000 Effects of distance on visual and audiovisual speech recognition. *Lang. Speech* **43**, 107–124.

Ludman, C. N., Summerfield, A. Q., Hall, D., Elliott, M., Foster, J., Hykin, J. L., Bowtell, R. & Morris, P. G. 2000 Lip-reading ability and patterns of cortical activation studied using fMRI. *Br. J. Audiol.* **34**, 225–230.

MacSweeney, M. et al. 2002 Neural systems underlying British Sign Language and audio-visual English processing in native users. *Brain* **125**, 1583–1593. (doi:10.1093/brain/awf153)

Massaro, D. W 1987 *Speech perception by ear and by eye.* Hillsdale, NJ: Lawrence Erlbaum Associates.

McGurk, H. & MacDonald, J. 1976 Hearing lips and seeing voices. *Nature* **264**, 746–748. (doi:10.1038/264746a0)

Meltzoff, A. N. & Moore, M. K. 1983 Newborn infants imitate adult facial gestures. *Child Dev.* **54**, 702–709. (doi:10.2307/1130058)

Miller, L. M. & D'Esposito, M. D. 2005 Perceptual fusion and stimulus coincidence in the crossmodal integration of speech. *J. Neurosci.* **25**, 5884–5893. (doi:10.1523/JNEUROSCI.0896-05.2005)

Möttönen, R., Schurmann, M. & Sams, M. 2004 Time course of multisensory interactions during audiovisual speech perception in humans: a magnetoencephalographic study. *Neurosci. Lett.* **363**, 112–115. (doi:10.1016/j.neulet.2004.03.076)

Möttönen, R., Järveläinen, J., Sams, M. & Hari, R. 2005 Viewing speech modulates activity in the left S1 mouth cortex. *Neuroimage* **24**, 731–737. (doi:10.1016/j.neuroimage.2004.10.011)

Munhall, K. G., Jones, J. A., Callan, D. E., Kuratate, T. & Vatikiotis-Bateson, E. 2004 Visual prosody and speech intelligibility: head movement improves auditory speech perception. *Psychol. Sci.* **15**, 133–137. (doi:10.1111/j.0963-7214.2004.01502010.x)

Musacchia, G., Sams, M.,Nicol, T. & Kraus, N. (2006) Seeing speech affects acoustic information processing in the human brainstem. *Exp.Brain Res.* **168**, 1–10

Nishitani, N. & Hari, R. 2002 Viewing lip forms: cortical dynamics. *Neuron* **36**, 1211–1220. (doi:10.1016/S0896-6273(02)01089-9)

Ojanen, V., Möttönen, R., Pekkola, J., Jääskeläinen, I. P., Joensuu, R., Autti, T. & Sams, M. 2005 Processing of audiovisual speech in Broca's area. *Neuroimage* **25**, 333–338. (doi:10.1016/j.neuroimage.2004.12.001)

Pekkola, J., Ojanen, V., Autti, T., Jääskeläinen, I. P., Möttönen, R., Tarkiainen, A. & Sams, M. 2005 Primary auditory cortex activation by visual speech: an fMRI study at 3T. *Neuroreport* **16**, 125–128. (doi:10.1097/00001756-200502080-00010)

Pekkola, J., Laasonen, M., Ojanen, V., Autti, T., Jäskeläinen, I. P., Kujala, T. & Sams, M. 2006 Perception of matching and conflicting audiovisual speech in dyslexic and fluent readers: an fMRI study at 3T. *Neuroimage* **29**, 797–807. (doi:10.1016/j.neuroimage.2005.09.069)

Puce, A., Syngeniotis, A., Thompson, J. C., Abbott, D. F., Wheaton, K. J. & Castiello, U. 2003 The human temporal lobe integrates facial form and motion: evidence from fMRI and ERP studies. *Neuroimage* **19**, 861–869. (doi:10.1016/S1053-8119(03)00189-7)

Radeau, M. & Bertelson, P. 1974 The after-effects of ventriloquism. *Q. J. Exp. Psychol.* **26**, 63–71. (doi:10.1080/14640747408400388)

Reisberg, D., McLean, J. & Goldfield, A. 1987 Easy to hear but hard to understand: a lip-reading advantage with intact auditory stimuli. In *Hearing by eye: the psychology of lip-reading* (eds B. Dodd & R. Campbell), pp. 97–113. Hillsdale, NJ: Lawrence Erlbaum Associates.

Remez, R. E. 2005 Three puzzles of multimodal speech perception. In *Audiovisual speech* (eds E. Vatikiotis-Bateson, G. Bailly & P. Perrier), pp. 12–19. Cambridge, MA: MIT Press.

Rizzolatti, G. & Arbib, M. A. 1998 Language within our grasp. *Trends Neurosci.* **21**, 188–194. (doi:10.1016/S0166-2236(98)01260-0)

Rosenblum, L. D. & Saldaña, H. M. 1996 An audiovisual test of kinematic primitives for visual speech perception. *J. Exp. Psychol. Hum. Percept. Perform.* **22**, 318–331. (doi:10.1037/0096-1523.22.2.318)

Rosenblum, L. D., Johnson, J. A. & Saldaña, H. M. 1996 Point-light facial displays enhance comprehension of speech in noise. *J. Speech Hear. Res.* **39**, 1159–1170.

Rosenblum, L. D., Schmuckler, M. A. & Johnson, J. A. 1997 The McGurk effect in infants. *Percept. Psychophys.* **59**, 347–357.

Ross, L. A. et al. (2007) Impaired multisensory processing in schizophrenia: deficits in the visual enhancement of speech comprehension in noisy conditions. *Schizophrenia Res.* **97**, 173–183.

Sadato, N. et al. 2005 Cross modal integration and changes revealed in lipmovement, random-dot motion and sign languages in the hearing and deaf. *Cereb. Cortex* **15**, 1113–1122. (doi:10.1093/cercor/bhh210)

Sams, M., Aulanko, R., Hämäläinen, M., Hari, R., Lounasmaa, O. V., Lu, S.-T & Simola, J. 1991 Seeing speech: visual information from lip movements modifies activity in the human auditory cortex. *Neurosci. Lett.* **127**, 141–145. (doi:10.1016/0304-3940(91)90914-F)

Sams, M., Mottonen, R. & Sihvonen, T. 2005 Seeing and hearing others and oneself talk. *Cogn. Brain Res.* **23**, 429–435. (doi:10.1016/j.cogbrainres.2004.11.006)

Santi, A., Servos, P., Vatikiotis-Bateson, E., Kuratate, T. & Munhall, K. 2003 Perceiving biological motion: dissociating visible speech from walking. *J. Cogn. Neurosci.* **15**, 800–809. (doi:10.1162/089892903322370726)

Scott, S. K. 2005 Auditory processing—speech, space and auditory objects. *Curr. Opin. Neurobiol.* **15**, 197–201. (doi:10.1016/j.conb.2005.03.009)

Skipper, J. I., Nusbaum, H. C. & Small, S. L. 2005 Listening to talking faces: motor cortical activation during speech perception. *Neuroimage* **25**, 76–89. (doi:10.1016/j.neuroimage.2004.11.006)

Soto-Faraco, S., Navarra, J. & Alsius, A. 2004 Assessing automaticity in audiovisual speech integration: evidence from the speeded classification task. *Cognition* **92**, B13–B23. (doi:10.1016/j.cognition.2003.10.005)

Sumby, W. H. & Pollack, I. 1954 Visual contribution to speech intelligibility in noise. *J. Acoust. Soc. Am.* **26**, 212–215. (doi:10.1121/1.1907309)

Summerfield, A. Q. 1979 The use of visual information in phonetic perception. *Phonetica* **36**, 314–331.

Summerfield, A. Q. 1987 Some preliminaries to a theory of audiovisual speech processing. In *Hearing by eye* (eds B. Dodd & R. Campbell), pp. 58–82. Hove, UK: Erlbaum Associates.

Thompson-Schill, S. L. 2005 Dissecting the language organ: a new look at the role of Broca's area in language processing. In *Twenty-first century psycholinguistics: four cornerstones* (ed. A. Cutler), pp. 173–189. Hillsdale, NJ: Lawrence Erlbaum Associates.

Tiippana, K., Andersen, T. S. & Sams, M. 2004 Visual attention modulates audiovisual speech perception. *Eur. J. Cogn. Psychol.* **16**, 457–472. (doi:10.1080/09541440340000268)

van Linden, S. & Vroomen J. 2007 Recalibration of phonetic categories by lipread speech versus lexical information. *JEP Hum. Perc. Perf.* **33**, 1483–1494.

van Wassenhove, V., Grant, K. W. & Poeppel, D. 2005 Visual speech speeds up the neural processing of auditory speech. *Proc. Natl Acad. Sci. USA* **102**, 1181–1186. (doi:10.1073/pnas.0408949102)

Watkins, K. E., Strafella, A. P. & Paus, T. 2003 Seeing and hearing speech excites the motor system involved in speech production. *Neuropsychologia* **41**, 989–994. (doi:10.1016/S0028-3932(02)00316-0)

Wright, T. M., Pelphrey, K. A., Allison, T., McKeown, M. J. & McCarthy, G. 2003 Polysensory interactions along lateral temporal regions evoked by audiovisual speech. *Cereb. Cortex* **13**, 1034–1043. (doi:10.1093/cercor/13.10.1034)

Yehia, H. C., Kuratate, T. & Vatikiotis-Bateson, E. 2002 Linking facial animation, head motion and speech acoustics. *J. Phonet.* **30**, 555–568. (doi:10.1006/jpho.2002.0165)

7

Listening to speech in the presence of other sounds

C. J. Darwin

> Although most research on the perception of speech has been conducted with speech presented without any competing sounds, we almost always listen to speech against a background of other sounds which we are adept at ignoring. Nevertheless, such additional irrelevant sounds can cause severe problems for speech-recognition algorithms and for the hard of hearing, as well as pose a challenge to theories of speech perception. A variety of different problems are created by the presence of additional sound sources: detection of features that are partially masked, allocation of detected features to the appropriate sound sources, and the recognition of sounds on the basis of partial information. The separation of sounds is arousing substantial attention in psychoacoustics and in computer science. An effective solution to the problem of separating sounds would have important practical applications.
>
> **Keywords:** Auditory grouping, Auditory localization, Auditory masking, Auditory perception, Auditory scene analysis, Cocktail-party problem, Pitch perception, Speech perception

7.1 Perception of multiple sound sources

Despite early work by Colin Cherry (1953; 1954) drawing attention to the 'cocktail-party problem', all of the major strands of research on the perception of speech have largely ignored the problems caused by the fact that, most of the time, we listen to speech against a background of, often intense, irrelevant sounds. For example, acoustic phonetics (see the chapter by Diehl in this volume) has concentrated on mapping the relationship between acoustic features (such as formants) and linguistic categories (such as phonemic features or syllables), but has been much less concerned with how the acoustic features produced by the required talker might be extracted from a sound mixture. Similarly, psycholinguists have studied the processing of connected speech, but with little attention to the problem of how we follow the speech of a particular talker in a mixture of sounds. Such neglect is understandable given the complexity of the problem posed by recognizing the speech of even an isolated talker, yet the presence of additional sound sources – especially the speech of another talker – can have significant practical and theoretical implications.

The practical application of speech-recognition algorithms is limited by the problems raised by additional sounds. Speech-recognition algorithms, though increasingly successful in a naturally quiet environment (or one that has been artificially quietened with a noise-cancelling microphone), fail in the presence of background sound which does not trouble a human listener (Lippmann 1997; Sroka & Braida 2005). Additional sounds create problems for speech recognition in a number of ways. Simple strategies for automatic speech recognition, such as spectral template-matching, can be seriously disrupted by additional sounds, since the overall spectrum of the mixture differs substantially from that of the target sound. This disruption is especially marked for non-stationary masking sounds, which are difficult to remove by adaptive filtering. More analytical recognition strategies can also be disrupted, since parts of a sound can be masked by a background sound, and the background sound can also provide additional acoustic features that need to be discounted.

Another practical consequence of the presence of additional sounds involves hearing-impaired listeners who typically find noisy environments disproportionately difficult for understanding speech, whether they wear a hearing aid or not (Killion 1997; Moore 1998). A major reason for noise causing additional problems for hearing-impaired listeners is that their auditory-filter bandwidths are wider than normal as a result of outer hair cell loss (see Moore this volume). The wider bandwidths lead both to spectral smearing of auditory features and to a reduced signal-to-noise ratio in the presence of background noise. Similar problems are encountered by users of cochlear implants, again because of the limited frequency resolution available with implants (Clark 2003).

The problem of additional irrelevant sounds is not, of course, one that is specific to speech perception; it applies to the recognition of any sound in a natural environment. Our conscious experience of sound is typically of a number of separate sources, each with their appropriate location, pitch, and timbre, apparently unaltered by whatever sounds they occur with. These percepts are the result of a variety of remarkable processing strategies by the brain.

Two of the problems created by sound mixtures are detection and allocation. For speech masked by high-level noise the main problem is detection – actually hearing the component frequencies of the sound. For speech masked by, say, a single other talker, the additional problem of allocation arises: which detected components belong to which sound source. This chapter deals with the latter problem extensively in the section on auditory scene analysis. After both of these problems have been overcome, an additional problem arises – how to recognize a particular category of speech sound on the basis of only partial information (Cooke *et al.* 2001). This problem is not, of course, unique to speech; listeners' perception of timbre is influenced by implicit assumptions about what aspects of a target sound might be masked by concurrent sounds (McDermott & Oxenham 2008).

7.2 Listening to speech in noise

To return to Cherry's cocktail party: Plomp (1977) has calculated that, when everyone at a well-attended party talks at the same level, the speech of the attended talker at a distance of 0.7 m has a signal-to-noise (S/N) ratio of about 0 dB – the background is as intense as the target talker. Since speech fluctuates substantially in level, an average of 0 dB implies that many of the weaker sounds of the target speaker will be masked. Nonetheless, 0 dB is sufficient to give adequate intelligibility with normal pronunciation and redundancy for listeners with normal hearing (Miller 1947). With a steady masking sound in non-reverberant conditions, intelligibility can be well predicted using the Speech Intelligibility Index (SII) (ANSI 1997), which is calculated from a weighted sum of the contributions of different frequency bands according to their S/N ratio. If the speech environment is reverberant, the Speech Transmission Index (STI) (Houtgast & Steeneken 1973), which is based on the modulation transfer function, can be used to calculate the predicted intelligibility. However, neither the SII nor the STI has a principled basis for modelling the short-term properties of either speech or noise that are involved in perceptual separation, since these indices are based on the long-term average properties of the speech and noise (but see Rhebergen *et al.* 2005).

These measures all work best when the interfering noise is stationary, so that there are no pronounced temporal fluctuations in level or spectrum. But the speech of a single background talker is far from stationary. For masking sounds that are equated for their overall average level, speech is considerably more intelligible when in the presence of a different-sex talker than with a steady noise with the same overall level and spectral composition produced by adding together many different talkers (Miller 1947). Intermediate levels of intelligibility occur for noise that has a spectrum similar to the long-term average spectrum of speech, but whose instantaneous amplitude fluctuates like that of speech (Duquesnoy 1983; Festen & Plomp 1990; Peters *et al.* 1998) or for a competing speaker that is identical to or of the same sex as the target speaker (Drullman & Bronkhorst 2000; Stubbs & Summerfield 1990). Such differences between different maskers are partly attributable to the spectral and temporal gaps in speech and in modulated noise (for more details see the review by Bronkhorst (2000)), but also require an understanding of how the brain deals, in general, with the separation of simultaneous sound sources across frequency and time.

7.3 Auditory scene analysis

The general problem of how the brain is able to separate component sound sources in mixtures has received considerable attention from psychologists over the last 30 years under the general name of Auditory Scene Analysis – the title of an influential book summarizing the field (Bregman 1990). Bregman's book makes the case for the brain using two types of information to group together sound components that have originated from a common source: heuristics based on general properties of sound sources, and schematic knowledge about specific sounds. General properties such as the common onset and (for periodic sounds) the harmonic relations between frequency components from a single source can help partition sounds from different sources that occur simultaneously; other properties, such as continuity of pitch, timbre, overall level, and spatial location can help to track a single sound source across time. For a more recent review of such low-level auditory grouping, see Darwin and Carlyon (1995). Bregman also recognizes that, as well as general heuristics, the brain may employ schema-based grouping. Here, the knowledge about specific sounds that schemata contain can be used to select from a mixture those components that form a schematized sound. Schema-based selection might be particularly important in the perception of speech. For example, if two steady vowels are synthesized with the same fundamental frequency and played strictly simultaneously, listeners can still identify them to some extent. There are no general grouping cues to help them to separate the two sounds of the mixture, only the experience of hearing particular vowels, or perhaps more abstract knowledge about how vocal tracts can shape sound (Darwin 1984).

(a) Auditory scene analysis and speech

The relationship between auditory scene analysis and speech perception has been a contentious issue. One view builds on the idea that speech is perceived by special processing mechanisms (Liberman 1982; Liberman *et al.* 1967) which are quite separate from those involved in the perception of other sounds – the 'speech is special' view. Here, speech perception pre-empts auditory scene analysis, using speech schemata to cherry-pick from mixtures of sounds those components that can form speech (Mattingly & Liberman 1990; Whalen & Liberman 1987). By contrast, others (Bregman 1990; Darwin 1991) have argued that all sound input is indiscriminately subject to low-level grouping mechanisms in addition to more specialized schemata-driven mechanisms which, in the case of speech, could include language-specific phonetic or articulatory-based knowledge of the structure of speech sounds. A third view (Remez *et al.* 1994) specifically rejects Bregman's view of auditory scene analysis on the grounds that neither its low-level nor its schematic grouping mechanisms can account for the phonetic perception of sine-wave speech (see below). Rather, this view maintains that 'phonetic perceptual organization is achieved by specific sensitivity to the acoustic modulations characteristic of speech' (p. 129).

It is uncontentious that simple low-level grouping heuristics that are generally of some help in sound segregation are inadequate for putting together, into a single source, the varied sounds that make up the natural speech stream, as it rapidly switches between voiced, aspirated, and fricated sounds and silence. More extremely, they are also of limited, though not negligible (see Barker & Cooke 1999), use in explaining the phonetic perception of sine-wave speech (Remez *et al.* 1981; 1994). Sine-wave speech is synthesized by adding together three frequency-modulated (FM) and slowly amplitude-modulated (AM) sine waves that follow the frequencies and smoothed amplitudes of the first three formants (the resonant frequencies of the vocal tract – see Diehl's chapter in this volume), producing the speech equivalent of a line-drawing. Individually, each FM sine-wave component sounds like a whistle that rises and falls in pitch; it does not sound like speech. But together, the three whistles gain – in addition to their individual whistliness – a speech-like quality that renders them, after a little exposure or being told that it is actually speech, intelligible (Remez *et al.* 1981, 1994, 2008). Since sine-wave speech lacks harmonic structure and the frequency movements of the three sine waves are largely uncorrelated, additional criteria for grouping them are required. It may well be that both innate and acquired schematic knowledge about speech is mobilized in this task. Infants show a very early sensitivity to speech that is modified by their experience of their native language; see the chapter by Kuhl, this volume. Adults clearly differ in what they incorporate into their speech stream: speakers of Khoisan languages such as Xhosa hear clicks phonetically whereas English speakers tend not to.

The need for phonetically constrained knowledge in deciding what auditory features should be incorporated into the speech stream does not preclude more general low-level grouping principles from contributing. Indeed, as Bregman has remarked: 'There is no biological advantage in replicating all these processes for the private use of the speech system' (Bregman 1991, p. 606). A similar point is made by Barker: 'As speech developed, it was just another sound source in [the complex mixtures that arrive at the ear] to which the existing well-established listening systems could be applied' (Barker 2006, p. 298). This point is most clearly made when we switch our focus from the problem of uniting the elements of a single sound source to the more ecologically valid one of sorting mixtures of multiple sound sources. As we document in greater detail later (see also Darwin & Carlyon 1995), there is extensive evidence that simple properties, such as harmonicity and onset-time which are used by the auditory system to segregate non-speech sounds, are also helpful in the difficult task of separating simultaneous talkers.

The limitations of specifically phonetic processes to separate multiple sound sources, however, are demonstrated first by the relatively poor performance of listeners asked to perceive one sine-wave (rather than natural) speech utterance in a mixture with another such utterance (Barker & Cooke 1999), although, as with the well-known camouflaged dalmatian dog, when the nature of the beast is pointed out, all is clear. Similar difficulties, but with natural speech, are found for users of cochlear implants, whose coding gives impoverished harmonic structure. Second, sine-wave speech, like a visual cartoon, probably makes its perceptual impact because it makes explicit perceptually useful abstractions of speech (i.e. formant frequencies) which may be only implicit in the natural signal. The frequency of the first formant must, in a natural signal, be inferred from the relative amplitudes of the harmonics around the formant peak; this is a non-trivial operation with simultaneous talkers, and one that has been shown to be very substantially aided by grouping by harmonic structure (Culling & Darwin 1993; Bird & Darwin 1998) and grouping by onset-time (Darwin 1984). These examples are unlikely to be due to speech-specific mechanisms, since the low-level grouping cues can lead to the percept of a less natural speech sound than pre-emptive phonetic grouping would have provided. In addition, some apparent grouping effects such as those involving onset time may actually be due to auditory coding mechanisms at the level of the cochlear nucleus (Roberts & Holmes 2006; 2007). Whether these coding mechanisms exist solely to help the separation of sounds or have additional functional significance is an interesting question.

A rather different approach to the problem of the relationship between auditory scene analysis and speech perception comes from work on the difficult problem of automatically recognizing speech when it is present in a background of other sounds. Similar issues to those raised in the psychological literature also appear here with a contrast between model-based approaches (Varga & Moore 1991) to recognizing simultaneous speech and those to incorporating low-level grouping. A thoughtful and psychologically sophisticated examination of the complex issues that arise in trying to link an auditory scene analysis front-end with speech-recognition algorithms is given by Barker (2006). As such attempts mature, they should provide an intellectually rigorous framework for characterizing the way that different types of knowledge about sounds and about speech can usefully interact.

The related question of which areas of the brain respond selectively to speech sounds has been addressed by a number of imaging studies. For example, a recent review claims that 'speech perception emerges from the connectivity between (generic) auditory areas and ... frontal lobe regions' (Price *et al.* 2005). For more details of this flourishing new area see the chapter in this volume by Patterson and Johnsrude.

(b) Intermediate representations in speech

Speech perception is remarkably resistant to distortion. For instance, speech remains moderately intelligible under extremes of high-pass or low-pass filtering, infinite peak clipping, or the addition of high levels of noise. It is also remarkably intelligible when its spectral information is reduced either to three FM sine-waves, as described above for sine-wave speech, or to only four broad frequency channels – each excited by a noise that is slowly amplitude modulated to match the energy distribution in those bands of the original speech (Shannon *et al.* 1995) – a transformation that is similar to that produced over a larger number of frequency channels by modern cochlear implants. More bizarrely, thanks to the automatic separation of the contributions to speech of the larynx (source) and the vocal tract (filter) made possible by linear-predictive coding (Atal & Hanauer 1971), the buzz and hiss of the larynx can be replaced by an arbitrary broadband signal – such as an orchestra – and the speech articulated by the moving vocal tract can then be heard as a ghostly modulation of the orchestral sound (Hunt *et al.* 1989).

Such resistance to distortion arises through speech being redundant at many levels, and the perceptual system being adept at perceiving sound that has been filtered and masked by its environmental context. At the acoustic–phonetic level, there are many sufficient acoustic cues to a phoneme but no necessary ones (see Diehl, this volume). Moreover, in interpreting a particular sound, listeners are able to compensate for the way that a sound has been filtered in travelling from its source to the listener (Darwin 1990; Watkins & Makin 1994). The sheer variability of sounds that can be effectively perceived as speech makes it very difficult to discover what intermediate representations of sound are used in perceiving speech.

Although formants are used to describe the way that sounds are synthesized in speech-perception experiments, and so have become a *de facto* acoustic description of at least the vocal portion of speech, their status as a perceptual entity is surprisingly vague. On the one hand, formant frequencies do seem to be perceptually salient: vowels change their perceived category when formant frequencies are changed, but usually only change their general timbre (e.g. becoming more or less muffled) when formant amplitudes are changed (Klatt 1985). For example, the amplitude of the second formant in a two-formant synthesis of a front vowel can be reduced by as much as 28 dB without the phonetic identity of the vowel being changed (Ainsworth & Miller 1972). On the other hand, listeners find it surprisingly hard to match single-formant sounds according to their formant frequency when they differ in fundamental frequency (Dissard & Darwin 2000; 2001) and when the harmonics close to the formant peak are resolved (see Moore, this volume). In addition, formants are notoriously difficult to track automatically in a way that is reliable enough to be the sole metric for recognition, and so speech-recognition algorithms tend to use spectral metrics that are more global than formant frequencies, such as cepstral coefficients (derived from a fourier transform of the log power spectrum), despite the fact that formants are more resistant to the overall changes to the spectrum that occur in natural listening situations (Hunt 1987).

There is, thus, no consensus about the form of the auditory information that is used to access the brain's acoustic–phonetic knowledge in order to categorize the speech sounds that we hear. Whatever form it takes, it is likely to be in a source-specific form. It is very unlikely that the brain would store information about, say, the vowel /a/ mixed with each of the background sounds with which we have ever encountered it. Rather, the stored description of /a/ must capture salient properties that allow us to recognize it when it is mixed with other sounds. The frequencies of spectral peaks, such as formants, since they represent the locally highest energy, will be the most likely to escape masking by other sounds. It seems likely that one of the functions of the early stages of auditory processing is to get the incoming sound mixture into a form where it can sensibly make contact with stored source-specific information (Gutschalk *et al.* 2005).

7.4 Separating speech signals

(a) The nature of the problem

One interesting property of speech is that, on a frequency–time plot such as a spectrogram, it is a sparse signal, which mostly consists of discrete harmonics, whose amplitudes vary gradually with frequency, being maximal near the formant peaks. There are also periods of quieter noise and silence, or near-silence, e.g. when the vocal tract is closed during production of a stop-consonant. Consequently, when two different speech signals are mixed together with similar overall levels, any particular local frequency–time region is substantially dominated by one signal or the other. Such dominance is exaggerated by the fact that the auditory system codes intensity approximately logarithmically, since if a>>b then $\log(a + b) \sim \log(a)$. In other words, in a mixture of two speech signals, each local frequency–time region will predominantly reflect the value of one of the speech signals. Put another way, the log-amplitude spectrogram of the mixture is almost the same as the (frequency/time) element-wise maximum of the component spectrograms (Moore 1986; Varga & Moore 1991). If knowledge of the two signals is used to extract just those frequency–time regions which are dominated by one particular signal using an 'ideal binary mask' (Hu & Wang 2004), then a good, intelligible version of that signal can be resynthesized from just those regions (Cooke 2003; Roweis 2004). Such independent knowledge is not, of course, available to someone listening to a mixture of two voices, or to a computer program attempting to separate a mixture. Both the listener and the program must use whatever knowledge they have in order to attempt to identify which frequency/time regions come from the signal of interest (Wang 2004).

The problem of separating two talkers is, thus, very different from the problem of trying to listen to one talker against a background of steady noise. Such noise provides a level of masking that is constant over time (apart from statistical fluctuations) in each frequency band. Consequently, only the most intense parts of the signal will be detectable against this background, and the perceptual (or computational) problem is one of detection. Anything that is detectably different from the background belongs to the signal. Thus, with a background of noise the problem is one of detecting parts of the signal, whereas with two talkers the problem is one of assigning readily detectable frequency/time elements to the appropriate sound source. This difference has been characterized as the difference between energetic masking (see Moore, this volume) and 'informational' masking (Durlach et al. 2003; Watson 1987), although probably all the phenomena covered by the latter term are more enlighteningly described as failures of auditory scene analysis or attention. Experiments which have formally assessed the intelligibility of speech produced by the ideal binary mask from a multi-talker background (Brungart 2005) have shown that energetic masking plays a very small role when there is only a single competing talker, but that its role increases rapidly when additional interfering voices are added. In the limit, a multi-talker babble becomes very similar to noise that has the same average spectrum as speech; the sparseness of the single interfering talker is lost.

Although speech masked by a single talker can be convincingly resynthesized by attributing each time/frequency element to either the target or the masking speech, the auditory system does not always conform to this principle of 'disjoint' allocation. Listeners can 'conjointly' allocate the energy in a particular frequency channel to more than one sound source – an application of Bregman's (1990) 'Old + New' heuristic. This heuristic states that we try to maintain perceived continuity of an initial sound when a new sound starts. What is then heard as the new sound corresponds to the total sound that is now present *minus* the old sound. Complementing the additivity of sound mixtures, this Old + New heuristic is subtractive within a frequency channel, so that, e.g. when a noise burst alternates with a more intense one, not only does the less intense one sound continuous, but the loudness of the more intense one is reduced compared with its loudness in isolation (McAdams *et al.* 1998; Warren *et al.* 1972). A similar effect can be shown on the timbre of complex sounds when part of the complex precedes the remainder (Darwin, 1995). Although the precise metric used for such subtraction is not clear, what is clear is that a single frequency channel is not exclusively allocated to one source or another. A minimalist demonstration of non-exclusive allocation can be made with a single frequency. Play the same sine-wave at equal amplitude to each ear, and then put a pulsed increment just on the left ear. You hear a continuous steady tone in the middle of the head, with an additional pulsing tone at the same frequency on the left ear. Incidentally, this demonstration is not consistent with Kubovy's (1981) notion that for two sounds to be heard they must differ in frequency (an 'indispensible attribute' of a sound source).

(b) Auditory grouping and the perception of speech

The perception of noisy speech has been characterized as 'glimpsing' (Miller & Licklider 1950), where the listener is given occasional, relatively undistorted views of the speech scattered across the frequency–time plane (Assmann & Summerfield 2004; Cooke 2003). The auditory system seems to be well adapted to deal with such glimpses since it can tolerate the temporal interruption of speech by noise, even asynchronously in different frequency channels (Buss *et al.* 2004; Carlyon *et al.* 2002; Howard-Jones & Rosen 1993) as illustrated in Figure 7.1. Intelligibility is improved if the interrupting noise is intense enough to be capable of masking the speech that it replaces, especially if what is being said is relatively predictable (Cherry & Wiley 1967; Miller & Licklider 1950; Powers & Wilcox 1977; Verschuure & Brocaar 1983). The presence of such noise also leads to the illusion that the speech is continuous and to an inability to identify which speech sounds are physically present, and which are replaced by noise (Samuel 1981; Warren 1970, 1984; Warren *et al.* 1972).

As we saw in the previous section, the perceptual problem is more complex when speech is heard at the same time as a small number of other talkers. How are the 'glimpses' from a particular talker perceptually separated from those coming from other talkers? This complex field brings together the research areas of speech perception and auditory scene analysis. The present chapter can only deal with some of the issues; for a more detailed discussion, the interested reader is referred to two excellent recent reviews (Assmann & Summerfield 2004; Bronkhorst 2000).

It is convenient to follow Bregman's (1990) conceptual structure by distinguishing between simultaneous and sequential grouping of speech glimpses. Simultaneous grouping determines whether frequency–time elements that are presented at the same time belong to the same sound source, whereas sequential grouping refers to the process of following a particular sound source across time. Such a distinction may be a simplification, since it may be useful for the listener to track, across time, an individual element such as a harmonic, rather than an entire sound source.

(i) Harmonicity

Much of speech is voiced, and normal voicing results in a quasi-periodic signal that shows considerable harmonic structure with a perceptible pitch corresponding to the fundamental frequency (F0) (see chapter by Moore, this volume). This harmonic structure is used by the auditory system to help separate simultaneous sounds. For example, two steady, synthetic simultaneous vowels are more intelligible if they are played at slightly different F0s than if they are played at the same F0 (Assmann & Summerfield 1989, 1990;

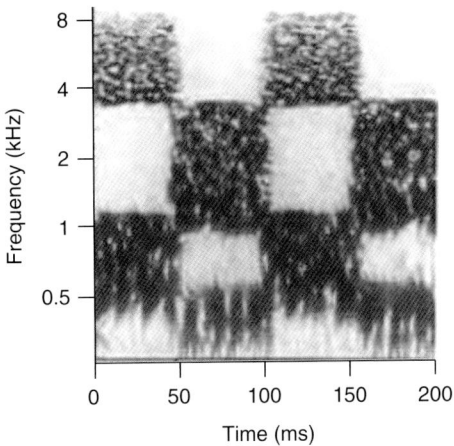

Figure 7.1 Spectrogram of 'checkerboard' noise with four channels logarithmically spaced in frequency. Redrawn from Howard-Jones & Rosen (1993).

Scheffers 1979, 1983). Although identification is well above chance when the sounds have the same F0, it increases by about 20% as the F0-difference increases to one semitone, but then asymptotes with further F0 increases. Although, this result stimulated subsequent research into both the psychology and physiology (Palmer 1988) of the perception of simultaneous sounds, the likely explanation of the intelligibility increase at very small F0s (~1 semitone) is not one that would be very useful for more natural sounds. When two harmonic sounds differ in F0 by, say, a semitone, corresponding pairs of harmonics are too close together in frequency (about 6% separation) to be resolved by the cochlea (see chapter by Moore, this volume). Since the two corresponding harmonics excite essentially the same region of the basilar membrane, its vibration reflects the physical addition of the sounds to give beats – relatively slow fluctuations in the level of the summed sound. It is likely that in a sufficiently long sound (Assmann & Summerfield 1994) these slow fluctuations allow the perceptual system to 'glimpse' one vowel or the other at different instants, as the spectral envelope systematically changes (Culling & Darwin 1993, 1994). Natural speech is not sufficiently stationary to benefit from this mechanism. Indeed, if the F0 of natural speech is manipulated to give consistent differences between the F0s of two competing speakers, intelligibility shows relatively little improvement at the 1-semitone F0 difference at which the double-vowel data asymptote (Bird & Darwin 1998; Brokx & Nooteboom 1982). However, there are two other reasons why a difference in F0 improves the intelligibility of natural speech in the presence of a competing talker (Bird & Darwin 1998).

One reason is that a difference in F0 improves the definition of the first-formant (F1) frequencies of the two speakers compared with when the F0s are the same. The upper panel of Figure 7.2 superimposes the spectra of two different vowels that have different F0s. The spectrum of each vowel consists of a harmonic series, and the peak of the amplitude envelope of each harmonic series corresponds to the vowel's F1 frequency. Provided that the perceptual mechanism is able to separate the two harmonic series in the upper panel sufficiently well, the lower F1 could be recovered. The lower panel of Figure 7.2 shows the spectrum of the sum of the two vowels when they have the same F0.

Notice that the spectral envelope of the summed sound does not have a peak corresponding to the lower of the two F1s and so it would be perceptually invisible.

The other reason, which at least for male voices only applies for F0 differences greater than about four semitones (Bird & Darwin 1998), is that a common harmonic series helps to group together the different formant regions that make up a voiced sound such as a vowel (as proposed by Broadbent & Ladefoged 1957) and to separate them from sounds on different F0s (Darwin 1981; Darwin & Gardner 1986; Gardner et al. 1989). When the F0 of the voice is high, as for the female voice, mechanisms based on the presence of harmonically related frequencies can group the lower formants (Darwin 1992), whereas lower-pitched male voices rely more on mechanisms that use the amplitude modulation that represents the pitch of unresolved harmonics (Schouten, 1940; see also chapter by Moore, this volume).

(ii) Onset-time
Frequency components from a single sound often share a common onset-time, and this property is used by listeners to group frequency components together in the perception

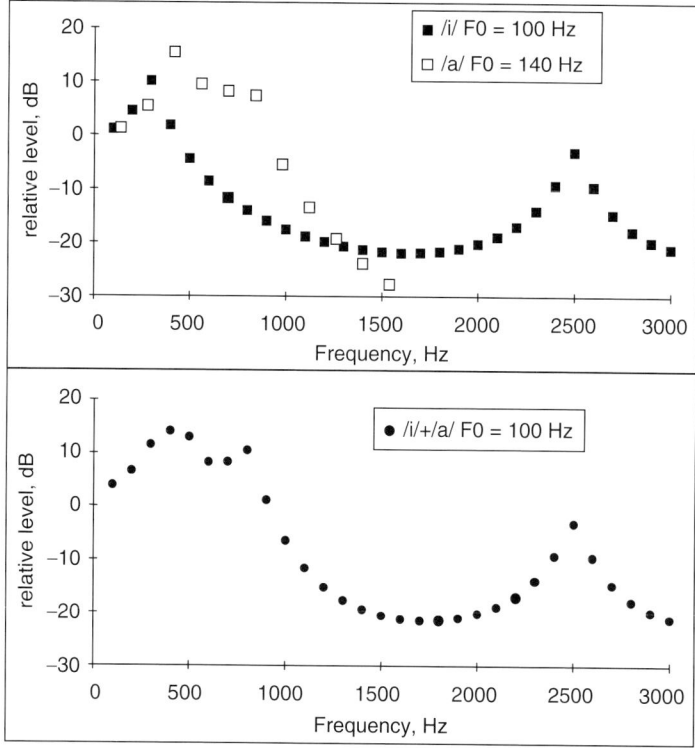

Figure 7.2 Schematic magnitude spectrum showing levels of individual harmonics in a mixture of vowel sounds. In the upper panel, the two vowels have different fundamental frequencies, forming two different harmonic series. In the lower panel, the two vowels are on the same fundamental frequency so only a single harmonic series is present. Note that the different first-formant frequencies of the two vowels are not well represented in the lower panel.

of timbre (Bregman & Pinker 1978) including vowel quality in steady vowels and simple syllables (Darwin 1981, 1984). For example, if the harmonics of a complex tone start at different times – say 100-ms apart, the frequencies of the individual harmonics are clear, but the timbre of the complex is less clear than if they had all started simultaneously. Similarly, if one of the harmonics of a steady vowel starts earlier than the others, it makes less of a contribution to the vowel quality than if it had been simultaneous. Such effects may be an advantageous consequence of the way that sounds are coded in auditory nerve fibres, which are known to adapt rapidly to steady sounds (Chimento & Schreiner 1991; Smith 1979; Yates *et al.* 1985). Rapid adaptation will reduce the neural response to a leading sound by the time the remaining sounds start. That this adaptation might not be a complete explanation was indicated first by offset asynchrony producing a similar, though somewhat smaller effect (Darwin & Sutherland 1984; Roberts & Moore 1991), and second by the fact that the contribution of the leading sound could be partly reinstated by making just its leading portion group with another sound that started with the leading sound but stopped when the remaining sounds started (Darwin & Sutherland 1984). However, this grouping explanation has itself been challenged recently, and an alternative explanation proposed, based on across-frequency interactions between sounds in the cochlear nucleus (Roberts & Holmes 2006, 2007). The relative contribution of relatively peripheral coding mechanisms and more central cognitive grouping mechanisms to the problem of sound-source separation is likely to remain a topic of interest for some time.

Onset-time also provides a useful grouping cue for connected speech. For example, it has been used in computational auditory scene analysis, along with harmonicity, to group together frequency components from one of two harmonic sound sources (Cooke 1993) and has also been found effective at grouping together the frequency-modulated sine waves of sine-wave speech (Barker & Cooke 1999). Its usefulness in connected speech is due to the interruptions to the continuous speech signal associated with stop and voiceless consonants. In the absence of onset-time differences between two voices, there is more scope for pitch-difference cues to be effective. This increased scope is probably the reason why the improvement in intelligibility with increasing difference in F0 of two simultaneous sentences is very much larger than normal when the sentences mainly contain continuant consonants (compare Brokx & Nooteboom 1982 and Bird & Darwin 1998).

(iii) Spatial direction

Different natural sound sources usually come from different directions in space, and localization cues have been used quite extensively for the machine segregation of different talkers (Bodden 1996). There are various different advantages that accrue to the listener from having sounds come from different directions (for a recent review, see Bronkhorst 2000); however, the way that the brain uses directional cues in auditory grouping is not straightforward, and reflects the problems that difficult listening environments provide for sound localization.

The intelligibility of speech masked by a single continuous noise is higher when the speech and the noise come from different directions. The two main reasons (Plomp 1976) are:

(1) for the higher frequency components, the head casts an acoustical shadow which can benefit the ear on the side of the speech and

(2) the different relative phases at the two ears of the low-frequency components of the speech and noise make the speech easier to detect (the so-called binaural masking level difference (Licklider 1948)).

The relative contribution of these two mechanisms to intelligibility depends on the type of speech material used – the former being more important for monosyllables, and the latter for spondees (Dirks & Wilson 1969). With speech, rather than noise, as the interfering sound, masking is reduced because of the sparsity of the speech signal (see above), but there are the additional problems of simultaneously grouping together the components that make up a particular speech sound, and of tracking one of the sources across time.

The dominant cue for localizing speech signals in the horizontal plane (or azimuth) is the inter-aural time differences (ITDs) of the sound's low-frequency (<1.5 kHz) components (Wightman & Kistler 1992), at least in non-reverberant environments. Curiously, though, most listeners are very poor at selectively grouping together simultaneous frequency components on the basis of common ITDs. Culling and Summerfield (1995) constructed four different vowel-like sounds by pairing in different combinations four different narrow-bandpass noises close in frequency to the first two formant frequencies of each vowel. For example, bands 1 and 4 together give a percept of /i/ while 2 and 3 together give /a/. The alternative grouping of 1 with 3 and 2 with 4 gives two different vowels. When one pair of noise bands (say 1 and 4) was played to one ear and the other pair to the other ear, listeners had no difficulty in identifying the vowel on the left ear. However, when the noise bands were all played to both ears but given different ITDs, listeners were, surprisingly, unable to identify the vowel heard on the left side.

We have, then, the apparently paradoxical result that the cue that is dominant for the localization of complex sounds is quite ineffective for the grouping of simultaneous sounds. This unexpected finding also seems to go against one's introspective experience of attending spatially to one sound source rather than another. Is this simply an illusion, although one supported by experimental evidence (Munte *et al.* 2001; Spence & Driver 1994)? Or is the relationship between spatial attention and auditory localization more complex than implied by the simple notion of attending to frequency channels that share the same ITD?

Evidence supports the more complex relationship. The basic idea is that auditory grouping (based, at least, on primitive cues such as harmonicity and onset-time) occurs prior to the localization of complex sounds. According to this idea, which was initially proposed by Woods and Colburn (1992) and subsequently pursued by Hill and Darwin (1996), the inter-aural time (and intensity) differences of individual frequency channels are computed independently, in parallel with a separate process which assigns these frequency channels to separate sound sources. The two types of information are then brought together so that the localization of an auditory object can be constructed from the ITDs of the frequency components that make up that auditory object. In the case of Culling and Summerfield's noise bands, there are no grouping cues other than ITD to pair off the bands, and consequently there is no segregation into two perceived vowels. By contrast, if the members of a pair of ordinary voiced vowels with a small difference in F0 are additionally given different ITDs, identification of the vowels improves (Shackleton *et al.* 1994).

Subsequent work on this lack of grouping by ITD has qualified the original conclusion. With practice, some listeners *can* learn to perform segregation by ITD (Darwin

2002; Drennan *et al.* 2003), though it is not clear how well this ability generalizes outside of the set of sounds that the listeners have been exposed to. Another important qualification is that if one uses more complex, natural sounds, rather than steady-state vowel-like sounds, listeners can much more easily identify two spatially separate sounds simply on the basis of ITD cues. Darwin and Hukin (1999) showed that natural sounds – modified to have the same F0, and that differed only in ITD – could be selectively attended very easily. For example, if two simultaneous monosyllabic words ('bead' and 'globe') are given ITDs of +90 μs and −90 μs, respectively, they are readily perceived as two spatially distinct auditory objects which can be readily attended to. For natural speech, there are many cues (e.g. harmonicity, onset-time differences) which can help the auditory system to allocate individual frequency channels to the two different sound sources. What is more surprising is that the impression of two separate objects survives when the two words are resynthesized on the same F0. For simpler sounds, such as steady vowels, the impression of two distinct sources with separate locations is destroyed by a common F0.

The upshot of these experiments (see also Best *et al.* (2007)) is to provide broad support for the model outlined by Woods and Colburn (1992). ITD is a surprisingly weak cue to segregation, at least for unpractised listeners and in the absence of other grouping cues. When the listener has some independent way of grouping together the frequency components that make up different sound sources, then ITD differences between the sources give improved identification. One reason why the auditory system may not use ITD as a strong primary grouping cue is that in normal, difficult listening situations (with multiple sound sources, sound reflections, and reverberation) the ITD information in a single frequency channel is not robust. The auditory system may be able to produce a more stable estimate of the location of sound sources by pooling ITD information across those frequencies, and only those frequencies (Hill & Darwin 1996), that make up a sound source. Bearing out this point is the interesting observation that the perceived location of sound sources is very stable: e.g. the high- and low-frequency parts of a sound never appear to come from different directions. Such spatial fragmentation might be a more common experience if the brain did use ITD as a primary grouping cue.

Once an auditory object is spatially localized, this location provides a powerful cue for tracking it across time. Spatial cues become particularly important if other cues such as semantic redundancy (Treisman 1960), pitch and voice differences (Darwin *et al.* 2003; Darwin & Hukin 2000), or level (Brungart 2001; Egan *et al.* 1954) are ineffective.

The perceptual separation of two interleaved melodies by spatial differences is strikingly illustrated in Bregman and Ahad's demonstration of West African xylophone (amadinda) music (1995, demonstration 41). Experimental data illustrating a similar effect of spatial segregation (using interleaved monotone rhythms) is provided by Sach and Bailey (2004), who also distinguish between the effects of perceived location and the effect of individual cues. They traded-off ITD and inter-aural level cues to show that perceived spatial separation, rather than an ITD difference itself, is responsible for the perceptual segregation of the two rhythmic streams.

An ingenious experiment by Freyman *et al.* (1999) led to a similar conclusion concerning the intelligibility of speech masked by similar speech. They exploited the precedence or Haas effect (Haas 1972) to produce the impression that sound sources were coming from different azimuthal directions without using either of the conventional binaural cues (ILD or ITD). They measured listeners' identification of nonsense sentences spoken by a female talker in the presence of either speech-spectrum noise or of similar sentences spoken by a second female talker. They used two spatial arrangements: one where the target and masker both came from straight ahead, and one in which the distractor also came from a second loudspeaker, 45° to the right, but its signal led the distractor signal in the straight-ahead loudspeaker by 4 ms. Thanks to the precedence effect, the latter configuration gives the clear impression that the distractor is coming from the right-hand loudspeaker, while the target remains straight ahead. This spatial separation substantially improved performance when the distractor was the other female talker. In the absence of other cues (the female talkers were similar, and the sentences were nonsense), the perceived spatial difference allowed the listener to selectively follow the target talker. However, the additional version of the distractor from straight ahead complicates the ITDs and intensity differences between the targets and the distractors so that there is little binaural masking level difference. Consequently, the impression of a large spatial difference produced by the precedence effect produced little intelligibility increase with the speech-spectrum noise. Freyman's experiment thus demonstrates that a perceived spatial difference between talkers can increase intelligibility under conditions where there is no energetic masking advantage from the spatial separation.

Although, as we have seen, the natural spatial separation of attended and unattended sources improves detection and tracking of the attended source, a spatially separated distractor sound *can* cause a remarkable amount of disruption to selective attention. Brungart and Simpson (2002) asked listeners to respond to a target speech signal spoken by one of two competing talkers in one ear while ignoring a simultaneous masking sound in the other ear. When the masking sound in the unattended ear was noise, listeners were able to segregate the competing talkers in the target ear nearly as well as they could with no sound in the unattended ear. But when the masking sound in the unattended ear was speech, speech segregation in the target ear was very substantially worse than with no sound in the unattended ear. The presence of speech-like (Brungart *et al.* 2005) sounds in the unattended ear makes the separation of sounds in the attended ear much harder.

7.5 Concluding remarks

The difficult problem of how speech can be recognized against a background of other sounds is receiving increased attention from both psychologists and computer scientists. Speech can be remarkably well recognized by human listeners under a wide variety of distortions and against both random and structured noise backgrounds in anechoic and reverberant surroundings. The brain uses a wide range of perceptual mechanisms to achieve a level of recognition that is presently beyond computer systems. Some of these mechanisms have probably arisen as a response to the general problem of recognizing sounds in noisy and reverberant environments, while others may rely on specific knowledge about speech. The types of mechanisms involved vary widely depending on the characteristics of the noise. For random noise, the problem is mainly one of detection and also requires recognition mechanisms that can operate on the basis of partial information, tolerating 'missing data'. For more structured noise, such as another talker, additional problems arise of allocating sensory fragments to one or other sound source, and of tracking an individual sound source over time. The effectiveness of various parameters in allowing this perceptual grouping has begun to be studied, although this knowledge has not yet been integrated into the mainstream of work on speech perception. One particular gap in our knowledge is an understanding of what the intermediate representations of sound between the sensory coding of the auditory nerve and the human brain's representation of phonetic knowledge could be. It is likely that those representations will be intimately linked to the problem of how the brain deals with multiple sound sources.

References

Ainsworth, W. A. & Miller, J. B. 1972 The effect of relative formant amplitude on the identity of synthetic vowels. *Lang. Speech* **15**, 328–341.
ANSI 1997 ANSI S3.5-1997, Methods for calculation of the Speech Intelligibility Index. New York: American National Standards Institute.
Assmann, P. F. & Summerfield, A. Q. 1989 Modelling the perception of concurrent vowels: vowels with the same fundamental frequency. *J. Acoust. Soc. Am.* **85**, 327–338.
Assmann, P. F. & Summerfield, A. Q. 1990 Modelling the perception of concurrent vowels: vowels with different fundamental frequencies. *J. Acoust. Soc. Am.* **88**, 680–697.
Assmann, P. F. & Summerfield, A. Q. 1994 The contribution of waveform interactions to the perception of concurrent vowels. *J. Acoust. Soc. Am.* **95**, 471–484.
Assmann, P. F. & Summerfield, Q. 2004 The perception of speech under adverse conditions. In *Speech Processing in the Auditory System* (eds S. Greenberg, W. A. Ainsworth, A. N. Popper & R. R. Fay), pp. 231–308. New York: Springer-Verlag, Inc.
Atal, B. S. & Hanauer, S. L. 1971 Speech analysis and synthesis by linear prediction of the acoustic wave. *J. Acoust. Soc. Am.* **50**, 637–655.
Barker, J. 2006 Robust automatic speech recognition. In *Computational auditory scene analysis* (eds D. L. Wang & G. J. Brown), pp. 297–350. Hoboken, New Jersey: Wiley.
Barker, J. & Cooke, M. 1999 Is the sine-wave speech cocktail party worth attending? *Speech Comm.* **27**, 159–174.
Best, V., Gallun, F. J., Carlile, S. & Shinn-Cunningham, B. G. 2007 Binaural interference and auditory grouping. *J. Acoust. Soc. Am.* **121**, 1070–1076.
Bird, J. & Darwin, C. J. 1998 Effects of a difference in fundamental frequency in separating two sentences. In *Psychophysical and physiological advances in hearing* (eds A. R. Palmer, A. Rees, A. Q. Summerfield & R. Meddis), pp. 263–269. London: Whurr.
Bodden, M. 1996 Auditory demonstrations of a cocktail-party processor. *Acustica* **82**, 356–357.
Bregman, A. S. 1978 Auditory streaming is cumulative. *J. Exp. Psychol. Hum. Perc. Perf.* **4**, 380–387.
Bregman, A. S. 1990 *Auditory Scene Analysis: the Perceptual Organisation of Sound*. Cambridge, MA: Bradford Books, MIT Press.
Bregman, A. S. & Ahad, P. A. 1995 *Compact Disc: Demonstrations of auditory scene analysis*. Montreal: Department of Psychology, McGill University.
Bregman, A. S., Liao, C. & Levitan, R. 1990 Auditory grouping based on fundamental frequency and formant peak frequency. *Canad. J. Psychol.* **44**, 400–413.
Bregman, A. S. & Pinker, S. 1978 Auditory streaming and the building of timbre. *Canad. J. Psychol.* **32**, 19–31.
Broadbent, D. E. & Ladefoged, P. 1957 On the fusion of sounds reaching different sense organs. *J. Acoust. Soc. Am.* **29**, 708–710.
Brokx, J. P. L. & Nooteboom, S. G. 1982 Intonation and the perceptual separation of simultaneous voices. *J. Phonetics* **10**, 23–36.
Bronkhorst, A. W. 2000 The cocktail party phenomenon: a review of speech intelligibility in multiple-talker conditions. *Acustica* **86**, 117–128.
Brungart, D. S. 2001 Informational and energetic masking effects in the perception of two simultaneous talkers. *J. Acoust. Soc. Am.* **109**, 1101–1109.

Brungart, D. S. 2005 A binary masking technique for isolating energetic masking in speech perception. *J. Acoust. Soc. Am.* **117**, 2484.

Brungart, D. S. & Simpson, B. D. 2002 Within-ear and across-ear interference in a cocktail-party listening task. *J. Acoust. Soc. Am.* **112**, 2985–2995.

Brungart, D. S., Simpson, B. D., Darwin, C. J., Arbogast, T. L. & Kidd, G. 2005 Across-ear interference from parametrically-degraded synthetic speech signals in a dichotic cocktail-party listening task. *J. Acoust. Soc. Am.* **117**, 292–304.

Buss, E., Hall, J. W., 3rd & Grose, J. H. 2004 Spectral integration of synchronous and asynchronous cues to consonant identification. *J. Acoust. Soc. Am.* **115**, 2278–2285.

Carlyon, R. P., Deeks, J., Norris, D. & Butterfield, S. 2002 The continuity illusion and vowel identification. *Acta Acustica - Acustica* **88**, 408–415.

Cherry, C. & Wiley, R. H. 1967 Speech communication in very noisy environments. *Nature* **214**, 1164.

Cherry, E. C. 1953 Some experiments on the recognition of speech, with one and with two ears. *J. Acoust. Soc. Am.* **25**, 975–979.

Cherry, E. C. & Taylor, W. K. 1954 Some further experiments upon the recognition of speech, with one and with two ears. *J. Acoust. Soc. Am.* **26**, 554–559.

Chimento, T. C. & Schreiner, C. E. 1991 Adaptation and recovery from adaptation in single fiber responses of the cat auditory-nerve. *J. Acoust. Soc. Am.* **90**, 263–273.

Clark, G. 2003 Cochlear Implants fundamentals and Applications. Berlin, London: Springer.

Cooke, M. 2003 Glimpsing speech. *J. Phonetics* **31**, 579–584.

Cooke, M. P. 1993 *Modelling Auditory Processing and Organisation.* Cambridge: Cambridge University Press.

Cooke, M. P., Green, P. D., Josifovski, L. & Vizinho, A. 2001 Robust automatic speech recognition with missing and unreliable acoustic data. *Speech Comm.* **34**, 267–285.

Culling, J. F. & Darwin, C. J. 1993 Perceptual separation of simultaneous vowels: within and across-formant grouping by F0. *J. Acoust. Soc. Am.* **93**, 3454–3467.

Culling, J. F. & Darwin, C. J. 1994 Perceptual and computational separation of simultaneous vowels: cues arising from low-frequency beating. *J. Acoust. Soc. Am.* **95**, 1559–1569.

Culling, J. F. & Summerfield, Q. 1995 Perceptual separation of concurrent speech sounds: absence of across-frequency grouping by common interaural delay. *J. Acoust. Soc. Am.* **98**, 785–797.

Darwin, C. J. 1981 Perceptual grouping of speech components differing in fundamental frequency and onset-time. *Q. J. Exp. Psychol.* **33A**, 185–208.

Darwin, C. J. 1984 Perceiving vowels in the presence of another sound: constraints on formant perception. *J. Acoust. Soc. Am.* **76**, 1636–1647.

Darwin, C. J. 1990 Environmental influences on speech perception. In *Advances in Speech and Language Processing* (ed. W.A.Ainsworth), pp. 219–241. London: JAI press.

Darwin, C. J. 1991 The relationship between speech perception and the perception of other sounds. In *Modularity and the Motor Theory of Speech Perception* (eds I. G. Mattingly & M. G. Studdert-Kennedy), pp. 239–259. Hillsdale,N.J: Erlbaum.

Darwin, C. J. 1995 Perceiving vowels in the presence of another sound: a quantitative test of the 'Old-plus-New' heuristic. In *Levels in Speech Communication: Relations and Interactions: a tribute to Max Wajskop* (eds C. Sorin, J. Mariani, H. Méloni & J. Schoentgen), pp. 1–12. Amsterdam: Elsevier.

Darwin, C. J. 2002 Auditory streaming in language processing. In *Genetics and the Function of the Auditory System: 19th Danavox Symposium* (eds L. Tranebjaerg, T. Andersen, J. Christensen-Dalsgaard & T. Poulsen), pp. 375–392. Denmark: Holmens Trykkeri.

Darwin, C. J., Brungart, D. S. & Simpson, B. D. 2003 Effects of fundamental frequency and vocal-tract length changes on attention to one of two simultaneous talkers. *J. Acoust. Soc. Am.* **114**, 2913–2922.

Darwin, C. J. & Carlyon, R. P. 1995 Auditory grouping. In *The Handbook of Perception and Cognition (2nd Edition) Volume 6, Hearing* (ed. B. C. J. Moore), pp. 387–424. London: Academic.

Darwin, C. J. & Gardner, R. B. 1986 Mistuning a harmonic of a vowel: grouping and phase effects on vowel quality. *J. Acoust. Soc. Am.* **79**, 838–845.

Darwin, C. J. & Hukin, R. W. 1999 Auditory objects of attention: the role of interaural time-differences. *J. Exp. Psychol. Hum. Perc. Perf.* **25**, 617–629.

Darwin, C. J. & Hukin, R. W. 2000 Effectiveness of spatial cues, prosody and talker characteristics in selective attention. *J. Acoust. Soc. Am.* **107**, 970–977.

Darwin, C. J. & Sutherland, N. S. 1984 Grouping frequency components of vowels: when is a harmonic not a harmonic? *Q. J. Exp. Psychol.* **36A**, 193–208.

Dirks, D. D. & Wilson, R. H. 1969 The effects of spatially separated sound sources on speech intelligibility. *J. Speech Hear. Res.* **12**, 5–38.

Dissard, P. & Darwin, C. J. 2000 Extracting spectral envelopes: formant frequency matching between sounds on different and modulated fundamental frequencies. *J. Acoust. Soc. Am.* **107**, 960–969.

Dissard, P. & Darwin, C. J. 2001 Formant frequency matching between sounds with different bandwidths and on different fundamental frequencies. *J. Acoust. Soc. Am.* **110**, 409–415.

Drennan, W. R., Gatehouse, S. & Lever, C. 2003 Perceptual segregation of competing speech sounds: the role of spatial location. *J. Acoust. Soc. Am.* **114**, 2178–2189.

Drullman, R. & Bronkhorst, A. W. 2000 Multichannel speech intelligibility and talker recognition using monaural, binaural, and three-dimensional auditory presentation. *J. Acoust. Soc. Am.* **107**, 2224–2235.

Duquesnoy, A. J. 1983 Perceptual segregation of competing speech sounds: the role of spatial location. *J. Acoust. Soc. Am.* **74**, 739–743.

Durlach, N. I., Mason, C. R., Kidd, G., Jr., Arbogast, T. L., Colburn, H. S. & Shinn-Cunningham, B. G. 2003 Note on informational masking. *J. Acoust. Soc. Am.* **113**, 2984–2987.

Egan, J. P., Carterette, E. C. & Thwing, E. J. 1954 Some factors affecting multi-channel listening. *J. Acoust. Soc. Am.* **26**, 774–782.

Festen, J. M. & Plomp, R. 1990 Effects of fluctuating noise and interfering speech on the speech-reception threshold for impaired and normal hearing. *J. Acoust. Soc. Am.* **88**, 1725–1736.

Freyman, R. L., Helfer, K. S., McCall, D. D. & Clifton, R. K. 1999 The role of perceived spatial separation in the unmasking of speech. *J. Acoust. Soc. Am.* **106**, 3578–3588.

Gardner, R. B., Gaskill, S. A. & Darwin, C. J. 1989 Perceptual grouping of formants with static and dynamic differences in fundamental frequency. *J. Acoust. Soc. Am.* **85**, 1329–1337.

Gutschalk, A., Micheyl, C., Melcher, J. R., Rupp, A., Scherg, M. & Oxenham, A. J. 2005 Neuromagnetic correlates of streaming in human auditory cortex. *J. Neurosci.* **25**, 5382–5388.

Haas, H. 1972 The influence of a single echo on the audibility of speech. *J. Aud. Eng. Soc*, **20**, 145–149.

Hill, N. I. & Darwin, C. J. 1996 Lateralisation of a perturbed harmonic: effects of onset asynchrony and mistuning. *J. Acoust. Soc. Am.* **100**, 2352–2364.

Houtgast, T. & Steeneken, H. J. M. 1973 The modulation transfer function in room acoustics as a predictor of speech intelligibility. *Acustica* **28**, 66–73.

Howard-Jones, P. A. & Rosen, S. 1993 Uncomodulated glimpsing in 'checkerboard' noise. *J. Acoust. Soc. Am.* **93**, 2915–2922.

Hu, G. & Wang, D. L. 2004 Monaural speech segregation based on pitch tracking and amplitude modulation. *Proceedings of IEEE Workshop on Applications of Signal Processing to Audio and Acoustics (WASPA)*. pp. 79–82.

Hunt, M. J. 1987 Delayed decisions in speech recognition – the case of formants. *Pattern Recognition Letters* **6**, 121–137.

Hunt, M. J., Zwierzynski, D. A. & Carr, R. C. 1989 Issues in high-quality LPC analysis and synthesis. *Proceedings of the European Conference on Speech Communication and Technology* (EuroSpeech-89).

Killion, M. C. 1997 Hearing aids: past, present, future: moving toward normal conversations in noise. *Brit. J. Audiol.* **31**, 141–148.

Klatt, D. H. 1985 The perceptual reality of a formant frequency. *J. Acoust. Soc. Am.* **78**, S81–82.

Kubovy, M. 1981 Concurrent pitch segregation and the theory of indispensible attributes. In *Perceptual Organization* (eds M. Kubovy & J. R. Pomerantz), pp. 55–98. Hillsdale, New Jersey: Erlbaum.

Liberman, A. M. 1982 On the finding that speech is special. *American Psychology*, **37**, 148–167.

Liberman, A. M., Cooper, F. S., Shankweiler, D. S. & Studdert-Kennedy, M. 1967 Perception of the speech code. *Psychol. Rev.* **74**, 431–461.

Licklider, J. C. R. 1948 The influence of interaural phase relations upon the masking of speech by white noise. *J. Acoust. Soc. Am.* **20**, 150–159.

Lippmann, R. P. 1997 Speech recognition by machines and humans. *Speech Comm.* **22**, 1–15.

McAdams, S., Botte, M. C. & Drake, C. 1998 Auditory continuity and loudness computation. *J. Acoust. Soc. Am.* **103**, 1580–1591.

McDermott, J. H. & Oxenham, A. J. 2008 Spectral completion of partially masked sounds. *Proceedings of the National Academy of Science, U. S. A.* **105**, 5939–5944.

Miller, G. A. 1947 The masking of speech. *Psychol. Bull.* **44**, 105–129.

Miller, G. A. & Licklider, J. C. R. 1950 The intelligibility of interrupted speech. *J. Acoust. Soc. Am.* **22**, 167–173.

Moore, B. C. J. 2007 *Cochlear Hearing Loss: Physiological, Psychological and Technical Issues*. 2nd Edn. Chichester: Wiley.

Moore, R. K. 1986 Signal decomposition using Markov modelling techniques. Royal Signals Research Establishment (UK). *Memorandum No.3931.*

Munte, T. F., Kohlmetz, C., Nager, W. & Altenmuller, E. 2001 Superior auditory spatial tuning in conductors. *Nature* **409**, 580.

Palmer, A. R. 1988 The representation of concurrent vowels in the temporal discharge patterns of auditory nerve fibers. In *Basic Issues in Hearing* (eds H. Duifhuis, J. W. Jorst & H. P. Wit), pp. 244–251. London: Academic.

Peters, R. W., Moore, B. C. J. & Baer, T. 1998 Speech reception thresholds in noise with and without spectral and temporal dips for hearing-impaired and normally hearing people. *J. Acoust. Soc. Am.* **103**, 577–587.

Plomp, R. 1976 Binaural and monaural speech intelligibility of connected discourse in reverberation as a function of a single competing sound source (speech or noise). *Acustica* **34**, 200–211.

Plomp, R. 1977 Acoustical aspects of cocktail parties. *Acustica* **38**, 186–191.

Powers, G. L. & Wilcox, J. C. 1977 Intelligibility of temporally-interrupted speech with and without intervening noise. *J. Acoust. Soc. Am.* **61**, 195–199.

Price, C., Thierry, G. & Griffiths, T. D. 2005 Speech-specific auditory processing: where is it? *Trends in Cogn. Sci.* **9**, 271–276.

Remez, R. E., Rubin, P. E., Pisoni, D. B. & Carrell, T. D. 1981 Speech perception without traditional speech cues. *Science* **212**, 947–950.

Remez, R. E., Rubin, P. E., Berns, S. M., Pardo, J. S. & Lang, J. M. 1994 On the perceptual organization of speech. *Psychol. Rev.* **101**, 129–156.

Remez, R. E., Ferro, D. F., Wissig, S. C. & Landau, C. A. 2008 Asynchrony tolerance in the perceptual organization of speech. *Psychol. Bull. Rev.* **15**, 861–865.

Rhebergen, K. S., Versfeld, N. J. & Dreschler, W. A. 2005 Release from informational masking by time reversal of native and non-native interfering speech. *J. Acoust. Soc. Am.* **118**, 1274–1277.

Roberts, B. & Holmes, S. D. 2006 Asynchrony and the grouping of vowel components: captor tones revisited. *J. Acoust. Soc. Am.* **119**, 2905–2918.

Roberts, B. & Holmes, S. D. 2007 Contralateral influences of wideband inhibition on the effect of onset asynchrony as a cue for auditory grouping. *J. Acoust. Soc. Am.* **121**, 3655–3665.

Roberts, B. & Moore, B. C. J. 1991 The influence of extraneous sounds on the perceptual estimation of first-formant frequency in vowels under conditions of asynchrony. *J. Acoust. Soc. Am.* **89**, 2922–2932.

Roweis, S. 2004 Automatic speech processing by inference in generative models. In *Speech Separation by Humans and Machines* (eds P. L. Divenyi), pp. 97–134. New York: Kluwer Academic Publishers.

Sach, A. J. & Bailey, P. J. 2004 Some characteristics of auditory spatial attention revealed using rhythmic masking release. *Perc. Psychophys.* **66**, 1379–1387.

Samuel, A. G. 1981 The role of bottom-up confirmation in the phonemic restoration illusion. *J. Exp. Psychol. Hum. Perc. Perf.* **7**, 1124–1131.

Scheffers, M. T. 1979 The role of pitch in perceptual separation of simultaneous vowels. Institute for Perception Research, Annual Progress Report **14**, 51–54.

Scheffers, M. T. 1983 Sifting vowels: auditory pitch analysis and sound segregation. Ph.D. thesis. The Netherlands: University of Groningen.

Schouten, J. F. 1940 The residue and the mechanism of hearing. *Proceedings of the Koninklijke Nederlandse Akademie van Wetenschappen* **43**, 991–999.

Shackleton, T. M., Meddis, R. & Hewitt, M. J. 1994 The role of binaural and fundamental frequency difference cues in the identification of concurrently presented vowels. *Q. J. Exp. Psychol.* **47A**, 545–563.

Shannon, R. V., Zeng, F.-G., Kamath, V., Wygonski, J. & Ekelid, M. 1995 Speech recognition with primarily temporal cues. *Science* **270**, 303–304.

Smith, R. L. 1979 Adaptation, saturation, and physiological masking in single auditory nerve fibers. *J. Acoust. Soc. Am.* **65**, 166–178.

Spence, C. J. & Driver, J. 1994 Covert spatial orienting in audition: exogenous and endogenous mechanisms. *J. Exp. Psychol. Hum. Perc. Perf.* **20**, 555–574.

Sroka, J. J. & Braida, L. D. 2005 Human and machine consonant recognition. *Speech Comm.* **45**, 401–423.

Stubbs, R. J. & Summerfield, A. Q. 1990 Algorithms for separating the speech of interfering talkers: evaluations with voiced sentences and normal-hearing and hearing-impaired listeners. *J. Acoust. Soc. Am.* **89**, 1383–1393.

Treisman, A. 1960 Contextual cues in selective listening. *Q. J. Exp. Psychol.* **12**, 242–248.

Varga, A. P. & Moore, Roger K. 1991 Simultaneous recognition of concurrent speech signals using hidden Markov model decomposition', In *EUROSPEECH-1991*, 1175–1178.

Verschuure, J. & Brocaar, M. P. 1983 Intelligibility of interrupted meaningful and nonsense speech with and without intervening noise. *Perc. Psychophys.* **33**, 232–240.

Wang, D. L. 2004 An ideal binary mask as the computational goal of auditory scene analysis. In *Speech Separation by Humans and Machines* (ed. P. L. Divenyi), pp. 181–197. New York: Kluwer Academic Publishers.

Warren, R. M. 1970 Perceptual restoration of missing phonemes. *Science* **167**, 392–393.

Warren, R. M. 1984 Perceptual restoration of obliterated sounds. *Psychol. Bull.* **96**, 371–383.

Warren, R. M., Obusek, C. J. & Ackroff, J. M. 1972 Auditory induction: perceptual synthesis of absent sounds. *Science* **176**, 1149–1151.

Watkins, A. J. and Makin, S. J. 1994 Perceptual compensation for speaker differences and for spectral-envelope distortion. *J. Acoust. Soc. Am.* **96**, 1263–1282.

Watson, C. S. 1987 Uncertainty, informational masking and the capacity of immediate auditory memory. In *Auditory Processing of Complex Sounds* (eds W. A. Yost & C. S. Watson), pp. 267–277. Hillsdale, New Jersey: Erlbaum.

Whalen, D. M. & Liberman, A. M. 1987 Speech perception takes precedence over nonspeech perception. *Science* **237**, 169–171.

Wightman, F. L. & Kistler, D. J. 1992 The dominant role of low-frequency interaural time differences in sound localization. *J. Acoust. Soc. Am.* **91**, 1648–1661.

Woods, W. A. & Colburn, S. 1992 Test of a model of auditory object formation using intensity and interaural time difference discriminations. *J. Acoust. Soc. Am.* **91**, 2894–2902.

Yates, G. K., Robertson, D. & Johnstone, B. M. 1985 Very rapid adaptation in the guinea-pig auditory nerve. *J. Speech Lang. Hear. Res.* **17**, 1–12.

8

Functional imaging of the auditory processing applied to speech sounds

Roy D. Patterson and Ingrid S. Johnsrude

In this paper, we describe domain-general auditory processes that we believe are prerequisite to the linguistic analysis of speech. We discuss biological evidence for these processes and how they might relate to processes that are specific to human speech and language. We begin with a brief review of (i) the anatomy of the auditory system and (ii) the essential properties of speech sounds. Section 4 describes the general auditory mechanisms that we believe are applied to all communication sounds, and how functional neuroimaging is being used to map the brain networks associated with domain-general auditory processing. Section 5 discusses recent neuroimaging studies that explore where such general processes give way to those that are specific to human speech and language.

Keywords: auditory anatomy; speech sounds; auditory processing; neuroimaging of pitch; neuroimaging of speech sounds

8.1 Introduction

Speech is a rich social signal that conveys a wealth of information. Not only is it a linguistic signal, used to communicate information and ideas, but it also contains non-linguistic information about the size, sex, background, social status and emotional state of the speaker. Finally, it is usually experienced as a multisensory and interactive signal; these are important aspects that also do not fall within the traditional realm of linguistic analysis. These non-linguistic aspects of communication are a reminder that speech shares characteristics with communication in other animals, including other primates. The initial stages of auditory processing, which rely on a neural organization that is evolutionarily conserved among many primate species, are probably general and apply to all communication sounds, not just to speech. Accordingly, we begin with a brief overview of primate anatomy. At the same time, the complexity of human communication indicates that it engages additional neural apparatus subserving linguistic and social cognition. The point in the system where the processing radiates out into divergent functions is the topic of §5.

8.2 A brief overview of auditory anatomy

(a) The subcortical auditory system in humans

In humans, the principal components of the subcortical auditory system lie in a frontal plane that extends from the ear canal to the upper surface of the central portion of the temporal lobe. Between the cochlea and the auditory cortex, there are four major centres of neural processing: the cochlear nucleus (CN); the superior olivary complex (SOC); the inferior colliculus (IC); and the medial geniculate body (MGB) of the thalamus. Work in other primates suggests that there are mandatory synapses for auditory processing in three of the four nuclei (CN, IC and MGB), which supports the view that these nuclei perform transformations that are applied to all sounds as they proceed up the pathway, much as the cochlea performs a mandatory frequency analysis on all sounds entering the auditory system. In the visual system, there is only one synapse between the retina and visual cortex in the lateral geniculate nucleus.

Information from the two ears is probably integrated in several nuclei in the subcortical auditory system. The CN projects to both the contralateral and the ipsilateral SOC, where minute differences in the timing of the versions of a sound at the two ears are correlated, permitting estimation of source location. The CN also projects to both contralateral and ipsilateral IC, and the two ICs are themselves densely interconnected. Thus, the subcortical auditory system does not maintain a clear segregation of information by the ear of entry. In contrast, in the visual system, there is no binocular processing prior to visual cortex. The complexity of the subcortical auditory system is probably due, at least in part, to the temporal precision of the neural representation of sound (Patterson *et al.* 1999). Auditory nerve fibres between the cochlea and the CN fire in phase with basilar membrane motion up to approximately 5000 Hz, and the nuclei that process this sub-millisecond information must be close to the source to minimize temporal distortion. The maximum rate of phase locking drops to approximately 500 Hz in the IC, and to approximately 50 Hz in the MGB and primary auditory cortex (PAC), which suggests that the form of the neural code changes at least twice as the information progresses from cochlea to cortex, once at the level of the IC and once at the level of the MGB.

(b) The anatomy of auditory cortex and its projections, in the macaque

In humans, the principal components of the cortical auditory system are not well understood. Microelectrode recordings, the cornerstone of non-human neurophysiology, can only be undertaken in rare circumstances (e.g. during neurosurgery; Howard *et al.* 2000; Brugge *et al.* 2003). Post-mortem histological material is scarce and of relatively poor quality (Hackett *et al.* 2001; Wallace *et al.* 2002), and *in vivo* tracer studies in humans are currently not possible. The rhesus macaque monkey (*Macaca mulatta*) provides an animal model for the organization of auditory cortex (Rauschecker *et al.* 1997; Rauschecker 1998; Kaas *et al.* 1999; Kaas & Hackett 2000), and this can be supplemented by the (relatively few) anatomical and neurophysiological studies that have been conducted in humans (Liegeois-Chauvel *et al.* 1991; Rivier & Clarke 1997; Howard *et al.* 2000; Hackett *et al.* 2001; Morosan *et al.* 2001; Rademacher *et al.* 2001; Wallace *et al.* 2002; see Hall *et al.* (2003) and Scott & Johnsrude (2003), for reviews).

A note of caution must be sounded in assuming anatomical and functional homologies between macaques and humans. Most obviously, functional specialization must diverge in the two species at, or before, the point where speech-specific processing begins in humans. Furthermore, unlike our own species, vocalization is not an important form of communication in macaques. Also, auditory research in the macaque has been largely restricted to experiments with very simple sounds such as clicks and pure tones, which may not require extensive cortical processing. As a result, the functional specialization of the core, belt and parabelt regions is simply not known, and macaque research provides only the most general indication of where to look for specific forms of processing in humans.

The organization in the macaque is shown in Figure 8.1a. Cortical afferents from the ventral division of the MGB project to three tonotopically organized fields on the superior temporal gyrus (STG; Rauschecker et al. 1997; Kaas et al. 1999; Kaas & Hackett 2000). This 'core' of primary areas projects to a surrounding 'belt' of anatomically distinguishable cortical fields which exhibit interconnections among adjacent regions (Merzenich & Brugge 1973; Pandya & Sanides 1973; Jones et al. 1995; Pandya 1995; Hackett et al. 1998; Rauschecker 1998; Kaas & Hackett 2000; Rauschecker & Tian 2000). Belt areas connect with lateral 'parabelt' fields, again through connections between physically adjacent regions. The hierarchical connections of the core, belt and parabelt areas suggest at least three discrete levels of processing in the macaque (Pandya 1995; Hackett et al. 1998; Rauschecker 1998; Kaas et al. 1999; Kaas & Hackett 2000; Rauschecker & Tian 2000).

Recent neuroimaging studies (reviewed in §5) indicate that the superior temporal sulcus (STS) region in humans is important for speech-sound perception. Drawing inferences from macaque cortical organization is problematic in the STS, since humans have a middle temporal gyrus (including the ventral bank of the STS) and macaques do not. Human homologies of the ventral bank regions that have been mapped in the macaque are particularly uncertain. Nevertheless, the anatomical organization of the upper bank of the macaque STS may be somewhat conserved in humans, and it is currently the best evidence we have as to what to expect in human STS.

The STS in the macaque is anatomically heterogeneous, but much of its upper bank, running the length of the STS, comprises a region (area TAa) that receives its input mainly from auditory cortex (Seltzer & Pandya 1978, 1989b). This region projects into adjacent polysensory cortex in the depth of the STS, as well as to the inferior parietal lobule and prefrontal cortex (Seltzer & Pandya 1989a, b). Furthermore, anterior STS regions project to ventral and anterior frontal regions, and more posterior STS regions project to more posterior and dorsal frontal regions (and to parietal cortex; Figure 8.1b). Similarly, as shown in Figures 8.1c, d anterior belt and parabelt also interconnect directly, and in a topographically organized way, with multiple sites within orbitofrontal, ventrolateral and dorsolateral frontal cortex including Brodmann areas 46, 12 and 45 (Petrides & Pandya 1984; Hackett et al. 1998, 1999; Romanski et al. 1999a, b). Importantly, area 45 in humans, located in the inferior frontal gyrus (IFG; pars triangularis), is considered as one of the architectonic constituents of Broca's area (Amunts et al. 1999). This distributed set of fields in STG, STS, parietal and prefrontal cortex constitutes a potential fourth stage of processing (Kaas et al. 1999; Figure 8.1a).

(c) Links between perception and production in humans

At the level of cortex, anatomical connectivity suggests that auditory perception and vocal production may be quite intimately linked. Auditory core, belt and parabelt regions all project into the dorsal caudate and putamen—components of the basal ganglia—which are traditionally considered to serve a primarily motor function (Yeterian & Pandya 1998). STS regions that receive projections from auditory cortices, in turn project to regions of the inferior parietal lobule that interconnect with motor cortex via premotor cortex (Pandya & Seltzer 1982; Seltzer & Pandya 1991; Petrides & Pandya 2002).

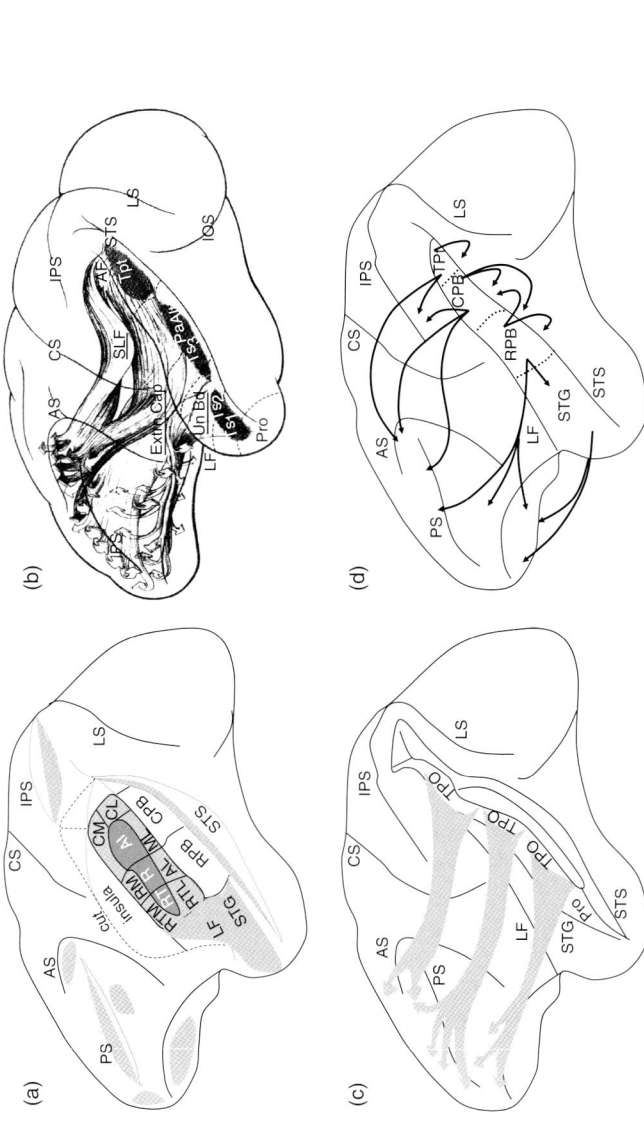

Figure 8.1 Four representations of the anatomical connections of the temporal lobe in the macaque brain. (a) The anatomical organization of the auditory cortex is consistent with at least four levels of processing, including core regions (darkest shading) belt regions (lighter shading), parabelt regions (stripes) and temporal and frontal regions that interconnect with belt and parabelt (lighter shading). (Adapted from Kaas *et al.* (1999) and Hackett & Kaas (2004)). Dotted lines indicate sulci that have been opened to show auditory regions. Regions along the length of (b) superior temporal gyrus and (c) dorsal bank of the superior temporal sulcus connect with prefrontal regions in a topographically organized anterior-to-posterior fashion. (b) Adapted from Petrides & Pandya (1988, p. 64); (c) adapted from Seltzer & Pandya (1989a). (d) Connectivity of auditory belt and parabelt; adapted from Hackett & Kaas (2004). AF, arcuate fasciculus; AS, arcuate sulcus; CS, central sulcus; Extm Cap, extreme capsule; IOS, inferior occipital sulcus; IPS, intraparietal sulcus; LF, lateral fissure; LS, lunate sulcus; PS, principal sulcus; SLF, superior longitudinal fasciculus; STG, superior temporal gyrus; STS, superior temporal sulcus; UnBd, uncinate bundle. (*Note*. Abbreviations are not spelt out if they are the conventional label for a microanatomically or physiologically defined area).

Finally, Brodmann areas 45 and 46 in frontal cortex, which receive auditory projections, interconnect with motor regions via area 44 and premotor cortex.

Physiological data are consistent with a link between auditory perception and vocal production, and they indicate that the coupling is quite rapid. Matt Howard, John Brugge and colleagues have used depth electrode stimulation and electrophysiological recording in neurosurgical patients to explore the evoked responses and connectivity in a circuit involving PAC, a posterolateral region of the STG which they call posterior lateral superior temporal (PLST), IFG (pars triangularis and opercularis) and orofacial motor cortex (Garell *et al.* 1998; Howard *et al.* 2000; Brugge *et al.* 2003; Greenlee *et al.* 2004). Evoked responses in PAC of Heschl's gyrus (HG) had response latencies ranging from 15 to 25 ms, which are compatible with the magnetoencephalography (MEG) data on click latency in PAC reported by Lütkenhöner *et al.* (2003). Then, when this region of HG was electrically stimulated, it resulted in an evoked potential in PLST (Howard *et al.* 2000; Brugge *et al.* 2003). The average onset latency for this evoked response was only 2 ms, consistent with an ipsilateral corticocortical connection between HG and PLST. PLST appears to make a functional connection with the IFG (Garell *et al.* 1998) with onset latencies of approximately 10 ms, and cortical stimulation of posterior IFG elicits responses in orofacial motor cortex with onset latencies of approximately 6 ms (Greenlee *et al.* 2004). Taken together, these results suggest that a sound in the environment could, in principle, have an impact on neural activity in orofacial motor cortex within 35 ms of stimulus onset, and most of that time is spent in the pathway from the cochlea to PAC.

In summary, this overview of the anatomy of auditory cortex suggests that, following the succession of nuclei in the subcortical pathway, the information in auditory cortex radiates out in parallel paths from core areas, and cascades into at least three spatially distributed sets of regions, comprising at least three further processing stages. Other sense information is integrated with auditory information early on in cortical processing, and prominent feedback routes connect adjacent regions at all levels. Perceptual processes must depend on this anatomical organization.

Now we turn to the characteristics of speech sounds and describe a model of the processes that we believe are applied to all communication sounds before speech-specific processing begins in cortex.

8.3 General auditory processes involved in speech perception

When a child and an adult utter the 'same' syllable, it is only the linguistic message of the syllable that is the same. The child has a shorter vocal tract and lighter vocal cords and, as a result, the waveforms carrying the message are quite different for the child and the adult. Although humans have no difficulty in understanding that a child and an adult have said the same word, evaluating the equivalence is far from trivial, as indicated by the fact that speech-recognition machines find this task difficult. Indeed, when trained on the speech of a man, recognition machines are notoriously bad at understanding the speech of a woman, let alone a child. The robustness of auditory perception has led Irino & Patterson (2002) to hypothesize that the auditory system possesses mechanisms that automatically assess the vocal-tract length (VTL) and glottal pulse rate (GPR) of the speaker. Moreover, since humans produce speech sounds in much the same way as all other mammals, it is assumed that such mechanisms are part of the processing applied to all sounds. The value of this analysis is that it helps to produce a size-invariant representation of the timbral cues that identify a species, and this greatly facilitates communication. In speech communication, such processes may be responsible for what is referred to as vowel normalization (e.g. Miller 1989).

(a) Communication sounds

At the heart of each syllable of speech is a vowel. Figure 8.2 shows four versions of the vowel /a/ as in 'hall'. From the auditory perspective, a vowel is a 'pulse-resonance' sound, that is, a stream of glottal pulses each with a resonance showing how the vocal tract responded to that pulse. From the speech perspective, the vowel contains three important components of the information in the larger communication (Irino & Patterson 2002). The first is the phonological 'message'; for the vowels in Figure 8.2, the message is that the vocal tract is currently in the shape that the brain associates with the phoneme /a/. This message is contained in the shape of the resonance which is the same in every cycle of all four waves. In Figures 8.2a, c one person has spoken two versions of /a/ using a high and a low GPR, respectively; the pulse rate determines the pitch of the voice. The resonances have the same form since it is the same person speaking the same vowel. In Figures 8.2b, d a short person and a tall person, respectively, have spoken versions of /a/ on the same pitch. The pulse rate and the shape of the resonance are the same, but the *rate* at which the resonance proceeds within the glottal cycle is slower in Figure 8.2d. This person has the longer vocal tract and so their resonances ring longer. Since the vocal tract connects the mouth and nose to the lungs, VTL is highly correlated with the height of the speaker. In summary, it is the shape of the resonance that corresponds to the message or content of the speech sound. The GPR, which corresponds to the pitch, and the resonance rate, which corresponds to VTL, are derived from the 'form' of the message.

(b) The auditory image model and auditory adaptation to GPR and VTL

The general transforms involved in analysing GPR and VTL will be presented in the context of the auditory image model (AIM; Patterson *et al.* 1992, 1995), a model that focuses on the internal 'auditory images' produced by communication sounds and how

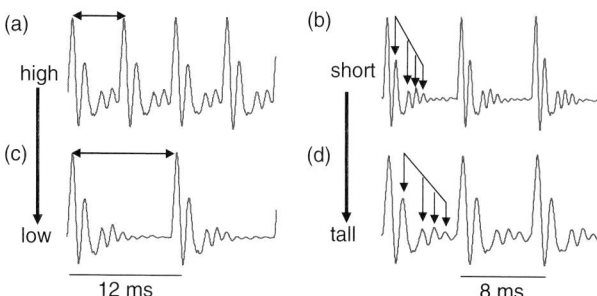

Figure 8.2 Internal structure of voiced sounds illustrating the size factors: pulse rate and resonance rate. (a, b) Glottal pulse rate and (c, d) vocal-tract length have a major effect on both the waveform and the spectrum of the sound, but human perception is extremely robust to changes in both of these factors.

these images can be produced with an ordered set of three transforms. The cochlea performs a spectral analysis of all incoming sounds. In AIM, this is simulated with an auditory filterbank in which the bandwidths of the filters are proportional to filter centre frequency. The filterbank converts an incoming sound wave into a multi-channel representation of basilar membrane motion. The most recent version of the auditory filter includes the fast-acting compression and two-tone suppression observed in the cochlea (Irino & Patterson 2006; Unoki et al. 2006). A 'transduction' mechanism involving half-wave rectification and low-pass filtering converts each channel of membrane motion into a simulation of the neural activity produced in the auditory nerve at that point on the basilar membrane. The result is a multi-channel, neural activity pattern (NAP) like that shown in Figure 8.3a; the dimensions of the NAP are time (the abscissa) and auditory-filter centre frequency on a quasi-logarithmic axis (the ordinate). The surface defined by the set of lines is AIM's simulation of the NAP produced in response to a short segment of this vowel. The channels cover the frequency range from 100 to 6000 Hz. The glottal pulses initiate activity in most of the channels every time they occur. The concentrations of energy in the mid-frequency region reveal the formants. Thus, the NAP of a vowel is a repeating pattern consisting of a warped vertical structure with triangular resonances on one side, which provide information about the shape of the vocal tract. The pattern repeats at the GPR which is heard as the voice pitch.

Whereas the activity in the NAP of a periodic sound oscillates on and off over the course of the glottal cycle, the percept evoked by such a stationary vowel does not flutter or wobble; indeed, periodic sounds produce the most stable of auditory perceptions. The contrast between the form of the NAP, which summarizes our understanding of the representation of sound in the early stages of the auditory pathway (CN and SOC), and the auditory image we hear indicate that there is some form of temporal integration between the NAP and the representation that is the basis of our initial perception of the sound. One process that could produce this stabilization is 'strobed' temporal integration (STI). It is assumed that there is a neural unit associated with each NAP channel that monitors its activity, to locate peaks like those produced by glottal pulses. The peaks cause the unit to 'strobe' the temporal integration process which (i) measures the time intervals from the strobe time to succeeding peaks in the decaying resonance and (ii) enters the time intervals into an interval histogram as they are generated (Patterson 1994). The histogram is dynamic; the information in it decays with a half-life of approximately 30 ms. The array of dynamic interval histograms across NAP channels is AIM's representation of the stabilized auditory image (SAI) that we hear in response to this kind of sound. The SAI of the NAP in Figure 8.3a is presented in Figure 8.3b. The rate of glottal cycles in speech is high relative to the rate of syllables; so, even for men who have GPRs in the range of 125 Hz, there are about four glottal cycles per 30 ms half-life. As a result, the level of the auditory image would be incremented four times during the time that it would otherwise take the image to decay to half its level, and so, a stable pattern builds up in the auditory image and remains there as long as the sound is on.

The process of stabilizing repeating patterns by calculating time intervals from NAP peaks applies to any periodic sound with amplitude modulation, that is, for sounds where one pulse of the period is somewhat larger than the others. This condition is common in the communication calls of animals and the sounds produced by most musical instruments (Fletcher & Rossing 1998; van Dinther & Patterson 2006). The normalization happens with the analysis; there is no need for a central pitch mechanism as in spectral models

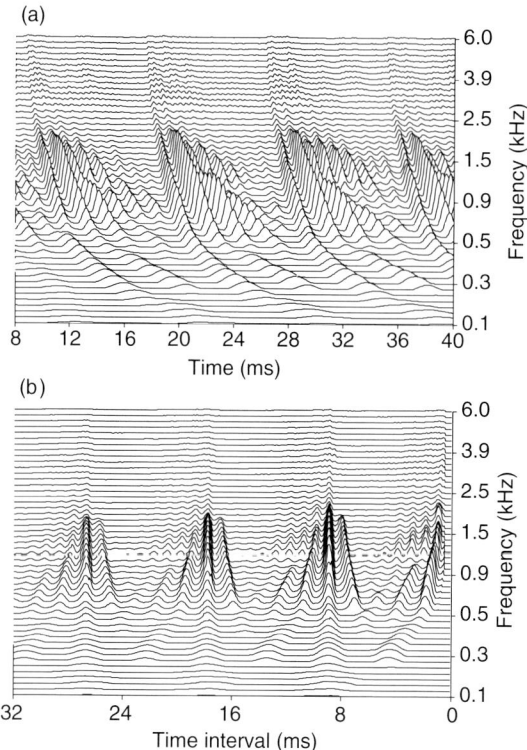

Figure 8.3 (a) The neural activity pattern and (b) the auditory image produced by the /a/ of 'hat'. Note that the abscissa of the auditory image (b) is 'time interval' rather than time itself.

of perception (e.g. Terhardt 1974), and no need for a central neural net to learn that tokens of a vowel with different pitches should all be mapped to the same vowel type. The STI does provide information about pitch and pitch strength (Yost et al. 1996; Patterson et al. 2000), but the pitch information arises as a by-product of image stabilization and adaptation to the sound's pulse rate.

Adaptation to resonance rate, and thereby to VTL, involves a mathematically straightforward affine-scaling transform (Cohen 1993; Irino & Patterson 2002). When the VTL becomes shorter, the formants move up in frequency *and* they shrink in time. If the time-interval dimension in each channel of the auditory image is stretched, by multiplying time interval by the centre frequency of the channel, then the upper formants are stretched horizontally relative to the lower formants, and the mathematics tells us that the image becomes scale covariant. That is, changes in VTL just cause the formants to move up or down, as a group, without changing shape, and the vertical position of the pattern is the size information.

In summary, temporal models of auditory perception like AIM suggest that, following the initial frequency analysis performed in the cochlea, two relatively simple transforms are applied to the internal representation of the sound, which extract the pulse rate and resonance rate of the sound, and produce a largely invariant representation of the linguistic message. The result is that three kinds of information relating directly to GPR,

VTL and to formant structure are available for processing. The model is helpful because it specifies a set of physiologically plausible processes and the order in which they occur. That is, the GPR and VTL adaptation mechanisms must occur, like frequency analysis and the binaural analysis of interaural phase and intensity cues, before the analysis of the sound's spectral structure (its timbre or formants). If the signature of one of the processes can be identified at a specific site in the auditory pathway, it places constraints on where the remaining processes are instantiated. This is the approach adopted by a loose consortium of auditory scientists to search initially for a 'hierarchy of pitch and melody processing' in the auditory system, and more recently to begin searching for the site of acoustic scaling in the auditory system. Neuroimaging research aimed at identifying and localizing the putative adaptation mechanisms is reviewed in §4.

8.4 Imaging methods and general auditory processing

We began by searching for evidence of processing of periodicity, such as would be used to extract GPR, within the auditory pathway using functional magnetic resonance imaging (fMRI). fMRI is a non-invasive technique that indirectly measures regional neural activity, by measuring regional changes in blood oxygenation level. Griffiths et al. (2001) conducted a study with regular-interval (RI) sounds (Patterson et al. 1996; Yost et al. 1996) that are spectrally matched stimuli with and without temporal regularity, which give rise to a noisy percept with and without a buzzy pitch, respectively (see Patterson et al. 2002; Figure 1). RI noise with pitch produces an auditory image with a vertical ridge at the pitch period, similar to the ridge produced by the glottal period in steady-state vowels, but without the formant structure. The study employed a 2-Tesla MR system, cardiac gating (Vlaardingerbroek & den Boer 1996; Guimaraes et al. 1998), sparse imaging (Edmister et al. 1999; Hall et al. 1999) and a magnet-compatible, high-fidelity sound system (Palmer et al. 1998). There were also 48 repetitions of each condition, which provided sufficient sensitivity to reveal activation in the four major subcortical nuclei of the auditory pathway. The contrast between the activation produced by sounds with and without temporal regularity revealed that the processing of temporal regularity begins in subcortical structures (CN and IC). A contrast with the same power, between sounds with a fixed pitch and sounds where the pitch was varied to produce a melody, revealed that changing pitch does not produce more activation than fixed pitch in these nuclei.

The cortical activation from this fMRI study was reported in Patterson et al. (2002); the main results are presented in Figure 8.4 (Figure 3, Patterson et al. 2002). The morphological landmark for PAC in humans is HG (Rademacher et al. 1993; Rivier & Clarke 1997; Morosan et al. 2001; Rademacher et al. 2001). The location of HG was identified in each of the subjects; the conjoint volume is shown in white in the central panels of Figure 8.4 and its location is in good agreement with the locations reported in other studies (Penhune et al. 1996; Leonard et al. 1998). The Figure shows that noise on its own (the blue regions in Figure 8.4) produced more activation than silence, bilaterally, in a large cluster of voxels centred on HG and planum temporale (PT) behind it. The same region is activated by the stimuli with temporal regularity, whether the regularity is fixed (so that pitch is fixed) or varying (as in the melodies). The activation peaks in this region are highly significant and they appear with remarkable consistency in all of the contrasts between sound and silence, and in all subjects.

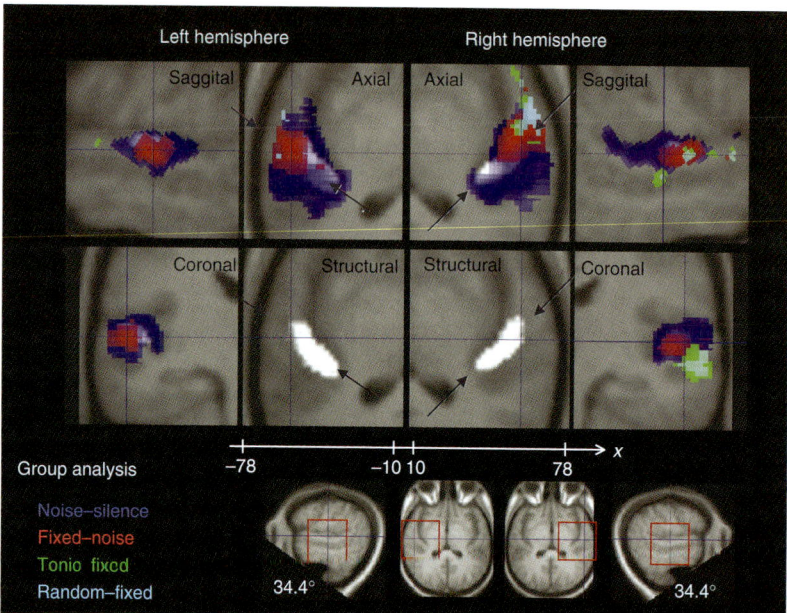

Figure 8.4 A summary of the results of Patterson *et al.* (2002). Group activation for four contrasts from Patterson *et al.* (2002), using a fixed-effects model, rendered onto the average structural image of the group (threshold $p < 0.05$, corrected for multiple comparisons across the whole brain). The position and orientation of the sections are shown in the bottom panels of the figure. The axial sections show the activity in a plane parallel to the surface of the temporal lobe and just below it. The highlighted regions in the structural sections show the average position of HG in the two hemispheres; they are replotted under the functional activation in the axial sections above. The functional activation shows that, as a sequence of noise bursts acquires the properties of melody (first pitch and then changing pitch), the region sensitive to the added complexity changes from a large area on HG and planum temporale (blue), to a relatively focused area in the lateral half of HG (red), and then on out into surrounding regions of the planum polare (PP) and STG (green and cyan mixed). The orderly progression is consistent with the hypothesis that the hierarchy of melody processing that begins in the brainstem continues in auditory cortex and subsequent regions of the temporal lobe. The activation is largely symmetric in auditory cortex and becomes asymmetric abruptly as it moves on to PP and STG with relatively more activity in the right hemisphere.

The region of the noise—silence contrast (blue) at the medial end of HG was particularly consistent. In this region, despite strong activation to all stimuli, there was no differential activity when activation to one sound condition was contrasted with that of any other. For example, Figure 8.4 shows that when the fixed-pitch condition was contrasted with noise (red), or when the changing-pitch conditions (melody) were contrasted with noise (cyan and green), there was no differential activation in medial HG. The obvious interpretation is that this is core auditory cortex (PAC) which is fully engaged by the processing of any complex sound, so the level of activation is the same for sounds with the same loudness.

When the fixed-pitch condition was contrasted with noise (red), it revealed differential activation, bilaterally, in anterolateral HG (al-HG), a region that could be auditory belt (Wallace *et al.* 2002) or core (Morosan *et al.* 2001). When activity in the melody conditions was contrasted with that in the fixed-pitch condition, it revealed differential activation in planum polare (PP) and in STG just below HG (cyan and green in Figure 8.4), and the activation was stronger in the right hemisphere. In bilateral al-HG, melody conditions produced roughly the same level of activity as fixed-pitch conditions. Taken together, these results suggest that neurons in the al-HG region are involved in the cross-frequency evaluation of pitch value and pitch strength, and subsequent temporal regions, particularly in the right hemisphere, evaluate changes in pitch over time.

Penagos et al. (2004) extended the results of Griffiths et al. (2001) and Patterson et al. (2002) using harmonic complex tones with and without resolved harmonics. They used a 3-Tesla MR system, cardiac gating and sparse imaging which enabled them to measure activation in CN, IC and HG, and show that the level of activation does not vary significantly with pitch salience, except in a small region of al-HG, bilaterally. There were no changing-pitch conditions in this study. Finally, Bendor & Wang (2005) have recently reported finding cells in marmoset (*Callithrix jacchus*) cortex sensitive to the low pitch of harmonic complex tones. The cells were in an auditory core area adjacent to A1 (area R) which, Bendor & Wang (2005) argue, is probably homologous to al-HG in humans.

The results of these experiments can be interpreted to indicate that there is a 'hierarchy of processing' in the auditory pathway. With regard to AIM and the transform that adapts to GPR, the results suggest that (i) the extraction of time-interval information from the firing pattern in the auditory nerve probably occurs in the brainstem, (ii) the construction of the time-interval histograms probably occurs in, or near, the thalamus (MGB), and the resulting SAI is in PAC, (iii) the cross-channel evaluation of pitch value and pitch strength probably occurs in al-HG, and (iv) assessment of pitch variation for the perception of melody, and perhaps prosody, appears to occur in regions beyond auditory cortex (anterior STG) particularly in the right hemisphere. It is this last process, requiring integration over long time periods that gives rise to the hemispheric asymmetries observed in neuropsychological and functional neuroimaging studies of pitch perception, rather than pitch extraction *per se* (Patterson et al. 2002).

(a) Imaging the auditory system with MEG

MEG measures the strength and direction of the magnetic dipole produced by activation in nerve fibres running parallel to the scalp; it is largely insensitive to radial sources. This is an advantage for measuring activity along the surface of the temporal lobe in the region of lateral belt and parabelt auditory cortex, and a disadvantage for imaging of activity in higher-level auditory areas like the STS. However, the main advantage of MEG for the investigation of auditory function is that it has millisecond temporal resolution which can be used to investigate the order of events in auditory cortex. It is also the case that recent MEG machines with hundreds of sensors make it possible to localize sources sufficiently well to associate them with regions of activation observed with fMRI.

The auditory evoked field (AEF) is dominated by a large negative deflection associated with stimulus onset; it appears in the interval between 80 and 130 ms post-stimulus onset, and when the source can be located, it is usually in PT just posterior to al-HG. It is referred to as N1 m or N100 m, and it is generally assumed to represent the aggregate activity of several sources involved in general auditory processing. There are several techniques for dissecting the components of this large, broad, negative deflection. One can simply gather sufficient MEG data to reveal smaller positive and negative peaks on the flanks of the N100 m by averaging. For example, Lütkenhöner *et al.* (2003) showed that the first cortical response in humans to transient sounds is a negative deflection in PAC 19 ms post-stimulus onset (N19 m), and Rupp *et al.* (2002) showed that short chirps produce an N19 m–P30 m complex from a source on HG near PAC. The P30 is the first of a set of positive field generators that appear in, or near, PAC between 30 and 60 ms post-onset, and which are collectively referred to as 'P1m'. Gutschalk *et al.* (2004) have shown that the P1m is related to stimulus onset but not the processing of temporal regularity.

Forss *et al.* (1993) have shown that the latency of the N100 m elicited by a regular click train is inversely related to the pitch of the sound, which led Crottaz-Herbette & Ragot (2000) to propose that the generators of the N100 m are involved in pitch processing. However, an earlier review of a wide range of studies (Näätänen & Picton 1987) concluded that an N100 m can be elicited by the onset of almost any kind of sound. So, while it is the case that the latency of the N100 m varies with pitch, the response is fundamentally confounded with the activation of other generators that reflect features like loudness and timbre rather than pitch. To isolate the pitch component of the N100 m, Krumbholz *et al.* (2003) developed a continuous stimulation technique in which the sound begins with a stationary noise and then, after a second or so, when the N100 m has passed and the AEF has settled into a sustained response, the fine structure of the noise is regularized to produce a RI sound (RIS) without changing the energy or spectral distribution of the energy. There is a marked perceptual change at the transition from noise to RIS, and it is accompanied by a prominent negative deflection in the magnetic field, referred to as the pitch onset response (POR). The inverse transition, from RIS to noise, produces virtually no deflection. Krumbholz *et al.* (2003) showed that the latency of the POR varies inversely with the pitch of the RIS, and the magnitude of the response increases with pitch strength. Gutschalk *et al.* (2004) constructed a continuous stimulus of alternating regular and irregular click trains, and used it to isolate the POR from an intensity-related response in PT. The source of the POR was located in al-HG very near the pitch centre identified by Patterson *et al.* (2002) and Penagos *et al.* (2004). The PORs were surprisingly late: approximately 120 ms *plus* four times the period of the click train. This is substantially longer than might be anticipated from temporal models of pitch perception (e.g. Patterson *et al.* 1995; Krumbholz *et al.* 2003). The results suggest that the POR reflects relatively late cortical processes involved in the cross-channel estimation of pitch and pitch strength. It stands in sharp contrast to the initial extraction of periodicity information with STI which is thought to be in the brainstem and thalamus.

The notes of music and the vowels of speech produce sustained pitch perceptions, and there is a sustained component of the AEF that appears to reflect the sustained perception of pitch. The advent of MEG systems with 125–250 sensors means that it is now possible to measure sustained fields. Gutschalk *et al.* (2002) contrasted the activity produced by regular and irregular click trains and performed the experiment at three intensities. This enabled them to isolate two sources in each hemisphere adjacent to PAC. The more anterior source was located in lateral HG in the pitch region identified with fMRI by Patterson *et al.* (2002) and Penagos *et al.* (2004). This source was particularly sensitive to regularity and largely insensitive to sound level. The second source was located just posterior to the first in PT; it was particularly sensitive to sound level and largely insensitive to regularity. This double dissociation provided convincing evidence that the source of the POR in al-HG also produces a sustained field that is related to the sustained perception of pitch. The posterior source in PT would appear to be more involved with the perception of loudness.

The studies discussed to this point are all related to the transforms that adapt auditory analysis to the GPR of the vowel and evaluate the pitch. There is only one very recent study (von Kriegstein et al. 2006) of the processes that adapt auditory analysis to the resonance rate of the vowel, that is, the VTL of the speaker. Subjects listened to sequences of syllables in which GPR and VTL either remained fixed or varied randomly in a 2×2 factorial design. The results are compatible with the model of hierarchical processing inasmuch as they indicate that the adaptation begins in the MGB and is not completed until regions of STG beyond auditory cortex. However, these initial results do not reveal one simple region for the processing of VTL information. The cortical activation arises from the interaction of GPR and VTL, which occurs naturally as children grow; however, it complicates the interpretation of the results.

8.5 The beginnings of voice-specific and speech-specific processing in the brain, and the implications for models of speech perception

As noted above, talker normalization is an important preliminary step to recovering the content of an utterance. Not until GPR and VTL have been taken into account, and a transformed auditory image reflecting the shape of the vocal apparatus achieved, can the linguistic content (phonemes/words/phrases) be analysed and interpreted. Until recently, research on speech perception (as a linguistic signal) assumed that talker-specific information, such as GPR and VTL, was simply stripped away from the linguistic content of the message relatively early in processing—it was thought that this was simply noise that did not contribute helpfully to interpreting speech. A growing body of research makes it clear, however, that form and content must, to some extent, be processed together (Goldinger 1998). Detailed information about an individual talker's voice is encoded and retained and can subsequently improve intelligibility of this familiar voice (Pisoni 1997; Nygaard & Pisoni 1998; Sheffert et al. 2002; Bent & Bradlow 2003). Thus, the transformations discussed in §3 must also permit talker-specific information such as GPR and VTL to contact the linguistic information processed in cortex.

The brain networks underlying both voice-specific processing and the transformation from general auditory to speech-specific processing have been studied using imaging methods, particularly fMRI. Voice-specific processing has not been extensively investigated, and speech-specific processing—particularly the question of where the transformation from auditory to linguistic processing occurs—has been neglected until recently. In fact, this is not one but many questions: Is the transformation localizable? If localizable, do we observe evidence for one neural locus or for several? Is such a transformation dependent only upon the acoustics of the signal or also upon the cognitive state of the individual? Is this a modular auditory process, or can it be influenced by other sensory modalities (i.e. visual; somatomotor)? Recent imaging studies have begun to answer these questions. First, however, we will briefly review the evidence supporting localized voice-specific processing in cortex.

Thierry et al. (2003) compared auditory processing of speech and environmental sound sequences matched for duration, rhythm, content and interpretability, in addition to identifying a network of areas activated during comprehension of both kinds of sound, several areas in the left anterior superior and middle temporal gyri, straddling the STS, were activated more by speech than by environmental sounds. As the authors point out (see Price *et al.* 2005), increased sensitivity to speech over environmental sounds could arise either because these STS areas are sensitive to speech *qua* linguistic signal or *qua* vocal signal. It is difficult to separate these two types of processing, but studies indicate that both types of processing appear to recruit similar regions, in both hemispheres. However, the processing of speech as a linguistic signal seems to recruit left-hemisphere STS areas preferentially, whereas processing of speech as a voice signal seems to recruit right-hemisphere STS areas preferentially.

Pascal Belin and colleagues (Belin *et al.* 2000, 2002; Fecteau *et al.* 2004; see Belin *et al.* 2004 for a review) used non-speech vocalizations such as laughs and cries to distinguish between processing of speech as voice and speech as linguistic content. They observed robust bilateral activity in the STS for human voices when compared with non-human sounds, irrespective of the linguistic content of the voice (Belin *et al.* 2000, 2002). In most listeners the peak of this activity appeared to be in the upper bank of the STS. Although anatomical homologies between humans and macaque monkeys in the STS are not yet known, this region in the macaque corresponds to a third or fourth stage of cortical auditory processing (Kaas *et al.* 1999, Figure 8.2*a*), and may similarly subserve late-stage auditory processing in humans. In a subsequent study with different listeners, similar, but somewhat more inferior, regions exhibited greater sensitivity to human than to animal vocalizations, suggesting that these STS regions are not just sensitive to voices, but are more sensitive to human voices. Von Kriegstein & Giraud (2004) used two different tasks with the same set of sentences spoken by different talkers to observe three regions in the right STS that are more active during recognition of target voices than during recognition of target sentence content, confirming a role for multiple right STS regions in voice perception. Unlike the studies by Belin and colleagues, differential activity in left STS regions was not observed in this contrast. Note however that voices were equally present in all conditions: if voice processing is more obligatory in the left hemisphere (perhaps because it is a necessary concomitant of the extraction of linguistic content), then it would be 'subtracted out' in the contrast.

In general, the evidence suggests that the transformation from an auditory signal to speech is localizable and is distributed across several neural loci, including PT (probable belt or parabelt) and STS but not HG (probable core and belt). Uppenkamp *et al.* (2006) scanned volunteers while they listened to natural and synthetic vowels, or to non-speech stimuli matched to the vowel sounds in terms of their long-term energy and spectro-temporal profiles. Vowels produced more activation than non-speech sounds in several regions along the STS bilaterally, in anterolateral PT as well as in premotor cortex, but not in any anatomically earlier auditory region. Other researchers observe multiple foci throughout the temporal lobe when speech and non-speech perceptions are compared, even despite the 8–12 mm smoothing that is common in imaging studies (e.g. Giraud *et al.* 2004; Rimol *et al.* 2005). Jacquemot *et al.* (2003) performed a study in which they examined the neural correlates of acoustic differences within, or across, phonetic categories. They exploited cross-linguistic differences in phonology between French and Japanese, to achieve a counterbalanced design in which stimuli that were perceived by one language group as belonging to the same phonetic category were perceived by the other group as belonging to different phonetic categories. When across- versus within-category stimuli were compared across groups, activation was observed in the supramarginal gyrus and in a region of anterior PT. Since linguistic stimuli were present in both conditions, any activation that was due to these stimuli being treated as speech would be subtracted out, and the activity could reflect some common, experience-dependent, magnification of acoustic differences across groups. The results are generally compatible with the hierarchical model of primate auditory processing, and with the idea that early cortical stages of processing respond indiscriminately to speech and non-speech sounds, and only regions at a higher stage of processing are specialized for speech perception. However, even though speech and non-speech stimuli were acoustically closely matched in the study by Uppenkamp *et al.* (2006), the synthetic speech stimuli had a perceptual coherence, eliciting a voice-like percept that the non-speech stimuli lacked. Thus, we cannot determine whether the activation foci observed in this study, and in many other studies comparing speech and non-speech stimuli (e.g. Demonet *et al.* 1992; Binder *et al.* 1997, 2000; Jancke *et al.* 2002), reflect voice or linguistic perception, or both.

Activity in speech-sensitive regions can be contingent upon the cognitive state of the individual. In several fMRI experiments (Liebenthal *et al.* 2003; Giraud *et al.* 2004; Dehaene-Lambertz *et al.* 2005; Möttönen *et al.* 2006) the physical identity of auditory stimuli was held constant between two conditions, but the cognitive state of the listeners was systematically manipulated so that they heard the stimuli as non-speech in one condition, but as speech in another. For example, in the study by Giraud *et al.* (2004), listeners heard sentences that had been noise-vocoded (divided into four frequency bands and then resynthesized onto a noise carrier) both before and after a period of training. These sentences were initially not heard as speech, but after they had been presented pairwise with their natural-speech homologues (training), listeners could understand them. A control stimulus set consisted of vocoded sentences that were degraded acoustically so that training did not increase comprehension. The post- versus pre-training contrast was essentially tested as an interaction with post-pre control stimuli, to remove systematic order-of-testing effects, and revealed areas that are selectively active during comprehension. Interestingly, this contrast did not reveal activity in the left STS, instead activity was observed in right STS and in left and right middle and inferior temporal areas. Dehaene-Lambertz *et al.* (2005) used sine-wave consonant–vowel syllables which can be perceived either as non-speech whistles or, following instructions and training, as syllables. Posterior left STS/STG was the only region that was more active when listeners heard the stimuli as speech compared to when they heard them as non-speech. Möttönen *et al.* (2006) also demonstrated left posterior STS activity when perceiving sine-wave processed non-word bisyllables as speech compared to non-speech. Again, whether such activation arises as a result of voice or linguistic perception is an open question.

Researchers have observed activity in premotor cortex or motor cortex during the perception of speech (words: Fadiga *et al.* 2002; monosyllables: Wilson *et al.* 2004, Uppenkamp *et al.* 2006; connected speech: Watkins *et al.* 2003, Watkins & Paus 2004). It is therefore possible that the acoustic-to-speech transformation relies on multiple regions in a distributed network including both temporal-lobe and motor–premotor regions, although the stimuli used in these studies may have engaged lexical, semantic and syntactic processes in addition to speech-sound processing, and further studies are required to determine whether premotor/motor activity reflects processes that are prerequisite to speech perception, or instead reflects some late process that is merely correlated with speech perception (e.g. semantically relevant imagery; cf. Hauk *et al.* 2004).

The anatomical connections between the auditory system and motor structures are highly compatible with a wealth of information attesting to speech perception as a sensorimotor phenomenon. Several models of speech perception, positing a basis in articulatory or gestural representations, have been formulated (Fowler 1996; Rizzolatti & Arbib 1998; Liberman & Whalen 2000; MacNeilage & Davis 2001). The motor theory of speech perception is unlikely to hold in its orthodox form (e.g. Liberman & Whalen 2000), but more moderate positions that acknowledge at least preliminary domain-general auditory processing of the speech signal also propose that speech perception and production are linked (Kluender & Lotto 1999). This is seen most clearly during development where imitation plays an important role in children's acquisition of spoken language (Kuhl 1994; Doupe & Kuhl 1999) but sensorimotor development would have an impact on the organization of speech perception in the adult brain.

Most authors have concluded that such motor activity during speech perception could reflect the activation of articulatory gestural representations which permit the listener to derive the *intended gesture* of the speaker; this is in line with the motor theory of speech perception (e.g. Liberman & Whalen 2000). Such access to gestural representations provides for parity; a shared code between speaker and listener, which is essential for speech communication (Rizzolatti & Arbib 1998; Liberman & Whalen 2000).

The highly parallel organization of cortical regions and their interconnections suggests a way that auditory and gestural accounts of speech perception can be reconciled (Scott & Johnsrude 2003). Different processing pathways may be differentially specialized to serve different processes or operate on complementary representations of speech (phonological versus articulatory). Multiple pathways, operating in parallel, would serve to make speech perception the efficient and robust communication system we know it to be.

8.6 Conclusions

The incoming auditory signal is extensively processed and recoded by the time it reaches auditory cortex, and it probably is not treated in any speech-specific way until relatively late—at least three or four cortical processing stages beyond PAC. Prior to that stage, processes are more domain general—more about the form of the speech (how the talker was talking) and less about the content of the speech (what the talker was saying). These two classes of processing work together to recover the speech content and information indicative of the size, sex and age of the talker among other indexical characteristics. The mammalian auditory system is organized hierarchically, and from PAC onwards the anatomy suggests multiple, parallel, processing systems with strong feedback connections suggesting that multiple aspects of the speech signal are processed more-or-less simultaneously with reference to the ongoing context. The fact that the processing is distributed means that functional imaging (fMRI and MEG) can assist in exploring both the subcortical and cortical networks involved in domain-general and speech-specific processing, and the interactions among them.

Acknowledgements

The writing of this paper was supported by a grant to the first author from the UK Medical Research Council (G0500221).

References

Amunts, K., Schleicher, A., Burgel, U., Mohlberg, H., Uylings, H. & Zilles, K. 1999 Broca's region revisited: cytoarchitecture and intersubject variability. *J. Comp. Neurol.* **412**, 319–341. (doi:10.1002/(SICI)1096-9861(19990920)412:2<319::AID-CNE10>3.0.CO;2-7)

Belin, P., Zatorre, R. J., Lafaille, P., Ahad, P. & Pike, B. 2000 Voice-selective areas in human auditory cortex. *Nature* **403**, 309–312. (doi:10.1038/35002078)

Belin, P., Zatorre, R. J. & Ahad, P. 2002 Human temporal-lobe response to vocal sounds. *Brain Res. Cogn. Brain Res.* **13**, 17–26. (doi:10.1016/S0926-6410(01)00084-2)

Belin, P., Fecteau, S. & Bedard, C. 2004 Thinking the voice: neural correlates of voice perception. *Trends Cogn. Sci.* **8**, 129–135. (doi:10.1016/j.tics.2004.01.008)

Bendor, D. & Wang, X. 2005 The neuronal representation of pitch in primate auditory cortex. *Nature* **436**, 1161–1165. (doi:10.1038/nature03867)

Bent, T. & Bradlow, A. R. 2003 The interlanguage speech intelligibility benefit. *J. Acoust. Soc. Am.* **114**, 1600–1610. (doi:10.1121/1.1603234)

Binder, J. R., Frost, J., Hammeke, T., Cox, R., Rao, S. & Prieto, T. 1997 Human brain language areas identified by functional magnetic resonance imaging. *J. Neurosci.* **17**, 353–362.

Binder, J. R., Frost, J., Hammeke, T., Bellgowan, P., Springer, J., Kaufman, J. & Possing, E. 2000 Human temporal lobe activation by speech and nonspeech sounds. *Cereb. Cortex* **10**, 512–528. (doi:10.1093/cercor/10.5.512)

Brugge, J. F., Volkov, I., Garell, P., Reale, R. & Howard III, M. 2003 Functional connections between auditory cortex on Heschl's gyrus and on the lateral superior temporal gyrus in humans. *J. Neurophysiol.* **90**, 3750–3763. (doi:10.1152/jn.00500.2003)

Cohen, L. 1993 The scale transform. *IEEE Trans. Acoust. Speech Signal Process.* **41**, 3275–3292.

Crottaz-Herbette, S. & Ragot, R. 2000 Perception of complex sounds: N1 latency codes pitch and topography codes spectra. *Clin. Neurophysiol.* **111**, 1759–1766. (doi:10.1016/S13882457(00)00422-3)

Dehaene-Lambertz, G., Pallier, C, Serniclaes, W., Sprenger-Charolles, L., Jobert, A. & Dehaene, S. 2005 Neural correlates of switching from auditory to speech perception. *NeuroImage* **24**, 21–33. (doi:10.1016/j.neuroimage.2004.09.039)

Demonet, J., Chollet, F., Ramsay, S., Cardebat, D., Nespoulous, J., Wise, R., Rascol, A. & Frackowiak, R. 1992 The anatomy of phonological and semantic processing in normal subjects. *Brain* **115**, 1753–1768. (doi:10.1093/brain/115.6.1753)

Doupe, A. & Kuhl, P. 1999 Birdsong and human speech: common themes and mechanisms. *Annu. Rev. Neurosd.* **22**, 567–631. (doi:10.1146/annurev.neuro.22.1.567)

Edmister, W., Talavage, T., Ledden, P. & Weisskoff, R. 1999 Improved auditory cortex imaging using clustered volume acquisitions. *Hum. Brain. Mapp.* **7**, 89–97. (doi:10.1002/(SICI)1097-0193(1999)7:2<89::AID-HBM2>3.0.CO;2-N)

Fadiga, L., Craighero, L., Buccino, G. & Rizzolatti, G. 2002 Speech listening specifically modulates the excitability of tongue muscles: a TMS study. *Eur. J. Neurosci.* **15**, 399–402. (doi:10.1046/j.0953-816x.2001.01874.x)

Fecteau, S., Armony, J., Joanette, Y. & Belin, P. 2004 Is voice processing species-specific in human auditory cortex? An fMRI study. *NeuroImage* **23**, 840–848. (doi:10.1016/j.neuroimage.2004.09.019)

Fletcher, N. H. & Rossing, T 1998 *The physics of musical instruments*. New York, NY: Springer.

Forss, N., Makela, J., McEvoy, L. & Hari, R. 1993 Temporal integration and oscillatory responses of the human auditory cortex revealed by evoked magnetic fields to click trains. *Hear. Res.* **68**, 89–96. (doi:10.1016/0378-5955(93)90067-B)

Fowler, C. A. 1996 Listeners do hear sounds, not tongues. *J. Acoust. Soc. Am.* **99**, 1730–1741. (doi:10.1121/1.415237)

Garell, P., Volkov, I., Noh, M., Damasio, H., Reale, R., Hind, J., Brugge, J. & Howard, M. 1998 Electrophysiologic connections between the posterior superior temporal gyrus and lateral frontal lobe in humans. *Soc. Neurosci. Abstr.* **24**, 1877.

Giraud, A. L., Kell, C., Thierfelder, C., Sterzer, P., Russ, M., Preibisch, C. & Kleinschmidt, A. 2004 Contributions of sensory input, auditory search and verbal comprehension to cortical activity during speech processing. *Cereb. Cortex* **14**, 247–255. (doi:10.1093/cercor/bhgl24)

Goldinger, S. D. 1998 Echoes of echoes? An episodic theory of lexical access. *Psychol. Rev.* **105**, 251–279. (doi:10.1037/0033-295X.105.2.251)

Greenlee, J. D., Oya, H., Kawasaki, H., Volkov, I., Kaufman, O., Kovach, C, Howard, M. & Brugge, J. 2004 A functional connection between inferior frontal gyrus and orofacial motor cortex in human. *J. Neurophysiol.* **92**, 1153–1164. (doi:10.1152/jn.00609.2003)

Griffiths, T. D., Uppenkamp, S., Johnsrude, I., Josephs, O. & Patterson, R. D. 2001 Encoding of the temporal regularity of sound in the human brainstem. *Nat. Neurosci.* **4**, 633–637. (doi:10.1038/88459)

Guimaraes, A. R., Melcher, J., Talavage, T., Baker, J., Ledden, P., Rosen, B., Kiang, N., Fullerton, B. & Weisskoff, R. 1998 Imaging subcortical auditory activity in humans. *Hum. Brain Mapp.* **6**, 33–41. (doi:10.1002/(SICI)1097–0193(1998)6:K33::AID-HBM3>3.0.CO;2-M)

Gutschalk, A., Patterson, R. D., Rupp, A., Uppenkamp, S. & Scherg, M. 2002 Sustained magnetic fields reveal separate sites for sound level and temporal regularity in human auditory cortex. *NeuroImage* **15**, 207–216. (doi:10.1006/nimg.2001.0949)

Gutschalk, A., Patterson, R. D., Scherg, M., Uppenkamp, S. & Rupp, A. 2004 Temporal dynamics of pitch in human auditory cortex. *NeuroImage* **22**, 755–766. (doi:10.1016/j.neuroimage.2004.01.025)

Hackett, T. A., & Kaas, J. H. 2004 Auditory cortex in primates: functional subdivisions and processing streams. In *The cognitive neurosciences III* (ed. M. Gazzaniga), ch. 16, pp. 215–232. Cambridge, MA: MIT Press.

Hackett, T. A., Stepniewska, I. & Kaas, J. 1998 Subdivisions of auditory cortex and ipsilateral cortical connections of the parabelt auditory cortex in macaque monkeys. *J. Comp. Neurol.* **394**, 475–495. (doi:10.1002/(SICI)1096-9861(19980518)394:4<475::AID-CNE6>3.0.CO;2-Z)

Hackett, T. A., Stepniewska, I. & Kaas, J. 1999 Prefrontal connections of the parabelt auditory cortex in macaque monkeys. *Brain Res.* **817**, 45–58. (doi:10.1016/S0006-8993(98)01182-2)

Hackett, T. A., Preuss, T. & Kaas, J. 2001 Architectonic identification of the core region in auditory cortex of macaques, chimpanzees, and humans. *J. Comp. Neurol.* **441**, 197–222. (doi:10.1002/cne.1407)

Hall, D. A., Haggard, M., Akeroyd, M., Palmer, A., Summerfield, A. Q., Elliott, M., Gurney, E. & Bowtell, R. 1999 'Sparse' temporal sampling in auditory fMRI. *Hum. BrainMapp.* **7**, 213–223. (doi:10.1002/(SICI)1097-0193(1999)7:3<213::AID-HBM5>3.0.CO;2-N)

Hall, D. A., Hart, H. & Johnsrude, I. 2003 Relationships between human auditory cortical structure and function. *Audiol. Neurootol.* **8**, 1–18. (doi:10.1159/000067894)

Hauk, O., Johnsrude, I. & Pulvermuller, F. 2004 Somatotopic representation of action words in human motor and premotor cortex. *Neuron* **41**, 301–307. (doi:10.1016/S0896-6273(03)00838-9)

Howard, M. *et al.* 2000 Auditory cortex on the human posterior superior temporal gyrus. *J. Comp. Neurol.* **416**, 79–92. (doi:10.1002/(SICI)1096-9861(20000103)416:1<79::AID-CNE6>3.0.CO;2-2)

Irino, T. & Patterson, R. D. 2002 Segregating information about the size and shape of the vocal tract using a time-domain auditory model: the stabilised wavelet-Mellin transform. *Speech Commun.* **36**, 181–203. (doi:10.1016/S0167-6393(00)00085-6)

Irino, T. & Patterson, R. D. 2006 A dynamic, compressive gammachirp auditory filterbank. *IEEE Audio, Speech Lang. Process (ASLP)* **14**, 2222–2232. (doi:10.1109/TASL.2006.874669)

Jacquemot, C., Pallier, C., LeBihan, D., Dehaene, S. & Dupoux, E. 2003 Phonological grammar shapes the auditory cortex: a functional magnetic resonance imaging study. *J. Neurosci.* **23**, 9541–9546.

Jancke, L., Wustenberg, T, Scheich, H. & Heinze, H. 2002 Phonetic perception and the temporal cortex. *NeuroImage* **15**, 733–746. (doi:10.1006/nimg.2001.1027)

Jones, E. G., Dell'Anna, M., Molinari, M., Rausell, E. & Hashikawa, T. 1995 Subdivisions of macaque monkey auditory cortex revealed by calcium-binding protein immunoreactivity. *J. Comp. Neurol.* **362**, 153–170. (doi:10.1002/cne.903620202)

Kaas, J. & Hackett, T. 2000 Subdivisions of auditory cortex and processing streams in primates. *Proc. Natl Acad. Sci. USA* **97**, 11 793–11 799. (doi:10.1073/pnas.97.22.11793)

Kaas, J., Hackett, T. & Tramo, M. 1999 Auditory processing in primate cerebral cortex. *Curr. Opin. Neurobiol.* **9**, 164–170. (doi:10.1016/S0959-4388(99)80022-1)

Kluender, K. R. & Lotto, A. 1999 Virtues and perils of an empiricist approach to speech perception. *J. Acoust. Soc. Am.* **105**, 503–511. (doi:10.1121/1.424587)

Krumbholz, K., Patterson, R. D., Seither-Preisler, A., Lammertmann, C. & Lütkenhöner, B. 2003 Neuromagnetic evidence for a pitch processing center in Heschl's gyrus. *Cereb. Cortex* **13**, 765–772. (doi:10.1093/cercor/13.7.765)

Kuhl, P. 1994 Learning and representation in speech and language. *Curr. Opin. Neurobiol.* **4**, 812–822. (doi:10.1016/0959-4388(94)90128-7)

Leonard, C, Puranik, C, Kuldau, J. & Lombardino, L. 1998 Normal variation in the frequency and location of human auditory cortex landmarks. Heschl's gyrus: where is it? *Cereb. Cortex* **8**, 397–406. (doi:10.1093/cercor/8.5.397)

Liberman, A. & Whalen, D. 2000 On the relation of speech to language. *Trends Cogn. Sci.* **4**, 187–196. (doi:10.1016/S1364-6613(00)01471-6)

Liebenthal, E., Binder, J., Piorkowski, R. & Remez, R. 2003 Short-term reorganization of auditory analysis induced by phonetic experience. *J. Cogn. Neurosci.* **15**, 549–558. (doi:10.1162/0898929 03321662930)

Liegeois-Chauvel, C., Musolino, A. & Chauvel, P. 1991 Localization of the primary auditory area in man. *Brain* **114**, 139–151.

Lütkenhöner, B., Krumbholz, K., Lammertmann, C., Seither-Preisler, A., Steinstrater, O. & Patterson, R. D. 2003 Localization of primary auditory cortex in humans by magnetoencephalography. *NeuroImage* **18**, 58–66. (doi:10.1006/nimg.2002.1325)

MacNeilage, P. F. & Davis, B. L. 2001 Motor mechanisms in speech ontogeny: phylogenetic, neurobiological and linguistic implications. *Curr. Opin. Neurobiol.* **11**, 696–700. (doi:10.1016/S0959-4388(01)00271-9)

Merzenich, M. & Brugge, J. 1973 Representation of the cochlear partition of the superior temporal plane of the macaque monkey. *Brain Res.* **50**, 275–296. (doi:10.1016/0006-8993(73)90731-2)

Miller, J. D. 1989 Auditory-perceptual interpretation of the vowel. *J. Acoust. Soc. Am.* **85**, 2114–2134. (doi:10.1121/1.397862)

Morosan, P., Rademacher, J., Schleicher, A., Amunts, K., Schormann, T. & Zilles, K. 2001 Human primary auditory cortex: cytoarchitectonic subdivisions and mapping into a spatial reference system. *NeuroImage* **13**, 684–701. (doi:10.1006/nimg.2000.0715)

Möttönen, R., Calvert, G., Jääskeläinen, I., Matthews, P., Thesen, T., Tuomainen, J. & Sams, M. 2006 Perceiving identical sounds as speech or non-speech modulates activity in the left posterior superior temporal sulcus. *NeuroImage* **30**, 563–569. (doi:10.1016/j.neuroimage.2005.10.002)

Näätänen, R. & Picton, T. 1987 The N1 wave of the human electric and magnetic response to sound: a review and an analysis of the component structure. *Psychophysiohgy* **24**, 375–425. (doi:10.1111/j.1469-8986.1987.tb00311.x)

Nygaard, L. & Pisoni, D. 1998 Talker-specific learning in speech perception. *Percept. Psychophys.* **60**, 355–376.

Palmer, A. R., Bullock, D. & Chambers, J. 1998 A high-output, high-quality sound system for use in auditory fMRI. *NeuroImage* **7**, S359.

Pandya, D. N. 1995 Anatomy of the auditory cortex. *Rev. Neurol. (Paris)* **151**, 486–494.

Pandya, D. N. & Sanides, F. 1973 Architectonic parcellation of the temporal operculum in rhesus monkey and its projection pattern. *Z. Anat. Entwicklungsgesch.* **139**, 127–161. (doi:10.1007/BF00523634)

Pandya, D. N. & Seltzer, B. 1982 Intrinsic connections and architectonics of posterior parietal cortex in the rhesus monkey. *J. Comp. Neurol.* **204**, 196–210. (doi:10.1002/cne.902040208)

Patterson, R. D. 1994 The sound of a sinusoid: time-interval models. *J. Acoust. Soc. Am.* **96**, 1419–1428. (doi:10.1121/1.410286)

Patterson, R. D., Robinson, K., Holdsworth, J., McKeown, D., Zhang, C. & Allerhand, M. 1992 Complex sounds and auditory images. In *Auditory physiology and perception, Proc. 9th Int. Symp. Hear*, (eds Y. Cazals, L. Demany & K. Horner), pp. 429–446. Oxford, UK: Pergamon.

Patterson, R. D., Allerhand, M. & Giguere, C. 1995 Time-domain modelling of peripheral auditory processing: a modular architecture and a software platform. *J. Acoust. Soc. Am.* **98**, 1890–1894. (doi:10.1121/1.414456)

Patterson, R. D., Handel, S., Yost, W. & Datta, A. 1996 The relative strength of the tone and noise components in iterated rippled noise. *J. Acoust. Soc. Am.* **100**, 3286–3294. (doi:10.1121/1.417212)

Patterson, R. D., Hackney, C. M. & Iversen, I. D. 1999 Interdisciplinary auditory neuroscience. *Trends Cognit. Neurosci.* **3**, 245–247. (doi:10.1016/S1364-6613(99)01347-9)

Patterson, R. D., Yost, W., Handel, S. & Datta, A. 2000 The perceptual tone/noise ratio of merged iterated rippled noises. *J. Acoust. Soc. Am.* **107**, 1578–1588. (doi:10.1121/1.428442)

Patterson, R. D., Uppenkamp, S., Johnsrude, I. & Griffiths, T. 2002 The processing of temporal pitch and melody information in auditory cortex. *Neuron* **36**, 767–776. (doi:10.1016/S0896-6273(02)01060-7)

Penagos, H., Melcher, J. & Oxenham, A. 2004 A neural representation of pitch salience in nonprimary human auditory cortex revealed with functional magnetic resonance imaging. *J. Neurosci.* **24**, 6810–6815. (doi:10.1523/JNEUROSCI.0383-04.2004)

Penhune, V. B., Zatorre, R. J., MacDonald, J. & Evans, A. 1996 Interhemispheric anatomical differences in human primary auditory cortex: probabilistic mapping and volume measurement from magnetic resonance scans. *Cereb. Cortex* **6**, 661–672. (doi:10.1093/cercor/6.5.661)

Petrides, M. & Pandya, D. N. 1984 Projections to the frontal cortex from the posterior parietal region in the rhesus monkey. *J. Comp. Neurol.* **228**, 105–116. (doi:10.1002/cne.902280110)

Petrides, M. & Pandya, D. N. 1988 Association fiber pathways to the frontal cortex from the superior temporal region in the rhesus monkey. *J. Comp. Neurol.* **273**, 52–66. (doi:10.1002/cne.902730106)

Petrides, M. & Pandya, D. N. 2002 Comparative cytoarchitectonic analysis of the human and the macaque ventrolateral prefrontal cortex and corticocortical connection patterns in the monkey. *Eur. J. Neurosci.* **16**, 291–310. (doi:10.1046/j.1460-9568.2001.02090.x)

Pisoni, D. 1997 Some thoughts on 'normalization' in speech perception. In *Talker variability in speech processing* (eds K. Johnson & J. W. Mullennix), pp. 9–32. San Diego, CA: Academic Press.

Price, C, Thierry, G. & Griffiths, T. 2005 Speech-specific auditory processing: where is it? *Trends Cogn. Sci.* **9**, 271–276. (doi:10.1016/j.tics.2005.03.009)

Rademacher, J., Caviness Jr, V., Steinmetz, H. & Galaburda, A. 1993 Topographical variation of the human primary cortices: implications for neuroimaging, brain mapping, and neurobiology. *Cereb. Cortex* **3**, 313–329. (doi:10.1093/cercor/3.4.313)

Rademacher, J., Morosan, P., Schormann, T., Schleicher, A., Werner, C, Freund, H. & Zilles, K. 2001 Probabilistic mapping and volume measurement of human primary auditory cortex. *NeuroImage* **13**, 669–683. (doi:10.1006/nimg.2000.0714)

Rauschecker, J. P. 1998 Parallel processing in the auditory cortex of primates. *Audiol. Neurootol.* **3**, 86–103. (doi:10.1159/000013784)

Rauschecker, J. P. & Tian, B. 2000 Mechanisms and streams for processing of 'what' and 'where' in auditory cortex. *Proc. Natl Acad. Sci. USA* **97**, 11800–11806. (doi:10.1073/pnas.97.22.11800)

Rauschecker, J. P., Tian, B., Pons, T. & Mishkin, M. 1997 Serial and parallel processing in rhesus monkey auditory cortex. *J. Comp. Neurol.* **382**, 89–103. (doi:10.1002/(SICI)1096-9861(19970526)382:K89::AID-CNE6>3.0.CO;2-G)

Rimol, L. M., Specht, K., Weis, S., Savoy, R. & Hugdahl, K. 2005 Processing of sub-syllabic speech units in the posterior temporal lobe: an fMRI study. *NeuroImage* **26**, 1059–1067. (doi:10.1016/j.neuroimage.2005.03.028)

Rivier, F. & Clarke, S. 1997 Cytochrome oxidase, acetyl-cholinesterase, and NADPH-diaphorase staining in human supratemporal and insular cortex. *NeuroImage* **6**, 288–304. (doi:10.1006/nimg.1997.0304)

Rizzolatti, G. & Arbib, M. 1998 Language within our grasp. *Trends Neurosci.* **21**, 188–194. (doi:10.1016/S0166-2236(98)01260-0)

Romanski, L., Bates, J. & Goldman-Rakic, P. 1999a Auditory belt and parabelt projections to the prefrontal cortex in the rhesus monkey. *J. Comp. Neurol.* **403**, 141–157. (doi:10.1002/(SICI)1096-9861(19990111)403:2<141::AID-CNE1>3.0.CO;2-V)

Romanski, L., Tian, B., Fritz, J., Mishkin, M., Goldman-Rakic, P. & Rauschecker, J. 1999b Dual streams of auditory afferents target multiple domains in the primate prefrontal cortex. *Nat. Neurosci.* **2**, 1131–1136. (doi:10.1038/16056)

Rupp, A., Uppenkamp, S., Gutschalk, A., Beucker, R., Patterson, R. D., Dau, T. & Scherg, M. 2002 The representation of peripheral neural activity in the middle-latency evoked field of primary auditory cortex in humans(1). *Hear. Res.* **174**, 19–31. (doi:10.1016/S0378-5955(02)00614-7)

Scott, S. K. & Johnsrude, I. 2003 The neuroanatomical and functional organization of speech perception. *Trends Neurosci.* **26**, 100–107. (doi:10.1016/S0166-2236(02)00037-1)

Seltzer, B. & Pandya, D. N. 1978 Afferent cortical connections and architectonics of the superior temporal sulcus and surrounding cortex in the rhesus monkey. *Brain Res.* **149**, 1–24. (doi:10.1016/0006-8993(78)90584-X)

Seltzer, B. & Pandya, D. N. 1989a Frontal lobe connections of the superior temporal sulcus in the rhesus monkey. *J. Comp. Neurol.* **281**, 97–113. (doi:10.1002/cne.902810108)

Seltzer, B. & Pandya, D. N. 1989b Intrinsic connections and architectonics of the superior temporal sulcus in the rhesus monkey. *J. Comp. Neurol.* **290**, 451–471. (doi:10.1002/cne.902900402)

Seltzer, B. & Pandya, D. N. 1991 Post-rolandic cortical projections of the superior temporal sulcus in the rhesus monkey. *J. Comp. Neurol.* **312**, 625–640. (doi:10.1002/cne.903120412)

Sheffert, S. M., Pisoni, D., Fellowes, J. & Remez, R. 2002 Learning to recognize talkers from natural, sinewave, and reversed speech samples. *J. Exp. Psychol. Hum. Percept. Perform.* **28**, 1447–1469. (doi:10.1037/0096-1523.28.6.1447)

Terhardt, E. 1974 Pitch, consonance, and harmony. *J. Acoust. Soc. Am.* **55**, 1061–1069. (doi:10.1121/1.1914648)

Thierry, G., Giraud, A. L. & Price, C. 2003 Hemispheric dissociation in access to the human semantic system. *Neuron* **38**, 499–506. (doi:10.1016/S0896-6273(03)00199-5)

Unoki, M., Irino, T., Glasberg, B., Moore, B. C. J. & Patterson, R. D. 2006 Comparison of the roex and gammachirp filters as representations of the auditory filter. *J. Acoust. Soc. Am.* **120**, 1474–1492. (doi:10.1121/1.2228539)

Uppenkamp, S., Johnsrude, I., Patterson, R. D., Norris, D. & Marslen-Wilson, W. 2006 Locating the initial stages of speech-sound processing in human temporal cortex. *NeuroImage* **31**, 1284–1296. (doi:10.1016/j.neuroimage.2006.01.004)

van Dinther, R. & Patterson, R. D. 2006 Perception of acoustic scale and size in musical instrument sounds. *J. Acoust. Soc. Am.* **120**, 2158–2176. (doi:10.1121/1.2338295)

Vlaardingerbroek, M. & den Boer, J. A. 1996. *Magnetic resonance imaging*. New York, NY: Springer.

von Kriegstein, K. & Giraud, A. L. 2004 Distinct functional substrates along the right superior temporal sulcus for the processing of voices. *NeuroImage* **22**, 948–955. (doi:10.1016/j.neuroimage.2004.02.020)

von Kriegstein, K., Warren, J. D., Ives, D. T., Patterson, R. D. & Griffiths, T. D. 2006 Processing the acoustic effect of size in speech sounds. *NeuroImage* **32**, 368–375. (doi:10.1016/j.neuroimage.2006.02.045)

Wallace, M. N., Johnston, P. W & Palmer, A. R. 2002 Histochemical identification of cortical areas in the auditory region of the human brain. *Exp. Brain Res.* **143**, 499–508. (doi:10.1007/s00221-002-1014-z)

Watkins, K. & Paus, T. 2004 Modulation of motor excitability during speech perception: the role of Broca's area. *J. Cogn. Neurosci.* **16**, 978–987. (doi:10.1162/0898929041502616)

Watkins, K. E., Strafella, A. P. & Paus, T. 2003 Seeing and hearing speech excites the motor system involved in speech production. *Neuropsychologia* **41**, 989–994. (doi:10.1016/S0028-3932(02)00316-0)

Wilson, S. M., Saygin, A. P., Sereno, M. I. & Iacoboni, M. 2004 Listening to speech activates motor areas involved in speech production. *Nat. Neurosci.* **7**, 701–702. (doi:10.1038/nn1263)

Yeterian, E. H. & Pandya, D. N. 1998 Corticostriatal connections of the superior temporal region in rhesus monkeys. *J. Comp. Neurol.* **399**, 384–402. (doi:10.1002/(SICI)1096-9861(19980928)399:3<384::AID-CNE7>3.0.CO;2-X)

Yost, W A., Patterson, R. & Sheft, S. 1996 A time domain description for the pitch strength of iterated rippled noise. *J. Acoust. Soc. Am.* **99**, 1066–1078. (doi:10.1121/1.414593)

9

Fronto-temporal brain systems supporting spoken language comprehension

Lorraine K. Tyler and William Marslen-Wilson

> The research described here combines psycholinguistically well-motivated questions about different aspects of human language comprehension with behavioural and neuroimaging studies of normal performance, incorporating both subtractive analysis techniques and functional connectivity methods, and applying these tasks and techniques to the analysis of the functional and neural properties of brain-damaged patients with selective linguistic deficits in the relevant domains. The results of these investigations point to a set of partially dissociable sub-systems supporting three major aspects of spoken language comprehension, involving regular inflectional morphology, sentence-level syntactic analysis and sentence-level semantic interpretation. Differential patterns of fronto-temporal connectivity for these three domains confirm that the core aspects of language processing are carried out in a fronto-temporo-parietal language system which is modulated in different ways as a function of different linguistic processing requirements. No one region or sub-region holds the key to a specific language function; each requires the coordination of activity within a number of different regions. Functional connectivity analysis plays the critical role of indicating the regions which directly participate in a given sub-process, by virtue of their joint time-dependent activity. By revealing these co-dependencies, connectivity analysis sharpens the pattern of structure–function relations underlying specific aspects of language performance.
>
> **Keywords:** fronto-temporal language system; connectivity analysis; neural language system

9.1 Introduction

In understanding normal spoken language, the listener is confronted with a flow of rapidly accumulating and dynamically varying acoustic–phonetic information. This needs to be broken down into its constituent words and morphemes so that the information carried by these primary linguistic units can be used to construct an interpretation of the message being transmitted, weaving together cues to the linguistic structure and the meaning. Over the last decade, a framework has begun to emerge for understanding these capacities from a cognitive neuroscience perspective. This cross-disciplinary perspective combines novel inputs from the neurobiology of primate auditory processing systems and from structural and functional neuroimaging of the intact and damaged human brain, with the older traditions of the neurological and neuropsychological study of language and the brain.

An outcome of this combination of sources is a renewed emphasis on the bi-hemispheric foundations of primate auditory and human speech communication systems, moving away from the classical view of language as a purely left hemisphere (LH) phenomenon, as exemplified in the standard Broca–Wernicke–Lichtheim diagram (Figure 9.1). Recent research with non-human primates highlights the underlying hemispheric symmetry of the auditory processing systems upon which human speech comprehension systems are presumably built, although several aspects of critically linguistic (as opposed to auditory) processing show clear asymmetries, as we describe below. These studies with non-human primates show that bilateral inputs to auditory processing areas in superior temporal cortex (core, belt and parabelt) link to major processing streams, running dorsally and ventrally to processing regions in frontal cortex, inferior parietal areas and other temporal lobe areas (e.g. Kaas & Hackett 1999; Rauschecker & Tian 2000).

Although this analysis has increasingly been adopted as a template for thinking about the organization of the speech and language processing system in the human brain (e.g. Hickok & Poeppel 2000; Scott & Johnsrude 2003), two major caveats are in order. The first is that the human brain diverges in many respects from the macaque brain, but most extensively in the anterior temporal lobe and frontal lobe areas that are critically involved in the systems postulated—for example, the macaque entirely lacks the middle temporal gyrus that is a prominent and functionally significant part of the human brain. The second is that a system designed to support spoken language will need to make different and additional functional demands to those served by the macaque system. Nonetheless, the emergence of a well-specified account of the neurobiological underpinnings of primate auditory processing has had important consequences. It provides a model for what a theory of these systems needs to look like, in terms of the specificity of both the functional and the neural account that is provided, and it suggests a very different approach to the characterization of human language function.

Classical cognitive and psycholinguistic approaches to the functional structure of the system for mapping from sound to meaning have always assumed that a single, unitary process (or succession of processes) is engaged to carry out this mapping. This is reflected in the focus in these models on a single neural system involving inferior frontal cortex (especially Broca's area) and posterior temporal cortex (Wernicke's area) and the major white matter tract (the arcuate fasciculus) that connects them (Figure 9.1). The neurobiological evidence suggests, however, that the underlying neural system is not organized

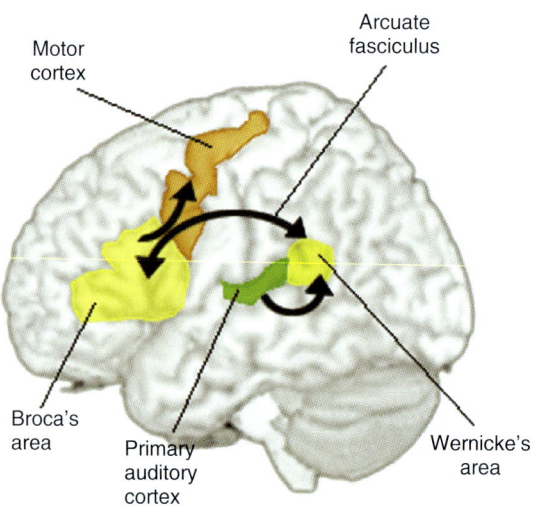

Figure 9.1 Classical model of the LH language system.

along these lines and that multiple parallel processing streams are involved, extending hierarchically outwards from auditory cortex in both posterior and anterior directions. Instead of language being processed primarily within a single dorsal stream, strong evidence is emerging that a substantial ventral stream is also involved (Hickok & Poeppel 2004). This is thought to extend from posterior temporo-parietal sites via the superior and middle temporal cortex to the anterior temporal and orbito-frontal cortex, by means of the white matter tracts of the inferior longitudinal fasciculus and the uncinate fasciculus. Although these dorsal and ventral streams potentially exist in both hemispheres, recent research suggests that there are considerable hemispheric asymmetries. For example, using new techniques such as diffusion tensor imaging (DTI), which examine the structure of white matter tracts, Buchel *et al.* (2004) have shown an increased white matter composition of the left arcuate fasciculus and inferior longitudinal fasciculus in healthy subjects. Other studies using DTI tractography have suggested that there may be even more marked hemispheric asymmetry. Parker *et al.* (2005), for example, have claimed that while the arcuate fasciculus is reliably observed in both the hemispheres across subjects, the ventral stream is only seen in the LH.

This anatomical asymmetry seems to be reflected in functional asymmetry. Although a combination of inputs from both novel and historical sources lend some support to both the active role of right hemisphere (RH) structures in language function and the existence of multiple processing streams (although these tend to be more left lateralized), the role of the RH appears to be limited. A number of functional imaging studies have shown that speech processing in humans activates bilateral temporal lobe structures in and around primary auditory processing areas (e.g. Zatorre *et al.* 1992; Zatorre & Gandour 2008; though see Scott & Wise (2003) for a critique of some earlier studies). Moreover, this bilateral activation is not only limited to low-level acoustic and phonetic analyses, but also implicates lexical analysis processes—mapping the speech input onto lexical representations (Binder *et al.* 2000; Scott & Wise 2004), as reflected both in studies using haemodynamic techniques such as PET and fMRI and those using electrophysiological techniques such as EEG and MEG (Marinkovic *et al.* 2003).

RH involvement at these levels is consistent with the neuropsychological evidence that patients with extensive damage to LH perisylvian language areas (L frontal and superior temporal regions) but spared RH can still recognize spoken words and access lexical meaning (Tyler *et al.* 1995a, 2002a; Dronkers *et al.* 2004). For example, such patients typically show semantic priming effects and reaction times within the normal range for spoken words—but only when they are morphologically simple, such as *desk*, *rabbit*, etc. (Tyler *et al.* 1995a,b, 2002a,b; Longworth *et al.* 2005). These normal priming effects suggest that the patients do not access semantic information more slowly than unimpaired listeners, contrary to earlier claims that patients with left perisylvian lesions are slow to access semantic information (Milberg *et al.* 1987). Moreover, given that the patients produce normal patterns of priming in the face of extensive LH damage, this suggests that the RH can support quite extensive processing of simple words.

Nonetheless, despite this evidence for a degree of bilateral parallelism in some aspects of language function, it is also clear that the most critical language functions depend on an intact left-dominant perisylvian core language system, linking left inferior frontal cortex (LIFC) with temporal and posterior parietal cortices. Damage to these regions can cause a permanent disruption of some key language functions while damage to parallel regions on the right generally does not. A particularly salient feature here is the disruption of the combinatorial aspects of language function—those processes which involve combining linguistic elements into more complex entities. Lexically based combinatorial processes typically combine morphemes into complex words through processes of derivation (*manage + ment = management*) or inflection (*jump + ed = jumped*), while syntactic combination involves combining words into syntactic phrases (e.g. noun phrases such as *the new book*, verb phrases such as *they walk carefully* or prepositional phrases such as *under the bricks*).

Patients with LH damage, especially involving LIFC, frequently have problems with syntax and inflectional morphology, both in production and in comprehension (Caplan & Futter 1986; Goodglass *et al.* 1993), even though the processing of simple, concrete words may remain relatively unimpaired. In a series of studies with such patients, we have shown that their comprehension of spoken inflected words (such as *blessed, jumped*) is significantly impaired (Marslen-Wilson & Tyler 1997, 1998; Tyler *et al.* 2002*a,b*; Longworth *et al.* 2005). These patients also typically have problems processing syntactic structure, although syntactic processing deficits are not confined to patients who only have damage to the LIFC. It is often difficult to attribute syntactic deficits solely to damage of the LIFC since damage to this region often also involves damage to proximate areas of the left superior temporal gyrus (STG). An important observation here is that patients who only have L posterior STG damage can also have problems with syntactic processing (Caplan & Hildebrant 1988; Caplan *et al.* 1996). We have observed the same kinds of behavioural deficits for syntactic processing in patients with intact LIFC but damage to L posterior STG/MTG, supporting the importance of both frontal and temporal cortex in spoken language function (see below and Figures 9.11 and 9.12). Evidence from neuroimaging confirms the salience of LH contributions to combinatorial aspects of language function, primarily located in inferior frontal and temporal cortex, and in regions around inferior parietal cortex (Demonet *et al.* 1992; Zatorre *et al.* 1996; Binder *et al.* 2000; Friederici *et al.* 2003; Scott & Johnsrude 2003; Stamatakis *et al.* 2005; Tyler *et al.* 2005*c*), together with some involvement of the RH, albeit to a lesser extent.

But while there is broad agreement on these general aspects of spoken language processing, there is still considerable disagreement about the detailed properties of the neural language system and how different language processes are instantiated within it. Even very basic functions, such as the fundamental process of mapping speech sounds onto semantic representations, are not well understood, perhaps because few studies are underpinned by theoretical claims about the cognitive processes involved. Attempts to characterize the sound-meaning mapping range from the claim that it takes place within a hierarchically organized speech processing stream within the STG/STS, possibly bilaterally, with posterior regions engaged in the processing of form and anterior regions engaged in the processing of meaning (Scott *et al.* 2000; Scott & Johnsrude 2003), to the view that L posterior superior temporal regions play the crucial role, with the emphasis varying between Wernicke's area (BA 22; Hillis *et al.* 2001; Mesulam *et al.* 2003), the junction between L posterior temporal cortex and inferior parietal cortex (Mummery *et al.* 1999; Binder *et al.* 2000), and L posterior MTG (Dronkers *et al.* 2004; Indefrey & Cutler 2004). Still other researchers identify the L posterior middle and inferior temporal cortex as the critical site (Hickok & Poeppel 2004).

Similarly, the neural instantiation of other basic language processes, such as those underpinning syntactic and morphological analyses, remains unclear and controversial.

Although there is convincing evidence that superior temporal cortex, possibly bilaterally, is engaged by processes involving morphological and syntactic combination (Friederici *et al.* 2003; Rodd *et al.* 2004; Tyler *et al.* 2005*b*), the role of other cortical regions in these processes is less certain. For example, while the LIFC, including Broca's area, is reliably activated when listeners are processing spoken language, its functions remain surprisingly contentious (see Kaan & Swaab 2002 for a review), despite its central role in most models of the neural language system since Paul Broca. On the one hand, sub-regions of LIFC are claimed to be functionally specialized for highly specific linguistic processes, such as syntactic parsing (Grodzinsky 2000; Friederici *et al.* 2003) or phonological analysis (Stromswold *et al.* 1996; Zatorre *et al.* 1996; Hagoort 2003), while on the other the LIFC is claimed to support general functions such as retrieval of linguistic information stored in posterior brain regions (e.g. Bokde *et al.* 2001; Gold & Buckner 2002), selection among competing alternatives (Thompson-Schill *et al.* 1997, 2005) or the maintenance of information in working memory (e.g. Gabrieli *et al.* 1998). Even the question of whether its role is specific to language is yet to be resolved, with several suggestions that LIFC supports processes shared by multiple cognitive domains (e.g. Miller 2000).

In summary, therefore, as this brief review indicates, there is still considerable uncertainty about the properties of the basic components of the neural language system, about the precise contribution of the regions themselves and how they operate together to support the dynamic processes of language comprehension and production. This state of affairs reflects, at least in part, the implicitly 'phrenological' assumptions underlying much current and historical research—namely a focus on delineating the functional specialization of specific brain areas, much as Broca and Wernicke originally attempted in their pioneering proposals nearly 150 years ago. To make progress, it is also necessary to focus on the nature of the functional relationships between the anatomically distinct regions which have been identified as constituting the neural language system, and how they are affected by different linguistic inputs. At the same time, however, it is necessary to do so in the context of a theory of the functional organization of the system as a cognitive process—in the current context, how speech inputs are mapped onto lexical representations and how these relate to processes of syntactic and semantic analysis. In the remainder of this paper, we outline recent research which attempts to address these issues.

In doing so, working with both intact and brain-damaged populations, we combine subtractive neuroimaging techniques with recent developments that provide analytic methods for examining the ways in which different regions within the neural language system interact with each other by analysing functional connectivity between cortical regions. This enables us to go beyond a description of the neural language system in terms of levels of activity in isolated regions, by determining the ways in which activity in one region modulates activity within another. These analyses capture time-dependent changes in the coupling or decoupling of anatomically remote brain areas, allowing us to study integration in the brain in the context of changing task conditions in a dynamic manner (Fletcher *et al.* 1999). We have exploited this type of approach, in combination with more conventional subtractive analyses of neuroimaging data, in order to determine the nature of the interactions between different cortical regions with respect to two core sets of linguistic functions, covering combinatorial operations in the lexical and in the grammatical domains.

9.2 Functional organization of the fronto-temporal language system: words and morphemes

We focus first on the lexical domain, examining the representational and processing consequences of inflectionally complex words such as *jumped* or *smiles*. These are forms, very common in English, made up of a noun or a verb stem and an inflectional morpheme ('jump' + '-ed') or ('smile' + '-s'). Inflectional (or grammatical) morphemes are particularly revealing because they link across the two fundamental combinatorial domains of lexical and syntactic combination. Forms like *jumped* or *smiles*, reflecting the operations of regular inflectional morphology in English, on the one hand engage and challenge the basic systems of lexical access, whereby phonological forms are mapped onto internal representations of lexical form and meaning. On the other hand, the information carried by these forms—especially the inflectional morphemes themselves—has a critical role to play in combining incoming words and morphemes into higher order structures. We therefore begin this analysis of functional connectivity by examining the processing of regularly inflected verbs and nouns. These enable us to probe basic processes of lexical access, the morpho-phonological parsing operations required by inflected forms and the structural processes that depend on the information carried by grammatical morphemes. This set of processes implicates key temporal and frontal lobe structures in these basic linguistic operations.

In considering these processes, we need to take the *morpheme*—the minimal meaning bearing element in human language—as a basic building block. Broadly speaking, morphemes can either carry semantic content or can function to communicate grammatical information of various types. In a language like English, semantic morphemes can almost always occur as monomorphemic 'free stems' as in words such as *dog*, *smile*, *tidy*, etc.—whereas grammatical morphemes are often 'bound' and only occur in conjunction with semantic morphemes. The plural morpheme {-s}, for example, can only appear in combination with a free stem such as {dog}, creating the form *dogs*. Similar constraints hold for the past tense morpheme {-d}, in forms like *smiled* or *jumped*, as well as for a large range of derivational morphemes (e.g. {-ness} combining with {tidy} to create the form *tidiness*).

Consistent with extensive neuropsychological evidence and at least some neuroimaging evidence, we assume that lexical access processes involving mono morphemic content words—the initial mapping of acoustic–phonetic information in the speech signal onto stored lexical representations of form and meaning—are mediated by brain regions in the superior and middle temporal lobes. As noted earlier, these lexical access processes seem to be supported bilaterally, although there is undoubtedly some degree of LH dominance. An increasing body of data, from both neuropsychological and neuroimaging sources, indicates that morphologically complex words involving regular inflectional morphology require these temporal lobe access processes to interact with inferior frontal areas, primarily via a so-called 'dorsal' route involving the arcuate fasciculus, likely to be critical for morpho-phonological parsing.

Perhaps the most direct evidence for the involvement of this dorsal route—as opposed to the ventral route likely to be active in more semantic aspects of language comprehension (as discussed later)—comes from a recent study using a lesion–behaviour correlational technique (Tyler *et al.* 2005*a*). This is a new methodology which correlates scores on two continuous variables—signal intensity of each voxel across the entire brain of brain-damaged patients and their continuous behavioural scores, in this case from a priming study. We used this method to determine whether disruption to the processing of regularly or irregularly inflected past tense forms is associated with damage to different brain regions. We correlated signal intensity across the brains of 22 right-handed brain-damaged patients, with the patients' behavioural scores on a priming study which tested their ability to process the phonological form, meaning and morphological structure of spoken words.

In the priming study, patients heard prime–target stimulus pairs and made a lexical decision to the second (target) stimulus in each pair. We compared word-pairs which were either regularly inflected past tense forms (e.g. *jumped-jump*) or irregularly inflected past tense forms (e.g. *slept-sleep*), or related only in phonological form (e.g. *pillow-pill*) or only in meaning (e.g. *card-paper*). Different neural regions correlated with behavioural priming scores in the four conditions. Priming for regularly inflected past tense words correlated most strongly with variations in signal intensity in the LIFG (BA 47, 45), as shown in Figure 9.2a. At a slightly lower threshold (Figure 9.2b), this cluster included a large

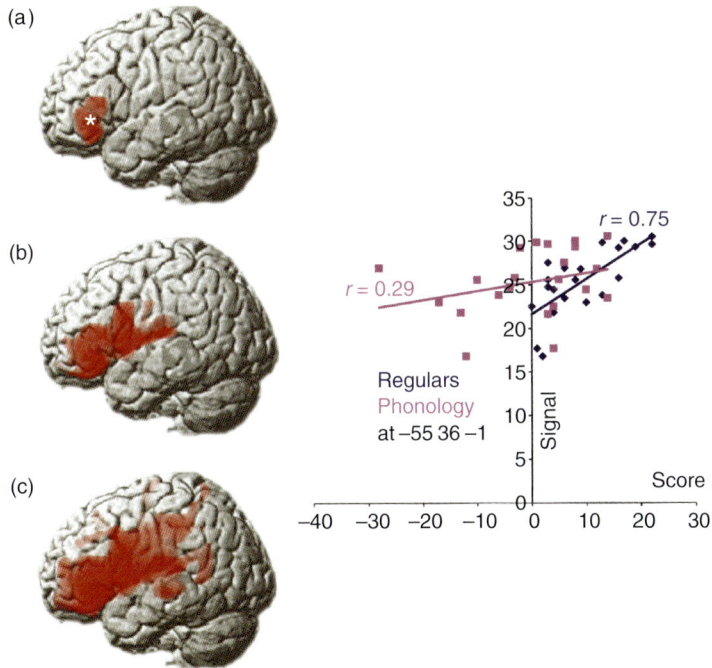

Figure 9.2 Structural correlates of regular inflection. Three-dimensional reconstructions of a T1-weighted MRI image showing brain areas where variations in signal density correlate with priming for regularly inflected words at: (a) $p < 0.001$, (b) $p < 0.01$ and (c) $p < 0.05$ voxel thresholds. The clusters shown survived correction at $p < 0.05$ cluster level adjusted for the entire brain. The statistical peak (–55, 36, –1) is in the left inferior frontal gyrus (BA 47), and the cluster extends superiorly into BA 45. At lower thresholds, the cluster extends from Broca's to Wernicke's areas and includes the arcuate fasciculus. The scatter plot shows the relationship between variations in signal density at the most significant voxel (see asterisk on (a)) and individual behavioural scores in the regular and the non-morphological phonological conditions. Adapted from Tyler *et al.* (2005a).

region of left superior temporal gyrus (LSTG) extending from the anterior extent of Wernicke's area (BA 41, 42) to the anterior LSTG. When the threshold was lowered still further (Figure 9.2c), all of Wernicke's area was included, looping around to include the arcuate fasciculus and Brodmann areas 47, 44 and 45 (Broca's area). This essentially replicates the classical Broca–Wernicke–Lichtheim model of language function, where the arcuate fasciculus connects superior temporal and inferior frontal regions in a neural language system (Figure 9.1), and is similar to the dorsal route identified in more recent neural accounts of the language system (e.g. Hickok & Poeppel 2000).

Priming for the irregularly inflected past tense forms, in contrast, correlated with signal intensity in completely different neural regions—the left superior parietal lobule, inferior parietal lobule and angular gyrus. These regions have been associated with irregular past tense processing in previous neuroimaging studies (Jaeger *et al*. 1996; Beretta *et al*. 2003) and are often reported as being activated in lexical processing tasks (Demonet *et al*. 1992; Celsis *et al*. 1999). The role of these regions in lexical processing is confirmed by the finding that when they are damaged, which typically occurs following lesions in Wernicke's area, the patient exhibits the speech comprehension deficits typical of Wernicke's aphasia (Selnes *et al*. 1985; Kertesz *et al*. 1993). Wernicke's area and the surrounding parietal regions are thought to be involved in the mapping between spoken forms and their meanings.

In a further series of behavioural and neuroimaging experiments, we used the contrast between words with regular and irregular past tense morphology to build up a more dynamic picture of the fronto-temporal systems underlying combinatorial morphological processes involving inflectional morphemes. The value of this contrast is that it provides sets of words that are matched for lexical and grammatical properties, but which differ in whether or not they are decomposable, via morpho-phonological parsing processes, into a stem morpheme plus an inflectional affix. This always holds for regularly inflected forms, like *smiled* or *jumped*, but not for irregular past tense forms, such as *bought* or *gave*, which cannot be decomposed into a stem plus an affix, and where the idiosyncratic and unpredictable nature of each form means that they have to be learnt and stored as unanalysable whole forms.

The importance of morpho-phonological parsing processes in the perceptual analysis of regular inflectional morphology was highlighted in a previous study with brain-damaged patients, where we found that patients with left perisylvian lesions had difficulties in processing stimuli (whether real words or non-words) that ended in either real or potential inflectional morphemes (Tyler *et al*. 2002*a, b*). These patients were impaired not only on regularly inflected past tense forms such as *played*, but also on real words like *trade* and non-words like *snade* which shared the specific phonological features that are diagnostic of the presence of a potential inflectional suffix. These features—a word-final coronal consonant (typically /d, t, s, z/) that agrees in voice with the preceding segment—we refer to as the *inflectional rhyme pattern*. The fact that the patients were impaired on all three types of words, but had no difficulty in processing words which did not contain these phonological properties, suggested that the processes of morpho-phonological parsing were disrupted in these patients because of damage to their L perisylvian language areas. These were patients where processing of monosyllabic content words was generally spared, based on relatively intact temporal lobe systems (LH and/or RH), but where the fronto-temporal system linking basic lexical access with combinatorial morpho-syntactic processing was in some way disrupted.

These hypotheses about patient performance made clear predictions about how the presence or absence of morpho-phonological complexity should affect the distribution of neural activity in the intact brain. We pursued these predictions in an fMRI study with healthy subjects (Tyler *et al.* 2005c), following this up with a re-analysis of the same data in a functional connectivity framework (Stamatakis *et al.* 2005). Listeners heard spoken word-pairs, such as *played-play* or *played-played*, and indicated, by means of a button-press, whether the pairs were the same or different. The same three types of real and pseudo-inflected forms were used (*played*, *trade* and *snade*), all sharing the inflectional rhyme pattern, as well as a matched set of real and non-word pseudo-irregulars (*thought*, *port* and *hort*), which do not end in potential suffixes. These were embedded in same–different pairs such as *thought-think*, *thought-thought*, etc. A third set consisted of simple words which have no morphological structure (e.g. *shelf-shell*) and which also did not end in potential suffixes. However, these pairs were similar to the regulars in sharing the same minimal (one phoneme) difference between word-pairs, controlling for the possibility that differential effects for the regulars might simply reflect the difficulty of making the same–different decision between stimuli which are perceptually very similar. If the neural language system is differentially sensitive to phonological cues which signal morphological decomposition, then we would expect a different pattern of activation for the regularly inflected sets compared with either the irregularly inflected or the simple sets, neither of which can be decomposed and must be accessed as full forms.

The fMRI analyses (Tyler *et al.* 2005c) showed that stimuli containing phonological cues to the presence of a potential suffix preferentially activated a fronto-temporal network, including anterior cingulate cortex (ACC), LIFG, bilateral STG, L inferior parietal lobule (LIPL) and bilateral MTG, over and above activation for the irregular sets (Figures 9.3a,b) or the simple words (Figure 9.3c). There were no regions that were significantly more activated for the irregulars when compared with the regulars, or for the simple words when compared with the regulars. This increased LIFG activation arguably reflects additional processes of morpho-syntactic analysis which are required for parsing regularly inflected forms into their stems and morphological affix. The finding that LIFG activation was obtained for inflected non-words as well as for real words, suggests that it is the morpho-phonological structure (real or potential) of stimuli containing the inflectional rhyme pattern that produces the additional activation. Further evidence for this comes from a comparison between two sets of non-words—regular non-words (e.g. *crade-cray*) and simple non-words (e.g. *blane-blay*). This contrast showed an increased activation for the regular non-words when compared with the simple non-words but only in the LIFG and not in the LMTG or STG. When neither a stem nor a whole word is accessed (as is the case for non-words), then there is no differential STG/MTG activity, but the inflectional rhyme pattern present in the regular non-words still triggers the LIFG.

We attribute the increased LMTG and STG activation (Figure 9.3) for real regular and pseudo-inflected forms to the special processing demands made by such forms. Although *jump*, or any other uninflected stem, can map straightforwardly onto temporal lobe lexical representations, the presence of a past tense affix, as in the form *jumped*, seems to place additional demands on this access process. To interpret *jumped* correctly, and to allow the process of lexical access to proceed normally, the recognition system requires the simultaneous access of the lexical content associated with the stem *play* and of the grammatical implications of the {-d} morpheme. This seems to require both an intact LIFC and intact links to left superior temporal cortex. Note that irregular past tense

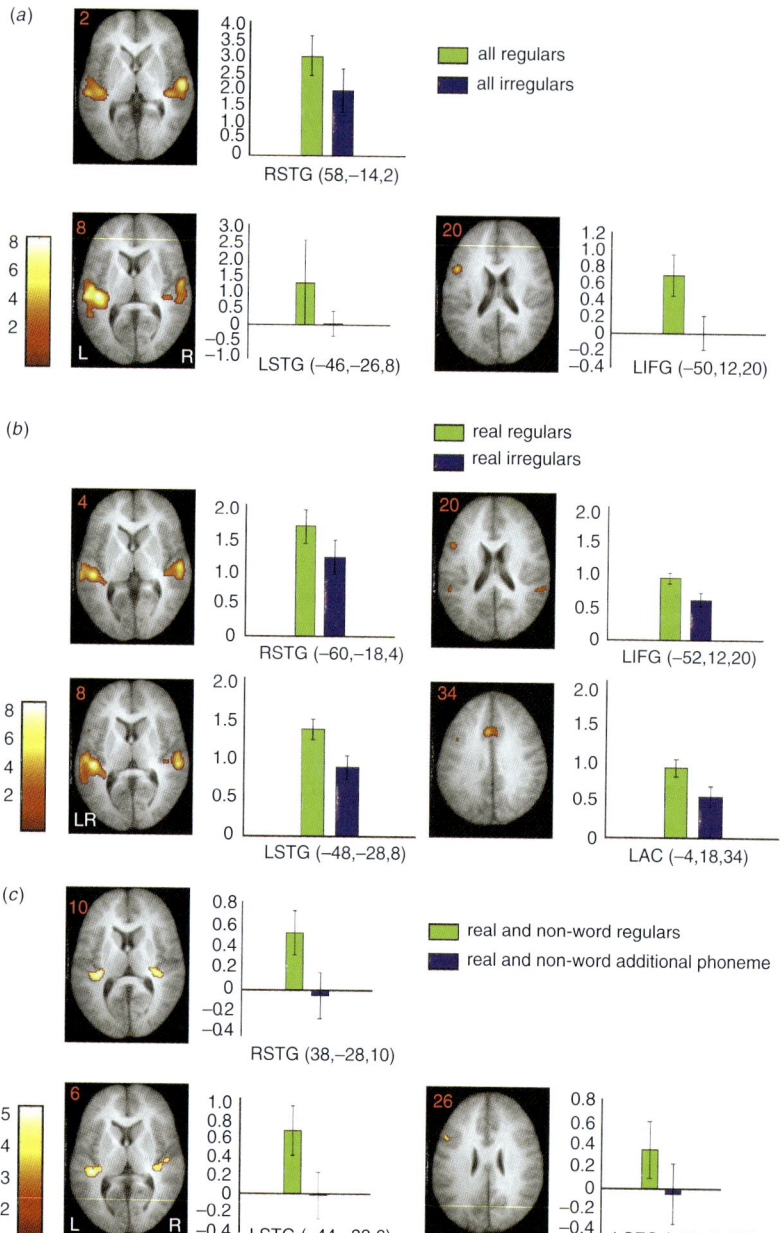

Figure 9.3 Functional correlates of regular inflection. The activations are superimposed on the mean T1 image of 18 volunteers. (a) Significant activations for the overall contrast of regulars (real, pseudo and non-word) minus irregulars (real, pseudo and non-word). Clusters were found in the RSTG, LSTG and LIFC. Activation peaks are given in brackets. (b) Significant activations for the contrast of real regulars minus real irregulars. Clusters were found in the RSTG, LSTG, LACC and LIFC. (c) Significant activations for the contrast of regulars (real and non-word) versus additional phoneme (real and non-word). Clusters were found in the RSTG, LSTG and LIFC. Adapted with permission from Tyler et al. (2005c). Copyright © Elsevier.

forms, which are never realized as an unchanged stem plus an affix, are not subject to the same additional processing requirement. They are assumed to be accessed as whole forms, exploiting the same temporal lobe systems as uninflected stems. Although irregular past tense forms will activate LIFC to some extent, owing to their morpho-syntactic implications, immediate access to their lexical meaning does not obligatorily require LIFC phonological parsing functions in the same way as regular past tense forms.

Independent evidence for this functional analysis comes from recent priming results (Longworth *et al.* 2005) demonstrating that patients with LIFC damage and difficulties with regular inflectional morphology also show deficits in semantic priming when the primes are regularly inflected forms, as in pairs like *jumped-leap*. At the same time, critically, they show normal performance both for pairs with stems as primes, as in *jump-leap*, and for pairs where the prime is an irregular past tense form, as in *shook-tremble*. Normal semantic priming performance in these auditory–auditory paired priming tasks requires rapid access to lexical semantic representations in the processing of both prime and target. The patients' preserved performance for stem and irregular spoken primes shows that the systems supporting fast access of meaning from speech are still intact for these types of input. The decrement in performance on the regular inflected forms means that these inflected forms make special processing demands and that an intact LIFC (and intact dorsal fronto-temporal links) are necessary to meet these demands.

A key component of this account is the claim that these special processing demands are automatically elicited by any input that shares the diagnostic properties of an inflectional affix, whether or not these forms correspond to existing lexical representations. Unless the system attempts the morpho-phonological segmentation of forms like *trade* or *snade*, it cannot rule out the possibility either that the pseudo-regular *trade* is actually the real regular *tray* in the past tense or that *snade* is the past tense of the potentially real stem *snay*. This, we argue, requires obligatory access to left inferior frontal regions. Additional evidence to support the across-the-board impact of the inflectional rhyme pattern comes from a recent behavioural study with intact young adults (Post *et al.* 2004), which not only replicates the finding that real, pseudo and non-word regulars group together against a range of control conditions, but also suggests that similar contrasts apply to English {-s} inflections, as in *jumps* or *yards*, which obey the same constraints of coronality and agreement in voice.

In summary, therefore, the increased activation for regulars (and pseudo-regulars) in temporal and inferior frontal areas reflects, on the one hand the specialized role of LIFC processes involved in analysing grammatical morphemes, and on the other the continuing STG/MTG activity involved in accessing lexical representations from the stems of regular and pseudo-regular inflected forms. The LIFC functions invoked here are likely to include support both for morpho-phonological parsing, segmenting complex forms into stems and affixes, and for syntactic processes triggered by the presence of grammatical morphemes such as the past tense marker.

A final consideration here is the potential control processes which regulate the proposed processing relationship between L frontal and temporal regions. Several lines of evidence suggest that the integration of information between superior temporal and L frontal areas may be modulated by anterior midline structures including the anterior cingulate, which both neuroanatomical and functional neuroimaging evidence suggest is well suited for this role. Work with non-human primates shows that the ACC projects to or receives connections from most regions of frontal cortex (Barbas 1995) and from superior temporal cortex (Pandya *et al.* 1981). Recent neuroimaging data not only implicate the ACC in the modulation of fronto-temporal integration (Fletcher *et al.* 1999), but also show it to be active in situations requiring the monitoring of interactions between different information processing pathways (Braver *et al.* 2001).

In this view, the increased activation of the ACC by real regular inflected forms (Figure 9.3b) may reflect the greater demands made on this monitoring function when complex forms such as *jumped* need to be parsed into a stem plus affix, with the bare stem then being able to act as a well-formed input to STG lexical access processes. The nature of this potential ACC contribution is examined in more detail in the connectivity analyses described below.

9.3 Functional connectivity in the intact and damaged brain: words and morphemes

The research described so far provides evidence for the activity of both frontal and temporal structures in the processing of morpho-phonologically complex words, combining evidence from behavioural studies with intact and brain-damaged populations with neuroimaging studies of the intact adult brain. We followed up the subtractive analyses reported above (Figure 9.3) with a series of connectivity analyses on the same data, in order to address more directly the functional relationship between the regions within the fronto-temporal language system. To do so, we used an approach which extended earlier proposals of Friston *et al.* (1997), aimed at identifying how the covariance between two regions could be modulated by a psychological variable or alternatively by the level of activity in a third region. The former was referred to as a psycho–physiological interaction and the latter as a physio–physiological interaction. In our study, we extended this approach to include both the psychological variable (morphological complexity) and the activity in a third region. This is therefore referred to as a psycho-physio-physiological interaction (Stamatakis *et al.* 2005).

The resulting connectivity analysis (Figure 9.4) shows that the LH regions identified in the subtractive analyses (Figures 9.3a, b), in LIFC and ACC, predict activity in L posterior middle temporal gyrus (MTG) for regularly inflected forms when compared with

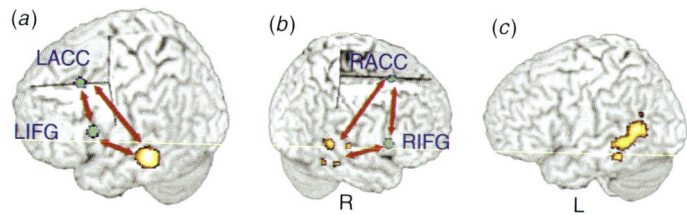

Figure 9.4 Functional connectivity analysis of regular inflection. Connectivity analysis in a group of healthy volunteers (20–40 years; based on data reported in Stamatakis et al. 2005). (a) The three-way interaction showing that the LACC predicts greater fronto-temporal interaction (LIFG and LMTG) in the context of regularly inflected when compared with irregularly inflected words, (b) The three-way interaction showing clusters in the RMTG that interact with activity in the RACC and RIFG in the context of regular versus irregular inflected forms, (c) The LMTG cluster predicted by the joint activity of RACC, RIFG for regular versus irregular inflected forms.

irregularly inflected forms (Figure 9.4a)—for example, *played* versus *taught*. A comparable analysis carried out on RH activations showed that the RACC and RIFC strongly predicted activity in LMTG (Figure 9.4c) and, to a lesser extent, in RMTG (Figure 9.4b).

This fronto-temporal interaction was reduced when the words were phonologically similar to the regular and irregular past tense but not themselves morphologically complex (e.g. for contrasts like *trade* versus *port*), suggesting that the modulatory effects we found for the regulars reflect the greater integration of the fronto-temporal language system required for processes of morpho-phonological decomposition and analysis rather than phonological differences between the regulars and the irregulars. The greater activation for real as opposed to pseudo-regulars reflects the likelihood that a form like *played* will trigger more activity than *trade*, both in terms of its consequences for the lexical access process and in terms of morpho-syntactic analysis processes. These latter processes will presumably be engaged more strongly when the evidence suggests that a grammatical morpheme is indeed present.

These results, showing connectivity between inferior frontal and middle temporal regions, are consistent with anatomical connectivity via the arcuate fasciculus between frontal and temporal regions, and between orbito-frontal and anterior temporal regions via ventral connections (Petrides & Pandya 1988; Morris *et al.* 1999). They are also consistent with recent analyses of the anatomical connections in the human brain, using DTI (Catani *et al.* 2005; Parker *et al.* 2005). As noted earlier, this work suggests that there may be important asymmetries in the anatomical connectivity between the cortical regions implicated in language function in the R and LH. In the DTI analyses of Parker *et al.* (2005), it is only in the LH that there is clear evidence for both a dorsal route, connecting Wernicke's and Broca's areas via the arcuate fasciculus, and a ventral route, connecting the LMTG to the LIFG via the uncinate fasciculus.

Connectivity studies with normal populations, as described above, form a valuable basis for investigating and interpreting the consequences of damage to the brain systems in question. Neuroimaging studies of patients with damage to the LH fronto-temporal system but preserved RH fronto-temporal cortex provide important additional information about the regions which are necessary for processing different types of linguistic inputs. Since, on the basis of the data described above, we claim that co-activation and modulation of LH fronto-temporal systems is integral to the processing of regularly inflected words, damage to this system should lead to greater difficulty in processing regularly when compared with irregularly inflected past tense forms. This should be revealed in abnormal patterns of L fronto-temporal connectivity.

To evaluate this hypothesis, we recently ran a chronic aphasic patient in the fMRI study described above (Tyler *et al.* 2005c). This patient, labelled here as patient P1, had extensive L perisylvian damage and showed persistent difficulties with the regular past tense which generalized to any speech token containing the diagnostic features of the inflectional rhyme pattern (Tyler *et al.* 2002a, b). In contrast to healthy age-matched controls, who showed increased activation for regular compared with irregular inflected forms in LIFG and bilateral STG/MTG (similar to the young normal patterns in Figure 9.3), P1 showed greater activity for the regulars in the RIFG and in the R insula. Note that the LMTG activations from the connectivity analysis carried out on unimpaired subjects (Figure 9.4a), which were predicted by the combined effects of the LIFG and the LACC, fell into damaged regions in the patient (Figure 9.5c). Owing to his extensive LH damage, connectivity analysis could only be carried out on the RH for this patient.

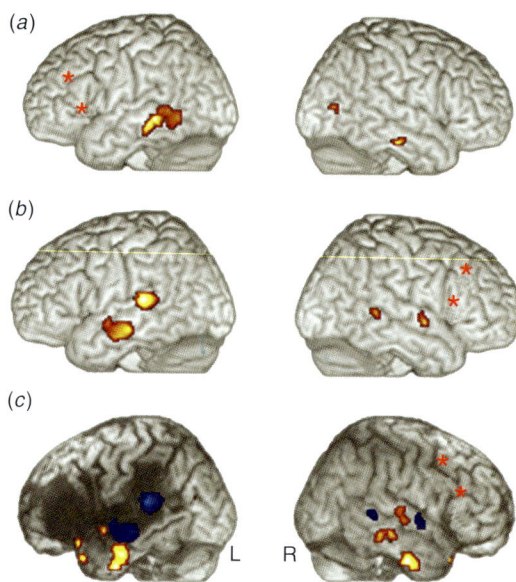

Figure 9.5 Functional connectivity for regular inflection following LH lesion. Connectivity analysis in (a, b) an age-matched control group and (c) patient P1, with extensive perisylvian damage. (a) The three-way interaction for a group of 40–60 year olds between the two time series derived from the LH peak voxels in the subtractive analysis and the experimental condition (regulars versus irregulars). Predictor time series, derived from maxima in group activation patterns, are shown with asterisks in the LIFG and LACC. These regions predict activity in LMTG in the context of the experimental condition (regulars versus irregulars). (b) RH connectivity (for the contrast regulars–irregulars) for the 40–60 year olds. Predictor time series are shown here with asterisks in the RIFG and RACC. These regions predict activity in LMTG as well as in the RH. (c) RH connectivity (i.e. three-way interaction between the two time series for the contrast between regulars and irregulars) for patient P1. Predictor time series, derived from maxima in P1's activation patterns (regulars–irregulars), are shown here with asterisks in the RIFG and RINS. The RH connectivity results from the controls (as in (b)) are shown in blue.

In order to compare P1 with the appropriate control group, we first carried out a connectivity analysis on a small group of healthy subjects age-matched to the patient. Figure 9.5 shows the pattern of connectivity for this group, which generated essentially the same pattern of connectivity as the young with one exception; the LIFG and LACC predicted activity in RMTG as well as LMTG (Figures 9.5a, b), perhaps indicating a degree of hemispheric reorganization with increasing age. The patient's connectivity analysis showed a stronger RH pattern of connectivity when compared with controls (Figure 9.5c) with increased activity in R inferior and RMTG, as well as the anterior temporal lobes in both the hemispheres (Figure 9.5c).

These anterior temporal areas are typically associated with semantic processing, consistent with such patients' greater reliance on semantic and pragmatic factors in language processing. P1's behavioural deficit in processing regularly inflected words (Tyler *et al.* 2002a,b; Longworth *et al.* 2005), and with syntax more generally (Tyler 1992), coupled with his extensive L perisylvian lesion and abnormal connectivity analyses, underlines both the importance of the LIFC and the apparent inability of the RIFC to take over the functions of the LIFC in these core domains of normal language processing. The RIFC activation that we observe for the patient, and its associated connectivity, may reflect functional reorganization where the patient evolves alternative strategies to meet the demands of language comprehension, building on residual functions of the normal network. This reorganization, nonetheless, notably fails to support the on-line processes of morpho-phonological segmentation and morpho-syntactic analysis required for successful processing of inflectionally complex words.

9.4 Functional organization in the intact and damaged brain: syntax and semantics

So far, our focus has been on how activity within the fronto-temporal language system is modulated as a function of the linguistic properties of individual words. The essence of language comprehension, however, involves combining words into structured sequences through processes of syntactic combination. Although the LIFG has long been considered to play a prominent role in these processes, there is a continuing disagreement about the nature of its contribution. At one extreme is the view that the LIFG supports general cognitive functions such as working memory and selection and is not specialized at all for syntactic processing (Thompson-Schill *et al.* 1997; Gabrieli *et al.* 1998; Miller 2000; Kaan & Swaab 2002). On the other hand, the LIFG is claimed to have a key role in syntactic processing, with Friederici (2004), for example, claiming that BA 44/45 in LIFG is involved in hierarchical structure-building, needed to capture long-distance dependencies between words and phrases, while phrasal level syntactic analyses—such as combining words into noun (e.g. *the dog*) and verb phrases (e.g. *he runs*)—involve the L frontal operculum (medial to BA 44). In contrast, Hagoort (2003) argues that the L posterior temporal cortex is important for the retrieval of syntactic frames stored in the lexicon whereas the LIFG binds this and other types of lexical information (phonology and semantics) together.

Accompanying these uncertainties about the nature of LIFG contributions is an equal degree of uncertainty about its relationship to other brain regions supporting language function, especially in the temporal lobes, as well as about the precise contribution that these regions themselves make to processes of language comprehension. While it is plausible that major dorsal and ventral processing streams, linking auditory processing areas in STG/STS to temporal, parietal and frontal regions, are involved in syntactic and sentential analyses (Hickok & Poeppel 2004), there are basic disagreements about the functional characterization of these pathways. In most accounts, the functional relationship between frontal and temporal areas is unspecified, and little attention is paid to the properties of parallel regions in the RH.

Our approach to these issues has been to explore, through connectivity analyses on fMRI data from both unimpaired and brain-damaged patients, the processing dependencies between frontal and temporal regions during the processing of spoken sentences. To understand LIFG function in the context of language processing requires an understanding of the functions it performs relative to the processing functions of other components of the neural language system.

To investigate the relationship between fronto-temporal systems in processing syntactic structure, we have carried out fMRI studies which differentiate semantic and syntactic sentential processing. In one recent study, we did this by presenting listeners with spoken sentences containing either semantic or syntactic ambiguities (Rodd *et al.* 2004). Ambiguity is a natural aspect of language; it occurs frequently and is rarely noticed by listeners because it is typically resolved almost immediately by the presence of a disambiguating context. For example, in 'She quickly learnt that injured calves...', the word *calves* has more than one meaning and is therefore momentarily ambiguous. However, this ambiguity is disambiguated by the following words '...moo loudly'. Sentences can also contain phrases which are syntactically ambiguous. For example, in the sentence 'Out in the open, flying kites...', 'flying kites' is syntactically ambiguous in that either *flying kites* can be a noun phrase in which *flying* modifies the noun *kite* or a verb phrase where flying is a progressive participle (as in 'I was flying kites'). This ambiguity can be immediately resolved by the inflection on the subsequent verb (e.g. '...*are*'/'...*is*'; Tyler & Marslen-Wilson 1977). Moreover, ambiguity is not a binary variable; words and sentences can vary in the degree to which they are ambiguous. We factored this into our study by obtaining 'dominance' ratings for each ambiguity. These provided an estimate of the extent to which one reading of a semantically ambiguous word or syntactically ambiguous phrase was preferred by listeners and were entered into the imaging analysis.

Using ambiguity as a way of manipulating syntactic and semantic structures avoids the criticisms that have been levied against previous studies, by minimizing overt working memory demands (Kaan & Swaab 2002). To reduce task requirements still further, we used a task which had been shown previously to produce patterns of activation which are indistinguishable from passive listening (Rodd *et al.* 2005). Listeners heard spoken sentences, and at the end of the sentence saw a visually presented probe word and made a judgement, indicated by a button-press, as to whether the word was related to the meaning of the sentence.

Syntactic ambiguity produced increased activation in LIFG (BA 44, 45, 47) and in a large swathe of LMTG, extending anteriorly into the anterior STG and posteriorly to the inferior parietal lobule (Figure 9.6). There was also a smaller cluster of activation which included the RSTG (Rodd *et al.* 2004). Activation in these regions increased as a function of increasing dominance, such that they were more strongly activated when the ambiguous phrase was followed by a continuation which was inconsistent with the strongly preferred syntactic interpretation. These regions are increasingly involved when listeners develop strong preferences for one particular syntactic reading, which is then overturned by the subsequent input, forcing a reinterpretation of the syntactic structure. Semantic ambiguity activated a subset of the same fronto-temporal regions as syntactic ambiguity. Although the LIFG activity overlapped considerably for both types of ambiguity, the LMTG activation for semantic ambiguity was confined to the mid portion of the MTG and did not extend posteriorly. Moreover, activity in the LMTG was substantially less than for syntactic processing and was only significant at a slightly lower threshold (Figure 9.6). In addition, the effect of semantic ambiguity was unaffected by the extent to which one meaning of a word was more strongly preferred over another, suggesting that both meanings are activated and listeners wait to make their choice until they hear the disambiguating information.

These results suggest that different cognitive strategies, seemingly rooted in separable underlying processing systems, govern the processing of the syntactic and semantic aspects of sentences. Younger listeners appear to handle syntactic ambiguity by choosing the most

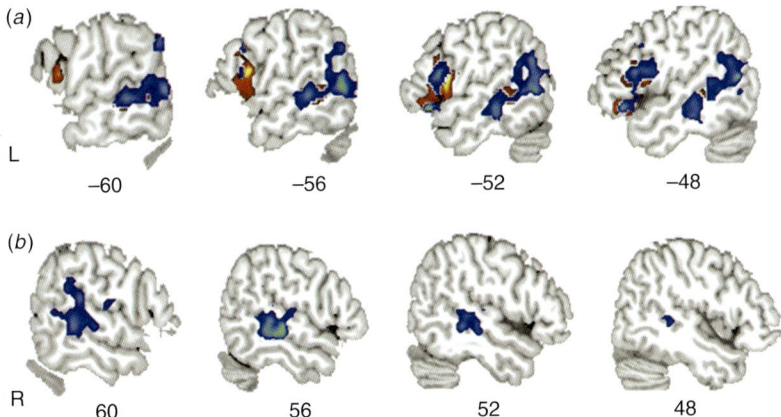

Figure 9.6 Contrasting effects of syntactic and semantic ambiguities. Significant activations (cluster threshold $p < 0.05$ corrected for the entire brain, voxel threshold $p < 0.01$ uncorrected) in LH and RH for (a) the contrast of semantically ambiguous–semantically unambiguous sentences (red) and (b) for the effect of syntactic dominance (blue; based on data reported in Rodd et al. 2004). The x coordinates are shown under each slice.

frequent reading and revising this interpretation when it fails to match the subsequent input. In contrast, they appear to delay their commitment to either reading of a semantic ambiguity until they have confirmatory information. These different sets of analysis processes affect the neural language system differentially, with only syntactic analysis engaging posterior temporal/parietal regions in the LH, perhaps indicating its particular involvement in combinatorial processing when working in concert with the LIFG.

9.5 Functional connectivity in the intact and damaged brain: syntax and semantics

Functional connectivity analyses can further sharpen these potential contrasts in the processing relationship between frontal and temporal cortices for syntactic and semantic aspects of sentential analysis. To explore this, we compared the activation patterns for sentences containing syntactic and semantic ambiguities with matched unambiguous sentences, using the peak frontal activations from the relevant subtractive analyses to predict activity elsewhere in the brain. The resulting functional connectivity analyses reveal distinct patterns of fronto-temporal connectivity for the two types of linguistic computation. For semantic processing (as shown in Figure 9.7a), activity in the LIFG positively predicts activation in the L temporal pole (BA 38), suggesting that this region and the LIFG co-modulate each other's activity during semantic processing.

The syntactic functional connectivity analysis (Figure 9.7b) showed the same co-modulation between LIFG and L temporal pole as for semantic processing, which is not surprising given that all sentences involved semantic analysis. However, in the syntactic analysis, this anterior STG activity was bilateral. Moreover, for the syntactic analysis only, the LIFG also predicts activity in LH posterior regions which included the L posterior MTG, L

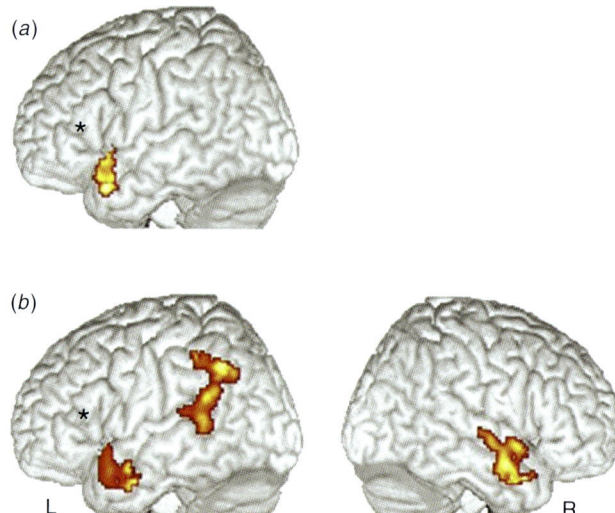

Figure 9.7 Functional connectivity analysis of syntactic and semantic ambiguity effects. Connectivity analysis using a predictor time series (marked by asterisks) found to be a statistical peak in the group (young normal) analysis. (a) The contrast of semantically ambiguous–unambiguous activity in the LIFG positively predicts activity in L anterior STG. (b) For syntactic dominance, activity in the LIFG positively predicts activity in bilateral anterior MTG/STG, L posterior MTG/STG and LIPL.

inferior parietal, angular gyrus and supramarginal gyrus (Figure 9.7b). These results suggest that syntactic combinatorial processes, revealed most strongly when the process is disrupted, involve the co-modulation of LIFG, bilateral anterior STG and left posterior temporal-parietal sites.

The left temporal areas that are active in these analyses of syntactic activity turn out to be adjacent to, but not overlapping with, the L posterior MTG region that showed a greater connectivity with the LIFG for regular when compared with irregular inflected words (Figure 9.8). The fact that activity in the LIFG during semantic processing is not correlated with activity in these more posterior temporal regions, whereas syntactic and morpho-phonological processing does seem to be, invites the inference that these adjacent regions of left posterior temporal cortex play related but different roles in mediating combinatorial linguistic processes.

Overall, these functional connectivity results suggest that successful syntactic processing requires the joint activity of an intact network of LH regions including the LIFC and regions of posterior temporal and parietal cortex. In contrast, semantic processing, while also involving the LIFC, engages a more anterior region of the LMTG/STG. Given these results, lesions which include LIFG and/or posterior temporal–parietal regions would be expected to impair syntactic processing. In a preliminary test of this hypothesis, using both subtractive neuroimaging methods and functional connectivity analyses, we studied two illustrative brain-damaged patients. One of these (patient P1) had extensive LIFC damage as well as damage which extended into temporal perisylvian language regions (Figure 9.5c), whereas the other (patient P2) had an intact LIFC but a lesion in L posterior temporal cortex, mostly involving the MTG (Figure 9.11a). Both had well-documented

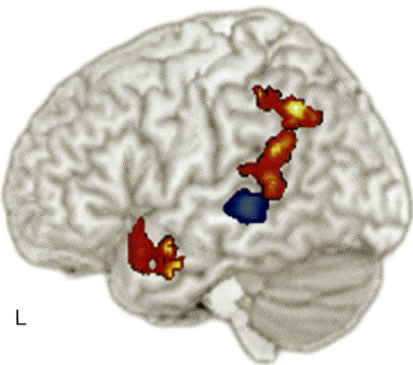

Figure 9.8 LH connectivity effects for regular inflection and for syntax. Results of connectivity analysis for syntactic dominance (red), from Figure 9.7b, contrasted with parallel results for real regulars versus real irregulars (blue), from Figure 9.4a, both for young controls ($p = 0.05$). Predictor time series for both analyses were located in the LIFG.

difficulties with syntactic processing in a variety of different tasks, while semantic processing was unimpaired (Tyler 2002*a*).

Patient P1 showed an abnormal pattern of activity for syntactic ambiguity, consistent with his behavioural deficit (Figure 9.9a). Syntactic ambiguities, when compared with unambiguous sentences, produced substantial perilesional activity in the L middle frontal gyrus and pre- and post-central gyrus, and in the right inferior parietal lobule, a region slightly more posterior than the comparable activations in the LH in healthy listeners. Connectivity analyses using the peak voxels in the LH from the subtractive analysis predicted activity in R posterior regions, including the R angular gyrus, supramarginal gyrus and inferior parietal lobule (Figure 9.9b). This anomalous network must reflect some degree of functional reorganization, given the destruction in this patient of so much of the left perisylvian network that supports syntactic function in the unimpaired brain. However, although this substitute sub-system seems capable of supporting some aspects of syntactic analysis—otherwise effects of syntactic ambiguity would not have been elicited—it is clearly unable to restore the key combinatorial functions underpinning normal performance.

For the same patient, semantic ambiguity extensively activated right frontal and bilateral parietal regions (Figure 9.10a), with the largest cluster in the RIFG. The exceptional extent of these activations may itself reflect another form of functional adaptation in this patient. Because normal syntactic constraints are not available, the processes of speech comprehension in such patients are heavily dependent on the semantic and pragmatic properties of the input. This means that processing is particularly strongly disrupted when these semantic expectations are violated, as we saw in earlier behavioural experiments (Tyler 1992) when this patient encountered semantic violations, as in 'John drank the guitar'. The functional connectivity analysis (Figure 9.10b) showed that activity in the RIFG predicted activation in the L posterior MTG and also in R anterior STG (Figure 9.10b), in regions similar to those activated in healthy subjects (Figure 9.7a), although here the LH anterior temporal activation is not seen. Given this relatively normal pattern and that this patient does not have a semantic deficit, it is clear that language-related semantic

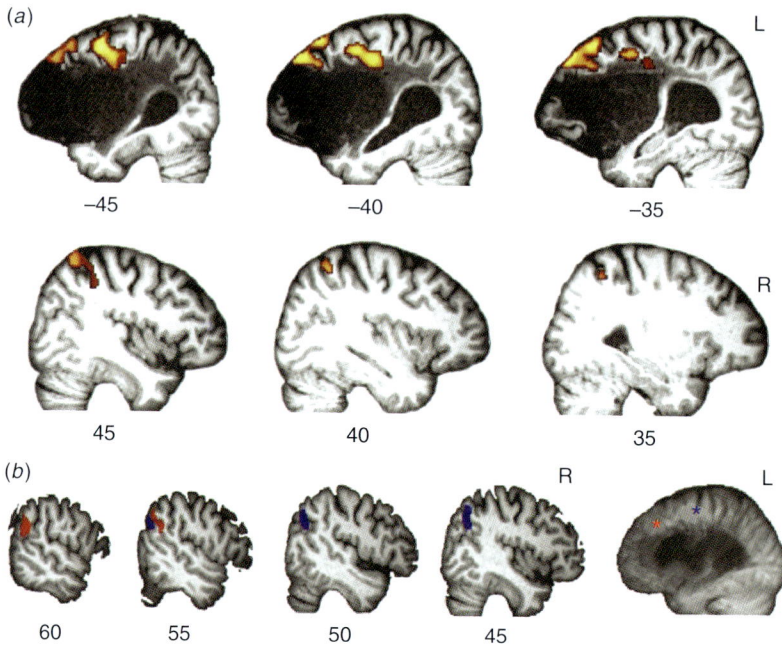

Figure 9.9 Syntactic ambiguity effects for patient P1. (a) LH and RH syntactic ambiguity activations, overlaid on sagittal slices of the patient's T1-weighted scan. The *x* coordinates are shown under each slice. (b) Connectivity analysis using predictors derived from P1's activation peaks (in L precentral G (blue asterisk) and LMFG (red asterisk)) for syntactic ambiguity, overlaid on the patient's RH. Activation in L precentral gyrus predicts activation in R angular gyrus (in blue); activation in LMFG predicts activation in R angular gyrus, extending to R supramarginal gyrus, RSTG and RIPL (in red).

processing can be achieved by means of a more distributed, more bilateral fronto-temporal system than is the case for syntax and does not seem to be dependent on the input from intact left perisylvian language areas to the same extent as syntactic processing.

Turning to patient P2, with damage restricted to L posterior temporal areas (Figure 9.11a), and with no LIFG involvement, here syntactically ambiguous sentences produced greater activation in the RIFC rather than the LIFC, even though the LIFC was not damaged. The fact that activation in the RIFC was nonetheless accompanied by a syntactic deficit is consistent with the view that the RIFC cannot play the same functional role as the LIFC in syntactic processing. In contrast, semantically ambiguous sentences produced a pattern in this patient similar to that of healthy subjects, with peak activation in the LIFC. We then carried out functional connectivity analyses on these data, using the peak activations from the subtractive analysis. Note that in the absence of significant LIFG activation in the syntactic conditions, these syntactic connectivity analyses are driven by seeds in the RIFG (Figure 9.11c).

In the semantic condition (Figure 9.11b), activity in the LIFG predicted activity in anterior LSTG/MTG (BA 21), a region close to that activated for healthy subjects (Figure 9.7a), as well as in the RSTG. In the syntactic analysis (Figure 9.11c), this same L anterior STG/LMTG region was modulated by activity in the RIFG. The RIFG also positively predicted

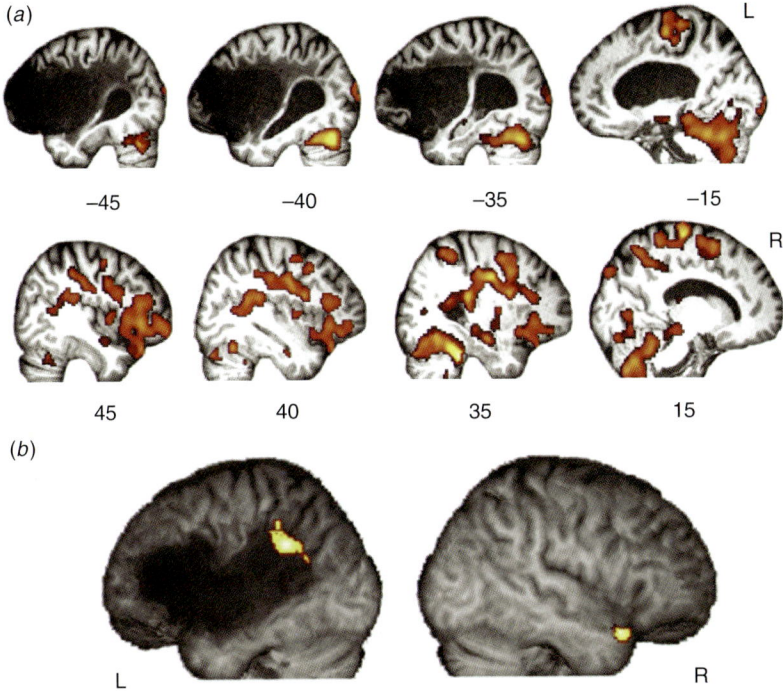

Figure 9.10 Semantic ambiguity effects for patient PI. (a) LH and RH semantic ambiguity activations for patient P1, overlaid on the patient's brain. The x coordinates are shown beneath each sagittal slice, (b) Connectivity analysis using predictors derived from P1's activation peak (see asterisk) for semantic ambiguity, overlaid on the patient's brain. Activation in RIFG, denoted by an asterisk, predicts activation in R anterior STG and L posterior MTG.

activity in bilateral posterior STG/MTG and IPL. The posterior LH activity was just perilesional to the patient's damage. These results suggest a degree of reorganization of function. Unlike in healthy subjects, semantic processing, which appears to be unimpaired, involves the co-activation of the LIFC and bilateral temporal cortex. Syntactic processing also involves a more bilateral system of connectivity than healthy subjects, with posterior temporal-parietal activity in the RH as well as in the LH, although with no LIFG activity detected. In spite of this additional RH involvement, syntactic processing is impaired, again consistent with the observation that this region cannot fully compensate for damage to critical LH regions and their connectivity (see below).

The patient's connectivity analysis reveals an abnormal pattern of connectivity for syntactic processing, which is associated with an abnormal behavioural profile. In contrast, semantic processing is normal, in terms of both functional connectivity and behaviour. We can unpack these contrasts still further, using recent developments in neuroimaging techniques, to ask whether the patient's syntactic deficit was due solely to grey matter damage in left posterior temporal cortex or whether white matter tracts connecting this region to other regions within the neural language system were also compromised. This is an important issue because patients with damage to posterior temporal cortex differ in the nature of their language deficits, with some showing evidence of a syntactic deficit

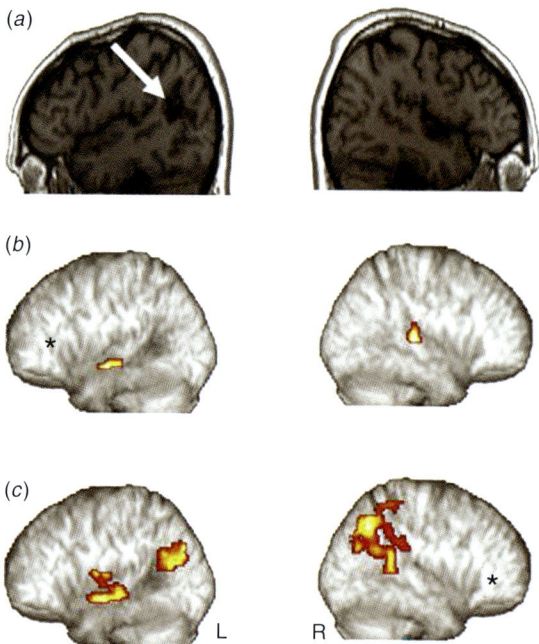

Figure 9.11 Semantic and syntactic connectivity effects for patient P2. (a) T1-weighted MR image for patient P2 (with lesion in L posterior MTG, indicated by a white arrow). (b) Connectivity analysis for semantically ambiguous words using predictors (see asterisk) derived from P2's activation peaks, overlaid on his three-dimensional reconstructed brain. Activity in the LIFG positively predicts activity in anterior regions of the LMTG and RSTG (BA 22, peak at MNI 62,–28, 4). (c) Connectivity analysis for syntactic dominance; activity in the RIFG, marked by an asterisk, positively predicts activity in anterior LMTG/STG and posteriorly in bilateral posterior MTG, and IPL.

and others not (Zurif *et al.* 1993; Wilson & Saygin 2004). One possible explanation for variation in the effect of left posterior temporal lesions may lie in the extent to which damage compromises the white matter connections between the lesion site and other anatomically distributed regions of the language system. Given that syntactic processing involves both posterior temporal and inferior frontal regions, syntactic deficits may be restricted to those patients whose damage includes the white matter tracts connecting these regions.

To determine whether there were any abnormalities in the patient's white matter tracts, we obtained DTI data and calculated fractional anisotrophy (FA), which provides a measure of the integrity of white matter tracts *in vivo*, by measuring directionality of water diffusion in each voxel (Basser & Pierpaoli 1996). In this analysis, we were primarily interested in the major white matter tracts which are thought to be of special importance in language function—the dorsal running arcuate fasciculus and the ventral running inferior longitudinal fasciculus—and therefore confined our analyses to these regions.

Figure 9.12 shows FA maps for patient P2 and, for comparison purposes, a healthy subject of a similar age. As the figure shows, the integrity of the patient's white matter tracts differs

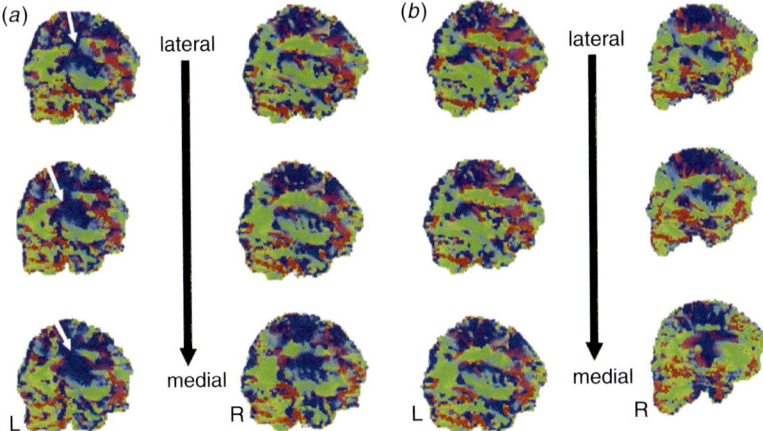

Figure 9.12 Disrupted white matter tracts in patient P2. Directional fractional anisotropy sagittal slices from (a) patient P2 (with L posterior temporal damage, see figure 11) and (b) an age-matched control. The colour maps are based on the principal diffusion directions: green, anterior to posterior; blue, inferior to superior; red, left to right. The arrows indicate a disruption in the LH arcuate fasciculus close to the patient's lesion.

markedly across the hemispheres, with greater integrity in the RH than in the LH. Comparing the FA values in the arcuate and inferior longitudinal fasciculi in the two hemispheres confirmed this pattern for the patient. The mean FA for the patient in the LH was 0.235, whereas it averaged 0.349 in the RH. In contrast, for the age-matched healthy control, there was no difference across the hemispheres, with FA averaging 0.377 in the LH and 0.367 in the RH. Moreover, when compared with the control subject, the patient showed a greater reduction in FA in the arcuate fasciculus than in the inferior longitudinal fasciculus. In fact, as Figure 9.12 shows (indicated by the white arrow), there is an apparent discontinuity in the left arcuate fasciculus which is neither present in the patient's RH nor in the healthy control.

This is an important observation for several reasons. First, it invites the inference that the disruption of this route between frontal and posterior temporal regions is a critical factor in the syntactic deficit shown by this patient. Second, it reinforces the significant theoretical and clinical point that the functional deficits associated with damage in particular locations needs to take into account white matter as well as grey matter damage. Finally, it underscores the critical role of connectivity between brain regions in characterizing the neural substrate for core linguistic functions.

9.6 Overview and conclusions

The research described in the preceding sections combines psycholinguistically well-motivated questions about different aspects of human language comprehension with behavioural and neuroimaging studies of normal performance, incorporating both subtractive analysis techniques and functional connectivity methods, and applying these same tasks and techniques to the analysis of the functional and neural properties of brain-damaged patients with selective linguistic deficits in the relevant domains.

The results of these investigations point to a set of partially dissociable sub-systems supporting three major aspects of spoken language comprehension, involving regular inflectional morphology, sentence-level syntactic analysis and sentence-level semantic interpretation. Differential patterns of fronto-temporal connectivity for these three domains confirm that the core aspects of language processing are carried out in a fronto-temporo-parietal language system which is modulated in different ways as a function of different linguistic processing requirements. No one region or sub-region holds the key to a specific language function; rather each requires the co-activation of activity within a number of different regions.

The use of functional connectivity analyses, in both intact and impaired systems, is critical to the ability to tease apart the wealth of overlapping activity associated with each function. While standard subtractive analyses delineate a range of regions potentially involved, functional connectivity analysis plays the critical role of indicating which regions directly participate in a given sub-process, by virtue of their joint time-dependent activity. By revealing these co-dependencies, connectivity analysis sharpens the pattern of structure–function relations underlying specific aspects of language performance.

Within the three aspects of language function addressed here, two of these, involving inflectional morphological and syntactic processes, clearly group together in distinction from the third, semantic function. Where the latter is concerned, the most salient outcome is the robustness of the ability to construct a semantic interpretation from linguistic inputs, even in the face of massive disruption to core LH language areas. A patient like P1 is able to use lexically derived semantic and pragmatic cues to meaning to drive an effective on-line interpretation process, with normal performance in semantic priming tasks (as long as inflectional morphology is not involved), and with normal sensitivity to semantic and pragmatic constraints in the speech input (e.g. Tyler 1992; Longworth *et al.* 2005). Functional connectivity analyses for P1 show considerable reorganization of functional networks, with additional recruitment of anterior temporal areas related to semantic function in the processing of isolated words (Figure 9.5c), and with greatly increased RH involvement in sentence-related semantic processing (Figure 9.10). Patient P2, with disruption of LH syntactic function, nonetheless shows normal performance on semantic tasks, in the context of stronger bilateral involvement (Figure 9.11b). Damage in other brain areas may well produce permanent impairment in semantic function, but for patients with L perisylvian damage it is clearly possible to retain, and perhaps to rebuild, the ability to semantically interpret spoken utterances, on the basis of functional reorganization of the neural substrates involved.

Both morphological and syntactic processes, in contrast, require an intact left-perisylvian language system—perhaps because they share a core language-specific combinatorial element (however this might be realized neuro-computationally). If the key LH regions (or the connections between them) are damaged, then the system seems to be unable to reorganize to restore effective morphological or syntactic function. Both P1 and P2 provide evidence for some degree of stable reorganization, with novel combinations of regions co-active in response to syntactic processing demands, but this had little impact on their continuing syntactic deficits.

Despite these core similarities, however, there are also substantial differences in the neural sub-systems linked together to support inflectional morphological processes on the one hand, and clausal and sentence-level syntactic interpretation on the other. The key network identified for regular inflectional morphology is relatively compact, and links LIFG, ACC and an area in LMTG (Figure 9.4a). This LMTG activation, likely to be implicated in basic lexical access processes (Dronkers *et al.* 2004), is adjacent to, but distinct from the more posterior temporo-parietal regions activated in the syntactic functional connectivity analyses (Figures 9.7b and 8), which extend into the L supramarginal gyrus, the angular gyrus and the IPL. The network implicated in syntactic processing also links to substantial areas of activation in bilateral anterior MTG/STG, showing some overlap with areas implicated in semantic processing (Figure 9.7), and presumably reflecting the involvement of processes along the STS (Scott & Johnsrude 2003). These results suggest differentiation in the anterior to posterior extent of the LSTG/MTG as a function of syntactic and semantic analysis processes (see also Caplan *et al.* 1996; Friederici *et al.* 2003; Hagoort 2003).

It is noteworthy that the areas implicated here in core language functions do not readily map onto the classical Broca and Wernicke regions (Figure 9.1). The inferior frontal activations were not confined to Broca's area but generally extended beyond it to include BA 46 and 47. Similarly, the posterior temporal activation, which was strongest for syntactic analysis, was not confined to Wernicke's area. Indeed, most of the posterior temporal activity we observed was centred around the posterior MTG and IPL which border the Wernicke's area. This adds to the growing evidence that the regions comprising the neural language system are more extensive than originally thought (e.g. Dronkers *et al.* 2004) and include these more posterior temporal and parietal sites. Moreover, they also highlight that the LMTG, and not only the LSTG, is important in sentence-level processing. Although previous studies have reported activity for spoken sentences solely in the STG (Davis & Johnsrude 2003; Friederici *et al.* 2003), we consistently found maximal activity in MTG.

In summary, the studies we report here suggest that spoken language comprehension involves a network of posterior and frontal regions, with posterior regions being especially important in syntactic processing. These posterior areas include L posterior STG/MTG, angular gyrus, supramarginal gyrus and inferior parietal cortex, regions initially identified as having a significant role in language comprehension, by Marie & Foix (1917). The task for twenty-first century neuro science is to use the imaging tools at our disposal in conjunction with well-developed cognitive models of language function to further elucidate the fine-grained structure of the neural language system.

Acknowledgements

We thank Ana Raposo, Emmanuel Stamatakis, Billi Randall and Jenni Rodd for their help with much of the research described here, and Marie Dixon for her help with the manuscript. A. R. and E. S. also provided the figures for this paper. This research was supported by an MRC programme grant to L.K.T.

References

Barbas, H. 1995 Anatomic basis of cognitive–emotional interactions in the primate prefrontal cortex. *Neurosci. Biobehav. Rev.* **19**, 499–510. (doi:10.1016/0149-7634(94)00053-4)

Basser, P. J. & Pierpaoli, C. 1996 Microstructural and physiological features of tissues elucidated by quantitative-diffusion-tensor MRI. *J. Magn. Reson. B* **111**, 209–219. (doi:10.1006/jmrb.1996.0086)

Beretta, A., Campbell, C, Carr, T. H., Huang, J., Schmitt, L. M., Christianson, K. & Cao, Y. 2003 An ER-fMRI investigation of morphological inflection in German reveals that the brain makes a distinction between regular and irregular forms. *Brain Lang.* **85**, 67–92. (doi:10.1016/S0093-934X(02)00560-6)

Binder, J. R., Frost, J. A., Hammeke, T. A., Bellgowan, P. S. R, Springer, J. A. & Kaufman, J. N. 2000 Human temporal lobe activation by speech and nonspeech sounds. *Cereb. Cortex* **10**, 512–528. (doi:10.1093/cercor/10.5.512)

Bokde, A. L. W., Tagamets, M.-A., Friedman, R. B. & Horwitz, B. 2001 Functional interactions of the inferior frontal cortex during the processing of words and word-like stimuli. *Neuron* **30**, 609–617. (doi:10.1016/S0896-6273(01)00288-4)

Braver, T. S., Barch, D. M., Gray, J. R., Molfese, D. L. & Snyder, A. 2001 Anterior cingulate cortex and response conflict: effects of frequency, inhibition and errors. *Cereb. Cortex* **11**, 825–836. (doi:10.1093/cercor/11.9.825)

Buchel, C., Raedler, T., Sommer, M., Sach, M., Weiller, C. & Koch, M. A. 2004 White matter asymmetry in the human brain: a diffusion tensor MRI study. *Cereb. Cortex* **14**, 945–951. (doi:10.1093/cercor/bhh055)

Caplan, D. & Futter, C. 1986 Assignment of thematic roles to nouns in sentence comprehension by an agrammatic patient. *Brain Lang.* **27**, 117–134. (doi:10.1016/0093-934X(86)90008-8)

Caplan, D. & Hildebrant, N. 1988 *Disorders of syntactic comprehension*. Cambridge, MA: MIT Press.

Caplan, D., Hildebrant, N. & Makris, N. 1996 Location of lesions in stroke patients with deficits in syntactic processing in sentence comprehension. *Brain* **119**, 933–949. (doi:10.1093/brain/119.3.933)

Catani, M., Jones, D. K. & Ffytche, D. H. 2005 Perisylvian language networks of the human brain. *Ann. Neurol.* **57**, 8–16. (doi:10.1002/ana.20319)

Celsis, P., Boulanouar, K., Doyon, B., Ranjeva, J. P., Berry, I., Nespoulous, J.-L. & Chollet, F. 1999 Differential fMRI responses in the left posterior superior temporal gyrus and left supramarginal gyrus to habituation and change detection in syllables and tones. *NeuroImage* **9**, 135–114. (doi:10.1006/nimg.1998.0389)

Davis, M. H. & Johnsrude, I. S. 2003 Hierarchical processing in spoken language comprehension. *J. Neurosci.* **28**, 3423–3431.

Demonet, J.-F., Chollet, F., Ramsay, S., Cardebat, D., Nespoulous, J.-L., Wise, R., Rascol, A. & Frackowiak, R. 1992 The anatomy of phonological and semantic processing in normal subjects. *Brain* **115**, 1753–1768. (doi:10.1093/brain/115.6.1753)

Dronkers, N. F., Wilkins, D. P., Van Valin Jr, R. D., Redfern, B. B. & Jaeger, J. J. 2004 Lesion analysis of the brain areas involved in language comprehension. *Cognition* **92**, 145–177. (doi:10.1016/j.cognition.2003.11.002)

Fletcher, P., McKenna, P. J., Friston, K. J., Frith, C. D. & Dolan, R. J. 1999 Abnormal cingulate modulation of fronto-temporal connectivity in schizophrenia. *NeuroImage* **9**, 337–342. (doi:10.1006/nimg.1998.0411)

Friederici, A. D. 2004 Processing local transitions versus long-distance syntactic hierarchies. *Trends Cogn. Sci.* **8**, 245–247. (doi:10.1016/j.tics.2004.04.013)

Friederici, A. D., Ruschemeyer, S.-A., Hahne, A. & Fiebach, C. J. 2003 The role of left inferior frontal and superior temporal cortex in sentence comprehension: localizing syntactic and semantic processes. *Cereb. Cortex* **13**, 170–177. (doi:10.1093/cercor/13.2.170)

Friston, K. J., Buechel, C, Fink, G. R., Morris, J., Rolls, E. & Dolan, R. J. 1997 Psychophysiological and modulatory interactions in neuroimaging. *NeuroImage* **6**, 218–229. (doi:10.1006/nimg.1997.0291)

Gabrieli, J. D. E., Poldrack, R. A. & Desmond, J. E. 1998 The role of left prefrontal cortex in language and memory. *Proc. Natl Acad. Sci. USA* **95**, 906–913. (doi:10.1073/pnas.95.3.906)

Gold, B. T. & Buckner, R. L. 2002 Common prefrontal regions coactivate with dissociable posterior regions during controlled semantic and phonological tasks. *Neuron* **35**, 803–812. (doi:10.1016/S0896-6273(02)00800-0)

Goodglass, H., Christiansen, J. & Gallagher, R. 1993 Comparison of morphology and syntax in free narrative and structured tests: fluent versus nonfluent aphasics. *Cortex* **29**, 377–407.

Grodzinsky, Y. 2000 The neurology of syntax: language use without Broca's area. *Behav. Brain Sci.* **23**, 1–21. (doi:10.1017/S0140525X00002399)

Hagoort, P. 2003 How the brain solves the binding problem for language: a neurocomputational model of syntactic processing. *NeuroImage* **20**, S18–S29. (doi:10.1016/j.neuroimage.2003.09.013)

Hickok, G. & Poeppel, D. 2000 Towards a functional neuroanatomy of speech perception. *Trends Cogn. Sci.* **4**, 131–138. (doi:10.1016/S1364-6613(00)01463-7)

Hickok, G. & Poeppel, D. 2004 Dorsal and ventral streams: a framework for understanding aspects of the functional anatomy of language. *Cognition* **92**, 67–99. (doi:10.1016/j.cognition.2003.10.011)

Hillis, A. E., Barker, P. B., Beauchamp, N. J., Winters, B. D., Mirski, M. & Wityk, R. J. 2001 Restoring blood pressure reperfused Wernicke's area and improved language. *Neurology* **56**, 670–672.

Indefrey, P. & Cutler, A. 2004 Pre-lexical and lexical processing in listening. In *The cognitive neurosciences* (ed. M. S. Gazzaniga), pp. 759–774. Cambridge, MA: MIT Press.

Jaeger, J. J., Lockwood, A. H., Kemmerer, D. L., Van Valin Jr, R. D., Murphy, B. W & Khalak, H. G. 1996 Positron emission tomographic study of regular and irregular verb morphology in English. *Language* **72**, 451–497. (doi:10.2307/416276)

Kaan, E. & Swaab, T. Y. 2002 The brain circuitry of syntactic comprehension. *Trends Cogn. Sci.* **6**, 350–356. (doi:10.1016/S1364-6613(02)01947-2)

Kaas, J. H. & Hackett, T. A. 1999 'What' and 'where' processing in auditory cortex. *Nat. Neurosci.* **2**, 1045–1047. (doi:10.1038/15967)

Kertesz, A., Lau, W K. & Polk, M. 1993 The structural determinants of recovery in Wernicke's aphasia. *Brain Lang.* **44**, 153–164. (doi:10.1006/brln.1993.1010)

Longworth, C., Marslen-Wilson, W D., Randall, B. & Tyler, L. K. 2005 Getting to the meaning of the regular past tense: evidence from neuropsychology. *J. Cogn. Neurosci.* **17**, 1087–1097. (doi:10.1162/0898929054475109)

Marie, P. & Foix, C. 1917 Les aphasies de guerre. *Rev. Neurol.* **24**, 53–87.

Marinkovic, K., Dhond, R. P., Anders, M. D., Glessner, M., Carr, V. & Halgren, E. 2003 Spatiotemporal dynamics of modality-specific and supramodal word processing. *Neuron* **38**, 487–497. (doi:10.1016/S0896-6273(03)001971)

Marslen-Wilson, W. D. & Tyler, L. K. 1997 Dissociating types of mental computation. *Nature* **387**, 592–594. (doi:10.1038/42456)

Marslen-Wilson, W. D. & Tyler, L. K. 1998 Rules, representations, and the English past tense. *Trends Cogn. Set.* **2**, 428–435. (doi:10.1016/S1364-6613(98)01239-X)

Mesulam, M.-M., Grossman, M., Hillis, A. E., Kertesz, A. & Weintraub, S. 2003 The core and halo of primary progressive aphasia and semantic dementia. *Ann. Neurol.* **54**, S11–S14. (doi:10.1002/ana.10569)

Milberg, W., Blumstein, S. & Dworetzky, R. 1987 Processing of lexical ambiguities in aphasia. *Brain Lang.* **31**, 138–150. (doi:10.1016/0093-934X(87)90065-4)

Miller, E. K. 2000 The prefrontal cortex and cognitive control. *Nat Rev. Neurosci.* **1**, 59–65. (doi:10.1038/35036228)

Morris, R., Pandya, D. N. & Petrides, M. 1999 Fiber system linking the mid-dorsolateral frontal cortex with the retro-splenial/presubicular region in the rhesus monkey. *J. Comp. Neurol.* **407**, 183–192. (doi:10.1002/(SICI)1096-9861(19990503)407:2<183::AID-CNE3>3.0.CO;2-N)

Mummery, C. J., Patterson, K., Wise, R. J. S., Vandenbergh, R., Price, C. J. & Hodges, J. R. 1999 Disrupted temporal lobe connections in semantic dementia. *Brain* **122**, 61–73. (doi:10.1093/brain/122.1.61)

Pandya, D. N., Hoesen, G. W & Mesulam, M.-M. 1981 Efferent connections of the cingulate gyrus in the rhesus monkey. *Exp. Brain Res.* **42**, 319–330. (doi:10.1007/BF00237497)

Parker, G. J. M., Luzzi, S., Alexander, D. C, Wheeler-Kingshott, C. A. M., Ciccarelli, O. & Lambon Ralph, M. A. 2005 Lateralization of ventral and dorsal auditory-language pathways in the human brain. *NeuroImage* **24**, 656–666. (doi:10.1016/j.neuroimage.2004.08.047)

Petrides, M. & Pandya, D. N. 1988 Association fiber pathways to the frontal cortex from the superior temporal region in the rhesus monkey. *J. Comp. Neurol.* **273**, 52–66. (doi:10.1002/cne.902730106)

Post, B., Randall, B., Tyler, L. K. & Marslen-Wilson, W. D. 2004 Morphological and phonological factors in the processing of English inflections. *Paper presented at Experimental Psychology Society Meeting, London.*

Rauschecker, J. P. & Tian, B. 2000 Mechanisms and streams for processing of 'what' and 'where' in auditory cortex. *Proc. Natl Acad. Sci. USA* **97**, 11800–11806. (doi:10.1073/pnas.97.22.11800)

Rodd, J. M., Longe, O. A., Randall, B. & Tyler, L. K. 2004 Syntactic and semantic processing of spoken sentences: an fMRI study of ambiguity. *J. Cogn. Neurosci.* **16**(Suppl. C), 89.

Rodd, J. M., Davis, M. H. & Johnsrude, I. S. 2005 The neural mechanisms of speech comprehension: fMRI studies of semantic ambiguity. *Cereb. Cortex* **15**, 1261–1269. (doi:10.1093/cercor/bhi009)

Scott, S. K. & Johnsrude, I. S. 2003 The neuroanatomical and functional organization of speech perception. *Trends Neurosci.* **26**, 100–107. (doi:10.1016/S0166-2236(02)00037-1)

Scott, S. K. & Wise, R. J. S. 2003 PET and fMRI studies of the neural basis of speech perception. *Speech Commun.* **41**, 23–34. (doi:10.1016/S0167-6393(02)00090-0)

Scott, S. K. & Wise, R. J. S. 2004 The functional neuroanatomy of prelexical processing in speech perception. *Cognition* **92**, 13–45. (doi:10.1016/j.cognition.2002.12.002)

Scott, S. K., Blank, C. C., Rosen, S. & Wise, R. J. S. 2000 Identification of a pathway for intelligible speech in the left temporal lobe. *Brain* **123**, 2400–2406. (doi:10.1093/brain/123.12.2400)

Selnes, O. A., Knopman, D. S., Niccum, N. & Rubens, A. B. 1985 The critical role of Wernicke's area in sentence repetition. *Ann. Neurol.* **17**, 549–557. (doi:10.1002/ana.410170604)

Stamatakis, E. A., Marslen-Wilson, W D., Tyler, L. K. & Fletcher, P. C. 2005 Cingulate control of fronto-temporal integration reflects linguistic demands: a three-way interaction in functional connectivity. *NeuroImage* **28**, 115–121. (doi:10.1016/j.neuroimage.2005.06.012)

Stromswold, K., Caplan, D., Alpert, N. & Rauch, S. 1996 Localization of syntactic comprehension by positron emission tomography. *Brain Lang.* **52**, 452–473. (doi:10.1006/brln.l996.0024)

Thompson-Schill, S. L., D'Esposito, M., Aguirre, G. K. & Farah, M. J. 1997 Role of left inferior prefrontal cortex in retrieval of semantic knowledge: a reevaluation. *Proc. Natl Acad. Sci. USA* **94**, 14 792–14 797. (doi:10.1073/pnas.94.26.14792)

Thompson-Schill, S. L., Bedny, M. & Goldberg, R. F. 2005 The frontal lobes and the regulation of mental activity. *Curr. Opin. Neurobiol.* **15**, 219–224. (doi:10.1016/j.conb.2005.03.006)

Tyler, L. K. 1992 *Spoken language comprehension: an experimental approach to the study of normal and disordered processing.* Cambridge, MA: MIT Press.

Tyler, L. K. & Marslen-Wilson, W D. 1977 The on-line effects of semantic context on syntactic processing. *J. Verbal Learn. Verbal Behav.* **16**, 645–659. (doi:10.1016/S0022-5371(77)80027-3)

Tyler, L. K., Moss, H. E. & Jennings, F. 1995a Abstract word deficits in aphasia: evidence from semantic priming. *Neuropsychology* **9**, 354–363. (doi:10.1037/0894-4105.9.3.354)

Tyler, L. K., Ostrin, R. K., Cooke, M. & Moss, H. E. 1995b Automatic access of lexical information in Broca's aphasies: against the automaticity hypothesis. *Brain Lang.* **48**, 131–162. (doi:10.1006/brln.1995.1007)

Tyler, L. K., de Mornay-Davies, P., Anokhina, R., Longworth, C., Randall, B. & Marslen-Wilson, W. D. 2002a Dissociations in processing past tense morphology: neuropathology and behavioural studies. *J. Cogn. Neurosci.* **14**, 79–94. (doi:10.1162/089892902317205348)

Tyler, L. K., Randall, B. & Marslen-Wilson, W. D. 2002b Phonology and neuropsychology of the English past tense. *Neuropsychologia* **40**, 1154–1166. (doi:10.1016/S0028-3932(01)00232-9)

Tyler, L. K., Marslen-Wilson, W. D. & Stamatakis, E. A. 2005a Differentiating lexical form, meaning, and structure in the neural language system. *Proc. Natl Acad. Sci. USA* **102**, 8375–8380. (doi:10.1073/pnas.0408213102)

Tyler, L. K., Marslen-Wilson, W. D. & Stamatakis, E. A. 2005b Dissociating neuro-cognitive component processes: voxel-based correlational methodology. *Neuropsychologia* **43**, 771–778. (doi:10.1016/j.neuropsychologia.2004.07.020)

Tyler, L. K., Stamatakis, E. A., Post, B., Randall, B. & Marslen-Wilson, W D. 2005c Temporal and frontal systems in speech comprehension: an fMRI study of past tense processing. *Neuropsychologia* **43**, 1963–1974. (doi:10.1016/j.neuropsychologia.2005.03.008)

Wilson, S. M. & Saygin, A. P. 2004 Grammaticality judgment in aphasia: deficits are not specific to syntactic structures, aphasic syndromes, or lesion sites. *J. Cogn. Neurosci.* **16**, 238–252. (doi:10.1162/089892904322984535)

Zatorre, R. J. & Gandour, J. T. 2008 Neural specializations for speech and pitch: moving beyond the dichotomies. *Phil. Trans. R. Soc. B* **363**, 1087–1104. (doi:10.1098/rstb.2007.2161)

Zatorre, R. J., Evans, A. C., Meyer, E. & Gjedde, A. 1992 Lateralization of phonetic and pitch discrimination in speech processing. *Science* **256**, 846–849. (doi:10.1126/science.1589767)

Zatorre, R. J., Meyer, E., Gjedde, A. & Evans, A. C. 1996 PET studies of phonetic processing of speech: review, replication and reanalysis. *Cereb. Cortex* **6**, 21–30. (doi:10.1093/cercor/6.1.21)

Zurif, E. B., Swinney, D., Prather, P., Solomon, J. & Bushell, C. 1993 An on-line analysis of syntactic processing in Broca's and Wernicke's aphasia. *Brain Lang.* **45**, 448–464. (doi:10.1006/brln.1993.1054)

10

The fractionation of spoken language understanding by measuring electrical and magnetic brain signals

Peter Hagoort

This paper focuses on what electrical and magnetic recordings of human brain activity reveal about spoken language understanding. Based on the high temporal resolution of these recordings, a fine-grained temporal profile of different aspects of spoken language comprehension can be obtained. Crucial aspects of speech comprehension are lexical access, selection and semantic integration. Results show that for words spoken in context, there is no 'magic moment' when lexical selection ends and semantic integration begins. Irrespective of whether words have early or late recognition points, semantic integration processing is initiated before words can be identified on the basis of the acoustic information alone. Moreover, for one particular event-related brain potential (ERP) component (the N400), equivalent impact of sentence- and discourse-semantic contexts is observed. This indicates that in comprehension, a spoken word is immediately evaluated relative to the widest interpretive domain available. In addition, this happens very quickly. Findings are discussed that show that often an unfolding word can be mapped onto discourse-level representations well before the end of the word. Overall, the time course of the ERP effects is compatible with the view that the different information types (lexical, syntactic, phonological, pragmatic) are processed in parallel and influence the interpretation process incrementally, that is as soon as the relevant pieces of information are available. This is referred to as the immediacy principle.

Keywords: speech; event-related brain potential; magnetoencephalography; N200; N400; P600/SPS

10.1 Introduction

Speed is one of the most remarkable characteristics of the human capacity for understanding spoken language. As listeners, we easily process three or four words per second. Although we do this seemingly without any effort, in fact a complex cascade of processes underlies our capacity for understanding. When we hear speech, numerous brain areas work together to analyse the acoustic information, select the proper words by mapping the sensory input onto stored lexical knowledge, extract the meaning of those words and integrate them into an ongoing sentential or discourse context (Marslen-Wilson 1973; Marslen-Wilson & Welsh 1978; Marslen-Wilson & Tyler 1980; Zwitserlood 1989; Norris 1994). All of these happen in a time span of only hundreds of milliseconds. An account of the neurobiology of spoken language processing can, therefore, only be adequate if the temporal dynamics is taken into consideration. In recent years, positron emission tomography (PET) and functional magnetic resonance imaging (fMRI) studies have given us important new insight into the network of brain areas involved in language processing. Diffusion tensor imaging has provided new data about the connectivity of language-related brain areas (Catani *et al.* 2004). However, none of these techniques has a temporal resolution of the order of milliseconds, which is necessary to study the time course of language processing. For a characterization of the temporal dynamics of spoken language comprehension, measuring electrical and magnetic brain responses is thus more appropriate. Here, I will focus on what has been learned from event-related brain potential (ERP) and magnetoencephalography (MEG) studies about the temporal and functional fractionation of the neurocognitive architecture for listening to language.

ERPs reflect the sum of simultaneous postsynaptic activity of a large population of mostly pyramidal neurons recorded at the scalp as small voltage fluctuations in the electroencephalography (EEG) time-locked to sensory, motor or cognitive processes. In a particular patch of cortex, excitatory input to the apical dendrites of pyramidal neurons will result in a net negativity in the region of the apical dendrite and a positivity in the area of the cell body. This creates a tiny dipole for each pyramidal neuron, which will summate with other dipoles, provided that there is simultaneous input to the apical dendrites of many neurons and a similar orientation of these cells. The cortical pyramidal neurons are all aligned perpendicular to the surface of the cortex, and thus share their orientation. The summation of the many individual dipoles in a patch of cortex is equivalent to a single dipole calculated by averaging the orientations of the individual dipoles (Luck 2005). This equivalent current dipole is the neuronal generator (or source) of the ERP recorded at the scalp. In many cases, a particular ERP component has more than one generator and contains the contribution of multiple sources. Mainly due to the high resistance of the skull, ERPs tend to spread laterally, blurring the voltage distribution at the scalp. An ERP generated locally in one part of the brain will therefore not only be recorded nearby but also at quite distant parts of the scalp. Recordings of magnetic fields instead of electrical potentials do not suffer from blurring as a consequence of high skull resistance. Since an electrical dipole is always surrounded by a magnetic field, and since these fields summate as well, a system for measuring the magnetic fields of the brain records the same activity from pyramidal neurons as EEG does. However, the spatial resolution of MEG is much greater than that of EEG, since the skull is transparent to magnetism and hence no blurring by the skull takes place. Another important difference is that MEG, in contrast to EEG, is not sensitive to dipoles with a radial orientation, i.e. those dipoles that are perpendicular to the skull. This holds mainly for the dipoles in the crowns of the gyri. The pyramidal cells in the banks of the sulci are oriented tangentially to the skull and their activity creates a recordable magnetic field. Thus, EEG and MEG measure overlapping but not identical overall contributions from pyramidal neurons in the cortical sheet.

In the remainder, I will focus mainly on what electrical and magnetic recordings reveal about spoken language understanding based on their high temporal resolution. In this context, I will not be able to discuss all the work that has been done, especially those on the mismatch negativity since these have mainly studied sublexical processing (Näätänen et al. 1997; Näätänen 2001; Pulvermüller et al. 2004). I will here focus on lexical processing and beyond.

10.2 Language-relevant event-related brain potentials

As it holds for psycholinguistics in general, ERP research on language has been done mostly with visual input. Language-relevant ERPs have been discovered almost without exception with visual instead of spoken language input. By way of introduction, I will thus have to refer mainly to ERP studies on reading. Once the basic language-related ERP effects have been introduced, I will discuss how they contributed to the fractionation of spoken language understanding.

The electrophysiology of language as a domain of study started with the discovery by Kutas & Hillyard (1980) of an ERP component that seemed especially sensitive to semantic manipulations. Kutas & Hillyard (1980) observed a negative-going potential with an onset at approximately 250 ms and a peak at approximately 400 ms (hence the N400), whose amplitude was increased when the semantics of the eliciting word (i.e. *socks*) mismatched with the semantics of the sentence context, as in 'He spread his warm bread with socks'. Since 1980, much has been learned about the processing nature of the N400 (for extensive overviews, see Kutas & Van Petten 1994 and Osterhout & Holcomb 1995). As Hagoort & Brown (1994) and many others have observed, the N400 effect does not depend on a semantic violation. Subtle differences in semantic expectancy as between *mouth* and *pocket* in the sentence context 'Jenny put the sweet in her *mouth/pocket* after the lesson' can modulate the N400 amplitude (Figure 10.1; Hagoort & Brown 1994).

The amplitude of the N400 is most sensitive to the semantic relations between individual words, or between words and their sentence and discourse context. The better the semantic fit between a word and its context, the more reduced the amplitude of the N400. Modulations of the N400 amplitude are generally viewed as directly or indirectly related

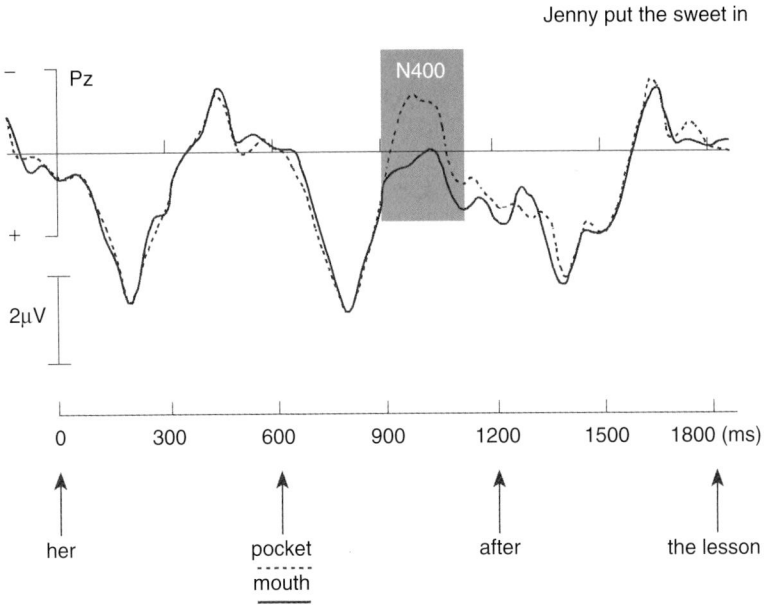

Figure 10.1 Modulation of the N400 amplitude as a result of a manipulation of the semantic fit between a lexical item and its sentence context. The grand average waveform is shown for electrode site Pz (parietal midline), for the best fitting word (high cloze; solid line), and a word that is less expected in the given sentence context (low cloze; dashed line). The sentences were visually presented word by word for every 600 ms. In the figure, the critical words (CWs) are preceded and followed by one word. The CW is presented at 600 ms on the time axis. Negativity is up.
(Reprinted with permission from Hagoort & Brown (1994). Copyright Erlbaum).

to the processing costs of integrating the meaning of a word into the overall meaning representation that is built up on the basis of the preceding language input (Osterhout & Holcomb 1992; Brown & Hagoort 1993). This holds equally when the preceding language input consists of a single word, a sentence or a discourse, indicating that semantic integration might be similar in word, sentence or discourse context (Van Berkum *et al.* 1999*b*). In addition, recent evidence indicates that sentence verification against world knowledge in long-term memory modulates the N400 in the same way (Hagoort *et al.* 2004).

In recent years, a number of ERP studies have been devoted to establishing ERP effects that can be related to the processing of syntactic information. These studies have found ERP effects to syntactic processing that are qualitatively different from the N400. Even though the generators of these effects are not yet well determined and not necessarily language specific (Osterhout & Hagoort 1999), the existence of qualitatively distinct ERP effects to semantic and syntactic processing indicates that the brain honours the distinction between semantic and syntactic processing operations. Thus, the finding of qualitatively distinct ERP effects for semantic and syntactic operations supports the claim that these two levels of language processing are domain specific. However, domain specificity should not be confused with modularity (Fodor 1983). The modularity thesis makes the much stronger claim that domain-specific levels of processing operate autonomously without interaction (informational encapsulation). Although domain specificity is widely assumed in models of language processing, there is much less agreement about the organization of the crosstalk between the different levels of processing (Boland & Cutler 1996).

ERP studies on syntactic processing have reported a number of ERP effects related to syntax (for an overview, see Hagoort *et al.* (1999)). The two most salient syntax-related effects are an anterior negativity, also referred to as LAN (left anterior negativity), and a more posterior positivity, here referred to as P600/SPS.

(a) LAN

A number of studies have reported negativities that are different from the N400, in that their voltage distribution over the scalp usually shows a more frontal maximum (but see Münte *et al.* 1997), sometimes larger over the left than the right hemisphere, although in many cases the distribution is bilateral (Hagoort *et al.* 2003). Moreover, the conditions that elicit these frontal negative shifts seem to be more strongly related to syntactic processing than to semantic integration. Usually, LAN effects occur within the same latency range as the N400, i.e. between 300 and 500 ms post stimulus (Osterhout & Holcomb 1992; Kluender & Kutas 1993; Münte *et al.* 1993; Rösler *et al.* 1993; Friederici *et al.* 1996). However, in some cases, and almost exclusively with spoken language input, the latency of a left frontal negative effect is reported to be much earlier, somewhere between approximately 100 and 300 ms (Neville *et al.* 1991; Friederici *et al.* 1993; Friederici 2002).

In some studies, LAN effects have been reported for violations of word-category constraints (Münte *et al.* 1993; Friederici *et al.* 1996; Hagoort *et al.* 2003). That is, if a word of a different syntactic class is presented (e.g. a verb) instead of the required one (e.g. a noun in the context of a preceding article and adjective), early negativities are observed. Friederici and colleagues (Friederici 1995; Friederici *et al.* 1996) have tied the early negativities specifically to the processing of word-category information. However, sometimes similar early negativities are observed with number, case, gender and tense mismatches in morphologically rich languages (Münte et al. 1993; Münte & Heinze 1994).

LAN effects have also been related to verbal working memory in connection with filler-gap assignment (Kluender & Kutas 1993). This working memory account of the LAN is compatible with the finding that lexical, syntactic and referential ambiguities seem to elicit very similar frontal negativities (Hagoort & Brown 1994; King & Kutas 1995; Van Berkum et al. 1999a; Kaan & Swaab 2003). Lexical and referential ambiguities are clearly not syntactic in nature, but can be argued to tax verbal working memory more heavily than sentences in which lexical and referential ambiguities are absent. Future research should indicate whether or not these two functionally distinct classes of LAN effects can be dissociated at a more fine-grained level of electrophysiological analysis.

(b) P600/SPS

A second ERP effect that has been related to syntactic processing is a later positivity, nowadays referred to as P600 or P600/SPS (Osterhout et al. 1997; Coulson et al. 1998; Hagoort et al. 1999). One of the antecedent conditions of P600/SPS effects is a violation of a syntactic constraint. If, for instance, the syntactic requirement of number agreement between the grammatical subject of a sentence and its finite verb is violated (see (1) below, with the critical verb form in italics; the * indicates the ungrammaticality of the sentence), a positive-going shift is elicited by the word that renders the sentence ungrammatical (Hagoort et al. 1993). This positive shift starts at approximately 500 ms after the onset of the violation and usually lasts for at least 500 ms. Given the polarity and the latency of its maximal amplitude, this effect was originally referred to as the P600 (Osterhout & Holcomb 1993) or, on the basis of its functional characteristics, as the syntactic positive shift (SPS; Hagoort et al. 1993).

(1) *The spoilt child *throw* the toy on the ground.

An argument for the independence of this effect from possibly confounding semantic factors is that it also occurs in sentences where the usual semantic/pragmatic constraints have been removed (Hagoort & Brown 1994). This results in sentences like (2a) and (2b) where one is semantically odd but grammatically correct, whereas the other contains the same agreement violation as in (1):

(2a) The boiled watering can *smokes* the telephone in the cat.
(2b) *The boiled watering can *smoke* the telephone in the cat.

If one compares the ERPs with the italicized verbs in (2a) and (2b), a P600/SPS effect to the ungrammatical verb form is visible (Figure 10.2). Despite the fact that these sentences do not convey any conventional meaning, the ERP effect of the violation demonstrates that the language system is nevertheless able to parse the sentence into its constituent parts.

Similar P600/SPS effects have been reported for a broad range of syntactic violations in different languages (e.g. English, Dutch, German), including phrase-structure violations (Neville et al. 1991; Osterhout & Holcomb 1992; Hagoort et al. 1993), subcategorization violations (Osterhout et al. 1994, 1997; Ainsworth-Darnell et al. 1998), violations in the agreement of number, gender and case (Hagoort et al. 1993; Osterhout & Mobley 1995; Münte et al. 1997; Osterhout 1997; Coulson et al. 1998), violations of subjacency

228   Peter Hagoort

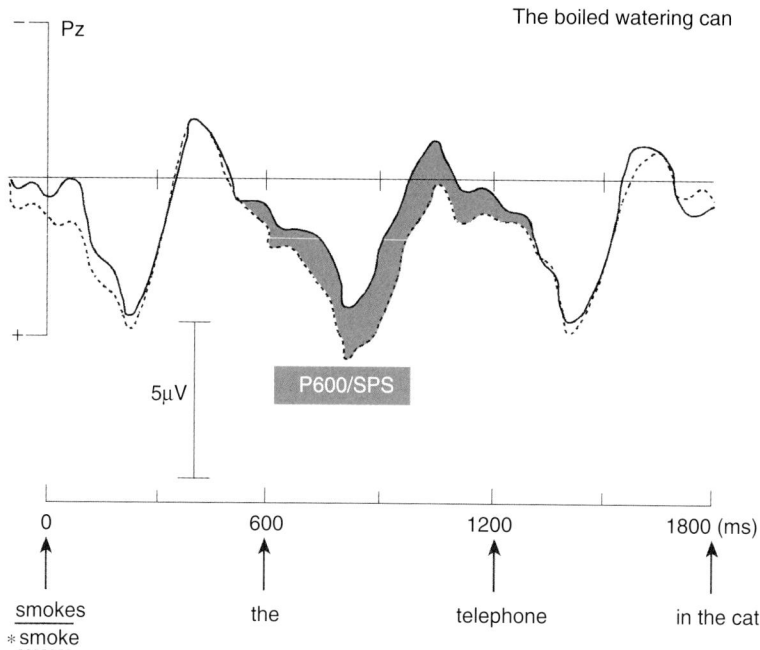

Figure 10.2 ERPs to visually presented syntactic prose sentences. These are sentences without a coherent semantic interpretation. A P600/SPS is elicited by a violation of the required number agreement between the subject noun phrase and the finite verb of the sentence. The averaged waveforms for the grammatically correct (solid line) and the grammatically incorrect (dashed line) words are shown for electrode site Pz (parietal midline). The word that renders the sentence ungrammatical is presented at 0 ms on the time axis. The waveforms show the ERPs to this and the following two words. Words were presented word by word, with an interval (stimulus onset asynchrony) of 600 ms. Negativity is plotted upwards; asterisk indicates the syntactic isolation.
(Reprinted with permission from Hagoort & Brown (1994). Copyright Erlbaum).

(Neville *et al.* 1991; McKinnon & Osterhout 1996) and of the empty-category principle (McKinnon & Osterhout 1996).

Recently, a P600/SPS is also reported in relation to thematic role assignment (Kuperberg *et al.* 2003; Van Herten *et al.* 2004; Kim & Osterhout 2005; Wassenaar & Hagoort 2007). In this case, a P600/SPS is elicited when constraints for grammatical role assignment are in conflict with thematic role biases. For instance, Kim & Osterhout (2005) report a P600/SPS to the verb *devouring* in the sentence 'The hearty meal was devouring...', where the first noun phrase (NP) is not a good *agent*, but would be fine as a *theme*. The fascinating possibility suggested by these results is that a strong thematic bias could induce a tendency to detect a grammatical error where there is none ('-ing should be -ed') or to assign the grammatical role of object to the first NP, whereas the syntactic cues indicate that it is the subject of the sentence. These opposing tendencies result in a P600/SPS.

In summary, two classes of syntax-related ERP effects have been consistently reported. These two classes differ in their polarity, topographic distribution and latency characteristics. In terms of latency, the first class of effects is an anterior negativity. Apart from LANs related to working memory, anterior negativities to syntactic violations are mainly seen.

In a later latency range, positive shifts occur that are not only elicited by syntactic violations but also in grammatically well-formed sentences that vary in complexity (Kaan et al. 2000), as a function of the number of alternative syntactic structures that are compatible with the input at a particular position in the sentence (syntactic ambiguity; Osterhout et al. 1994; Van Berkum et al. 1999a), or when constraints for grammatical role assignment are overwritten by thematic role biases (Kim & Osterhout 2005).

10.3 Speech-related event-related brain potential effects

Despite the fact that spoken language is the primary mode of language communication, and reading a recent invention that is usually only acquired with substantial effort and through formal instruction, the majority of psycholinguistic research is on reading rather than listening or speaking. Likewise, most neuroimaging and ERP/MEG studies on language have used visual instead of auditory input. Hence, the number of ERP studies on spoken language processing is relatively limited. Studies that present natural connected speech have often found the same basic semantic (N400) and syntactic (P600/SPS) ERP effects as in reading, although the overall morphology of the ERP waveforms is quite different in speech than in reading.

(a) Auditory N400/N200

Regarding the N400 effect in speech, a number of studies have found the effect to have an earlier onset latency than in reading and usually the effect has a longer duration for speech than for written input (Holcomb & Neville 1990, 1991; Hagoort & Brown 2000a). For instance, Holcomb & Neville (1990) reported a very early effect to semantic anomalies in sentences spoken at a normal rate. They characterized this as an N400 effect. Over occipital sites, the onset latency of the auditory N400 effect was as early as 50–100 ms, and 150 ms over left and right parietal sites. The earlier onset of N400 effects on connected speech is surprising since, in contrast to written words, spoken words are encoded in a signal that is extended over time.

However, there is some doubt as to whether the auditory N400 effect is indeed a similar monophasic negative shift as the visual N400 effect is. The alternative is that the so-called auditory N400 effect actually is composed of at least two separate negative polarity effects, of which only the second negativity is an N400. In one of the first ERP studies on spoken language processing, McCallum et al. (1984) already reported that in their data the auditory N400 effect seemed to be preceded by a separate N200 effect. This early effect reached its maximal amplitude for the semantically incongruous sentence endings between 208 and 216 ms. The possible separation of the overall negative shift into a functionally different early and late negative effect was tested in a series of studies by Connolly et al. (1990, 1992, 1995). These authors compared ERPs with sentence-final words in highly constraining sentence contexts (e.g. 'The king wore a golden crown.') with ERPs in sentence contexts with low constraints (e.g. 'The woman talked about the frogs.'). A negative shift to words of low constraining sentences was observed (i.e. frogs) relative to words in highly constraining sentences (i.e. crown). The authors reported (Connolly et al. 1990) that individual difference waveforms (but not the grand averages), obtained by subtracting for each subject from each other the waveforms to the critical words (CWs) in the two context conditions, showed two distinct peaks, an early one with a central distribution (N200 effect) and a later one with a centroparietal distribution (N400 effect). The authors suggest that the N200 effects reflect acoustic/phonological word processing. That is, if the initial phoneme of the CW mismatches with the onset of the expected word, an N200 effect emerges. The N400 amplitude is claimed to be modulated by semantic expectancy.

Connolly & Phillips (1994) have attempted a more direct test of their account of an early and a later negativity in the ERP for words that do not allow a straightforward semantic fit with the context. In their study, next to semantically correct sentences, they presented sentences that ended in a semantically anomalous way. The anomalous word either started with the same phoneme as the most expected word given the sentence context (phoneme match condition), or its onset was different from the expected ending (phoneme mismatch condition). One negative peak (the N400) was reported for the phoneme match condition, whereas two negative peaks were observed for the phoneme mismatch condition. The authors attributed the earlier negativity to the phonemic deviation from the expected lexical form. Therefore, they called this effect the phonological mismatch negativity (PMN). The account of this early negative shift as a PMN is based on the idea that in spoken word recognition, word-initial sounds activate a cohort (Marslen-Wilson & Tyler 1980) or a shortlist (Norris 1994) of possible lexical candidates. In the process of recognizing a word, further incoming sensory information and top-down contextual information result in a reduction of the cohort or shortlist of possible candidates to one. This is the word that is actually perceived. Since in the phoneme mismatch condition the expected word is not a member of the cohort of lexical candidates, the mismatch can be detected early. In contrast, in the phoneme match condition, the expected word is a member of the cohort of lexical candidates that is instantiated by the onset of the sentence-final anomalous word. Therefore, the mismatch supposedly can be detected only later, when context information contributes to the pruning of the cohort of lexical candidates.

Hagoort & Brown (2000*a*) reported two experiments on semantic violations in connected speech that both resulted in substantial N400 effects time-locked to the word in the sentence that was semantically at odds with the preceding context. Unlike the prototypical visual N400 effect, which tends to be slightly larger over the right hemisphere, the auditory N400 effect was either symmetrical or larger over the left than the right hemisphere. However, just as in the visual modality, the auditory N400 effect had a clear posterior distribution. Both functionally and topographically, there is a strong correspondence between the visual and the auditory N400 effects. The hemispheric differences suggest that probably there is also a contribution from non-overlapping neuronal generators for the two input modalities.

In the Hagoort & Brown (2000*a*) study, the onset of N400 effects in the auditory modality was similar to that in the visual modality. However, it was found that, in addition, the anomalies elicited another effect that preceded the N400 effect in time. The onset of this effect (150 ms) was quite comparable to what Holcomb & Neville (1991) reported for their parietal sites, and to the PMN of Connolly & Phillips (1994). Van den Brink *et al.* (2001) found further evidence for an early N200 effect and a later N400 effect. The early effect occurred in the latency range of 150–250 ms and had an even distribution over the scalp. The N400 effect between 300 and 500 ms had the usual posterior distribution (Figure 10.3). Moreover, in contrast to the N400 effect, the N200 effect disappeared

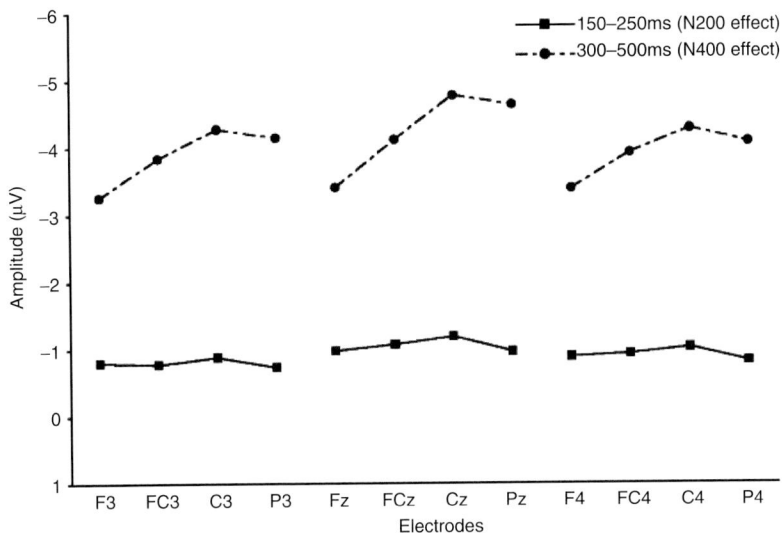

Figure 10.3 Connected speech. Distribution of the N200 and N400 effects for four left hemisphere sites (F3, FC3, C3 and P3), four midline sites (Fz, FCz, Cz and Pz) and four right hemisphere sites (F4, FC4, C4 and P4). The N200 effect was determined by subtracting the mean amplitude in the 150–250 ms latency window of the grand average ERP for the semantically congruent sentence-final words from the mean amplitudes of the grand average ERP for the semantically anomalous sentence-final words that did not share the same initial phonemes as the semantically congruous words. The N400 effect was determined in the same manner on the basis of the mean amplitudes in the 300–500 ms latency window
Van den Brink *et al.* (2001).

when the semantic anomaly shared the initial CV cluster with the semantically expected word.

The design of the study by Van den Brink *et al.* (2001) was very similar to that of the study by Van Petten *et al.* (1999). These authors presented sentences that ended with a pause followed by a target word that was either (i) the word with the highest cloze probability (e.g. 'It was a pleasant surprise to find that the car repair bill was only seventeen *dollars*.'), or a semantically incongruous ending that consisted of (ii) a word that shared the initial phonemes with the congruous target (e.g. *dolphins*), (iii) a word that had a different onset but rhymed with the highest cloze probability word (e.g. *scholars*), or (iv) a semantically incongruous sentence-final word without any form overlap with the congruous target (e.g. *bureau*). All three semantic anomaly conditions resulted in a significantly larger N400 than the congruous (highest cloze probability) ending. However, the onset of the N400 effect differed between conditions, as a function of the phonological overlap with the congruous ending. The onset of the N400 effect was later for the anomaly with a word-initial form overlap with the congruous ending, compared with onset latencies for the fully anomalous and rhyme word conditions. This suggests that the onset of the N400 effects was determined by the moment at which the acoustic input became inconsistent with a congruous sentence completion.

In §3*b*, how N200 and N400 effects can be characterized in terms of a functional account of spoken word recognition is discussed.

(b) Aspects of spoken word recognition

In models of spoken word recognition, usually a distinction is made between access, selection and integration (Marslen-Wilson 1987). The mental lexicon is the crucial interface between language form and content, two fundamentally distinct knowledge domains (Marslen-Wilson 1989). Lexical access refers to the mapping of the input signal onto word form representations in the mental lexicon. Selection refers to the discrimination among competing alternatives for a match between the acoustic input and lexical form representations (Frauenfelder & Tyler 1987). Through the mapping dynamics of access and selection, the retrieval of information associated with the word form is achieved, including the syntactic properties (e.g. gender, word class) and the meaning of a lexical item. However, since we normally perceive words not in isolation, but in the context of other words, the retrieved syntactic and semantic information have to be integrated with the higher-level context representation of the preceding part of utterance. We will refer to this process as integration (cf. Marslen-Wilson 1987). The nature of spoken word recognition is co-determined by the specifics of the speech signal. Mapping the spoken signal onto information in the mental lexicon is in many ways very different from mapping a visual signal onto lexical information. A central aspect of processing spoken words is that it occurs from left to right, starting from word onset (Marslen-Wilson & Welsh 1978). This left-to-right processing of spoken words allows the identification of the moment in time at which a particular word is recognized. The recognition point (RP) is defined as that part of the signal where the actual word becomes uniquely different from all other words in the listener's mental lexicon (Marslen-Wilson 1987). For instance, when presented in isolation, the RP of the word *captain* occurs after the /t/, since it is at this point that the sensory information excludes the only remaining alternative word candidate *captive*. For most multisyllabic words, the RP is located well before the end of the word (Marslen-Wilson 1984, 1987).

Electrophysiological evidence supporting the behavioural evidence for the concept of a RP in spoken word recognition was recently found in a study by O'Rourke & Holcomb (2002). These authors used the well-established fact that each content word elicits an N400 component. They had subjects listen to words with an early RP (e.g. pupil) and words with a late RP (e.g. carriage), without any additional task. When measured from acoustic word onset, the N400 component had an earlier peak latency for words with early RPs compared with words with late RPs. However, when the ERPs were averaged time-locked to the individual RPs, this difference disappeared, and the waveforms were identical following the RP.

In sentence context, it has been found that context information can speed up word processing (Tyler & Wessels 1983; Zwitserlood 1989). On the basis of experimental evidence, it is estimated that for selecting the word forms of one- and two-syllable content words, subjects need to hear an average of 200 ms in sentence context and more than 300 ms in isolation (Grosjean 1980; Marslen-Wilson 1984).

Although the amplitude of the N400 component is larger for content words than function words, and smaller for high frequency than low frequency words, modulations of its amplitude (i.e. the N400 effect) seem to be especially sensitive to integration—that is, to match the content specifications of a lexical candidate against the content specifications of the word, sentence or discourse context (Brown & Hagoort 1993; Holcomb 1993; Kellenbach & Michie 1996; Hagoort *et al.* 2004).

In contrast to the N400, the N200 effect seems specific for spoken language. The findings of Van den Brink *et al.* (2001) suggest a different functional account for the N200 effect than for the N400 effect. They found no N200 when the semantic anomaly shared its phonological onset with the contextually expected word candidate. Based on the initial form overlap, the expected candidate might initially be co-activated as a member of the cohort or shortlist of activated lexical candidates. This cohort/shortlist therefore contained the semantic features that are supported by the context. The authors propose that the amplitude modulation of the early negativity preceding the N400 effect reflects the lexical selection process that occurs at the interface of lexical form and contextual meaning. The word-initial speech segment activates a cohort of lexical candidates that are compatible with the initial stretch of the speech signal. This is a purely form-driven bottom-up process. Once the initial cohort (or shortlist) of lexical candidates is instantiated, top-down context information can start to have its effect. The interaction of form-based activation and content-based modulatory influences on the activational status of the available lexical candidates results in the selection of the lexical candidate that is optimally compatible with both form and content constraints. The N200 effect might reflect the lexical selection process that occurs at the interface of lexical form and contextual meaning. That is, if the contextual specifications do not support the form-based activation of a lexical candidate, an N200 effect is visible relative to a situation in which form-based activation is supported by contextual specifications. A large N200 would then indicate that the cohort of activated candidates does not contain semantic features that fit the preceding sentence frame well.

In contrast to the N200 effects, the N400 effect is claimed to arise at the content level only. Once a word's meaning is activated, the language processing system tries to incorporate its content specifications into the overall higher-order representation of the preceding utterance part. Even a clear mismatch between the meaning of a word and the semantics of the context does not prevent the mandatory process of matching the semantics of a word against the semantics of the context. It is this purely content matching and integration process of the most highly activated lexical candidates against their context that is reflected in the amplitude of the N400. The better the semantic fit, the more reduced the N400 amplitude.

One caveat has to be made with respect to these functional accounts. Not always is the N200 observed when it should be expected on the basis of the above account (Van Petten *et al.* 1999). This might be due to the overlapping component problem. The N200 and the N400 effects tend to overlap in time. This overlap implies that it is very hard to disentangle the neuronal generators of these two effects, and hence to find solid evidence, showing that indeed the N200 and N400 effects are qualitatively distinct.

(c) Lexical selection versus integration

The temporal resolution of EEG/MEG also allows one to investigate whether lexical selection is a prerequisite for the process of integration. This issue was addressed in a recent study by Van den Brink *et al.* (2006), which was designed to investigate the temporal relationship between lexical selection and semantic integration in auditory sentence processing. The authors investigated whether there is a discrete moment when lexical selection ends and semantic integration begins, or whether these two processes are of a cascading nature with semantic integration starting before lexical selection is completed. Information about the RP was used to investigate the onset of the N400 effect. Preceding the ERP experiment, a gating study was done on 522 spoken words. The gating method (Grosjean 1980; Tyler & Wessels 1983) allows the presentation of incremental portions of the acoustic signal until the full word is presented. For each gate, subjects specify which word they believe to be listening to and how confident they are that their response is correct. In this way, the RP can be established empirically. This study revealed that whereas the duration of the 522 words was on average 516 ms, their mean RP was well before word offset, namely at 286 ms. These words served as congruent and incongruent completions of spoken sentences. The results of the gating study were used to divide the CWs into two groups. One group contained contextually congruent and incongruent words with early RPs (with a mean of 230 ms), and the other group consisted of congruent and incongruent completions with late RPs (mean of 340 ms). If integration is dependent on word identification, then the N400 effect should set in at or soon after the RP and should therefore differ between the early and late RP words. However, if semantic integration starts before the word is uniquely recognized on the basis of bottom-up acoustic information, then the onset of the N400 could occur prior to the RP and should not have to differ as a function of RP latency. The results revealed that, despite a mean difference in RP of at least 100 ms between words with early and late RPs, the factor of early or late RPs neither affected the onset nor the peak of the N400. Incongruent completions in both groups elicited an N400 before the RP, which had an onset latency at approximately 200 ms (Figure 10.4). This indicates that irrespective of whether the selection process had successfully singled out one candidate, integration processing was started up nonetheless. Moreover, integration starts very early, well before the full word has been heard and in many cases even before the RP. These results favour a cascading account of spoken word processing in context. It seems that the semantic integration process does not wait until one appropriate candidate has been selected on the basis of a phonological analysis (cf. Marslen-Wilson (1989) for a similar view).

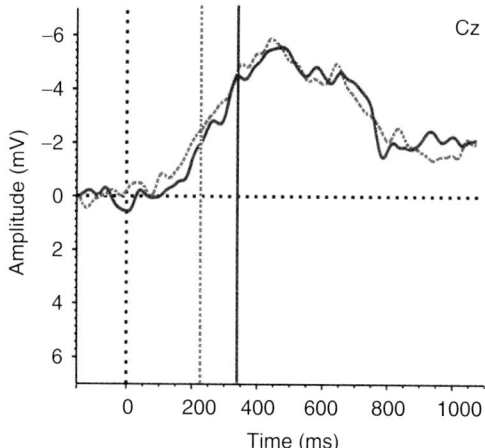

Figure 10.4 Connected speech. Difference waveforms from a representative electrode site (Cz), for semantically congruent versus semantically anomalous words with early RPs (dotted line; mean 229 ms) and late RPs (solid black line; mean 343 ms). The time axis is in milliseconds. Negative polarity is plotted upwards.

Up until recently, many researchers have shared the view that language processing is continuous and incremental, but that each incoming word is processed with bottom-up priority, and a number of spoken word processing models, such as the cohort model and shortlist, have incorporated this view (Marslen-Wilson & Welsh 1978; Marslen-Wilson 1987, 1993; Norris 1994). The findings of Van den Brink *et al.* (2006) could be explained in terms of bottom-up priority of spoken word processing. However, a number of recent ERP studies have investigated the possibility that expectancies for an upcoming word form are being generated on the basis of the preceding context in combination with the comprehender's knowledge about the world (Wicha *et al.* 2004; DeLong *et al.* 2005; Van Berkum *et al.* 2005). The results of these studies revealed that, in highly constraining contexts, words are not only rapidly integrated into the higher-order meaning representation of the preceding sentence or discourse context but also that the constraining context is used to form probabilistic predictions of which specific word will be presented next. This idea of anticipation is not new. Several models of spoken word recognition such as the Logogen model by Morton (1969) and TRACE by McClelland & Elman (1986) have already allowed for lexical preactivation of words based on the context. However, compelling evidence for preactivation during online sentence processing is limited, whereas a number of behavioural studies have provided evidence for an initial bottom-up priority based on the acoustic input (for a review of the literature relevant to this issue, see Van Berkum *et al.* (2005)).

In the case of a violation of anticipation of specific words, it is not surprising to find that the N400 effect sets in before the RP. Analysis of the first phonemes reveals that they do not match with those of any of the words anticipated. However, in light of the majority of behavioural evidence favouring bottom-up priority for acoustic processing of initial phonemes of the perceived word, Van den Brink *et al.* (2006) propose the following scenario. During sentence processing, there is a certain time frame in which lexical selection on the basis of a combination of the acoustic analysis of a word's first phonemes and context-based specifications can take place, and presumably it happens in the case of congruent words in highly to moderately constraining sentences. However, when selection of one appropriate candidate is difficult, as would be the case for anomalous words, or even congruent words in low-constraining contexts, integration as reflected by the N400 seems to be attempted for those candidates that match the acoustic input.

In conclusion, when words are spoken in context, there is no 'magic moment' when lexical selection ends and semantic integration begins. Irrespective of whether words have early or late RPs, semantic integration processing is initiated before words can be identified on the basis of the acoustic information alone.

(d) Neuronal generators of the N400

In an MEG version of the Van den Brink and Hagoort study (Jensen *et al.* in preparation), it was found that the N400 m effect is strongly left lateralized, with maximal activity over temporo-frontal areas (see Figure 10.5). MEG studies have identified a strong and a weak source of the N400 m in the left and right superior temporal sulci, respectively (Helenius *et al.* 1998; Halgren *et al.* 2002; Helenius *et al.* 2002). Using distributed currents source modelling, Halgren *et al.* (2002) identified an additional source in the left prefrontal cortex. The sources in the superior temporal sulci are consistent with the topography of the N400 m effect in Figure 10.5. It is interesting to note that the ERP

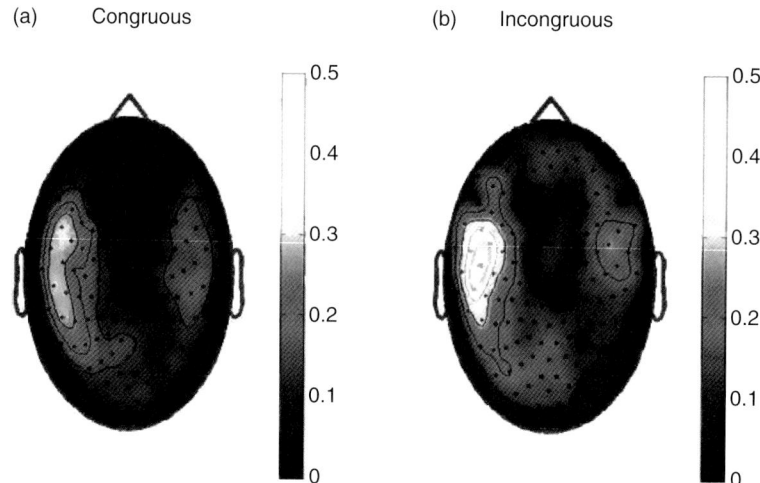

Figure 10.5 MEG recordings of the N400. The N400 m component is observed from 300 to 600 ms. The topography of the evoked fields (planar gradients of the evoked fields) for (a) the congruent nouns and (b) the incongruent nouns in sentence context. For the incongruous nouns, an increased N400 m is visible over the left hemisphere.

study (Van den Brink *et al.* 2006) using the same materials did not show a left lateralization of the N400 effect. Nevertheless, the topographies of the N400 effect are consistent for the EEG and MEG studies. The topography of the N400 ERP effect can be explained by dipolar sources in the left and right temporal sulci, given that these dipoles are approximately oriented towards the vertex of the head. Since both dipoles produce negative potentials over the midline, no lateralization is observed.

An MEG component with a similar time course and distribution is the M350 (Embick *et al.* 2001). The latency of the M350 has been found to be sensitive to word frequency. The M350 has been recorded mainly for written words. For the N400, it is the amplitude rather than the latency that is responsive to frequency. It is currently unclear if and, if so, to what degree the M350 and the N400m are based on overlapping neuronal generators.

10.4 Beyond the single utterance

A major task of psycholinguistics is to find out how syntactic, semantic and referential analyses are orchestrated as language comprehension unfolds in time. Research on this topic has primarily focused on the relative timing and the informational dependency of these various analyses. An important unresolved question, for example, is whether the different types of analysis are conducted in some principled sequential order, with some theorists assigning a fundamental priority to syntactic analysis (Frazier 1987; Mitchell *et al.* 1995) and others instead arguing for a simultaneous evaluation of syntactic, semantic and referential aspects of the input (Tyler & Marslen-Wilson 1977; Marslen-Wilson & Tyler 1980; MacDonald *et al.* 1994; Tanenhaus & Trueswell 1995; Jackendoff 1999; Hagoort 2005). A closely related question is whether the results of semantic and referential processing affect the initial syntactic analysis of the input (Crain & Steedman 1985; Altmann 1988) or not (Ferreira & Clifton 1986).

Van Berkum *et al.* (2003) explored the possibility that the ERP method can also be used to selectively track some of the *referential* aspects of language comprehension, while people listen to a piece of discourse. This is a relatively unexplored territory in electrophysiological research. Although ERPs have been used to address issues in referential processing before (Osterhout & Mobley 1995; Osterhout *et al.* 1997; Streb *et al.* 1999; Schmitt *et al.* 2002; see Kutas *et al.* 2000, for a brief review), no studies have directly looked for an ERP signature of referential analysis in discourse-level spoken language comprehension.

In the Van Berkum *et al.* (2003) study, subjects were asked to listen to several ministories, such as the one below (translated from Dutch, boldface and italics added):

(3) David had told *the boy and the girl* to clean up their room before lunchtime. But the boy had stayed in bed all morning, and the girl had been on the phone all the time. David told **the girl** that had been on the phone to hang up.

Following earlier research on this topic (Crain & Steedman 1985), the stories were varied such that they provided either a *single* unique referent for the NP 'the girl', as in (3), or *two* equally eligible referents, as in (4):

(4) David had told *the two girls* to clean up their room before lunchtime. But one of the girls had stayed in bed all morning, and the other had been on the phone all the time. David told **the girl** that had been on the phone to hang up.

The authors found that referentially ambiguous spoken nouns elicited a negative shift that emerged in the ERPs at approximately 300–400 ms after acoustic onset, and that, although widely distributed, was most prominent and most sustained at anterior recording sites (Figure 10.6). These findings show that very quickly the processing system has determined whether a singular definite noun has a single unique referent in the earlier discourse or not (in italics in (3) and (4), respectively).

The higher-level processes associated with establishing reference, particularly those that require consulting one's model of the prior discourse, are frequently assumed to be rather slow, as compared with the lower-level syntactic and sentence-semantic aspects of comprehension (Kintsch 1998). However, listeners needed at most only 300–400 ms at some level to detect a difference between, say, 'the girl' in a discourse that had introduced a single girl or that had introduced *two* girls. Given that all known sentence-semantic and sentence-syntactic ERP effects emerge within some 500 ms after CW onset (see Brown *et al.* 2000 for review), this discourse-related ERP effect occurs within the same temporal window of opportunity. Of course, this does not mean that referential ambiguity is *always* detected within some 300–400 ms. For instance, the moment at which an unfolding noun reveals its number depends on a wide variety of factors, including where the language at hand codes a noun for its number (e.g. suffix, prefix or other); the duration of the spoken noun at hand; and the way the stem of a noun changes with pluralization. In sum, the early onset of the referential ERP effect should primarily be taken as an indication that discourse-dependent referential ambiguity *can* be detected within that short duration.

Apart from its rapid emergence, the referentially induced ERP effect observed with spoken language is also 'immediate' in suggesting that reference is established *incrementally*, i.e. at each relevant word coming in. In Dutch, a referentially ambiguous NP can always be extended by a post-nominal modifier that supplies additional information,

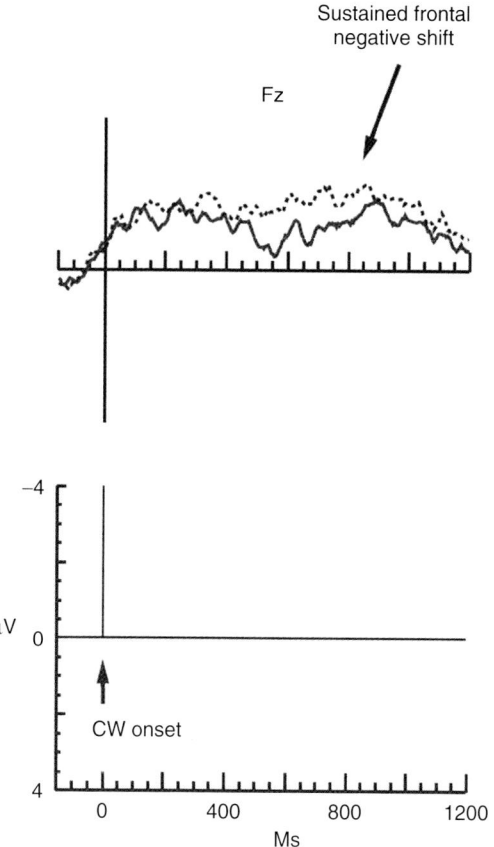

Figure 10.6 Grand average ERP waveform elicited by spoken singular nouns presented in a one-referent (solid line) and a two-referent (dotted line) context. The acoustic onset of the noun is at 0 ms, and negative polarity is plotted upwards. The waveforms are shown for a representative frontal midline site (Fz).

such as a relative clause ('the girl that was waiting'). In principle, the comprehension system might thus delay its attempt to establish reference until later sentential input signals that the NP is unequivocally complete. What these and earlier (Van Berkum *et al.* 1999*a*) findings suggest is that the comprehension system does *not* do this, and instead initiates sufficient referential processing at the head noun to at least determine, within some 300–400 ms after noun onset, whether it is referentially unique or not.

In addition to timing, there is also information on the *nature* of the referentially induced ERP effect, a frontally dominant and sustained negative shift. The frontally sustained negative shift reported for spoken referentially ambiguous nouns is very similar to sustained ERP effects observed under conditions of increased memory load in language comprehension (Kluender & Kutas 1993; King & Kutas 1995; Friederici *et al.* 1996; Kutas 1997; Münte *et al.* 1998; Fiebach *et al.* 2001; Vos *et al.* 2001) as well as in non-linguistic processing tasks (Rösier *et al.* 1993; Donaldson & Rugg 1999; Rugg & Allan 2000).

It is not difficult to imagine why referential ambiguity might be a memory-demanding situation. For one, referential ambiguity may trigger additional retrieval from episodic discourse memory, associated with a search for less obvious clues that might help to infer the most plausible referent (Myers & O'Brien 1998). Alternatively, referential ambiguity may require the system to actively maintain two candidate fillers for an unresolved single referential slot in working memory (see Gibson (1998) for an account of referentially induced working memory load in sentence comprehension). The latter would explain why the ERP effect of referential ambiguity resembles the ERP effect elicited by various other types of expressions that impose a higher load on working memory, such as (i) object-relative clauses (King & Kutas 1995; Kutas 1997), (ii) temporal expressions like '*Before* the psychologist submitted the article, the journal changed its policy' (Münte *et al.* 1998) in which the information supplied in the first phrase does not describe what actually happened first, or (iii) expressions like 'The *pitcher* fell down and broke/cursed' that contain a lexically ambiguous word (Hagoort & Brown 1994).

In the domain of language processing, memory-related sustained frontal negativities are sometimes referred to as left anterior negativities or LAN effects, owing to their frequent left anterior maximum over the scalp (Kluender & Kutas 1993). However, some particularly early left anterior negativities have also been claimed to directly reflect aspects of early syntactic processing (Friederici *et al.* 1996; see Friederici (1998) for a review). It is yet unclear as to what extent the early and later LAN effects reflect the same set of neuronal generators (see Friederici (1998), Hagoort *et al.* (1999), Brown & Hagoort (2000) and Brown *et al.* (2000) for discussions). The exact relationship between the LAN family of effects and the referentially induced sustained negative shift thus remains to be established.

In an additional study by Van Berkum *et al.* (2003), further evidence was obtained for the claim that the discourse context can affect the interpretation of an unfolding sentence extremely rapidly. This study compared discourse-dependent N400 effects with the standard N400 effect observed in response to a semantic anomaly. Subjects listened to short Dutch stories, in which the final sentence contained the CW. For each story, two CW alternatives were available: a discourse-coherent CW that was a good continuation of the earlier discourse and a discourse-anomalous CW that did not continue the discourse in a semantically plausible way (see the English translations of an example of the Dutch materials in (5); CWs are in bold). The difference in coherence between the two conditions hinged on considerable inferencing about the discourse topic and the situation it described. In the example, 'quick' and 'slow' are equally compatible with the sentence context, but only 'quick' is licensed by the discourse.

(5) As agreed upon, Jane was to wake her sister and her brother at five o'clock in the morning. But the sister had already washed herself, and the brother had even got dressed. Jane told the brother that he was exceptionally **quick/slow**.

When the final sentences were presented in isolation, no differences were obtained between the ERPs elicited by the CWs in the two conditions. However, in their discourse context, the discourse-anomalous CWs elicited a large and widely distributed negative deflection that emerged at approximately 150–200 ms after their acoustic onset, peaked at approximately 400 ms, lasted for approximately 800–1000 ms and reached its maximum over centroparietal electrode sites. In other words, a standard N400 effect was obtained that is indistinguishable from an N400 effect elicited by a semantic anomaly in a local sentence context (Figure 10.7).

These findings show that the incoming words of an unfolding spoken sentence not only very rapidly make contact with 'global' discourse-level semantic information, but also do so in a way that is indistinguishable from how they make contact with 'local' sentence-level semantic information. The equivalent impact of sentence- and discourse-semantic contexts on the N400 elicited by a spoken word suggests that the process at work does not care where the semantic context originally came from, and evaluates the incoming words relative to the widest interpretive domain available. Moreover, these findings show that in natural spoken language comprehension, an unfolding word can be mapped onto discourse-level representations extremely rapidly, after only about three phonemes, and in many cases well before the end of the word.

10.5 Conclusions

In addition to discussing a number of relevant ERP/MEG studies on the nature of spoken language processing, some final remarks and conclusions will be made to highlight important aspects of electrical and magnetic measurements for this domain of research.

One particularly noteworthy aspect of a substantial number of ERP studies that were discussed above is that subjects were only engaged in the natural task, which in the case of speech is listening for understanding. An advantage of ERP/MEG recordings is that reliable effects can be obtained in the absence of potentially intrusive secondary tasks.

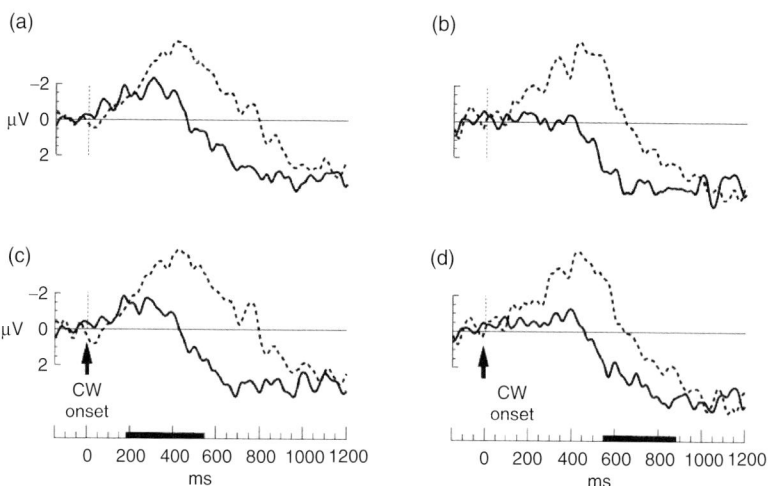

Figure 10.7 Discourse-dependent semantic anomaly effects by word position and length. Grand average ERPs, at Pz, elicited by spoken words that were coherent (solid line) or anomalous in discourse (dotted line), for words in (a) sentence-medial and (b) sentence-final position, as well as for (c) short words (below 550 ms and a mean duration of 451 ms) and (d) long words (over 550 ms and a mean duration of 652 ms). The horizontal bars mark the range of acoustic word offsets for short and long CWs.
(Reprinted with permission from Van Berkum *et al.* (2003). Copyright Elsevier B. V.)

Secondary tasks such as, for instance, lexical decision and word monitoring create a dual-task situation that might impact the primary process under investigation or the response characteristics (Norris *et al.* 2000). Getting reliable measurements in the absence of a secondary task is especially helpful in studies with language-impaired subjects, such as aphasic patients. For these patients, a dual-task situation might tax the processing resources beyond their limits. It is, nevertheless, reassuring that many of the ERP findings are consistent with the chronometric research on spoken word recognition and sentence processing.

Apart from the N200, all major effects discussed above are not only observed with spoken input, but also when subjects are reading. Since all these effects are triggered by processes beyond the single word level, such as integration, syntactic parsing or referential binding, this suggests that the functional architecture of language processing might be modality specific at the single word level, but beyond that spoken and written language comprehension seems to share processing principles and neural architecture. A recent series of ERP and fMRI studies on the integration of speech and co-speech gestures even suggests that similar principles apply beyond the domain of core language operations, with the left inferior frontal and the left superior temporal cortices as important nodes in the neural network for understanding (Hagoort 2005; Özyürek *et al.* 2007; Willems *et al.* 2007).

Even more remarkable is that, in general, the time course of the auditory N400 and P600/SPS (Hagoort & Brown 2000*a*, *b*), although often somewhat more extended, is similar compared with the N400 and P600/SPS elicited in reading. However, this similarity does not necessarily imply that the underlying processes have the same time course in both modalities. The crucial difference between reading and the processing of speech is the difference in the time at which word information is made available. In reading, words are essentially instantaneously available, whereas in speech the information accrues in a left-to-right order. Therefore, relative to the availability of information in reading, the N400 and P600/SPS effects as well as the referential negativity in speech are actually remarkably early. The time course of the ERP effects is compatible with the view that the different information types (lexical, syntactic, phonological, pragmatic) are processed in parallel and influence the interpretation process incrementally, i.e. as soon as the relevant pieces of information are available (Marslen-Wilson 1989; Zwitserlood 1989; Jackendoff 2002). I refer to this as the immediacy principle.

Furthermore, the ERP results clearly show that there does not seem to be a separate stage during which word meaning is exclusively integrated at the sentence level. Incremental interpretation is, for the most part, done by an immediate mapping onto a discourse model. Moreover, listeners relate the incoming acoustic signal to the semantic context not just before they have heard the complete word but before they can actually know exactly what the unfolding word itself is going to be. The presented ERP evidence is consistent with chronometric data, in showing that discourse- and sentence-semantic information differentially affects the processing of word candidates before the acoustic input itself uniquely specifies a word.

Finally, a cognitive neuroscience account of human language processing should not only specify the crucial network of brain areas supporting listening, reading and speaking. In addition, the temporal dynamics of the neurophysiological processes that instantiate these language functions has to be specified as well. EEG and MEG recordings are superior over any other non-invasive method in providing a window onto exactly these dynamics. As I have shown here, the limited number of studies on speech that are currently available already have provided interesting insights into the organization of spoken language understanding.

References

Ainsworth-Darnell, K., Shulman, H. & Boland, J. 1998 Dissociating brain responses to syntactic and semantic anomalies: evidence from event-related potentials. *J. Mem. Lang.* **38**, 112–130. (doi:10.1006/jmla.1997.2537)
Altmann, G. T. M. 1988 Ambiguity, parsing strategies, and computational models. *Lang. Cogn. Process.* **3**, 73–97.
Boland, J. E. & Cutler, A. 1996 Interaction with autonomy: multiple output models and the inadequacy of the great divide. *Cognition* **58**, 309–320. (doi:10.1016/0010-0277 (95)00684-2)
Brown, C. M. & Hagoort, P. 1993 The processing nature of the N400: evidence from masked priming. *J. Cogn. Neurosci.* **5**, 34–44.
Brown, C. M. & Hagoort, P. 2000 On the electrophysiology of language comprehension: implications for the human language system. In *Architectures and mechanisms for language processing* (eds M. W. Crocker, M. Pickering & C. Clifton), pp. 213–237. Cambridge, UK: Cambridge University Press.
Brown, C. M., Hagoort, P. & Chwilla, D. J. 2000 An event-related brain potential analysis of visual word priming effects. *Brain Lang.* **72**, 158–190. (doi:10.1006/brln.1999.2284)
Catani, M., Jones, D. K. & Ffytche, D. H. 2004 Perisylvian language networks of the human brain. *Ann. Neurol.* **57**, 8–16. (doi:10.1002/ana.20319)
Connolly, J. F. & Phillips, N. A. 1994 Event-related potential components reflect phonological and semantic processing of the terminal words of spoken sentences. *J. Cogn. Neurosci.* **6**, 256–266.
Connolly, J. F., Stewart, S. H. & Phillips, N. A. 1990 The effects of processing requirements on neurophysiological responses to spoken sentences. *Brain Lang.* **39**, 302–318. (doi:10.1016/0093-934X(90)90016-A)
Connolly, J. F., Phillips, N. A., Stewart, S. H. & Brake, W. G. 1992 Event-related potential sensitivity to acoustic and semantic properties of terminal words in sentences. *Brain Lang.* **43**, 1–18. (doi:10.1016/0093-934X(92)90018-A)
Connolly, J. F., Phillips, N. A. & Forbes, K. A. 1995 The effects of phonological and semantic features of sentence-ending words on visual event-related brain potentials. *Electroencephalogr. Clin. Neuropsychol.* **94**, 276–287. (doi:10.1016/0013-4694(95)98479-R)
Coulson, S., King, J. W. & Kutas, M. 1998 Expect the unexpected: event-related brain response to morphosyn-tactic violations. *Lang. Cogn. Process.* **13**, 21–58. (doi:10.1080/016909698386582)
Crain, S. & Steedman, M. 1985 On not being led up the garden path: the use of context by the psychological parser. In *Natural language parsing* (eds D. R. Dowty, L. Karttunen & A. M. N. Zwicky), pp. 320–358. Cambridge, UK: Cambridge University Press.
DeLong, K. A., Urbach, T. P. & Kutas, M. 2005 Probabilistic word pre-activation during language comprehension inferred from electrical brain activity. *Nat. Neurosci.* **8**, 1117–1121. (doi:10.1038/nn1504)
Donaldson, D. I. & Rugg, M. D. 1999 Event-related potential studies of associative recognition and recall: electrophysiological evidence for context dependent retrieval processes. *Brain Res. Cogn. Brain Res.* **8**, 1–16. (doi:10.1016/S0926-6410(98)00051-2)
Embick, D., Hackl, M., Schaeffer, J., Kelepir, M. & Marantz, A. 2001 A magnetoencephalographic component whose latency reflects lexical frequency. *Brain Res. Cogn. Brain Res.* **10**, 345–348. (doi:10.1016/S0926-6410(00)00053-7)
Ferreira, F. & Clifton, C. J. 1986 The independence of syntactic processing. *J. Mem. Lang.* **25**, 348–368. (doi:10.1016/0749-596X(86)90006-9)
Fiebach, C. J., Schlesewsky, M. & Friederici, A. D. 2001 Syntactic working memory and the establishment of filler-gap dependencies: insights from ERPs and fMRI. *J. Psycholinguist. Res.* **30**, 321–338. (doi:10.1023/A:1010447102554)

Fodor, J. D. 1983 *The modularity of mind.* Cambridge, MA: MIT Press.

Frauenfelder, U. H. & Tyler, L. K. 1987 The process of spoken word recognition: an introduction. *Cognition* **25**, 1–20. (doi:10.1016/0010-0277(87)90002-3)

Frazier, L. 1987 Sentence processing: a tutorial review. In *Attention and performance XII* (ed. M. Coltheart), pp. 559–585. Hillsdale, NJ: Lawrence Erlbaum.

Friederici, A. D. 1995 The time course of syntactic activation during language processing: a model based on neuropsy-chological and neurophysiological data. *Brain Lang.* **50**, 259–281. (doi:10.1006/brln.1995.1048)

Friederici, A. D. 1998 The neurobiology of language comprehension. In *Language comprehension: a biological perspective* (ed. A. D. Friederici), pp. 263–301. Berlin, Germany: Springer.

Friederici, A. D. 2002 Towards a neural basis of auditory sentence processing. *Trends Cogn. Sci.* **6**, 78–84. (doi:10.1016/S1364-6613(00)01839-8)

Friederici, A. D., Pfeifer, E. & Hahne, A. 1993 Event-related brain potentials during natural speech processing: effects of semantic, morphological and syntactic violations. *Cogn. Brain Res.* **1**, 183–192. (doi:10.1016/0926-6410(93)90026-2)

Friederici, A. D., Hahne, A. & Mecklinger, A. 1996 Temporal structure of syntactic parsing: early and late event-related brain potential effects. *J. Exp. Psychol. Learn.* **22**, 1219–1248. (doi:10.1037/0278-7393.22.5.1219)

Gibson, E. 1998 Linguistic comlexity: locality of syntactic dependencies. *Cognition* **68**, 1–76. (doi:10.1016/S0010-0277(98)00034-1)

Grosjean, F. 1980 Spoken word recognition processes and the gating paradigm. *Percept. Psychophys.* **28**, 267–283.

Hagoort, P. 2005 On Broca, brain, and binding: a new framework. *Trends Cogn. Sci.* **9**, 416–423.

Hagoort, P. & Brown, C. 1994 Brain responses to lexical ambiguity resolution and parsing. In *Perspectives on sentence processing* (eds C. Clifton, L. Frazier & K. Rayner), pp. 45–81. Hillsdale, NJ: Erlbaum.

Hagoort, P. & Brown, C. M. 2000a ERP effects of listening to speech: semantic ERP effects. *Neuropsychologia* **38**, 1518–1530. (doi:10.1016/S0028-3932(00)00052-X)

Hagoort, P. & Brown, C. M. 2000b ERP effects of listening to speech compared to reading: the P600/SPS to syntactic violations in spoken sentences and rapid serial visual presentation. *Neuropsychologia* **38**, 1531–1549. (doi:10.1016/S0028-3932(00)00053-1)

Hagoort, P., Brown, C. M. & Groothusen, J. 1993 The syntactic positive shift (SPS) as an ERP measure of syntactic processing. *Lang. Cogn. Process.* **8**, 439–483.

Hagoort, P., Brown, C. M. & Osterhout, L. 1999 The neurocognition of syntactic processing. In *Neurocognition of language* (eds C. M. Brown & P. Hagoort), pp. 273–317. Oxford, UK: Oxford University Press.

Hagoort, P., Wassenaar, M. & Brown, C. M. 2003 Syntax-related ERP-effects in Dutch. *Cogn. Brain Res.* **16**, 38–50. (doi:10.1016/S0926-6410(02)00208-2)

Hagoort, P., Hald, L., Bastiaansen, M. & Petersson, K M. 2004 Integration of word meaning and world knowledge in language comprehension. *Science* **304**, 438–441. (doi:10.1126/science.1095455)

Halgren, E., Dhond, R. P., Christensen, N., Van Petten, C. Marinkovic, K., Lewine, J. D. & Dale, A. M. 2002 N400-like magnetoencephalography responses modulated by semantic context, word frequency, and lexical class in sentences. *Neuroimage* **17**, 1101–1116. (doi:10.1006/nimg.2002.1268)

Helenius, P., Salmelin, R., Service, E. & Connolly, J. F. 1998 Distinct time courses of word and context comprehension in the left temporal cortex. *Brain* **121**, 1133–1142. (doi:10.1093/brain/121.6.1133)

Helenius, P., Salmelin, R., Service, E., Connolly, J. F., Leinonen, S. & Lyytinen, H. 2002 Cortical activation during spoken-word segmentation in nonreading-impaired and dyslexic adults. *J. Neurosci.* **22**, 2936–2944.

Holcomb, P. J. 1993 Semantic priming and stimulus degradation: implications for the role of the N400 in language processing. *Psychophysiology* **30**, 47–61.

Holcomb, P. J. & Neville, H. J. 1990 Semantic priming in visual and auditory lexical decision: a between modality comparison. *Lang. Cogn. Process.* **5**, 281–312.

Holcomb, P. J. & Neville, H. J. 1991 Natural speech processing: an analysis using event-related brain potentials. *Psychobiology* **19**, 286–300.

Jackendoff, R. 1999 The representational structures of the language faculty and their interactions. In *The neurocognition of language* (eds C. M. Brown & P. Hagoort), pp. 37–79. Oxford, UK: Oxford University Press.

Jackendoff, R. 2002 *Foundations of language: brain, meaning, grammar, evolution*. Oxford, UK: Oxford University Press.

Jensen, O., Weder, N., Bastiaansen, M., van den Brink, D., Dijkstra, T. & Hagoort, P. In preparation. Modulation of the beta rhythm in a language comprehension task.

Kaan, E. & Swaab, T. Y. 2003 Repair, revision and complexity in syntactic analysis: an electrophysiological differentiation. *J. Cogn. Neurosci.* **15**, 98–110. (doi:10.1162/089892903321107855)

Kaan, E., Harris, A., Gibson, E. & Holcomb, P. J. 2000 The P600 as an index of syntactic integration difficulty. *Lang. Cogn. Process.* **15**, 159–201. (doi:10.1080/016909600386084)

Kellenbach, M. I. & Michie, P. T. 1996 Modulation of event-related potentials by semantic priming: effects of color-dued selective attention. *J. Cogn. Neurosci.* **8**, 155–173.

Kim, A. & Osterhout, L. 2005 The independence of combinatory semantic processing: evidence from event-related potentials. *J. Mem. Lang.* **52**, 205–225. (doi:10.1016/j.jml.2004.10.002)

King, J. W. & Kutas, M. 1995 Who did what and when? Using word- and clause-level ERPs to monitor working memory usage in reading. *J. Cogn. Neurosci.* **7**, 376–395.

Kintsch, W. 1998 *Comprehension: a paradigm for cognition*. Cambridge, UK: Cambridge University Press.

Kluender, R. & Kutas, M. 1993 Subjacency as a processing phenomenon. *Lang. Cogn. Process.* **8**, 573–633.

Kuperberg, G. R., Sitnikova, T., Caplan, D. & Holcomb, P. J. 2003 Electrophysiological distinctions in processing conceptual relationships within simple sentences. *Cogn. Brain Res.* **17**, 117–129. (doi:10.1016/S0926-6410(03)00086-7)

Kutas, M. 1997 Views on how the electrical activity that the brain generates reflects the functions of different language structures. *Psychophysiology* **34**, 383–398. (doi:10.1111/j.1469-8986.1997.tb02382.x)

Kutas, M. & Hillyard, S. A. 1980 Reading senseless sentences: brain potentials reflect semantic anomaly. *Science* **207**, 203–205. (doi:10.1126/science.7350657)

Kutas, M. & Van Petten, C. K. 1994 Psycholinguistics electrified: event-related brain potential investigations. In *Handbook of psycholinguistics* (ed. M. A. Gernsbacher), pp. 83–143. New York, NY: Academic Press.

Kutas, M., Federmeier, K. D., Coulson, S., King, J. W. & Münte, T. F. 2000 Language. In *Handbook of psychophysiology* (eds J. T. Cacioppo, L. G. Tassinary & G. G. Berntson), pp. 576–601. New York, NY: Cambridge University Press.

Luck, S. J. 2005 *An introduction to the event-related potential technique*. Cambridge, MA: The MIT Press.

MacDonald, M. C., Pearlmutter, N. J. & Seidenberg, M. S. 1994 Lexical nature of syntactic ambiguity resolution. *Psychol. Rev.* **101**, 676–703. (doi:10.1037/0033-295X.101.4.676)

Marslen-Wilson, W. D. 1973 Linguistic structure and speech shadowing at very short latencies. *Nature* **244**, 522–523. (doi:10.1038/244522a0)

Marslen-Wilson, W. D. 1984 Function and process in spoken word-recognition. In *Attention and performance X* (eds H. Bouma & D. Bouwhuis), pp. 125–150. Hillsdale, NJ: Erlbaum.

Marslen-Wilson, W. D. 1987 Functional parallelism in spoken word-recognition. *Cognition* **25**, 71–102. (doi:10.1016/0010-0277(87)90005-9)

Marslen-Wilson, W. D. 1989 Access and integration: projecting sound onto meaning. In *Lexical representation and process* (ed. W. D. Marslen-Wilson), pp. 3–24. Cambridge, MA: MIT Press.

Marslen-Wilson, W. D. 1993 Issues of process and representation in lexical access. In *Cognitive models of speech processing: the second sperlonga meeting* (eds G. T. M. Altmann & R. Shillock), pp. 187–210. Hillsdale, NJ: Erlbaum.

Marslen-Wilson, W. D. & Tyler, L. K. 1980 The temporal structure of spoken language understanding. *Cognition* **8**, 1–71. (doi:10.1016/0010-0277(80)90015-3)

Marslen-Wilson, W. D. & Welsh, A. 1978 Processing interactions and lexical access during word-recognition in continuous speech. *Cogn. Psychol* **10**, 29–63. (doi:10.1016/ 0010-0285(78)90018-X)
McCallum, W.C., Farmer, S. F. & Pocock, P. V. 1984 The effects of physical and semantic incongruities on auditory event-related potentials. *Electroencephalogr. Clin. Neuropsychol.* **59**, 477–488. (doi:10.1016/0168-5597(84)90006-6)
McClelland, J. L. & Elman, J. L. 1986 The TRACE model of speech perception. *Cogn. Psychol.* **18**, 1–86. (doi:10.1016/0010-0285(86)90015-0)
McKinnon, R. & Osterhout, L. 1996 Constraints on movement phenomena in sentence processing: evidence from event-related brain potentials. *Lang. Cogn. Process.* **11**, 495–523. (doi:10.1080/016909696387132)
Mitchell, D. C., Cuetos, F., Corley, M. M. B. & Brysbaert, M. 1995 Exposure-based models of human parsing: evidence for the use of coarse-grained (nonlexical) statistical records. *J. Psycholinguist. Res.* **24**, 469–488. (doi:10.1007/BF02143162)
Morton, J. 1969 Interaction of information in word recognition. *Psychol. Rev.* **76**, 165–178. (doi:10.1037/h0027366)
Münte, T. F. & Heinze, H. J. 1994 ERP negativities during syntactic processing of written words. In *Cognitive electro-physiology* (eds H. J. Heinze, T. F. Münte & H. R. Mangun), pp. 211–238. Boston, MA: Birkhauser.
Münte, T. F., Heinze, H. J. & Mangun, G. R. 1993 Dissociation of brain activity related to syntactic and semantic aspects of language. *J. Cogn. Neurosci.* **5**, 335–344.
Münte, T. F., Matzke, M. & Johannes, S. 1997 Brain activity associated with syntactic incongruities in words and pseudo-words. *J. Cogn. Neurosci.* **9**, 300–311.
Münte, T. F., Schiltz, K. & Kutas, M. 1998 When temporal terms belie conceptual order. *Nature* **395**, 71–73. (doi:10.1038/25731)
Myers, J. L. & O'Brien, E. J. 1998 Accessing the discourse representation during reading. *Discourse Process.* **26**, 131–157.
Näätänen, R. 2001 The perception of speech sounds by the human brain as reflected by the mismatch negativity (MMN) and its magnetic equivalent (MMNm). *Psycho-physiology* **38**, 1–21.
Näätänen, R. et al. 1997 Language-specific phoneme representations revealed by electric and magnetic brain responses. *Nature* **385**, 432–434. (doi:10.1038/385432a0)
Neville, H. J., Nicol, J. L., Barss, A., Forster, K. I. & Garrett, M. F. 1991 Syntactically based sentence processing classes: evidence from event-related brain potentials. *J. Cogn. Neurosci.* **3**, 151–165.
Norris, D. 1994 Shortlist: a connectionist model of continuous speech recognition. *Cognition* **52**, 189–234. (doi:10.1016/0010-0277(94)90043-4)
Norris, D., McQueen, J. M. & Cutler, A. 2000 Merging information in speech recognition: feedback is never necessary. *Behav. Brain Sci.* **23**, 299–370. (doi:10.1017/S0140525X00003241)
O'Rourke, T. B. & Holcomb, P. J. 2002 Electrophysiological evidence for the efficiency of spoken word processing. *Biol. Psychol.* **60**, 121–150. (doi:10.1016/S0301-0511(02)00045-5)
Osterhout, L. 1997 On the brain response to syntactic anomalies: manipulations of word position and word class reveal individual differences. *Brain Lang.* **59**, 494–522. (doi:10.1006/brln.1997.1793)
Osterhout, L. & Hagoort, P. 1999 A superficial resemblance doesn't necessarily mean you're part of the family: counterarguments to Coulson, King, and Kutas (1998) in the P600/SPS debate. *Lang. Cogn. Process.* **14**, 1–14. (doi:10.1080/016909699386356)
Osterhout, L. & Holcomb, P. J. 1992 Event-related brain potentials elicited by syntactic anomaly. *J. Mem. Lang.* **31**, 785–806. (doi:10.1016/0749-596X(92)90039-Z)
Osterhout, L. & Holcomb, P. J. 1993 Event-related potentials and syntactic anomaly: evidence of anomaly detection during the perception of continuous speech. *Lang. Cogn. Process.* **8**, 413–438.
Osterhout, L. & Holcomb, P. J. 1995 Event-related potentials and language comprehension. In *Electrophysiology of mind* (eds M. D. Rugg & M. G. H. Coles), pp. 171–215. Oxford, UK: Oxford University Press.
Osterhout, L. & Mobley, L. A. 1995 Event-related brain potentials elicited by failure to agree. *J. Mem. Lang.* **34**, 739–773. (doi:10.1006/jmla.1995.1033)

Osterhout, L., Holcomb, P. J. & Swinney, D. A. 1994 Brain potentials elicited by garden-path sentences: evidence of the application of verb information during parsing. *J. Exp. Psychol. Learn.* **20**, 786–803. (doi:10.1037/0278-7393.20.4.786)

Osterhout, L., Bersick, M. & McLaughlin, J. 1997 Brain potentials reflect violations of gender stereotypes. *Mem. Cognition* **25**, 273–285.

Özyürek, A., Willems, R. M., Kita, S. & Hagoort, P. 2007 On-line integration of semantic information from speech and gesture: insights from event-related brain potentials. *J. Cogn. Neurosci.* **4**, 605–616. (doi:10.1162/089892903322370807)

Pulvermüller, F., Shtyrov, Y., Kujala, T. & Näätänen, R. 2004 Word-specific cortical activity as revealed by the mismatch negativity. *Psychophysiology* **41**, 106–112. (doi:10.1111/j.1469-8986.2003.00135.x)

Rösler, F., Friederici, A. D., Pütz, P. & Hahne, A. 1993 Event-related brain potentials while encountering semantic and syntactic constraint violations. *J. Cogn. Neurosci.* **5**, 345–362.

Rugg, M. D. & Allan, K. 2000 Event-related potential studies of long-term memory. In *The Oxford handbook of memory* (eds E. Tulving & F. I. M. Craik), pp. 521–537. Oxford, UK: Oxford University Press.

Schmitt, B. M., Lamers, M. J. A. & Münte, T. F. 2002 Electrophysiological estimates of biological and syntactic gender violation during pronoun. *Cogn. Brain Res.* **14**, 333–346. (doi:10.1016/S0926-6410(02)00136-2)

Streb, J., Rösler, R & Hennighausen, E. 1999 Event-related responses to pronoun and proper name anaphors in parallel and nonparallel discourse structures. *Brain Lang.* **70**, 273–286. (doi:10.1006/brln.1999.2177)

Tanenhaus, M. K. & Trueswell, C. 1995 Sentence comprehension. In *Speech, language, and communication* (eds J. L. Miller & P. D. Eimas), pp. 217–262. San Diego, CA: Academic Press.

Tyler, L. K. & Marslen-Wilson, W. D. 1977 The on-line effects of semantic context on syntactic processing. *J. Verbal Learn. Verbal Behav.* **16**, 683–692. (doi:10.1016/S0022-5371(77)80027-3)

Tyler, L. K. & Wessels, J. 1983 Quantifying contextual contributions to word-recognition processes. *Percept. Psychophys.* **34**, 409–420.

Van Berkum, J. J. A., Brown, C. M. & Hagoort, P. 1999a Early referential context effects in sentence processing: evidence from event-related brain potentials. *J. Mem. Lang.* **41**, 147–182. (doi:10.1006/jmla.1999.2641)

Van Berkum, J. J. A., Brown, C. M. & Hagoort, P. 1999b When does gender constrain parsing? Evidence from ERPs. *J. Psycholinguist. Res.* **28**, 555–566. (doi:10.1023/A:1023224628266)

Van Berkum, J. J. A., Brown, C. M., Hagoort, P. & Zwitserlood, P. 2003 Event-related brain potentials reflect discourse-referential ambiguity in spoken language comprehension. *Psychophysiology* **40**, 235–248. (doi:10.1111/1469–8986.00025)

Van Berkum, J. J. A., Brown, C. M., Zwitserlood, P., Kooijman, V & Hagoort, P. 2005 Anticipating upcoming words in discourse: evidence from ERPs and reading times. *J. Exp. Psychol. Learn.* **31**, 443–467. (doi:10.1037/0278-7393.31.3.443)

Van den Brink, D., Brown, C. M. & Hagoort, P. 2001 Electrophysiological evidence for early contextual influences during spoken-word recognition: N200 versus N400 effects. *J. Cogn. Neurosci.* **13**, 967–985. (doi:10.1162/089892901753165872)

Van den Brink, D., Brown, C. M. & Hagoort, P. 2006 The cascaded nature of lexical selection and integration in auditory sentence processing. *J. Exp. Psychol. Learn.* **32**, 364–372. (doi:10.1037/0278-7393.32.3.364)

Van Herten, M., Kolk, H. H. J. & Chwilla, D. J. 2004 An ERP study of P600 effects elicited by semantic anomalies. *Cogn. Brain Res.* **22**, 241–255. (doi:10.1016/j.cogbrainres.2004.09.002)

Van Petten, C., Coulson, S., Rubin, S., Plante, E. & Parks, M. 1999 Time course of word identification and semantic integration in spoken language. *J. Exp. Psychol. Learn.* **25**, 394–417. (doi:10.1037/0278-7393.25.2.394)

Vos, S. H., Gunter, T. C., Kolk, H. H. & Mulder, G. 2001 Working memory constraints on syntactic processing: an electrophysiological investigation. *Psychophysiology* **38**, 41–63. (doi:10.1111/1469-8986.3810041)

Wassenaar, M. & Hagoort, P. 2007 Thematic role assignment in patients with Broca's aphasia: sentence–picture matching electrified. *Neuropsychologia.* **45**, 716–740. (doi:10.1016/j.neuropsychologia.2006.08.016)

Wicha, N. Y. Y., Moreno, E. M. & Kutas, M. 2004 Anticipating words and their gender: an event-related brain potential study of semantic integration, gender expectancy, and gender agreement in Spanish sentence reading. *J. Cogn. Neurosci.* **16**, 1272–1288. (doi:10.1162/0898929041920487)

Willems, R. M., Özyürek, A. & Hagoort, P. 2007 When language meets action: the neural integration of gesture and speech. *Cerebral Cortex* **17**, 2322–2333. (doi:10.1093/cercor/bhL141)

Zwitserlood, P. 1989 The locus of the effects of sentential-semantic context in spoken-word processing. *Cognition* **32**, 25–64. (doi:10.1016/0010-0277(89)90013-9)

11

Speech perception at the interface of neurobiology and linguistics

David Poeppel, William J. Idsardi, and Virginie van Wassenhove

Speech perception consists of a set of computations that take continuously varying acoustic waveforms as input and generate discrete representations that make contact with the lexical representations stored in long-term memory as output. Because the perceptual objects that are recognized by the speech perception enter into subsequent linguistic computation, the format that is used for lexical representation and processing fundamentally constrains the speech perceptual processes. Consequently, theories of speech perception must, at some level, be tightly linked to theories of lexical representation. Minimally, speech perception must yield representations that smoothly and rapidly interface with stored lexical items. Adopting the perspective of Marr, we argue and provide neurobiological and psychophysical evidence for the following research programme. First, at the implementational level, speech perception is a multi-time resolution process, with perceptual analyses occurring concurrently on at least two time scales (approx. 20–80 ms, approx. 150–300 ms), commensurate with (sub) segmental and syllabic analyses, respectively. Second, at the algorithmic level, we suggest that perception proceeds on the basis of internal forward models, or uses an 'analysis-by-synthesis' approach. Third, at the computational level (in the sense of Marr), the theory of lexical representation that we adopt is principally informed by phonological research and assumes that words are represented in the mental lexicon in terms of sequences of discrete segments composed of distinctive features. One important goal of the research programme is to develop linking hypotheses between putative neurobiological primitives (e.g. temporal primitives) and those primitives derived from linguistic inquiry, to arrive ultimately at a biologically sensible and theoretically satisfying model of representation and computation in speech.

Keywords: multi-time resolution; temporal coding; analysis-by-synthesis; predictive coding; forward model; distinctive features

11.1 Introduction

We take speech perception to be the set of computations that entail as their 'endgame' and optimal result the identification of words, either presented in isolation or in spoken discourse. This—almost banal—presupposition, that speech perception is primarily about finding words in ecologically natural contexts (and not, say, about spotting phonemes or indicating intelligibility in experimental contexts; see Cleary & Pisoni (2001) for a related perspective), provides an important boundary condition on a programme of research; because words (or syllables or morphemes), once identified, must enter into *subsequent* linguistic computation (phonological, morphological, syntactic) to permit successful language comprehension, the internal representation of words generated by the speech perception processes must be suitable for the range of linguistic operations performed with these words. In short, it is a critical requirement that the output of the processes that constitute speech perception are representations that permit *using* and *manipulating* these representations in specific ways. Such a requirement implies that research on speech perception must interface closely with theories of lexical representation. It is, in our view, not a sufficient answer to state that a word has been recognized without specifying rather explicitly what the format of the representation is. More colloquially, if the neural code for lexical representation is written in, say, Brain + +, speech perception must transform the input signal, a continuously varying waveform, into Brain + + objects. On this view, any theory of speech perception thus requires making commitments to theories of lexical representation.

Based on this perspective, we outline a research programme on speech perception that is strongly influenced by Marr's (1982) approach to understanding visual perception. Marr's suggestion to distinguish between computational, algorithmic and implementational levels of description when investigating computational systems in cognitive neuroscience seems to be very helpful to us in fractionating the problem and organizing the set of questions one faces in the study of speech perception. We adopt the taxonomic organizational principles outlined by Marr and discuss from that perspective three major properties of speech perception that we take to require a principled explanation. *First*, at the *implementational* level of description, speech perception is a *multi-time resolution process*, with signal analysis occurring concurrently on (at least) two time scales relevant to speech, syllabic-level (approx. 5 Hz) and segmental-level (approx. 20 Hz) temporal analyses. Naturally, multiresolution processing is but one of many relevant implementational issues, but it has received recent empirical support in both human and animal studies (Boemio *et al.* 2005; Narayan *et al.* 2006) and has interesting consequences for the architecture of the system; consequently we focus on that issue here. Multiresolution processing is widely observed in other systems (e.g. vision) and can, we suggest, be used profitably in engineering approaches to speech recognition. *Second*, at the *algorithmic* level of description, the central algorithm we invoke is *analysis-by-synthesis*. This constitutes a set of operations first discussed in the 1950s and 1960s (and specifically for the speech case by Halle & Stevens (1959, 1962) and Stevens & Halle (1967)) that provide an approach to bottom-up processing challenges in perception by using 'hypothesize-and-test' methods. Based on minimal sensory information, the perceptual system generates knowledge-based 'guesses' (hypotheses) about possible targets and internally synthesizes these targets. Matching procedures between the synthesized candidate targets and the input signal ultimately select the best match; in other words, the analysis is guided by internally synthesized candidate representations. In the terminology of contemporary cognitive neuroscience, analysis-by-synthesis as we develop it here is closely related to the concept of internal forward models. In the terminology of automatic speech recognition and statistics, analysis-by-synthesis is also conceptually related to Bayesian classification approaches. *Third*, at the *computational* level of description, we commit to a specific representational theory, that of *distinctive features* as the primitives for lexical representation and phonological computation. Our proposal contrasts with views that argue for *strictly* episodic (acoustic) representations—although we are sympathetic to the fact that the rich evidence for episodic effects must be accommodated, and we articulate a proposal in §5. In our view, words are represented in the mind/brain as a series of segments each of which is a bundle of distinctive features that indicate the articulatory configuration underlying the phonological segment. As decades of research show, phonological generalizations are stated over features (neither holistic phonemes nor *a fortiori* 'epiphones'), reflecting their epistemological primacy. Given the importance of features for the organization of linguistically significant sounds and given the fact that their articulatory implementation results in specific acoustic correlates (Stevens 1998, 2002), we assume that one of the central aspects of speech perception is the extraction of distinctive features from the signal. The fact that the elements of phonological organization can be interpreted as articulatory gestures with distinct acoustic consequences suggests a tight and efficient architectural organization of the speech system in which speech production and perception are intimately connected through the unifying concept of distinctive features.

The ideas we raised earlier provide a new perspective on some challenges in speech recognition that we take to be fundamental: the problem of linearity (segmentation); the problem of invariance; and the problem of perceptual constancy. These three problems are, of course, closely related and constitute irritating stumbling blocks for automatic speech recognition research as well as accounts of human speech perception. We are in no position to provide answers to these foundational challenges, and the paper is not focused on segmentation and invariance. However, the three properties of speech perception that we argue for here may provide a wedge into dealing with these challenges to the recognition process. For example, we argue that multi-time resolution processing—in the context of which segmentation occurs concurrently on segmental and syllabic time scales—relates closely to the 'landmarks' approach advocated by Stevens (2002). In particular, our approach allows for a 'quick and coarse' sample of the input that can subsequently be refined by the further analysis in a parallel stream. This concept is very similar to Stevens' notion of looking for informative landmarks and then verifying and testing the information around these landmarks to specify the speech information at that time point in the waveform. If this model is on the right track, and if Stevens and we are on the right track in hypothesizing that distinctive features are both the basis for speech representation and have acoustic realizations, one can begin to formulate models that try to link *acoustic information on multiple time scales* to featural information. Secondly, once one has such featural hypotheses, one can generate internal guesses that can then guide further perceptual processing. That is, guesses based on coarsely represented spectro-temporal representations would constitute a way to ignite the analysis-by-synthesis algorithm that we take to be particularly useful to rapidly recognize incoming speech based on predictions that are conditioned by both the prior speech context and higher-order linguistic knowledge. It is conceivable that the linearity and invariance problems are not the principled limitations they are now if one adopts a multi-time resolution perspective, because it is possible that when one looks at the information on multiple scales, there is more robustness in the acoustic-to-feature mapping than when looking only at processing on one time scale (as is typical in hidden Markov models for automatic speech recognition systems). Certainly this is no solution to invariance, but it does provide one new perspective on how to approach this highly important and vexing issue in the perception of spoken language.

Figure 11.1 schematizes the operative representations in speech perception in the context of the present proposal. Based on a continuously varying signal at the periphery (Figure 11.1a), the afferent auditory pathway constructs a detailed spectro-temporal representation (Figure 11.1b), by hypothesis based on operations well described by the Fourier transform (at the periphery) and the Hilbert transform (in cortex, to extract envelope information). These assumptions are not particularly controversial and follow from extensive research in auditory theory and neurophysiology. Assuming that the representation of lexical items is a discrete series of segments composed of distinctive features

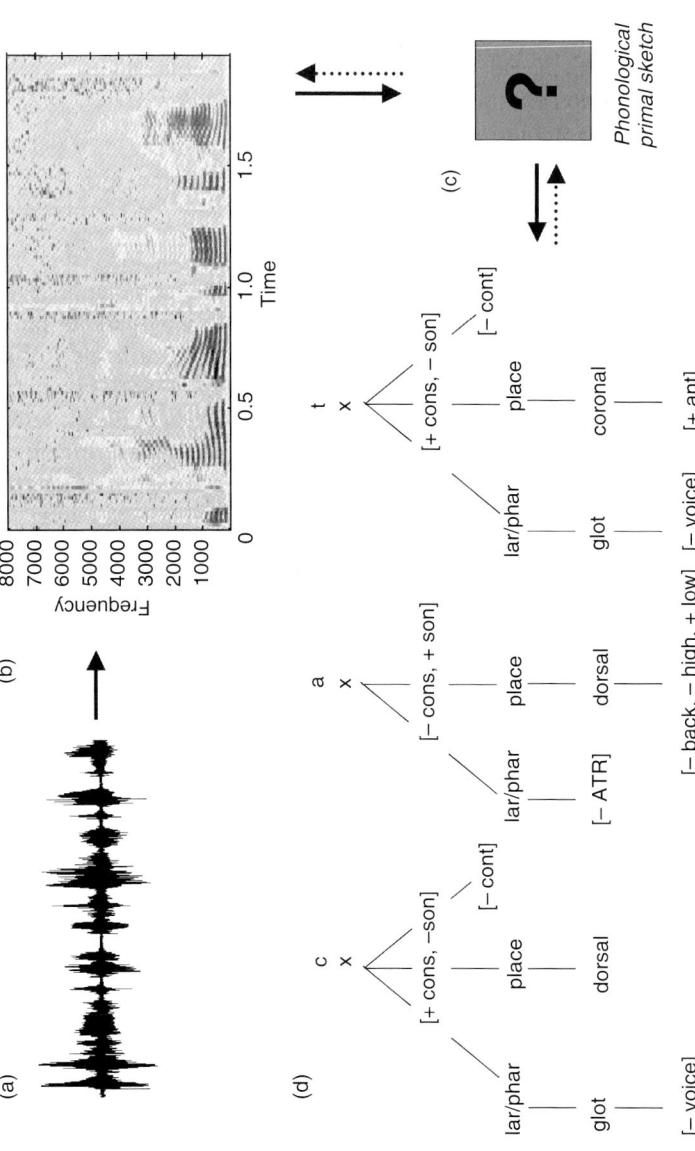

Figure 11.1 Representations and transformations from input signal to lexical representation. Solid arrows represent logically required steps and dotted arrows reflect hypothesized top-down mappings. (a) At the auditory periphery, the listener has to encode a continuously varying acoustic waveform (x-axis, time; y-axis, amplitude). (b) The afferent auditory pathway analyses the input signal in time and frequency. A neural 'analogue' of the spectrogram is generated to highlight both spectral and temporal variations in the signal. (cf. STRFs in auditory cortex.) (c) An intermediate representation may be necessary to map from a spectro-temporal representation of the acoustic input signal to the putative abstract representation of the word. The intermediate representation may be a PPS, built on temporal primitives (temporal windows of specific sizes) and spectral primitives. (d) The hypothesized representation of the word *cat* in the mind/brain of the speaker/listener. Each of the three segments of this consonant-vowel-consonant word is built from distinctive features that as a bundle are definitional of the segment.

(Figure 11.1d)—a view that is also uncontroversial insofar as one accepts the last few decades of phonological research—a central question is how to accomplish the mapping from a spectro-temporal, acoustic, representation to the lexical–phonological one. This mapping may—as suggested in Figure 11.1c—or may not involve further intermediate representations. Note that it *is* controversial as to what extent the processing steps can feedback to previous stages (cf. Norris *et al.* 2000). From our perspective that incorporates both multi-time resolution processing and analysis-by-synthesis, it follows that there is an intermediate representation (Figure 11.1c; say, the auditory equivalent of Marr's 2 1/2-dimensional sketch), namely (minimally) temporal windows of different sizes that represent different attributes of the signal, the *phonological primal sketch* (PPS). The properties of this putative intermediate representation are largely unknown, for the moment. (For a discussion of the related concept *auditory primal sketch*, see Todd 1994) For example, calling it 'phonological' implies a categorical representation, but the extent to which the information in each of the two windows is categorical is unclear because it is untested. One could call the representation 'phonetic primal sketch', as well, if the information remains graded. Crucial is that *something* must mediate the mapping from spectro-temporal signal configurations to lexical entries (on our view of lexical representation). Such an intermediate (and fleeting) multi-time resolution representation will retain acoustic properties, but they will differ depending on whether one is looking at the shorter (segmental) or longer (syllabic) temporal primitive. We see the primal sketch as related to Stevens' (2002) notion of landmarks. It is not yet worked out in what way a PPS relates to the representations stipulated by models, such as TRACE, NAM, shortlist or distributed cohort. Because such models do not make explicit reference to multi-time resolution processing, it is not obvious whether short, segmental or long, syllabic temporal primitives can be accommodated best within such theories. Finally, it stands to reason that the locus of computation is the superior temporal cortex, but any claim beyond that must remain speculative. If permitted such speculation in the context of our proposed functional anatomy (see Figure 11.2), one might argue that posterior superior temporal gyrus (STG) and superior temporal sulcus (STS) in the ventral pathway (Hickok & Poeppel 2004) are the relevant part of cortex to construct an interface representation, given that middle temporal gyrus (MTG) and STG are argued to be the substrates for lexical representation and auditory analysis, respectively.

The approach to the speech perception that we advocate here differs from many current proposals in explicitly (re)incorporating linguistic and psycholinguistic considerations, particularly considerations of lexical and phonological representations. Much research makes the implicit (and sometimes explicit) presupposition that the best route for understanding the major challenges to speech recognition comes from trying to bridge auditory theory with auditory neuroscience (e.g. the papers in Greenberg & Ainsworth (2006)), while dismissing the (often largely representational) issues raised by phonological theory. While we are sympathetic to what can be learned from such a research programme, we are convinced that one cannot do without the constraints derived from linguistics, and particularly phonology. Obviously, auditory theory (say, with regard to the importance of critical bands, modulation transfer functions, masking, pitch extraction, stream segregation, the modulation spectrum and so on) is crucial to an understanding of how the incoming signal is analysed and transformed into *representations that form the basis for speech recognition*. Similarly, (i) cellular and systems neuroscience teaches us essential facts about how acoustic signals are analysed in the afferent auditory pathway and (ii) the distributed cortical functional anatomy associated with speech recognition suggests that various dimensions are processed in a segregated manner. (Note that our own work often focuses on these issues, i.e. we are not just sympathetic to auditory neuroscience and auditory theory, we are also practitioners; e.g. Hickok & Poeppel 2000, 2004). These domains of investigation provide critical knowledge about the *construction* of the representations that constitute speech.

However, the nature of the speech representations *as they enter speech-related computation* is rarely, if ever, spelt out. For example, the field is very comfortable talking about how acoustic signals can be characterized by spectro-temporal receptive fields (STRFs) of auditory cortical neurons (e.g. Shamma 2001). This is a terrific set of results—but such a characterization tells us nothing about how such a (neuronal) representation allows for *further* computation with that token. Suppose one has recognized the acoustic realization of the word '*caterpillar*' using only the machinery of auditory theory and the neuronal concept of STRFs. What is now owed is a set of linking hypotheses from auditory-based representations of that type to whatever machinery or representational structure underlies further, language-based processing. Why? Because the recognized item typically enters into phonological and morphological operations (say, pluralization) as well as syntactic ones (say, subject-predicate agreement, *viz.* '*caterpillar-s change-Ø into butterflies*'). To connect with that aspect of processing, the representations in play must be in the same 'code', a rather straightforward conjecture. Now, if we assume (for us, uncontroversially; for some, shockingly) that there are abstract internal representations that form the basis for linguistic representation and processing, there must be some stage at which auditory signals are translated into such representations. If one is disinclined to invoke linguistically motivated representations early in the processing stream, then one owes a statement of linking hypotheses that connect the different formats (unless one does not, categorically, believe in any internal abstract representations for language processing). Alternatively, perhaps the representations of speech that are motivated by linguistic considerations are in fact active in the analysis process itself and therefore active throughout the subroutines that make up the speech perception process. Unsurprisingly, we adopt the latter view.

A slightly different way to characterize the programme of research is to ask: what are the representational and computational primitives in auditory cortex; what are the primitives for speech; and how can we build defensible linking hypotheses that bridge these domains (cf. Poeppel & Embick 2005)? Here, we will discuss three steps that we take to be essential in the process of transforming signals to interpretable internal representations: (i) multi-time resolution processing in auditory cortex as a computational strategy to fractionate the signal into appropriate 'temporal primitives' commensurate with processing the auditory input concurrently on a segmental and a syllabic scale; (ii) analysis-by-synthesis as a computational strategy linking top-down and bottom-up operations in auditory cortex; and (iii) the construction of abstract representations (distinctive features) that form the computational basis for both lexical representation and transforming between sensory and motor coordinates in speech processing.

In our view, the three attributes of speech representation (features) and processing (multi-time resolution, analysis-by-synthesis) that we raised provide a way to (begin to) think about how one might more explicitly link the acoustic signal to the internal abstractions that are words. Building on the intuitions of Marr (1982), we see the perception and recognition processes as having a number of bottom-up and top-down steps. It is not a 'subtle interplay' of feed-forward and feedback steps that we have in mind, though, but a rather unromantic, mechanical (forward) calculation of perceptual candidates based on very precisely guided synthesis steps. In a first pass, the system attempts a quick reduction (primal sketch) of the total search space for lexical access by finding the—somewhat coarsely specified—*landmarks* (Stevens 2002) through the *articulator-free* (major-class, place-less) features (Halle 2002). That is, the initial pass defines a neighbourhood on broad-class and manner features (e.g. stop-fricative-nasal-approximant; the term 'approximant' covers both glides and vowels). These initial guesses are based on minimal spectro-temporal information (say, two or three analysis windows) and can be stepwise refined in small time increments (approx. 30 ms or so) owing to the multi-time resolution nature of the process. Subsequent to the initial hypotheses triggered by the construction of the PPS, a cohort-type selection is elicited from the articulator-bound (place) features. In this way, we try to have our cake and eat it too—trying to capture both (gross) neighbourhood and (gross) cohort-model effects. Overall, our proposal is similar in spirit (if not in details) to the featurally underspecified lexicon (FUL) model of speech recognition (Lahiri & Reetz 2002).

11.2 Functional anatomic background

Importantly, once again from the perspective of the implementational level of Marr (1982), the research we outline is consistent with (and in part explicitly motivated by) cognitive neuroscience approaches to speech perception, specifically considerations of the cortical functional architecture of speech perception (Figure 11.2). Several recent reviews develop large-scale models of the cortical basis of speech perception and, despite some disagreement on various details, there is also considerable convergence among these proposals (Binder *et al.* 2000; Hickok & Poeppel 2000, 2004; Scott & Johnsrude 2003; Boatman 2004; Indefrey & Levelt 2004; Poeppel & Hackl 2007). We briefly outline the cortical architecture here.

The initial cortical analysis of speech occurs bilaterally in core and surrounding superior auditory areas (see Hackett *et al.* (2001) for relevant human auditory cortex anatomy). Subsequent computations (typically involving lexical-level processing) are largely left lateralized (with the exception of the analysis of pitch change; the analysis of voice; and the analysis of syllable-length signals), encompassing the STG, anterior and posterior aspects of the STS as well as inferior frontal, temporo-parietal and inferior temporal structures (see Poeppel *et al.* (2004) for arguments and imaging evidence that speech is bilaterally mediated). This listing shows that practically all classical, peri-Sylvian language areas are implicated in some aspects of the perception of speech. Therefore, one goal has to be to begin to specify what the computational contribution of each cortical field might be.

With regard to the input signal travelling up the afferent pathway, there are notable asymmetries in the brain stem and even at the cochlear level of sound analysis (e.g. Sininger & Cone-Wesson 2004), but it is not well understood whether these subcortical asymmetries condition the processing in a way that is sufficiently rich to account for the compelling asymmetries that emerge at the cortical level. Imaging studies show very convincingly that the processing of speech at the initial stages is robustly bilateral, at least at the level of core and surrounding STG (Mummery *et al.* 1999; Binder *et al.* 2000;

Norris & Wise 2000; Poeppel *et al.* 2004). The fact that imaging studies show bilateral activation does, of course, not imply that the computations executed in left and right core auditory cortices are identical—there are, presumably, important differences in local computation. Nevertheless the processing is bilateral as assessed by haemodynamic and electrophysiological methods. We hypothesize that the STRFs of neurons in bilateral core auditory cortex generate high-resolution neuronal representations of the input signal (which of course is already highly pre-processed in subcortical areas, say, the inferior colliculus).

A growing body of neuroimaging research deals with the question of what exactly is computed in left and right auditory areas during speech and non-speech processing. Zatorre and colleagues have argued on the basis of neuropsychological and neuroimaging data that right hemisphere superior temporal areas are specialized for the analysis of spectral properties of signals, in particular spectral change, and the analysis of pitch, specifically pitch change. In contrast, they argue that left hemisphere areas are better suited to the processing of rapid temporal modulation (Zatorre *et al.* 2002). Their view converges with that of Poeppel (2001, 2003), where it is suggested that the spectral versus temporal right–left asymmetry is a consequence of the size of the temporal integration windows of the neuronal ensembles in these areas. Neuronal ensembles in left (non-primary) temporal cortex are associated with somewhat shorter integration constants (say, 20–50 ms) and therefore left hemisphere cortical fields preferentially reflect temporal properties of acoustic signals. Right hemisphere (non-primary) cortex houses neuronal ensembles, a large proportion of which have longer (150–300 ms) integration windows, and therefore are better suited to analyse spectral change. These ideas are discussed in more detail below, but they build on a long history and literature that investigates hemispheric asymmetry in the auditory cortex related to spectral versus temporal processing (Schwartz & Tallal 1980; Robin *et al.* 1990). In summary, we hypothesize that primary (core) auditory cortex builds high-fidelity representations of the signal, and surrounding non-primary areas differentially 'elaborate' this signal by analysing it on different time scales.

Beyond this initial analysis of sounds that is robustly bilateral and may involve all the steps involved in the acoustic-to-phonetic mapping, there is wide agreement that speech perception is lateralized. The right STG and STS have been shown to play a critical role in the analysis of voice information (Belin *et al.* 2004) and dynamic pitch. The analysis of prosodic features of speech has also been suggested to be lateralized to right STG. The processing of speech *per se*, i.e. that aspect of processing that permits lexical access and further speech-based computation, however, is lateralized to left temporal, parietal and frontal cortices (Binder *et al.* 2000; Hickok & Poeppel 2000; Boatman 2004; Indefrey & Levelt 2004; Scott & Wise 2004).

Beginning in the STG, research in the last few years has identified the emergence of two processing streams. The idea of segregated and parallel pathways is closely related to vision research, where the concept of a 'what' versus a 'where'/'how' pathway is very firmly established (Ungerleider & Mishkin 1982). In the auditory domain, one can also think of a what (ventral) pathway (the pathway responsible for the 'sound-to-meaning mapping'; Hickok & Poeppel 2004) that involves various aspects of the temporal lobe that are apparently dedicated to sound identification. Both more anterior parts of the STS (Scott *et al.* 2000) and more posterior parts of the STG/STS (Binder *et al.* 2000) as well as MTG (Indefrey & Levelt 2004) have been implicated in speech-sound processing.

Scott *et al.* (2000) were the first to show that anterior STS plays a crucial role in speech intelligibility. Binder and colleagues have suggested that posterior STG and STS are critical for the transformation from acoustic-to-phonetic information; and, based on a large meta-analysis, Indefrey & Levelt (2004) suggest that the interface of phonetic and lexical information is at least in part mediated by posterior MTG. Note that it is not at all clear which aspect of the so-called 'what' pathway in auditory processing is responsible for lexical access. There are some suggestions that middle and inferior temporal gyri and basal temporal cortex reflect lexical processing, but the fractionation of speech processing and lexical processing is, perhaps unsurprisingly, not straightforwardly reflected in imaging studies. Nevertheless, there is consensus that the STG from rostral to caudal fields and the STS constitute the neural tissue in which many of the critical computations for speech recognition are executed. It is worth bearing in mind that the range of areas implicated in speech processing go well beyond the classical language areas typically mentioned for speech; the vast majority of textbooks still state that this aspect of perception and language processing occurs in Wernicke's area (the posterior third of the STG).

In analogy to the visual 'what'/'where' distinction, evidence from auditory anatomy and neurophysiology (e.g. Romanski *et al.* 1999) as well as imaging suggests that there is a dorsal pathway that plays a role—not just in 'where'-type computations but also in speech processing (the pathway responsible for the 'sound-to-articulation mapping'; Hickok & Poeppel 2004). The dorsal pathway implicated in auditory tasks includes temporo-parietal, parietal and frontal areas. The specific computational contribution of each area is not yet understood for either 'where'/'how' tasks in hearing or speech perception tasks. However, there is evidence, from the domain of speech processing, that a temporo-parietal area plays an important role in the (hypothesized) coordinate transformation from auditory to motor coordinates. This Sylvian parieto-temporal area has been studied by Hickok and colleagues (Hickok *et al.* 2003) and is argued to be necessary to maintain parity between input- and output-based speech tasks. Furthermore, aspects of Broca's area (Brodmann areas 44 and 45) are also regularly implicated in speech processing (see Burton 2001 for review). It is of considerable interest that frontal cortical areas are involved in perceptual tasks. Such findings have (i) challenged the view that Broca's area is principally responsible for production tasks or syntactic tasks and (ii) reinvigorated the discussion of a 'motor' contribution to speech perception. In part, these discussions are reflected in the debates surrounding mirror neurons and the renewed interest in the motor theory of speech perception. Figure 11.2 shows the functional anatomy of speech perception derived from neuropsychological and neuroimaging data. A central challenge to the field is to begin to formulate much more detailed hypotheses about what computations are executed in each of these areas, first for speech perception and second for other linguistic and non-linguistic operations in which the computations mediated by these areas participate in causal ways. We now turn to the three hypothesized attributes of speech perception introduced above and, when possible, relate them to the sketch of the anatomy outlined here.

11.3 Multi-time resolution processing

It is an intuitively straightforward observation that visual signals are processed on multiple spatial scales. For example, faces can be analysed at a detailed, featural level, but also at a coarser, configural level, and these correspond to different spatial frequencies in

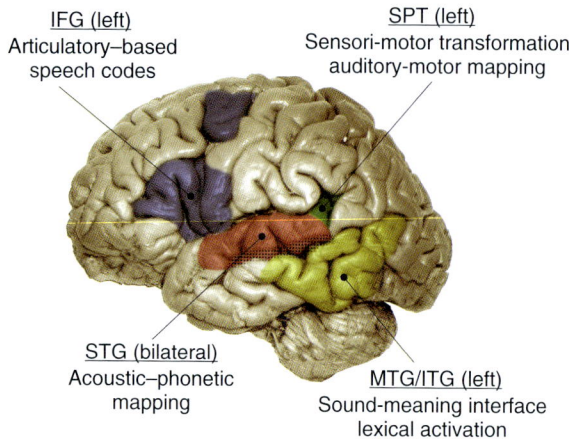

Figure 11.2 Functional anatomy of speech-sound processing. In the mapping from input to lexical representation, the initial steps are bilateral, mediated by various cortical fields on the STG; subsequent computation is typically left lateralized and extends over many left per-Sylvian areas. IFG, inferior frontal gyrus; SPT, Sylvian parieto-temporal area; MTG, middle temporal gyrus; ITG, inferior temporal gyrus; STG, superior temporal gyrus. (Adapted from Hickok & Poeppel (2004)).

the visual image. The information carried in these different channels is not identical— different spatial frequencies are associated with differential abilities to convey emotional information (low spatial frequencies) versus image details (high spatial frequencies; e.g. Vuilleumier *et al.* 2003). An alternative way to conceptualize this distinction is to think of it as the tension between global versus local information. Processing on multiple spatial scales in the visual domain has been studied psycho-physically and harnessed for provocative analyses of contemporary art (Pelli 1999). Whereas in the visual case the image can be fractionated into different spatial scales, in the auditory case both frequency and time can be thought of as dimensions along which one could fractionate the signal. We pursue the hypothesis that auditory signals are processed in time windows of different sizes, or durations. (For data supporting this conjecture from the domain of neural coding in bird song, see Narayan *et al.* (2006)). The idea that time windows of different sizes are relevant for speech analysis and perception derives from several phenomena. In particular, acoustic as well as articulatory-phonetic phenomena occur on different time scales. Investigation of a waveform and spectrogram of a spoken sentence reveal that at the scale of roughly 20–80 ms, segmental and subsegmental cues are reflected, as well as local segmental *order* (i.e. the difference between 'pe*st*' and 'pe*ts*'). In contrast, at the scale of roughly 150–300 ms (corresponding acoustically to the envelope of the waveform), suprasegmental and syllabic phenomena are reflected. One way to reconcile the tension between local (fast modulation frequency) and global (slower modulation frequency) information is to assume hierarchical processing such that higher-order, longer representations are constructed on the basis of smaller units. Alternatively, perhaps information on multiple time scales is processed concurrently. We explore the latter possibility here. In particular, we discuss the hypothesis that there are two principal time windows within which a given auditory signal (speech or non-speech) is processed, with the mean durations as above. Although we are not 'two time window imperialists' and recognize the importance of processing on the (sub-)millisecond scale in the brainstem and the 1000+ millisecond scale for phrases, we argue that the two windows we identify play a privileged role in the analysis and perceptual interpretation of auditory signals, and that these two time windows have special consequences for speech perception.

Multi-time resolution processing, as we develop the hypothesis here, is built on the concept of temporal integration windows (Figure 11.3). Both psychophysical and neurobiological evidence suggest that physically continuous information is broken apart and processed in temporal windows. The claim that there is temporal integration is rather uncontroversial. More controversial is the hypothesis that there is not just integration but discretization (Saberi & Perrott 1999; VanRullen & Koch 2003). Either way, it is clear that signals are analysed in a discontinuous fashion.

We hypothesize that there are two integration windows and that their implementation occurs in *non-primary* auditory cortex. As stated above, the auditory signal up to primary auditory cortex is processed in a predominantly symmetric manner (although there are notable asymmetries at the subcortical level). The STRFs in core auditory cortex permit the construction of a relatively high-fidelity representation of the signal. Based on this initial representation, there is a temporally asymmetric elaboration of the signal by 'sampling' the

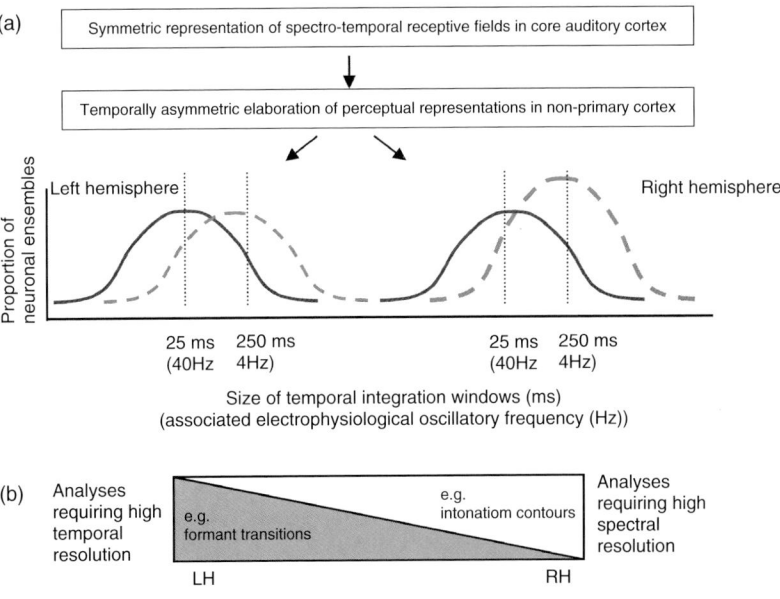

Figure 11.3 Temporal integration in auditory and speech analysis. (a) Temporal integration and multi-time resolution analysis: quantization and lateralization. Both left and right auditory cortices have populations of neurons (wired as neuronal ensembles) that have preferred integration constants of two types. By hypothesis, one set of neurons prefers approximately 25 ms integration, another 250 ms. In electrophysiological studies, such integration windows may be reflected as activity in the gamma and theta bands, respectively. The evidence for a rightward asymmetry of slow integration is growing and the evidence for a leftward asymmetry of rapid integration is unsettled. Minimally, both hemispheres are equipped to deal with subtle temporal variation (Boemio *et al.* 2005). (b) Functional lateralization as a consequence of temporal integration. From asymmetric temporal integration of this type, it follows that different auditory tasks will recruit the two populations differentially owing to sensitivity differences and lead to hemispheric asymmetry.

output of core auditory cortex using two window sizes. One window is of the order of 20–50 ms, another of the order of 200 ms. One way to develop a visual intuition for this idea is to imagine two spectrograms of the same signal, one highlighting the rapid temporal changes and representing the glottal pulse, the other representing the narrowband frequency variation (say, formants). What is the purpose of such a proposed temporal quantization of the input waveform? We hypothesize that this sampling serves as a logistical or administrative device to generate auditory representations of the appropriate granularity to interface with higher-order, abstract representations. The model is outlined in more detail in Poeppel (2001, 2003), and is consistent with functional magnetic resonance imaging (fMRI) data in Boemio *et al.* (2005); Schonwiesner *et al.* (2005) and others.

If this type of multi-time resolution model is on the right track, evidence is owed for the model's constituent claims. In particular, one needs to show that there is (i) integration on the 25–80 ms time scale, (ii) integration on the 150–300 ms time scale, (iii) a perceptually relevant interaction between representations constructed on these two time scales, and (iv) lateralization of function associated with processing on these time scales.

Evidence for temporal integration on the short time scale is relatively abundant and will not be discussed further here (see Poeppel 2003; Wang *et al.* 2003). Both for speech and non-speech signals, it has been shown psychophysically and electrophysiologically that integration on a 20–50 ms time scale has compelling perceptual consequences. In the non-speech case, three clear examples of processing on this time scale are provided by the psychophysical order threshold (Hirsh & Sherrick 1961), by frequency modulation (FM) direction discrimination (Gordon & Poeppel 2002; Luo *et al.* 2007) and click-train integration studies (Lu *et al.* 2001; A. Boemio 2002, unpublished dissertation). Experiments testing the minimum stimulus onset asynchrony (SOA) at which the *order* of two concurrently presented signals can be reliably indicated show that, across sensory modalities, approximately 20–30 ms are the relevant time scale (Hirsh & Sherrick 1961). Experiments testing the minimum signal duration at which one can reliably discriminate between upward and downward moving FM tones (glides) consistently show that the stimulus duration at which one performs robustly is 20 ms (Gordon & Poeppel 2002; Luo *et al.* 2007). Psychophysical and electrophysiological experiments on the processing of click trains also highlight the relevance of processing on this time scale, in both humans and non-human primates (Lu *et al.* 2001; A. Boemio 2002, unpublished dissertation). As one varies the SOA between clicks from 1 to 1000 ms, subjects experience categorically different auditory percepts: at SOAs above 50 ms, each click is perceived as an individuated event; at SOAs below 10 ms, the click train is perceived as a tone with a well-defined pitch. However, there is a sharp perceptual transition between roughly 10 and roughly 50 ms in which subjects have neither clearly discrete nor clearly continuous judgements of click trains; and, importantly, electrophysiological recordings show that this perceptual transition region (associated with the sensation of roughness) is associated with a particular neurophysiological response profile that reflects the transitory nature of this temporal regime at which one begins to construct perceptual pitch. Animal (Lu *et al.* 2001) and human (A. Boemio 2002, unpublished dissertation) studies show that response properties change at precisely that perceptual boundary, possibly associated with a transition from temporal to rate coding schemes.

A compelling example from speech perception comes from the work of Saberi & Perrott (1999), who took spoken sentences, cut them into time slices and locally reversed the direction of each time window. The intelligibility function they reported was strongly conditioned by the size of the window. For segment durations of up to 50 ms, intelligibility was not significantly affected, supporting the notion of the special nature of integration, and perhaps even discretization, on this time scale.

Evidence for temporal integration on a longer, 150–300 ms time scale also comes from physiological and psychophysical studies. For non-speech signals, loudness integration has been shown to require roughly 200 ms of signal to reach asymptotic psychophysical loudness judgement (Moore 1989; Green 1993). Electrophysiological mismatch negativity studies testing temporal integration of non-speech signals also point to 150–300 ms as the relevant time scale (Yabe *et al.* 1997, 2001 *a*, *b*). There are two observations from the domain of speech that we consider critical in this context. First, cross-linguistic measurements of mean syllable duration have revealed that although there is tremendous variability in syllable structure across languages, mean acoustic duration is remarkably stable, peaking between 100–300 ms. These values are commensurate with measurements of the modulation spectrum of spoken language (Greenberg 2005). Peaks in the modulation spectrum occur between 2 and 6 Hz and are argued to reflect the underlying syllabic structure of the spoken utterance, which determines its envelope.

A second source of evidence from the speech domain comes from studies on audio-visual (AV) integration. Several studies have replicated the following surprising finding (Massaro *et al.* 1996; Munhall *et al.* 1996; Grant *et al.* 2004; van Wassenhove *et al.* 2007). When one presents listeners with AV syllables and desynchronizes the audio and video tracks, one might expect severe perceptual disturbances given that one has disrupted the temporal alignment of the auditory and the visual information. Surprisingly, listeners tolerate enormous temporal asynchronies when viewing asynchronous AV speech. For example, it has been shown (Grant *et al.* 2004; van Wassenhove *et al.* 2007) that both McGurk auditory–visual stimuli and congruent auditory–visual stimuli are judged to be interpretable (with little performance degradation) despite AV asynchronies of up to 200 ms. In other words, the perceptual system interprets as simultaneous AV speech signals that are within a 200 ms window, the mean duration of a syllable.

Does the information carried on these two time scales interact in perceptually relevant ways? If an auditory signal is indeed analysed concurrently on two time scales, is there a binding of information that modulates speech perception? This question has, to our knowledge, not previously been addressed experimentally. Whereas there are studies exploiting signal-processing techniques to highlight the contributions of higher-or lower-modulation frequencies (Drullman *et al.* 1994*a*,*b*), the interaction between temporal information on different scales has been taken for granted (i.e. it is an implicit presupposition that segmental and syllabic information are 'congruent' in some sense, permitting successful comprehension). A new study by Chait *et al.* (submitted) tests the idea directly. They created signals in which either low-modulation frequency information was maintained across the spectrum (0–4 Hz, corresponding to long temporal window analysis) or higher-modulation frequency information was selectively retained (22–40 Hz, corresponding to the shorter integration constants). Subjects heard sentences (binaurally) that had only low- or high-modulation frequency and were tested for intelligibility. Consistent with previous work (Drullman *et al.* 1994*a*), low-modulation frequency signals, despite being extremely impoverished relative to a normal speech signal, allow for surprisingly good comprehension, with subjects showing intelligibility scores of well over 40% despite a severely degraded signal. In contrast, signals that retain only higher-modulation frequencies (above 20 Hz) generated low-intelligibility scores (below 20%, which is still remarkable given the restricted nature of the signal). But, crucially, what happens when both signals are presented at the same time (dichotically)? Many patterns *could* be obtained. The two signals could destructively interfere with each other, yielding low-intelligibility scores; the signals could be processed independently, yielding no net gain overall; the two signals could interact and yield an additive or even a supra-additive effect. Interestingly, it is the last possibility that is obtained: dichotically presented signals presented concurrently were apparently bound to generate representations that allowed for intelligibility scores that were significantly larger (68%) than a predicted linear additive effect (approx. 50%). This observation argues for the view that information extracted and analysed on two time scales interacts synergistically and in a perceptually relevant manner, supporting the hypothesis of multi-time resolution processing.

A final point concerns the hypothesis that processing on different time scales is actually associated with different cortical areas and possibly lateralized. In particular, it has been proposed that more slowly modulated signals—on the 150–300 ms scale—preferentially drive right hemisphere (non-primary) auditory areas whereas rapidly modulated signals—say, 20–80 ms—drive left cortical areas. If distinct auditory cortical fields are found to be differentially sensitive to temporal information, such data would support the model that different time scales are processed in parallel. In a recent fMRI study, Boemio *et al.* (2005) tested this hypothesis using non-speech signals that were constructed to closely match certain properties of speech signals. They observed that STG, bilaterally, is exquisitely sensitive to rapid temporal signals; however, right STS was preferentially driven by longer-duration signals, supporting the hypothesis that signals are not only processed on multiple time scales but also that the processing is partially lateralized, with slower signals differentially associated with right hemisphere mechanisms in higher-order auditory (and the canonical multisensory) area, STS. Lateralization of auditory analysis as a function of temporal signal properties has been observed in a number of studies now and can be viewed as a well-established finding (Hesling *et al.* 2005; Meyer *et al.* 2005; Schonwiesner *et al.* 2005). Cumulatively, the psychophysical and neurobiological data are most consistent with the conjecture of multi-time resolution processing, with the two critical time scales being a short (perhaps segmental) temporal integration window of 20–80 ms and a longer (syllabic) window of 150–300 ms. Regardless of the specifics of the values, the multi-time resolution nature of the processing seems like a very solid hypothesis based on a large body of convergent evidence.

11.4 Analysis-by-synthesis—internal forward models

Models of speech perception and lexical access tend to come in two types. Either the processing is rather strictly bottom-up (e.g. Norris *et al.* 2000) or there is feed-forward and feedback processing during perceptual analysis and lexical access, as is typical in most connectionist-style models. An attribute that tends to be common among such models is that the analysis is relatively 'passive'. That is to say, features percolate up the processing hierarchy in bottom-up models, or activation spreads in interactive models. The approach to recognition that we advocate here differs from such proposals. In particular, analysis-by-synthesis, or perception driven by predictive coding based on internal forward models, is a decidedly active stance towards perception that has been characterized as a 'hypothesize-and-test' approach. A minimal amount of signal triggers internal guesses about the perceptual target representation; the guesses (hypotheses) are recoded, or synthesized, into a format that permits comparison with the input signal. It is this 'forward' synthesis of candidate representations that is the central property of the approach and makes it a completely active process. Hypothesize-and-test models for perception were discussed in the 1950s and 1960s, for example, by Miller et al. (1960). Halle & Stevens (1959, 1962) and Stevens & Halle (1967) first developed the idea of analysis-by-synthesis for speech perception and an updated model, very much in line with our thinking, is provided in Stevens (2002).

Anticipating somewhat the discussion of distinctive features from §5, we see analysis-by-synthesis employed as a general computational architecture common to the whole recognition process. Acoustic measurements yield guesses about distinctive feature values in the string. 'Mini-lexicons' of valid syllable types are consulted, and a space of possible parses is constructed as a first-pass analysis. Frequency information is encoded throughout the system, so that the search can proceed on a 'best-first' basis, with more probable parses assigned greater weight in the system. More specifically, we see the landmarks of Stevens (2002), which correspond to the articulator-free features of Halle (2002), and which define the 'major' classes of phonemes (stop, fricative, nasal and approximant), as defining a PPS of the segmental time scale. This primal sketch gives a neighbourhood of words matching the detected landmark sequence. (Note that the speech perceptual model we outline is very sympathetic to lexical access models that emphasize the role of lexical neighbourhoods in processing). The primal sketch includes enough information to broadly classify certain prosodic characteristics as well (such as the approximate number of moras or syllables in the word).

Of course, the various feature detectors are fallible and return probabilistic information, which can then be inverted using Bayes's rule.

Bayes's rule: $p(H|E) = p(E|H) \times p(H)/p(E)$

(here 'H' can be read as 'hypothesis' and 'E' as 'evidence').

The quantity $p(H|E)$ represents the likelihood of the analysis, $p(E|H)$ is the likelihood of the synthesis of the data given the analysis. It is rather astonishing how seldom this connection between Bayesian methodology and analysis-by-synthesis has been drawn in the literature. Note that there are a small number of precedents linking analysis-by-synthesis with a Bayesian perspective, especially Hinton & Nair (in press) on handwriting recognition, and Bayesian perspectives have become much more common in perception (e.g. Knill & Richards 1996) and signal analysis (e.g. Bretthorst 1988) as well as in functional neuroimaging analysis (e.g. Friston et al. 2002). Moreover, much work in automatic speech recognition has a Bayesian orientation through the use of inverse-probability techniques such as hidden Markov models. However, much of the work in perception, signal analysis and speech recognition is of the empirical Bayes variety, using non-informative priors. In contrast, our view is that much of the interest is in discovering informative priors, that is, part of the content of universal grammar, in addition to incorporating the priors derived from the online perceptual processing.

The PPS is then specified by identifying the articulator-bound features within the detected landmarks. That is, the 2–1/2-dimensional analogue is constructed using probabilistic information about features such as [labial], [coronal], etc. (the *articulator-bound* features, see below) within the major class defined by the landmark primal sketch.

For example, the detection of [labial] or [coronal] place is different within nasals in English (which lacks a velar nasal phoneme) than it is for stops (which contrast all three categories). That is, we see this layer of analysis as evaluating conditional probabilities of the sort p([labial]|[+ nasal]). The hypothesized temporally synchronized feature sequences are then matched against the main lexicon and a list of candidates is generated, and then the rules of the phonology of the language are used to resynthesize the predictable features, again using Bayes's rule, and thus allowing a straightforward inclusion of variable rules (Labov 1972, 2001). Figure 11.4 shows some of the hypothesized set of computations, adapted and extended from Klatt (1979), to make clear where the synthesis process sits within the larger architecture.

One source of evidence for internal forward models of this type comes from studies on AV integration in speech. Several investigators have tested temporal constraints on AV speech integration and, as mentioned above, it is now reasonably well established that AV syllables tolerate signal desynchronization of up to 250 ms (Massaro *et al.* 1996; Munhall *et al.* 1996; Grant *et al.* 2004; van Wassenhove *et al.* 2007). When subjects are presented with AV syllables in which either the audio or the video signals lead or lag by up to 200 ms,

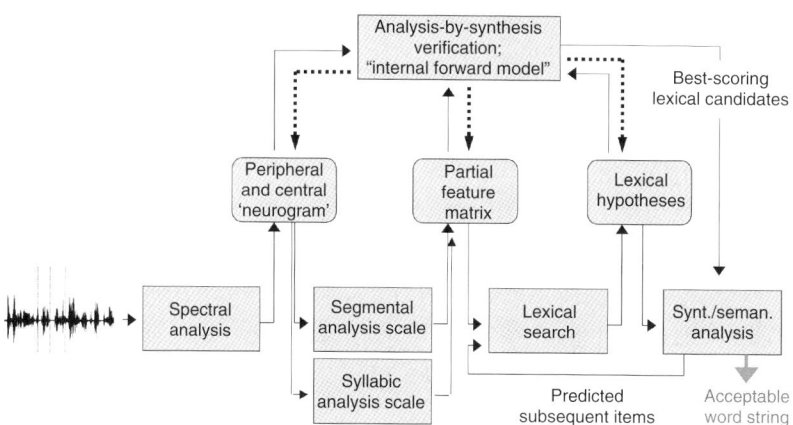

Figure 11.4. Possible processing steps in an analysis-by-synthesis model. The bottom tier incorporates distinct levels of representation in the mapping from sound to word (spectral analysis–segmental analysis–lexical hypotheses). The intermediate tier shows possible representations and computations that interact with the bottom and top (analysis-by-synthesis) levels to generate the correct mappings. The internal forward model can synthesize the candidates for matching at each level (neuronal, featural decomposition, lexical hypotheses) depending on how much information the forward model has to guide the internal synthesis. We hypothesize that the internal model is updated approximately every 30 ms, i.e. with each new sample that is available. Segmental and syllabic-level analyses of the signal are concurrent (multi-time resolution). Spectro-temporal analysis and the construction of a high-resolution auditory representation that is performed in the afferent pathway and core auditory cortex. Segmental-and syllabic-size analyses are hypothesized to occur in STG and STS (bilaterally), respectively; the mapping from hypothesized featural information to lexical entries may be mediated in STS, the lexical processes (search, activation) in middle temporal gyrus (while the conceptual information associated with lexical entries is likely to be much more distributed). The syntactic and compositional semantic representations further constraining lexical hypotheses are, perhaps, executed in frontal areas. The top-down forward model signals feed to temporal lobe from all connected areas, with a strong contribution from frontal articulatory cortical fields. (Adapted and extended from Klatt (1979)).

subjects apparently integrate the desynchronized signal successfully and interpret the AV signal as coherent and bound. One way to verify this is to use McGurk tokens in which listeners are presented, for example, with an auditory /pa/ and a visual /ka/. In synchronous presentation, subjects typically report perceiving the syllable /ta/. However, what happens during signal desynchronization is not clear—subjects could report perceiving either the audio or the video or the fused representation. As it turns out, over a time interval of 200 ms or more of desynchronization, listeners reliably perceive the fused syllable /ta/. This evidence is of course consistent with the claim that there is a temporal integration window of roughly 250 ms in AV speech (van Wassenhove et al. 2006).

Interestingly, the audio versus video lead or lag is not symmetric. Whereas visual leads are tolerated very well in perceptual experiments, auditory leads are more detrimental. Is such a result plausible from a more ecological perspective? We believe such a psychophysical result follows from the fact that movement of the articulators naturally precedes auditory speech output. In spoken language, auditory and visual onsets are not actually simultaneous in the physical sense and therefore incorporate a natural SOA. Such observations suggest that a tolerance for visual leads is not only preferable but also natural in AV speech perception. Unclear, however, is the issue of whether the information associated with the visual signal plays any specific role in auditory speech perception.

van Wassenhove et al. (2005) conducted a combined psychophysical and ERP study to investigate this issue. Subjects listened to and viewed congruent (audio /pa/ and video /pa/ or audio /ta/ and video /ta/ or audio /ka/ and video /ka/) syllables or incongruent (audio /pa/ and video /ka/) McGurk stimuli. In a three-alternative forced choice test, subjects had to categorize what they perceived; concurrently, ERPs were recorded. Based on the multisensory literature, which derives primarily from single-unit studies (Stein & Meredith 1993) or haemodynamic imaging studies (Calvert 2001), the most straightforward prediction was to observe supra-additivity on one of the major auditory evoked responses. In particular, because previous imaging work had observed that responses to AV (non-speech) signals could be supra-additive, it was predicted that the auditory N1 or P2 responses should reflect this multisensory interaction.

In contrast to the supra-additivity prediction, the experiment showed that the auditory-evoked N1 and P2 responses were actually reduced in amplitude in the AV case. The *amplitude reduction* of both the N1 and the P2 were independent of the stimulus. A very interesting pattern was obtained when looking at the response latency. The *peak latency* of the N1 and the P2 responses varied systematically as a function of the correctly identified viseme. In this study, subjects were asked to identify, based only on the visual signal, spoken syllables. For bilabials, correct identification was, predictably, very high (more than 90%); for alveolars, identification was at an intermediate level (approx. 70–85% correct); for velars, the correct identification was typically approximately 60–65%. When plotting electrophysiological response peak latency as a function of correct identification of the visual signal, significant temporal facilitation of the auditory evoked responses (N1 and P2) was observed as a function of how informative the face was for the listeners. In particular, the visual /ka/ was associated with a temporal facilitation of 5–10 ms. In comparison, the visual /pa/, which is almost always correctly identified, was associated with much greater temporal facilitation (up to 25 ms at the P2). In other words, the rate of correct identification of the visual-alone signal predicted the degree of temporal savings of the auditory N1 and P2 components. van Wassenhove et al. (2005) interpret these findings in the following way: the visual speech input, which typically precedes the auditory signal, elicits a (broad class of) internal abstract representations (the hypothesis space).

These internal representations predict the possible audio targets. The internally synthesized and predicted targets are compared against the auditory speech input and the residual error is calculated and fed back for correction. The more informative the facial information is, the more specific the prediction can be, and the more temporal savings is observed. The (abstract) internal representation that is rapidly elicited by the leading visual signal (which may, of course, be somewhat coarse) elicits a candidate set of possible targets; the fewer targets there are, as in the case of bilabials, the more rapid and precise the synthesis is, and the more temporal facilitation is observed. Cumulatively, these data are most consistent with an analysis-by-synthesis model, an internal forward model in which perceptual analysis is guided by the predictions made based on the internally synthesized candidates that are compared against input signals. The multimodality sensory-motor integration is thus in the articulation that underlies the various outputs, not in a sensory-to-sensory mapping between, say, vision and audition. That is, the integration is in abstract, amodal articulation space.

11.5 Lexical representation and distinctive features

We adopt the idea (probably first expressed in Bell (1867); cited in Halle 2002, pp. 3–4, see also pp. 97–100) that the mental representation of speech sounds is not as segment-sized (alphabetic) units, but is decomposed into *distinctive features*. Following Jakobson *et al.* (1952; see also Halle 2002, pp. 108–110), the features have dual definitions and provide the fundamental connection between action (articulation) and perception (audition). Each feature is defined by its effective motoric gesture(s), and also by the auditory patterns that trigger its detection. Though the search for phonetic invariants for features has a long and controversial history (see Stevens & Blumstein (1981), in particular, and Perkell & Klatt (1986), in general), we are not ready to abandon the search just yet. As a fairly clear example, the feature [+round] defines the connection between the motor gesture of lip rounding (the enervation of the *orbicularis oris* muscle) and the perceptual pattern of a down sweep in frequencies across the whole spectral range. For the derivation of the acoustic effect from the motor gesture of lip protrusion through the physics of resonant tube models, see Stevens (1998). Thus, in our view, distinctive features are the sort of representational primitives that allow us to talk both about action and perception and about the connection between action and perception in a principled manner. We can do this because distinctive features are stated in both articulatory (i.e. as gestures performed in a motor coordinate system) and acoustic/auditory terms (i.e. as events statable in an acoustic coordinate system). From this it follows that one must be able to translate between acoustic and motor representations of words: there must be some type of coordinate transformations, and, indeed, there is evidence for thinking about the problem in that way (cf. Hickok & Poeppel 2000, 2004; Hickok *et al.* 2003).

One reason we believe in features as primitives rather than segments is that generalizations in phonology typically affect or involve more than one segment of the language. That is, rules of pronunciation traffic in *natural classes* of sounds rather than individual sounds. To give just one example, the rule of coronal palatalization in Tohono O'odham (formerly known as Papago) given in many introductory phonology texts (e.g. Halle & Clements 1990) changes the *set* of alveolar stops /t d/ to the set of alveopalatal affricates /c j/ when they occur before any vowel in the set of [+high] vowels /i u _/. Likewise, in word-final position Korean neutralizes all coronal obstruents /t tʰ c cʰ s s'/ to the plain coronal stop /t/ (Martin 1951). The sets of sounds triggering and undergoing changes receive perspicuous description with distinctive features; in an alphabetic–phoneme world they are just arbitrary subsets of the language's sounds. In addition, the changes the sets undergo are also typically simple in terms of features—in Tohono O'odham adding [–back] and [+strident]; in Korean removing [+spread glottis] and [+strident]. Psycholinguistic evidence for features in speech perception has been available since the pioneering study by Miller & Nicely (1955, p. 338) who found evidence for distinctive features in that 'the perception of any one of these five features (which are [±voice], [±nasal], [±strident], duration and place of articulation DP/WJI) is relatively independent of the perception of the others, so that it is as if five separate, simple channels were involved rather than one complex channel.' Neuroimaging evidence for the psychological reality of these sets has been provided by Phillips *et al.* (2000), who found mismatch negativity responses in English subjects between the set of voiced stops /b d g/ and the set of voiceless stops /p t k/.

The rate of change of feature values in running speech varies, but often reaches the level of one feature per segment. That is, therefore, some features must be detected within the segmental time frame (approx. 20–80 ms). However, in addition, there is also abundant phonological evidence for syllable-level generalizations in phonology (a recent survey of such effects is Féry & van de Vijver (2004)). Almost all languages display various (approx. one-per-syllable) phenomena, such as stress, tone and vowel harmony (Archangeli & Pulleyblank 1994). The importance of the syllable in speech perception has also been emphasized by many researchers (e.g. Greenberg (2005) for an important perspective). For example, one study (Kabak & Idsardi submitted) shows differential sensitivity in cases of perceptual vowel epenthesis (the illusion of hearing vowels that are not present in the signal) to the syllabic status of consonants. They found that the onset-only set of consonants in Korean (e.g. [+strident] ones such as /c/ in 'pachma') induce perceptual epenthesis (i.e. they are indistinguishable from 'pachima' by Korean listeners), whereas other syllable contact violations of Korean (such as the /k.m/ in the impossible Korean form 'pakma') are ignored and do not result in the percept of an illusory vowel. More intriguingly, however, languages seem to organize their features so as to minimize the number of features used in the language to distinguish among both consonants and vowels (see especially the discussion of combinatorial specification in Archangeli & Pulleyblank (1994)). That is, for example [+round] is typically a feature only for vowels; much less often is [+round] used contrastively to distinguish different consonants (though one such example is Ponapean and others are documented in Ladefoged & Maddieson (1996)). In the case of canonical CV syllables, this allows the features such as [+round] to 'flow' at the syllable resolution rate (integration time scale) as well as at the segmental resolution rate. Furthermore, processes of vowel–consonant assimilation (such as nasalization of vowels and rounding or palatalization of consonants) serve to further 'smear' featural information into the syllabic time scale.

Thus for speakers to use segments to produce speech requires the segments to be decomposed into features in order to adequately account for rules of pronunciation, and listeners also construct representations using the same features, as shown by various psycholinguistic and neurolinguistic tests. Moreover, the features seem to organize along both a slower (a syllabic-level analysis) and a faster rate of change (a segment-level analysis). We find the convergence to a multiresolution analysis on two time scales of approximately the same size particularly intriguing and very much worth pursuing as one of the fundamental principles for speech recognition. For a related multi-tier framework on speech that also engages the multiple time scale, spectral integration and representational challenges to recognition, see Greenberg (2005).

11.6 Discussion

The research programme we have outlined has implications for some of the issues in speech perception research that tend to elicit high blood pressure. We briefly mention some of the consequences of our proposal for two major issues here. First, there have been many discussions on the question of whether speech is 'special'. In its most pointed form, the concept that speech is special presumably means that the cerebral machinery we have to analyse speech is specialized for speech signals in many or most parts of the auditory pathway. It is not obvious whether very much can be learned by focusing on whether or not there is this kind of extreme specialization. Presumably, what everybody is actually interested in is to try to understand how speech recognition works. However, some things do need to be said, since we advocate such a strong linguistically oriented position. As has been argued recently, for example by Price *et al*. (2005), all the cortical machinery that is used for speech is also used for other tasks. It seems like a reasonable proposition that in the difficult case of having to analyse complex signals extremely rapidly, you use whatever is available to you. However, there is a point at which there must be specialization, and that is the point at which the auditory representation interfaces with lexical representation. Lexical representations are *sui generis*: they *may* share properties with other cognitive representations, but they have a number of extremely specialized properties that seem to be restricted to the representation of lexical items in the human brain. So, for example, lexical items do not, at least to our knowledge, look like the internal representations of jingling keys, faces, melodies or odours. Therefore, there is a stage in speech perception at which this format must be constructed, and if that format is of a particular type, there is necessarily specialization. As mentioned above, a critical requirement of lexical representation is that the representations enter into subsequent computation, for example of the morphological or syntactic flavour. One could imagine that lexical roots share properties with the mental/neural representations of non-linguistic sounds, but *some* formal attribute of the representations must be such that they can participate in formal operations ranging from pluralization to compound generation to phrase structure construction. Second, rather than adopt episodic models, we propose that speech is processed in featural and syllabic (categorical) terms. But listeners also pay attention to the location of individual tokens in the acoustic 'clouds' defined by the categories (or types). That is, they detect, know and remember something about the speaker's speech, as compared with the listener's statistical summary of the acoustic variation in the categories that they have already encountered (Goldinger *et al.* 1992). Parallel to Labov (1972) who found *accommodation* by speakers in a variable speech community to the traits of other conversational partners (dropping more /r/'s in 'fourth floor' when their conversational partner dropped their /r/'s), we see the episodicists' findings as showing the willingness of speakers to track and accommodate their low-level speech traits to those of their conversational partners, presumably for sociological reasons (as argued by Labov 1972, 2001). Strong confirmation of the sociolinguistic mediation of speech accommodation is given by the work of Howard Giles (e.g. Giles 1973) and colleagues. In particular, Bourhis & Giles (1977) were able to produce accent divergence by Welsh speakers to an English-speaking authority figure by having the authority figure profess derogatory attitudes towards the Welsh language and culture. The speech of the Welsh speakers showed more Welsh characteristics after demeaning questions than after neutral questions. Similar results were also found by Bourhis *et al.* (1979) in a study on trilingual Flemish students. Thus, accommodation is not a mechanical exemplar-driven process but is rather mediated by the attitude of the listener to the speaker. That is, we believe that the listener constructs a (statistical) model of the speaker and then decides whether to (temporarily) move the speech targets towards the speakers when a sociological message of convergence is desired, or to move them away when wanting to convey an attitude of divergence with the speaker. That is, the speaker's *knowledge OF language* (in the sense of Chomsky (1986)) serves as the basis for the collection of *knowledge ABOUT language*. We believe that the episodic evidence is best understood as the statistical collection of knowledge *about* language, the sort of knowledge we draw on to complete crossword puzzles and knowledge similar to the statistics collected by all animals in various domains (Gallistel 1990). These concepts are not mutually exclusive—rather, knowledge about language is built upon knowledge of language.

A related way to address this tension comes from the cognitive psychology of concepts. In particular, the concepts literature has struggled with the tension between a range of well-documented surface effects (typically perceptual similarity effects) and the necessary and sufficient conditions that are definitional of the 'classical' accounts of concepts (see for review Murphy 2002). The disagreement has been, principally, about how to account for categorical versus gradient phenomena in conceptual processing. For many inferential psychological processes (say, deductive reasoning), a categorical perspective on concepts has been more successful; in contrast, many other processing effects have been best described by gradient, non-categorical representations of some type. How has this conflict been addressed? One approach has been to appeal to a 'theory' view of concepts in which concepts have, constitutively, a 'core' and a 'periphery', where the core corresponds, roughly, to the necessary and sufficient conditions for category membership and the periphery corresponds to the representational architecture that permits gradient characterization. (Naturally, it can be weighted to what extent core versus periphery are more important for various tasks using that concept.) The kind of phenomena that motivated this relatively complex view of concepts include experiments in which it must be the case that both kinds of information are consulted. For example, Armstrong *et al.* (1983) showed that a concept such as 'odd number', surely a classical concept given its formal definition for category membership, nevertheless is also subject to interesting non-categorical effects; i.e. subjects reliably judge '7' to be a better odd number than '237' despite the fact that both are, for computational inferential purposes, totally non-gradient.

This debate on concepts in cognitive psychology seems to us quite analogous to the conflict between abstractionist versus episodic models of speech perception and lexical representation. We advocated an abstractionist model (distinctive features as the representational primitives for phonology and lexicon) that, we argued, can be linked in principled ways to acoustic implementation and also holds hope for developing spectro-temporal primitives (Stevens 2002). But we are, of course, appreciative that there are gradient effects in speech recognition that require explanation. We are, therefore, not at all hostile to all episodic effects in models of speech perception. However, we are against episodic models insofar as they are not just episodic but also explicitly anti-abstractionist. It seems to us an unnecessary consequence to discard abstraction because there is evidence for episodic encoding. From the important demonstration of gradient episodic effects in recognition (Goldinger *et al.* 1992), it does not follow that categorical-type abstract representations do not exist. Instead, we believe that we can learn from the concepts literature and accommodate both types of effects. How can the disagreement be resolved? Perhaps lexical representation is like conceptual representation of the type discussed above: the mind/brain representation of lexical items is made up of a core (abstract, categorical, symbolic) and a periphery (close to the signal, gradient, statistical), both of which are essential for successful representation and are responsible for different aspect of lexical processing. Some speech or language tasks can be (or must be) driven by the *type*—say, morphological computation—and some tasks can be or must be conditioned by the *token* of the type. Either way, we see no logical reason why episodic and abstractionist models are mutually exclusive since they, for the most part, are designed to account for very different sets of phenomena.

Acknowledgements

Supported by NIH DC 05660 to D.P. and a Canada–US Fulbright Program award to W.J.I. We thank Norbert Hornstein for numerous insightful and provocative comments on these issues.

References

Archangeli, D. & Pulleyblank, D. 1994 *Grounded phonology*. Cambridge, MA: MIT Press.
Armstrong, S. L., Gleitman, H. & Gleitman, L. 1983 What some concepts might not be. *Cognition* **13**, 263–308. (doi:10.1016/0010-0277(83)90012-4)
Belin, P., Fecteau, S. & Bédard, C. 2004 Thinking the voice: neural correlates of voice perception. *Trends Cogn. Sci.* **8**, 129–135. (doi:10.1016/j.tics.2004.01.008)
Binder, J. R., Frost, J. A., Hammeke, T. A., Bellgowan, P. S. F., Springer, J. A., Kaufman, J. N. & Possing, E. T. 2000 Human temporal lobe activation by speech and nonspeech sounds. *Cereb. Cortex* **10**, 512–528. (doi:10. 1093/cercor/10.5.512)
Boatman, D. 2004 Cortical bases of speech perception: evidence from functional lesion studies. *Cognition* **92**, 47–65. (doi:10.1016/j.cognition.2003.09.010)
Boemio, A., Fromm, S., Braun, A. & Poeppel, D. 2005 Hierarchical and asymmetric temporal sensitivity in human auditory cortices. *Nat. Neurosci.* **8**, 389–395. (doi:10.1038/nnl409)
Bourhis, R. Y. & Giles, H. 1977 The language of intergroup distinctiveness. In *Language, ethnicity and intergroup relations* (ed. H. Giles), pp. 119–135. London, UK: Academic Press.
Bourhis, R. Y., Giles, H., Leyens, J. & Tajfel, H. 1979 Psycholinguistic distinctiveness: language divergence in Belgium. In *Language and social psychology* (eds H. Giles & R. St Clair), pp. 158–185. Oxford, UK: Blackwell.
Bretthorst, G. L. 1988 *Bayesian spectrum analysis and parameter estimation*. Berlin, Germany: Springer.
Burton, M. W 2001 The role of inferior frontal cortex in phonological processing. *Cogn. Sci.* **25**, 695–709. (doi:10.1016/S0364-0213(01)00051-9)
Calvert, G. A. 2001 Crossmodal processing in the human brain: insights from functional neuroimaging studies. *Cereb. Cortex* **11**, 1110–1123. (doi:10.1093/cercor/11.12. 1110)

Chait, M.j Greenberg, S., Arai, T, Simon, J. Z. & Poeppel, D. Submitted. Multi-time resolution analysis of speech.

Chomsky, N. 1986 *Knowledge of language: its nature, origin, and use*. New York, NY: Praeger.

Cleary, M. & Pisoni, D. 2001 Speech perception and spoken word recognition: research and theory. In *Handbook of perception* (ed. E. B. Goldstein), pp. 499–534. Cambridge, UK: Blackwell.

Drullman, R., Festen, J. M. & Plomp, R. 1994*a* Effect of temporal envelope smearing on speech reception. *J. Acoust. Soc. Am.* **95**, 1053–1064. (doi:10.1121/1.408467)

Drullman, R., Festen, J. M. & Plomp, R. 1994*b* Effect of reducing slow temporal modulations on speech reception. *J. Acoust. Soc. Am.* **95**, 2670–2680. (doi:10.1121/1.409836)

Féry, C. & van de Vijver, R. 2004 *The syllable in optimality theory*. Cambridge, MA: Cambridge University Press.

Friston, K. J., Penny, W, Phillips, C, Kiebel, S., Hinton, G. & Ashburner, J. 2002 Classical and Bayesian inference in neuroimaging: theory. *Neuroimage* **16**, 465–483. (doi:10.1006/nimg.2002.1090)

Gallistel, C. R. 1990 *The organization of learning*. Cambridge, MA: MIT Press.

Giles, H. 1973 Accent mobility: a model and some data. *Anthropol. Linguist.* **15**, 87–105.

Goldinger, S. D., Luce, P. A., Pisoni, D. B. & Marcario, J. K. 1992 Form-based priming in spoken word recognition: the roles of competition and bias. *J. Exp. Psychol. Learn. Mem. Cogn.* **18**, 1210–1238.

Gordon, M. & Poeppel, D. 2002 Inequality in identification of direction of frequency change (up vs. down) for rapid frequency modulated sweeps. *ARLO J. Acoust. Soc. Am.* **3**, 29–34.

Grant, K. W, van Wassenhove, V. & Poeppel, D. 2004 Detection of auditory (cross-spectral) and auditory-visual (cross-modal) synchrony. *Speech Commun.* **44**, 43–53. (doi:10.1016/j.specom.2004.06.004)

Green, D. M. 1993 Auditory intensity discrimination. In *Human psychophysics* (eds W A. Yost, A. N. Popper & R. R Fay), pp. 13–55. New York, NY: Springer.

Greenberg, S. 2005 A multi-tier theoretical framework for understanding spoken language. In *Listening to speech: an auditory perspective* (eds S. Greenberg & W A. Ainsworth), pp. 411–433. Mahwah, NJ: Erlbaum.

Greenberg, S. & Ainsworth, W. A. 2006 *Listening to speech: an auditory perspective*. Mahwah, NJ: Erlbaum.

Hackett, T. A., Preuss, T. M. & Kaas, J. H. 2001 Architectonic identification of the core region in auditory cortex of macaques, chimpanzees, and humans. *J. Comp. Neurol.* **441**, 197–222. (doi:10.1002/cne.1407)

Halle, M. 2002 *From memory to speech and back: papers on phonetics and phonology* 1954–2002. Berlin, Germany: Mouton de Gruyter.

Halle, M. & Clements, G. N. 1990 *Problem book in phonology: a workbook for courses in introductory linguistics and modern phonology*. Cambridge, MA: MIT Press.

Halle, M. & Stevens, K. N. 1959 Analysis by synthesis. In *Proc. Seminar on Speech Compression and Processing*, vol. 2 (eds W Wathen-Dunn & L. E. Woods), paper D7.

Halle, M. & Stevens, K. N. 1962 Speech recognition: a model and program for research, Reprinted in Halle. 2002.

Hesling, I., Dilharreguy, B., Clement, S., Bordessoules, M. & Allard, M. 2005 Cerebral mechanisms of prosodic sensory integration using low-frequency bands of connected speech. *Hum. Brain Mapp.* **26**, 157–169. (doi:10.1002/hbm.20147)

Hickok, G. & Poeppel, D. 2000 Towards a functional neuroanatomy of speech perception. *Trends Cogn. Sci.* **4**, 131–138. (doi:10.1016/S1364-6613(00)01463-7)

Hickok, G. & Poeppel, D. 2004 Dorsal and ventral streams: a framework for understanding aspects of the functional anatomy of language. *Cognition* **92**, 67–99. (doi:10.1016/j.cognition.2003.10.011)

Hickok, G., Buchsbaum, B., Humphries, C. & Muftuler, T. 2003 Auditory-motor interaction revealed by fMRI: speech, music, and working memory in area Spt. *J. Cogn. Neurosd.* **15**, 673–682.

Hinton, G. & Nair, V. In press. Inferring motor programs from images of handwritten digits. In *Proc. 2005 Neural Information Processing Systems*.

Hirsh, I. J. & Sherrick Jr, C. E. 1961 Perceived order in different sense modalities. *J. Exp. Psychol.* **62**, 423–432. (doi:10.1037/h0045283)

Indefrey, P. & Levelt, W. J. 2004 The spatial and temporal signatures of word production components. *Cognition* **92**, 101–144. (doi:10.1016/j.cognition.2002.06.001)

Jakobson, R., Fant, G. & Halle, M. 1952 *Preliminaries to speech analysis*. Cambridge, MA: MIT Press.

Kabak, B. & Idsardi, W. J. Submitted. Speech perception is not isomorphic to phonology: the case of perceptual epenthesis.

Klatt, D. H. 1979 Speech perception: a model of acoustic–phonetic analysis and lexical access. In *Perception and production of fluent speech* (ed. R. A. Cole), pp. 243–288. Hillsdale, NJ: Erlbaum.

Knill, D. C. & Richards, W. 1996 *Perception as Bayesian inference*. Cambridge, UK: Cambridge University Press.

Labov, W. 1972 *Sociolinguistic patterns*. Oxford, UK: Basil Blackwell.

Labov, W. 2001 *Principles of linguistic change. Social factors*, vol. 2. Cambridge, MA: Blackwell.

Ladefoged, P. & Maddieson, I. 1996 *The sounds of the World's languages*. Cambridge, MA: Blackwell.

Lahiri, A. & Reetz, H. 2002 *Laboratory phonology. Phonology and phonetics*, vol. 7. Berlin, Germany: Mouton de Gruyter.

Lu, T, Liang, L. & Wang, X. 2001 Temporal and rate representations of time-varying signals in the auditory cortex of awake primates. *Nat. Neurosci.* **4**, 1131–1138. (doi:10.1038/nn737)

Luo, H., Boemio, A., Gordon, M. & Poeppel, D. 2007 The perception of FM sweeps by Chinese and English listeners. *Hearing Res.* **224**, 75–83. (doi:10.1016/j.heares.2006.11.007)

Marr, D. 1982 *Vision*. San Francisco, CA: Freeman.

Martin, S. E. 1951 Korean phonemics. *Language* **27**, 519–533. (doi:10.2307/410039)

Massaro, D. W., Cohen, M. M. & Smeele, P. M. 1996 Perception of asynchronous and conflicting visual and auditory speech. *J. Acoust. Soc. Am.* **100**, 1777–1786. (doi:10.1121/1.417342)

Meyer, M., Zaehle, T., Gountouna, V. E., Barron, A., Jancke, L. & Turk, A. 2005 Spectro-temporal processing during speech perception involves left posterior auditory cortex. *NeuroReport* **16**, 1985–1989. (doi:10.1097/00001756-200512190-00003)

Miller, G. A. & Nicely, P. E. 1955 An analysis of perceptual confusions among some English consonants. *J. Acoust. Soc. Am.* **27**, 338–352. (doi:10.1121/1.1907526)

Miller, G. A., Galanter, E. & Pribram, K. H. 1960 *Plans and the structure of behavior*. New York, NY: Henry Holt & Co.

Moore, B. C. J. 1989 *An introduction to the psychology of hearing*. San Diego, CA: Academic Press.

Mummery, C. J., Ashburner, J., Scott, S. K. & Wise, R. J. 1999 Functional neuroimaging of speech perception in six normal and two aphasic subjects. *J. Acoust. Soc. Am.* **106**, 449–457. (doi:10.1121/1.427068)

Munhall, K., Gribble, P., Sacco, L. & Ward, M. 1996 Temporal constraints on the McGurk effect. *Percept. Psychophys.* **58**, 351–362.

Murphy, G. 2002 *Big book of concepts*. Cambridge, MA: MIT Press.

Narayan, R., Grana, G. & Sen, K 2006 Distinct time scales in cortical discrimination of natural sounds in songbirds. *J Neurophysiol.* **96**, 252–258. (doi:10.1152/jn.01257.2005)

Norris, D. & Wise, R. 2000 The study of prelexical and lexical processes in comprehension: psycholinguistics and functional neuroimaging. In *The new cognitive neurosciences* (ed. M. Gazzaniga), pp. 867–880. Cambridge, MA: MIT Press.

Norris, D., McQueen, J. M. & Cutler, A. 2000 Merging information in speech recognition: feedback is never necessary. *Behav. Brain Sci.* **23**, 299–325. (doi:10.1017/S0140525X00003241)

Pelli, D. G. 1999 Close encounters: an artist shows that size affects shape. *Science* **285**, 844–846. (doi:10.1126/science.285.5429.844)

Perkell, J. S. & Klatt, D. 1986 *Invariance and variability of speech processes*. Hillsdale, NJ: Lawrence Erlbaum Associates, Inc.

Phillips, C, Pellathy, T, Marantz, A., Yellin, E., Wexler, K., Poeppel, D., McGinnis, M. & Roberts, T. P. L. 2000 Auditory cortex accesses phonological categories: an MEG mismatch study. *J. Cogn. Neurosci.* **12**, 1038–1055. (doi:10.1162/08989290051137567)

Poeppel, D. 2001 Pure word deafness and the bilateral processing of the speech code. *Cogn. Sci.* **21**, 679–693. (doi:10.1016/S0364-0213(01)00050-7)

Poeppel, D. 2003 The analysis of speech in different temporal integration windows: cerebral lateralization as 'asymmetric sampling in time'. *Speech Commun.* **41**, 245–255. (doi:10.1016/S0167-6393(02)00107-3)

Poeppel, D. & Embick, D. 2005 The relation between linguistics and neuroscience. In *Twenty-first century psycho-linguistics: four cornerstones* (ed. A. Cutler), pp. 103–120. Hillsdale, NJ: Lawrence Erlbaum Associates, Inc.

Poeppel, D. & Hackl, M. 2007 The architecture of speech perception. In *Topics in integrative neuroscience: from cells to cognition* (ed. J. Pomerantz). Cambridge, UK: Cambridge University.

Poeppel, D., Wharton, C, Fritz, J., Guillemin, A., San Jose, L., Thompson, J., Bavelier, D. & Braun, A. 2004 FM sweeps, syllables, and word stimuli differentially modulate left and right non-primary auditory areas. *Neuropsychologia* **42**, 183–200. (doi:10.1016/j.neuropsychologia.2003.07.010)

Price, C. J., Thierry, G. & Griffiths, T. 2005 Speech specific neuronal processing: where is it? *Trends Cogn. Sci.* **9**, 271–276. (doi:10.1016/j.tics.2005.03.009)

Robin, D. A., Tranel, D. & Damasio, H. 1990 Auditory perception of temporal and spectral events in patients with focal left and right cerebral lesions. *Brain Lang.* **39**, 539–555. (doi:10.1016/0093-934X(90)90161-9)

Romanski, L. M., Tian, B., Fritz, J., Mishkin, M., Goldman-Rakic, P. S. & Rauschecker, J. P. 1999 Dual streams of auditory afferents target multiple domains in the primate prefrontal cortex. *Nat. Neurosci.* **2**, 1131–1136. (doi:10.1038/16056)

Saberi, K. & Perrott, D. R. 1999 Cognitive restoration of reversed speech. *Nature* **398**, 760. (doi:10.1038/19652)

Schonwiesner, M., Rubsamen, R. & von Cramon, D. Y. 2005 Hemispheric asymmetry for spectral and temporal processing in the human antero-lateral auditory belt cortex. *Eur. J. Neurosd.* **22**, 1521–1528. (doi:10.1111/j.l460-9568. 2005.04315.x)

Schwartz, J. & Tallal, P. 1980 Rate of acoustic change may underlie hemispheric specialization for speech perception. *Science* **207**, 1380–1381. (doi:10.1126/science. 207.4431.665)

Scott, S. K. & Johnsrude, I. S. 2003 The neuroanatomical and functional organization of speech perception. *Trends Neurosci.* **26**, 100–107. (doi:10.1016/S0166-2236(02)00037-1)

Scott, S. K. & Wise, R. J. 2004 The functional neuroanatomy of prelexical processing in speech perception. *Cognition* **92**, 13–45. (doi:10.1016/j.cognition.2002.12.002)

Scott, S. K., Blank, C. C, Rosen, S. & Wise, R. J. S. 2000 Identification of a pathway for intelligible speech in the left temporal lobe. *Brain* **123**, 2400–2406. (doi:10.1093/brain/123.12.2400)

Shamma, S. 2001 On the role of space and time in auditory processing. *Trends Cogn. Sci.* **5**, 340–348. (doi:10.1016/S1364-6613(00)01704-6)

Sininger, Y. S. & Cone-Wesson, B. 2004 Asymmetric cochlear processing mimics hemispheric specialization. *Science* **305**, 1581. (doi:10.1126/science.1100646)

Stein, B. & Meredith, A. 1993 *The merging of the senses*. Cambridge, MA: MIT Press.

Stevens, K. N. 1998 *Acoustic phonetics*. Cambridge, MA: MIT Press.

Stevens, K. N. 2002 Toward a model for lexical access based on acoustic landmarks and distinctive features. *J. Acoust. Soc. Am.* **111**, 1872–1891. (doi:10.1121/1.1458026)

Stevens, K. N. & Halle, M. 1967 Remarks on analysis by synthesis and distinctive features. In *Models for the perception of speech and visual form: proceedings of a symposium* (ed. W Wathen-Dunn), pp. 88–102. Cambridge, MA: MIT Press.

Stevens, K. N. & Blumstein, S. E. 1981 The search for invariant acoustic correlates of phonetic features. In *Perspectives in the study of speech* (eds P. D. Eimas & J. L. Miller), pp. 1–39. Hillsdale: Lawrence Erlbaum.

Todd, N. P. 1994 The auditory primal sketch: a multi-scale model of rhythmic grouping. *J. New Music Res.* **23**, 25–70.

Ungerleider, L. G. & Mishkin, M. 1982 Two cortical visual systems. In *Analysis of visual behavior* (eds D. J. Ingle, M. A. Goodale & R. J. W Mansfield), pp. 549–586. Cambridge, MA: MIT Press.

Van Rullen, R. & Koch, C. 2003 Is perception discrete or continuous? *Trends Cogn. Sci.* **7**, 207–213. (doi:10.1016/S1364-6613(03)00095-0)

van Wassenhove, V., Grant, K. & Poeppel, D. 2005 Visual speech speeds up the neural processing of auditory speech. *Proc. Natl Acad. Sci. USA* **102**, 1181–1186. (doi:10.1073/pnas.0408949102)

van Wassenhove, V, Grant, K. W & Poeppel, D. 2007 Temporal window of integration in auditory-visual speech perception. *Neuropsychologia* **45**, 598–607. (doi:10.1016/j.neuropsychologia.2006.01.001)

Vuilleumier, P., Armony, J. L., Driver, J. & Dolan, R. J. 2003 Distinct spatial frequency sensitivities for processing faces and emotional expressions. *Nat. Neurosci.* **6**, 624–631. (doi:10.1038/nnl057)

Wang, X., Lu, T. & Liang, L. 2003 Cortical processing of temporal modulations. *Speech Commun.* **41**, 107–121. (doi:10.1016/S0167-6393(02)00097-3)

Yabe, H., Tervaniemi, M., Reinikainen, K. & Näätänen, R 1997 Temporal window of integration revealed by MMN to sound omission. *NeuroReport* **8**, 1971–1974. (doi:10.1097/00001756-199705260-00035)

Yabe, H., Koyama, S., Kakigi, R., Gunji, A., Tervaniemi, M., Sato, Y & Kaneko, S. 2001*a* Automatic discriminative sensitivity inside temporal window of sensory memory as a function of time. *Cogn. Brain Res.* **12**, 39–48. (doi:10.1016/S0926-6410(01)00027-l)

Yabe, H., Winkler, I., Czigler, I., Koyama, S., Kakigi, R., Sutoh, T, Hiruma, T. & Kaneko, S. 2001*b* Organizing sound sequences in the human brain: the interplay of auditory streaming and temporal integration. *Brain Res.* **897**, 222–227. (doi:10.1016/S0006-8993(01)02224-7)

Zatorre, R. J., Belin, P. & Penhune, V. B. 2002 Structure and function of auditory cortex: music and speech. *Trends Cogn. Sci.* **6**, 37–46. (doi:10.1016/S1364-6613(00)01816-7)

12

Neural specializations for speech and pitch: moving beyond the dichotomies

Robert J. Zatorre and Jackson T. Gandour

The idea that speech processing relies on unique, encapsulated, domain-specific mechanisms has been around for some time. Another well-known idea, often espoused as being in opposition to the first proposal, is that processing of speech sounds entails general-purpose neural mechanisms sensitive to the acoustic features that are present in speech. Here, we suggest that these dichotomous views need not be mutually exclusive. Specifically, there is now extensive evidence that spectral and temporal acoustical properties predict the relative specialization of right and left auditory cortices, and that this is a parsimonious way to account not only for the processing of speech sounds, but also for non-speech sounds such as musical tones. We also point out that there is equally compelling evidence that neural responses elicited by speech sounds can differ depending on more abstract, linguistically relevant properties of a stimulus (such as whether it forms part of one's language or not). Tonal languages provide a particularly valuable window to understand the interplay between these processes. The key to reconciling these phenomena probably lies in understanding the interactions between afferent pathways that carry stimulus information, with top-down processing mechanisms that modulate these processes. Although we are still far from the point of having a complete picture, we argue that moving forward will require us to abandon the dichotomy argument in favour of a more integrated approach.

Keywords: hemispheric specialization; functional neuroimaging; tone languages

12.1 Introduction

The power of speech is such that it is often considered nearly synonymous with being human. It is no wonder, then, that it has been the focus of important theoretical and empirical science for well over a century. In particular, a great deal of effort has been devoted to understanding how the human brain allows speech functions to emerge. Several distinct intellectual trends can be discerned in this field of research, of which two particularly salient ones will be discussed in this paper. One important idea proposes that speech perception (and production) depends on specialized mechanisms that are dedicated exclusively to speech processing. A contrasting idea stipulates that speech sounds are processed by the same neural systems that are also responsible for other auditory functions.

These two divergent ideas will be referred to here as the domain-specific and the cue-specific models, respectively. Although they are most often cast as mutually exclusive, it is our belief, and the premise of this piece, that some predictions derived from each of these views enjoy considerable empirical support, and hence must be reconciled. We shall therefore attempt to propose a few ideas in this regard, after first reviewing the evidence that has been adduced in favour of each model. In particular, we wish to review the evidence pertaining to patterns of cerebral hemispheric specialization within auditory cortices, and their relation to the processing of speech signals and non-speech signals, with particular emphasis on pitch information. We constrain the discussion in this way in order to focus on a particularly important aspect of the overall problem, and also one that has generated considerable empirical data in recent years; thus, it serves especially well to illustrate the general question. Also, owing to the development of functional neuroimaging over the past few years, predictions derived from these two models have now been tested to a much greater extent than heretofore feasible. We therefore emphasize neuroimaging studies insofar as they have shed light on the two models, but also mention other sources of evidence when pertinent.

First, let us consider some of the origins of the domain-specificity model. Much of the impetus for this idea came from the behavioural research carried out at the Haskins laboratories by Alvin Liberman and his colleagues in the 1950s and 1960s (for reviews, see Liberman & Mattingly (1985) and Liberman & Whalen (2000)). These investigators took advantage of then newly-developed techniques to visualize speech sounds (the spectrograph), and were struck by the observation that the acoustics of speech sounds did not map in a one-to-one fashion to the perceived phonemes. These findings led to the development of the motor theory of speech perception, which proposes that speech sounds are deciphered not by virtue of their acoustical structure, but rather by reference to the way that they are articulated. The lack of invariance in the signal was explained by proposing that invariance was instead present in the articulatory gestures associated with a given phoneme, and that it was these representations which were accessed in perception. More generally, this model proposed that speech bypassed the normal pathway for analysis of sound, and was processed in a dedicated system exclusive to speech. This view therefore predicts that specialized left-hemisphere lateralized pathways exist in the brain which are unique to speech. A corollary of this view is that low-level acoustical features are not relevant for predicting hemispheric specialization, which is seen instead to emerge only from abstract, linguistic properties of the stimulus. Thus, a strong form of this model would predict, for instance, that certain left auditory cortical regions are uniquely specific to speech and would not be engaged by non-speech signals. An additional prediction is that the linguistic status of a stimulus will change the pattern of neural response (e.g. a stimulus that is perceived as being speech or not under different circumstances, or by different persons, would be expected to result in different neural responses).

An alternative approach to this domain-specific model proposes that general mechanisms of auditory neural processing are sufficient to explain speech perception (for a recent review, see Diehl *et al.* (2004)). Several different approaches are subsumed within what we refer to as the cue-specific class of models, but all of them would argue that speech-unique mechanisms are unnecessary and therefore unparsimonious. In terms of the focus of the present discussion, a corollary of this point of view is that low-level acoustical features of a stimulus can determine the patterns of hemispheric specialization that may be observed; hence, it would predict that certain features of non-speech sounds should reliably recruit the left auditory cortex, and that there would be an overlap between speech- and non-speech-driven neural responses under many circumstances. That is, to the extent that certain left-hemisphere auditory cortical regions are involved with the analysis of speech, this is explained on the basis that speech sounds happen to have particular acoustical properties, and it is the nature of the processing elicited by such properties that is at issue, not the linguistic function to which such properties may be related.

The problem that faces us today is that both models make at least some predictions which have been validated by experimental findings. There is thus ample room for theorists to pick and choose the findings that support their particular point of view, and on that basis favour one or the other model. Many authors seem to have taken the approach that if their findings favour one model, then this disproves the other. In the context of allowing the marketplace of ideas to flourish, this rhetorical (some would say argumentative) approach is not necessarily a bad thing. But at some point a reckoning becomes useful, and that is our aim in the present contribution. Let us therefore review some of the recent findings that are most pertinent before discussing possible resolutions of these seemingly irreconcilable models.

12.2 Evidence that simple acoustic features of sounds can explain patterns of hemispheric specialization

One of the challenges in studying speech is that it is an intricately complex signal and contains many different acoustical components that carry linguistic, paralinguistic and non-linguistic information. Many early behavioural studies focused on particular features of speech (such as formant transitions and voice-onset time) that distinguish stop consonants from one another, and found that these stimuli were perceived differently from most other sounds because they were parsed into categories. Furthermore, discrimination within categories was much worse than across categories, which violates the more typical continuous perceptual function that is observed with non-speech sounds (Eimas 1963). This categorical perception was believed to be the hallmark of the speech perception 'mode'. A particularly compelling observation, for instance, was made by Whalen & Liberman (1987), who noted that the identical formant transition could be perceived categorically or not depending on whether it formed part of a sound complex perceived as a speech syllable or not. This sort of evidence suggested that the physical cues themselves are insufficient to explain the perceptual categorization phenomena.

However, much other research showed that, at least in some cases, invariant acoustical cues did exist for phonetic categories (Blumstein 1994), obviating the need for a special speech-unique decoding system. Moreover, it was found that categorical perception was not unique to speech because it could be elicited with non-speech stimuli that either emulated certain speech cues (Miller *et al.* 1976; Pisoni 1977), or were based on learning of arbitrary categories, such as musical intervals (Burns & Ward 1978; Zatorre 1983). Eimas *et al.* (1971), together with other investigators (Kuhl 2000), also demonstrated that infants who lacked the capacity to articulate speech could nonetheless discriminate speech sounds in a manner similar to adults, casting doubt on the link between perception and production. The concept that speech may not depend upon a unique human mechanism but rather on general properties of the auditory system was further supported by findings that chinchillas and quail perceived phonemes categorically (Kuhl & Miller 1975; Kluender *et al.* 1987). Thus, from this evidence, speech came to be viewed as one class of sounds with certain particular properties, but not requiring some unique system to handle it.

A parallel trend can be discerned in the neuroimaging literature dealing with the neural basis of speech: particular patterns of neural engagement have been observed in many studies, which could be taken as indicative of specialized speech processors, but other studies have shown that non-speech sounds can also elicit the same patterns. For example, consider studies dealing with the specialization of auditory cortical areas in the left hemisphere. Many of the first functional neuroimaging studies published were concerned with identifying the pathways associated with processing of speech sounds. Most of these studies did succeed in demonstrating greater response from left auditory cortical regions for speech sounds as compared with non-speech controls, such as tones (Binder *et al.* 2000*a*), amplitude-modulated noise (Zatorre *et al.* 1996) or spectrally rotated speech (Scott *et al.* 2000). However, as noted above, it is somewhat difficult to interpret these responses owing to the complexity of speech; thus, if speech sounds elicit a certain activity pattern that is not observed with some control sound, it is not clear whether to attribute it to the speech *qua* speech, or to some acoustical feature contained within the speech signal but not in the control sound. Conversely, if a non-speech sound that is akin to speech elicits left auditory cortical activation, then one can always argue that it does so by virtue of its similarity to speech. Such findings can therefore comfort adherents of both models.

More recently, however, there has been greater success among functional imaging studies that have focused on the specific hypothesis that rapidly changing spectral energy may require specialized left auditory cortical mechanisms independently of whether they are perceived as speech. One such study was carried out by Belin *et al.* (1998) who used positron emission tomography (PET) to examine processing of formant transitions of different durations in pseudospeech syllables. The principal finding was that whereas left auditory cortical response was similar to both slower and faster transitions, the right auditory cortex responded best to the slower transitions, indicating a differential sensitivity to speed of spectral change. Although the stimuli used in this study were not speech, they were derived from speech signals and therefore one could interpret the result in that light. Such was not the case with a PET experiment by Zatorre & Belin (2001), who used pure-tone sequences that alternated in pitch by one octave at different temporal rates. As the speed of the alternation increased, so did the neural response in auditory cortices in both hemispheres, but the magnitude of this response was significantly greater on the left than on the right. Although this observation is different from the Belin *et al.* (1998) result, which showed little difference in left auditory cortex between faster and slower transitions, it supports the general conclusion of greater sensitivity to temporal rate on the left, using stimuli that bear no relationship to speech sounds. The findings of Zatorre & Belin were recently replicated and extended by Jamison *et al.* (2006), who used functional magnetic resonance imaging (fMRI) with the same stimuli, and found very consistent results even at an individual subject level. A further test of the general proposition is provided in a recent fMRI study by Schönwiesner *et al.* (2005), who used a sophisticated stimulus manipulation consisting of noise bands that systematically varied in their spectral width and temporal rate of change (Figure 12.1). The findings paralleled those of the prior studies to the extent that increasing rate of change elicited a more consistent response from lateral portions of Heschl's gyrus on the left when compared with the right. Although the precise cortical areas identified were somewhat different from those of the other studies, no doubt related to the very different stimuli used, the overall pattern of lateralization was remarkably similar.

A series of additional studies have also recently been carried out supporting this general trend. For example, Zaehle *et al.* (2004) carried out an fMRI study comparing the activation associated with speech syllables that differed in voice-onset time, and non-speech noises that differed in gap duration. Both classes of stimuli vary in terms of the

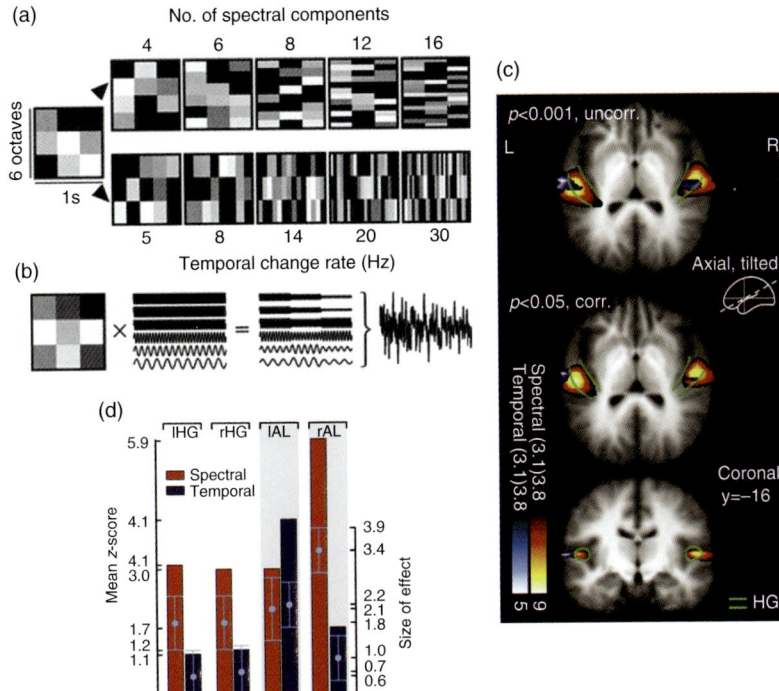

Figure 12.1 Hemispheric differences in auditory cortex elicited by noise stimuli. (a, b) Illustration of how noise stimuli were constructed; each matrix illustrates stimuli with different bandwidths (on the ordinate) and different temporal modulation rates (on the abscissa). (c) fMRI results indicating bilateral recruitment of distinct cortical areas for increasing rate of temporal or spectral modulation. (d) Effect sizes in selected areas of right (r) and left (l) auditory cortices. Note significant interaction between left anterolateral region, which responds more to temporal than to spectral modulation, and right anterolateral region, which responds more to spectral than to temporal modulation. HG, Heschl's gyrus (Schonwiesner et al. 2005).

duration between events, but in one case it cues a speech-relevant distinction, and the stimuli are perceived as speech, whereas the other stimuli are merely broadband noises with a certain size gap inserted. The findings indicated that there was a substantial overlap within left auditory cortices in the response to both speech syllables and noises (Figure 12.2). Thus, the physical cues present in both stimuli, regardless of linguistic status, seemed to be the critical factor determining recruitment of left auditory cortex, as predicted by the cue-specific hypothesis. Further consistent findings were reported by another research group (Joanisse & Gati 2003) who contrasted speech versus non-speech along with slow versus fast changes in a 2 × 2 factorial design. The most relevant finding was that certain areas of the superior temporal gyrus (STG) in both hemispheres responded similarly to speech and non-speech tone sweeps that both contained rapidly changing information; however, the response was substantially greater in the left hemisphere. No such response was observed to stimuli with more slowly changing temporal information, whether speech (vowels) or non-speech (steady-state tones). Thus, in this study too, the physical cues

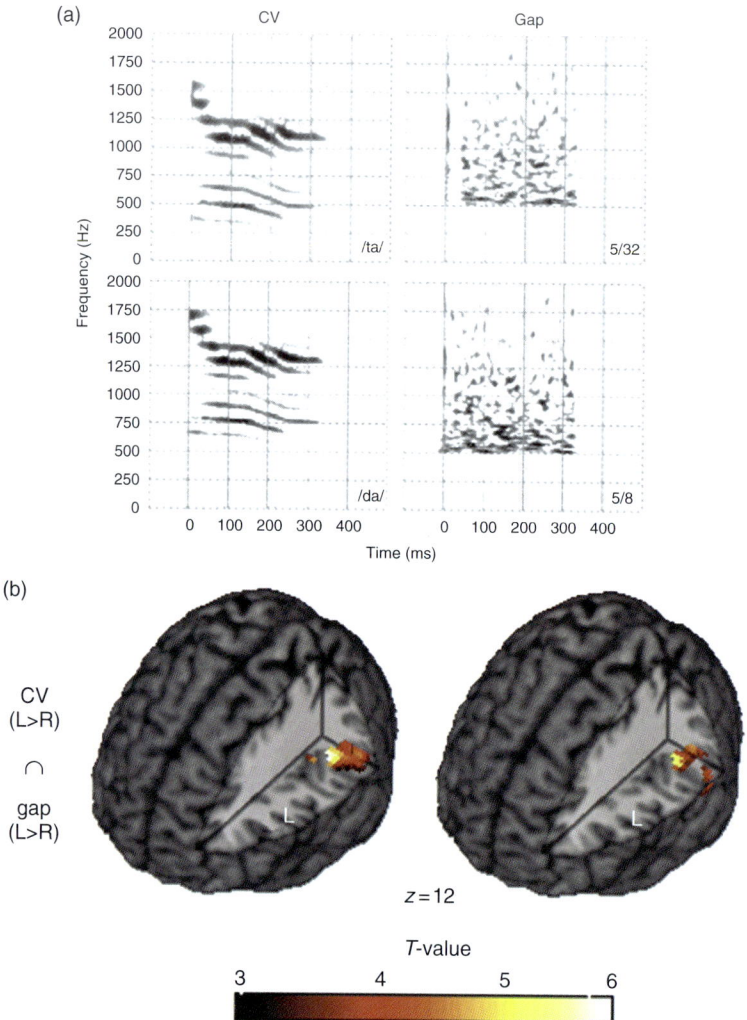

Figure 12.2 (a) Illustrations of consonant–vowel (CV) speech and non-speech sounds containing similar acoustical properties. (b) fMRI brain images illustrating an overlap between left auditory cortex responses to syllables and to non-speech gap stimuli in left auditory cortex. Similar results are obtained at two distinct time points (TPA1 and TPA2; Zaehle et al. 2004).

present in the stimuli seem to predict left auditory cortex activation, regardless of whether the stimuli were speech or not.

A final, recent, example of hemispheric asymmetries in auditory cortices arising from low-level features is provided by Boemio et al. (2005), who parametrically varied the segment transition rates in a set of concatenated narrowband noise stimuli, such that the segment durations varied from quickly changing (12 ms) to more slowly changing (300 ms). Sensitivity to this parameter was demonstrated bilaterally in primary and adjacent auditory cortices and, unlike the studies just reviewed, the strength of the response was essentially identical on both sides. However, in more downstream auditory regions within the superior temporal sulcus (STS), a clear asymmetry was observed, with the more slowly modulated signals preferentially driving the regions on the right side. The authors conclude that '...there exist two timescales in STG...with the right hemisphere receiving afferents carrying information processed on the long time-scale and the left hemisphere those resulting from processing on the short time-scale.' and that their findings are '...consistent with the proposal suggesting that left auditory cortex specializes in processing stimuli requiring enhanced temporal resolution, whereas right auditory cortex specializes in processing stimuli requiring higher frequency resolution' (Boemio *et al.* 2005, p. 394). Thus, the conclusion reached by an impressive number of independent studies is that these low-level features drive hemispheric differences.

One might be concerned with the substantial differences across these studies in terms of precisely which cortical zones are recruited, and which ones show hemispheric differences: core or adjacent belt cortices in some studies (Zatorre & Belin 2001; Schönwiesner *et al.* 2005; Jamison *et al.* 2006), as opposed to belt or parabelt areas in STS or anterior STG in others (Joanisse & Gati 2003; Boemio *et al.* 2005). Also, one might point out discrepancies of the specific circumstances under which one observes either enhanced left-hemisphere response to rapidly changing stimuli (Zaehle *et al.* 2004) or, instead, preferential right-hemisphere involvement for slowly varying stimuli (Belin *et al.* 1998; Boemio *et al.* 2005) or, in several cases, both simultaneously (Schönwiesner *et al.* 2005; Jamison *et al.* 2006). Indeed, these details remain to be specified in any clear way, especially in terms of which pathways in the auditory processing stream are being engaged in the various studies (Hickok & Poeppel 2000; Scott & Johnsrude 2003). One might also object, as Scott & Wise (2004) have, that not every study that has examined hemispheric differences based on low-level cues has succeeded in finding them (Giraud *et al.* 2000; Hall *et al.* 2002), or have found them, but not in auditory cortex (Johnsrude *et al.* 1997). There are many reasons that one could generate why a particular study may not have yielded significant hemispheric differences or observed them in predicted regions, and this too remains to be worked out in detail. But what is striking in the papers reviewed above is the consistency of the overall pattern, as observed repeatedly by a number of different research groups despite the very different stimuli and paradigms used. In particular, we call attention to the clear, replicable evidence that stimuli which are not remotely like speech either perceptually or acoustically (e.g. Boemio *et al.* 2005; Schönwiesner *et al.* 2005; Jamison *et al.* 2006) can and do elicit patterns of hemispheric differences that are consistent and predictable based on acoustical cues, and that brain activation areas are often, though not exclusively, overlapping with those elicited by speech within the left auditory cortex.

Furthermore, the conclusions from these imaging studies fit in a general way with earlier behavioural–lesion studies that focused on the idea that temporal information processing may be especially important for speech. Among the earliest to argue in favour of this idea were researchers working with aphasic populations, who noted associations between aphasia and temporal judgement deficits (Efron 1963; Swisher & Hirsh 1972; Phillips & Farmer 1990; von Steinbüchel 1998), and from research on children with specific language impairment, who seem to demonstrate global temporal processing deficits (Tallal *et al.* 1993, 1996). A similar conclusion regarding the perceptual deficits of patients with pure word deafness was reached by Phillips & Farmer (1990), who commented that the critical problem in these patients relates to a deficit in processing of sounds with temporal content in the milliseconds to tens of milliseconds range. Complementary data pointing in the same direction are provided by depth electrode recordings from human auditory cortex (Liégeois-Chauvel *et al.* 1999). These authors observed that responses from left but not right Heschl's gyrus distinguished differences related to the voice-onset time feature in speech; critically, a similar sensitivity was present for non-speech analogues that contained similar acoustic features, supporting the contention that it is general temporal acuity which is important, regardless of whether the sound is speech or not. Thus, the neuroimaging findings, which we focus on here, are, broadly speaking, also compatible with a wider literature.

The studies reviewed above, which argue for the importance of low-level temporal properties of the acoustical signal in defining the role of left auditory cortices, have also led to some theoretical conclusions that have implications beyond the debate about speech. In particular, a major advantage of the cue-specific hypothesis is that because it is neutral about the more abstract status of the stimulus, it can make predictions about all classes of signals, not just speech. Another advantage of the cue-specific model is that it leads to more direct hypotheses about the neural mechanisms that may be involved.

For example, both Zatorre *et al.* (2002) and Poeppel (2003) have independently proposed models whereby differences in neural responses between the left and right auditory cortices are conceptualized as being related to differences in the speed with which dynamically changing spectral information is processed. The Poeppel model emphasizes differential time integration windows, while Zatorre and colleagues emphasize the idea of a relative trade-off in temporal and spectral resolution between auditory cortices in the two hemispheres. One useful feature of these models is that they generate testable predictions outside the speech domain. The domain of tonal processing is especially relevant in this respect.

12.3 Evidence for right auditory cortex processing of pitch information

A large amount of evidence has accumulated, indicating that right auditory cortex is specialized for the processing of pitch information under specific circumstances. In neuroimaging studies, asymmetric responses favouring right auditory cortices have been reported in tasks that require pitch judgments within melodies (Zatorre *et al.* 1994) or tones (Binder *et al.* 1997); maintenance of pitch while singing (Perry *et al.* 1999); imagery for familiar melodies (Halpern & Zatorre 1999), or for the timbre of single tones (Halpern *et al.* 2004); discrimination of pitch and duration in short patterns (Griffiths *et al.* 1999); perception of melodies made of iterated ripple noise (Patterson *et al.* 2002); reproduction of tonal temporal patterns (Penhune *et al.* 1998); timbre judgments in dichotic stimuli (Hugdahl *et al.* 1999); perception of missing fundamental tones (Patel & Balaban 2001); and detection of deviant chords (Tervaniemi *et al.* 2000), to cite a few.

What is of interest is that these findings can be accounted for at least in part by the same set of assumptions of a cue-specific model in terms of differential temporal integration. The concept is that encoding of pitch information will become progressively better the longer the sampling window; this idea follows from straightforward sampling considerations—the more cycles of a periodic signal that can be integrated, the better the frequency representation should be. Hence, if the right auditory cortex is proposed to be a slower system—a disadvantage presumably when it comes to speech analysis—it would conversely have an advantage in encoding information in the frequency domain. Several of the studies reviewed in §2 explicitly tested this idea as a means of contrasting the response of the two auditory cortices. For example, Zatorre & Belin (2001) and Jamison *et al.* (2006) demonstrated that neural populations in the lateral aspect of Heschl's gyrus bilaterally responded to increasingly finer pitch intervals in a stimulus, and that this response was greater in the right side. Schönwiesner *et al.* (2005) found a similar phenomenon, but notably their stimuli contained no periodicity, since they used only noise bands of different filter widths. Nonetheless, the right auditory cortical asymmetry emerged as bandwidth decreased, supporting the idea that spectral resolution is greater in those regions.

Additional evidence in favour of the importance of right auditory cortical systems in the analysis and encoding of tonal information comes from a number of other sources, notably behavioural-lesion studies which have consistently reported similar findings. Damage to superior temporal cortex on the right affects a variety of tonal pitch processing tasks (Milner 1962; Sidtis & Volpe 1988; Robin *et al.* 1990; Zatorre & Samson 1991; Zatorre & Halpern 1993; Warrier & Zatorre 2004). More specifically, lesions to right but not left auditory cortical areas within Heschl's gyrus specifically impair pitch-related computations, including the perception of missing fundamental pitch (Zatorre 1988), and direction of pitch change (Johnsrude *et al.* 2000). The latter study is particularly relevant for our discussion for two reasons. First, the effect was limited to lesions encroaching upon the lateral aspect of Heschl's gyrus—the area seen to respond to small frequency differences by Zatorre & Belin (2001)—and not seen after more anterior temporal lobe damage. Second, the lesion resulted not in an abolition of pitch-discrimination ability, but rather in a large increase in the discrimination threshold. This result fits with the idea of a *relative* hemispheric asymmetry related to resolution. Auditory cortices on both sides must therefore be sensitive to information in the frequency domain, but the right is posited to have a finer resolution; hence, lesions to this region result in an increase in the minimum frequency needed to indicate the direction of change.

Taken together, the neuroimaging findings and the lesion data just reviewed point to a clear difference favouring the right auditory cortex in frequency processing, a phenomenon which can be explained based on a single simplifying assumption that differences exist in the capacity of auditory cortices to deal with certain types of acoustical information. The parsimony of making a single assumption to explain a large body of data concerning processing of frequency information as well as temporal information relevant to speech is attractive. Based on the foregoing empirical data, it seems impossible to support the proposition that low-level acoustical features of sounds have *no* predictive power with respect to patterns of hemispheric differences in auditory cortex. Yet, if one accepts this conclusion, it need not necessarily follow that higher-order abstract features of a stimulus would have no bearing on the way that they are processed in the brain, nor that low-level features can explain all aspects of neural specialization. In fact, despite the clear evidence just mentioned regarding tonal processing, the story becomes much more complex (and interesting) when tonal cues become phonemic in a speech signal, as we shall see in §§7–12. Before reviewing information on use of tonal cues in speech, however, we review some examples of findings that cannot be explained purely on the basis of the acoustic features present in a stimulus, and show that more abstract representations, or context effects, play an important role as well.

12.4 Evidence that linguistic features of a stimulus can influence patterns of hemispheric specialization

In a fundamental way, a model that does not take into account the linguistic status of a speech sound at all in predicting neural processing pathways cannot be complete; indeed, consider that if a sound which is speech was processed *identically* to a sound which is not speech at all levels in the brain, then we should not ever be able to perceive a difference! The fact that a speech sound is interpreted as speech, and may lead to phonetic recognition, or to retrieval of meaning, immediately implies contact with memory traces; hence, on those grounds alone different mechanisms would be expected. For instance, studies by Scott *et al*. (2000) and Narain *et al*. (2003) demonstrate that certain portions of the left anterior and posterior temporal lobes respond to intelligible speech sentences, whether they are produced normally or generated via a noise vocoding algorithm which results in an unusual, unnatural timbre but can still be understood. This result can be understood as revealing the brain areas in which meaning is processed based not upon the acoustical details of the signal, but rather on higher-order processes involved with stored phonetic and/or semantic templates. However, this effect need not indicate that auditory cortices earlier on in the processing stream are not sensitive to low-level features. Rather, it indicates a convergence of processing for different stimulus types at higher levels of analysis where meaning is decoded.

As noted above, adjudicating between the competing models is not always easy because differences in neural activity that may be observed for speech versus non-speech sounds are confounded by possible acoustic differences that are present in the stimuli. One clever way around this problem is to use a stimulus that can be perceived as speech or not under different circumstances. Sine-wave speech provides just such a stimulus, because it is perceived by naive subjects as an unusual meaningless sound, but, after some training, it can usually be perceived to have linguistic content (Remez *et al.* 1981). It has been shown that sine-wave speech presented to untrained subjects does not result in the usual left auditory cortex lateralization seen with speech sounds (Vouloumanos *et al.* 2001), a result taken as evidence for speech specificity. However, sine-wave stimuli are not physically similar to real speech, and thus the differences may still be related to acoustical variables. Two recent fMRI studies exploit sine-wave speech by comparing how these sounds are perceived before and after training sessions, which resulted in a subject being able to hear speech content (Dehaene-Lambertz *et al.* 2005; Möttönen *et al.* 2006). In both studies, the principal finding was that cortical regions in the left hemisphere showed enhanced activity after training. In the Dehaene-Lambertz study, however, significant left-sided asymmetries were noted even before training. In the Möttönen study, the enhanced lateralization was seen only in those subjects who were able to learn to identify the stimuli as speech. Thus, these findings demonstrate with an elegant paradigm that identical physical sounds are processed differently when they are perceived as speech than when they are not, although there is also evidence that even without training they may elicit lateralized responses.

A similar but more specific conclusion was reached in a recent study directly examining the neural correlates of categorical perception (Liebenthal *et al.* 2005). These authors compared categorically perceived speech syllables with stimuli containing the same acoustical cues, but which are perceived neither as speech nor categorically. Their finding of a very clear left STS activation exclusively for the categorically perceived stimuli indicates that this region responds to more than just the acoustical features, since these were shared across the two types of sounds used. The authors conclude that this region performs an intermediate stage of processing linking early processing regions with more anteroventral regions containing stored representations. More generally, these results indicate once again that experience with phonetic categories, and not just physical cues, influences patterns of activity, a conclusion which is also consistent with cross-language studies using various methodologies. For example, Näätänen and colleagues showed that the size of the mismatch negativity (MMN) response in the left hemisphere is affected by a listener's knowledge of vowel categories in their language (Näätänen *et al.* 1997). Golestani & Zatorre (2004) found that several speech-related zones, including left auditory cortical areas, responded to a greater degree after training with a foreign speech sound than they had before; since the sound had not changed, it is clearly the subjects' knowledge that caused additional activation.

12.5 Evidence that context effects modulate early sensory cortical response

There is a related body of empirical evidence, deriving from studies of task-dependent modulation, or context effects, that also indicates that the acoustical features of a stimulus by themselves are insufficient to explain all patterns of hemispheric involvement. Interestingly, such effects have been described for both speech and non-speech stimuli. Thus, they are not predicted by a strong form of either model, since, according to a strict cue-specific model, only stimulus features and not task demands are relevant, while the domain-specific model focuses on the idea that speech is processed by a dedicated pathway, but makes no predictions for non-speech sounds.

For instance, a recent study (Brechmann & Scheich 2005) using frequency-modulated tones shows that an asymmetry favouring the right auditory cortex emerges for these stimuli only when subjects are actively judging the direction of the tone sweep, and not under passive conditions. In a second condition of this study, these investigators contrasted two tasks, one requiring categorization of the duration of the stimulus, while the other required categorization of the direction of pitch change, using identical stimuli in both conditions. One region of posterior auditory cortex was found to show sensitivity to the task demands, such that more activation was noted in the left than the right for the duration task, while the opposite lateralization emerged for the pitch task. While this result is not predicted by a strict bottom-up model, it is, broadly speaking, consistent with the cue-specific model described above according to which pitch information is better processed in the right auditory cortex while temporal properties are better analysed in the left auditory cortex. However, this model would need to be refined in order to take into account not only the nature of the acoustical features present in the stimulus, but also their relevance to a behavioural demand.

Another recent example demonstrating interactions between physical features of a stimulus and context effects comes from a study using magnetoencephalography and a passive-listening MMN paradigm (Shtyrov *et al.* 2005). These investigators used a target sound containing rapidly changing spectral energy, which in isolation is heard as non-speech, in different contexts: one where it is perceived as a speech phoneme and another as a non-speech sound. In addition, the context itself was either a meaningful word or a meaningless but phonetically correct pseudoword. Left-lateralized effects were only observed when the target sound was presented within a word context, and not when it was placed within a pseudoword context, even though it was perceived as speech in the latter case (Figure 12.3). This result is important because, once again, it is not predicted by a strong form of either model. A domain-specific prediction would be that as soon as the sound is perceived as speech, it should result in recruitment of left auditory cortical speech zones, but this was not observed. Similarly, since the target sound is always the same, a strict cue-specific model would have to predict similar patterns regardless of context. Instead, the findings of this study point to the interaction between acoustical features and learned representations.

12.6 Evidence that not all language processing involves acoustical cues: sign language studies

One of the most dramatic demonstrations of the independence of abstract, symbolic processing involved in language from the low-level specializations relevant to speech comes from the study of sign language. Since perception of sign languages is purely visual, then whatever results one obtains with them must perforce pertain to general linguistic properties and be unrelated to auditory processing. The existing literature on sign language aphasia in deaf persons suggests that left-hemisphere damage is associated with

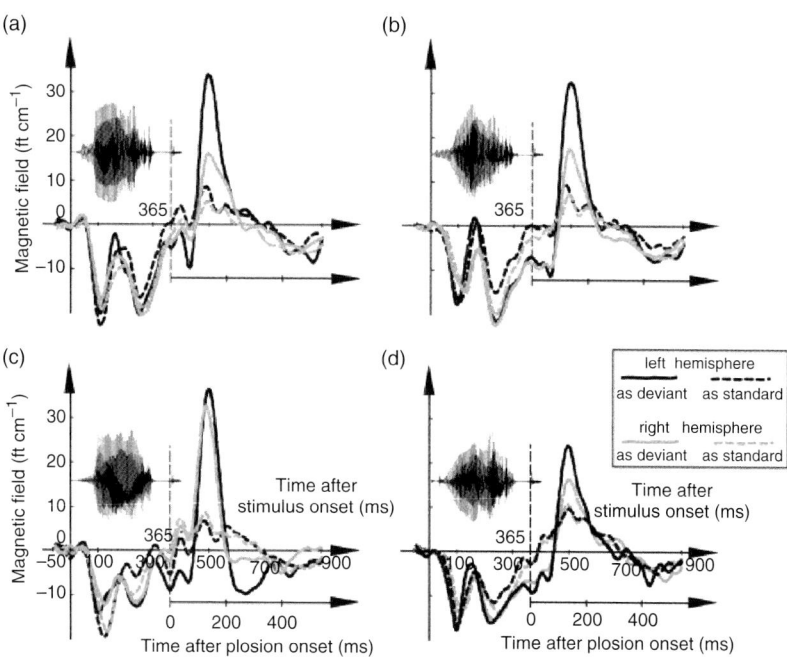

Figure 12.3 Magnetic-evoked potentials to a non-speech sound presented in different contexts. A left-hemisphere advantage is observed only when the target sound is embedded in (a, b) a real word (verb or noun, respectively), but not when it is presented in either (c) a non-speech context or (d) a pseudoword context, where it is perceived as speech but not as a recognizable word (Shtyrov *et al.* 2005).

aphasia-like syndromes, manifested as impairments in signing or in perceiving sign (Hickok *et al.* 1998). More recently, this literature has been enhanced considerably by functional neuroimaging studies of neurologically intact deaf signers. A complete review of this expanding literature is not our aim here, but there are some salient findings which are very relevant to our discussion. In particular, there is a good consensus on the conclusions that (i) what would typically be considered auditory cortex can be recruited for visual sign language processing in the deaf, and (ii) under many circumstances, sign language processing recruits left-hemisphere structures. Moreover, these findings are consistent across a range of distinct sign languages, including American, Quebec, British and Japanese sign languages.

The finding that auditory cortex can be involved in the processing of visual sign can perhaps best be interpreted in light of cross-modal plasticity effects. Many studies have now shown that visual cortex in the blind is functional for auditory (Kujala *et al.* 2000; Weeks *et al.* 2000; Gougoux *et al.* 2005) and tactile tasks (Sadato *et al.* 1996). It is therefore reasonable to conclude that the cross-modal effects in the deaf are not necessarily related to language, a conclusion strengthened by the finding that superior temporal areas respond to non-linguistic visual stimuli in the deaf (Finney *et al.* 2001). But the recruitment of left-hemisphere language areas for the performance of sign language tasks would appear to be strong evidence in favour of a domain-specific model (Bavelier *et al.* 1998). For example, Petitto *et al.* (2000) showed that when deaf signers are performing a verb generation task, neural activity is seen in the left inferior frontal region, comparable with what is observed with speaking subjects (Petersen *et al.* 1988). Similarly, MacSweeney *et al.* (2002) demonstrated that left-hemisphere language areas in frontal and temporal lobes are recruited in deaf signers during a sentence comprehension task, and that these regions overlap with those of hearing subjects performing a similar speech-based task. However, the latter study also points out that left auditory cortical areas are more active for speech than for sign in hearing persons who sign as a first language; therefore, the left auditory cortex would seem to have some privileged status for processing speech.

In any case, these findings and others would seem to spell trouble for a cue-specific model that predicts hemispheric differences based on acoustical features. The logic goes something like this: if visual signals can result in recruitment of left-hemisphere speech zones, then how can rapid temporal processing of sound be at all relevant? At first glance, this would appear to be strong evidence in favour of a domain-specific model. However, this reasoning does not take into account an evolutionary logic. That is, anyone born deaf still carries in his or her genome the evolutionary history of the species, which has resulted in brain specializations for language. What has changed in the deaf individual is the access to these specializations, since there is a change in the nature of the peripheral input. But it was processing *of sound* that presumably led to the development of the specializations during evolutionary history, assuming that the species has been using speech and not sign during its history (indeed, there is no evidence of any hearing culture anywhere not using speech for language communications). What the literature on deaf sign language processing teaches us is certainly something important that higher-order linguistic processes can be independent of specializations for acoustical processing. But, as with the literature discussed above, this conclusion does not mean that the cue-specific model is incorrect; instead, it tells us, as do many of the other findings we have reviewed, that there are interesting and complex interactions between low-level sensory processing mechanisms and higher-order abstract processing mechanisms.

12.7 Evidence for an interaction between processing of pitch information and higher-order linguistic categories: tonal languages

As we have seen, there is consistent evidence for a role of right-hemisphere cortical networks in the processing of tonal information. But tonal information can also be part of a linguistic signal. The question arises whether discrimination of *linguistically relevant* pitch patterns would depend on processing carried out in right-hemisphere networks, as might be predicted by the cue-specific hypothesis, or, alternatively, whether it would engage left-hemisphere speech zones as predicted by the domain-specific hypothesis. Pitch stimuli that are linguistically relevant at the syllable level thus afford us a unique window for investigating the case for the domain-specific hypothesis.

Tone languages exploit variations in pitch at the syllable level to signal differences in word meaning (Yip 2003). The bulk of information on tonal processing in the brain comes primarily from two tone languages: Mandarin Chinese, which has four contrastive tones, and Thai, which has five (for reviews, see Gandour 1998, 2006a, b). Similar arguments in support of domain-specificity could be made for those languages in which duration is used to signal phonemic oppositions in vowel length (Gandour *et al.* 2002a, b; Minagawa-Kawai *et al.* 2002; Nenonen *et al.* 2003). However, cross-language studies of temporal processing do not provide a comparable window for adjudicating between cue- and domain-specificity, because both temporal and language processes are generally considered to be mediated primarily by neural mechanisms in the left hemisphere.

12.8 Evidence for left-hemisphere involvement in processing of linguistic tone

As reviewed in §2 and 3, in several functional neuro-imaging studies of phonetic and pitch discrimination in speech, right prefrontal cortex was activated in response to pitch judgments for English-speaking listeners, whereas consonant judgments elicited activation of the left prefrontal cortex (Zatorre *et al.* 1992, 1996). In English, consonants are phonemically relevant; pitch patterns at the syllable level are not. What happens when pitch patterns are linguistically relevant in the sense of cuing lexical differences? Several studies have now demonstrated a left-hemisphere specialization for tonal processing in native speakers of Mandarin Chinese. For example, when asked to discriminate Mandarin tones and homologous low-pass filtered pitch patterns (Hsieh *et al.* 2001), native Chinese-speaking listeners extracted tonal information associated with the Mandarin tones via left inferior frontal regions in both speech and non-speech stimuli. English-speaking listeners, on the other hand, exhibited activation in homologous areas of the right hemisphere. Pitch processing is therefore lateralized to the left hemisphere only when the pitch patterns are phonologically significant to the listener; otherwise, to the right hemisphere. It appears that left-hemisphere mechanisms mediate processing of linguistic information irrespective of acoustic cues or type of phonological unit, segmental or suprasegmental. No activation occurred in left frontal regions for English listeners on either the tone or pitch task because they were judging pitch patterns that are not phonetically relevant in the English language.

Given these observations, one might then ask whether knowledge of a tone language changes the hemispheric dynamics for processing of tonal information in general, or whether it is specific to one's own language. The answer appears to be that it is tied closely to knowledge of a specific language. When asked to discriminate Thai tones, Chinese listeners fail to show activation of left-hemisphere regions (Gandour *et al.* 2000, 2002*a*). This finding is especially interesting insomuch as the tonal inventory of Thai is similar to that of Mandarin in terms of number of tones and type of tonal contours. In spite of similarities between the tone spaces of Mandarin and Thai, the fact that we still observe cross-language differences in hemispheric specialization of tonal processing suggests that the influence of linguistically relevant parameters of the auditory signal is specific to experience with a particular language. Other cross-language neuroimaging studies of Mandarin have similarly revealed activation of the left posterior prefrontal and insular cortex during both perception and production tasks (Klein *et al.* 1999, 2001; Gandour *et al.* 2003; Wong *et al.* 2004).

An optimal window for exploring how the human brain processes linguistically relevant temporal and spectral information is one in which both vowel length and tone are contrastive. Thai provides a good test case, insomuch as it has five contrastive tones in addition to a vowel length contrast. To address the issue of temporal versus spectral processing directly, it is imperative that the same group of subjects be asked to make perceptual judgments of tone and vowel length on the same stimuli. To test this idea in a cross-language study (Gandour *et al.* 2002*a*), Thai and Chinese subjects were asked to discriminate pitch and timing patterns presented in the same auditory stimuli under speech and non-speech conditions. If acoustic in nature, effects due to this level of processing should be maintained across listeners regardless of language experience. If, on the other hand, effects are driven by higher-order abstract features, we expect brain activity of Thai and Chinese listeners to vary depending on language-specific functions of temporal and spectral cues in their respective languages. The question is whether language experience with the same type of phonetic unit is sufficient to lead to similar brain activation patterns as those of native listeners (cf. Gandour *et al.* 2000). In a comparison of pitch and duration judgments of speech relative to non-speech, the left inferior prefrontal cortex was activated for the Thai group only (Figure 12.4). No matter whether the phonetic contrast is signalled primarily by spectral or temporal cues, the left hemisphere appears to be dominant in processing contrastive phonetic features in a listener's native language. Both groups, on the other hand, exhibited similar fronto-parietal activation patterns for spectral and temporal cues under the non-speech condition. When the stimuli are no longer perceived as speech, the language-specific effects disappear. Regardless of the neural mechanisms underlying lower-level processing of spectral and temporal cues, hemispheric specialization is clearly sensitive to higher-order information about the linguistic status of the auditory signal.

The activation of posterior prefrontal cortex in the above-mentioned studies of tonal processing raises questions about its functional role. Because most of these studies used discrimination tasks, there were considerable demands on attention and memory. The influence of higher-order abstract features notwithstanding, the question still remains whether activation of this subregion reflects tonal processing, working memory in the auditory modality or other mediational, task-specific processes that transcend the cognitive domain. In an attempt to fractionate mediational components that may be involved in phonetic discrimination, tonal matching was compared with a control condition in which whole syllables were matched to one another (Li *et al.* 2003). The only difference between conditions is the focus of attention, either to a subpart of the syllable or to the

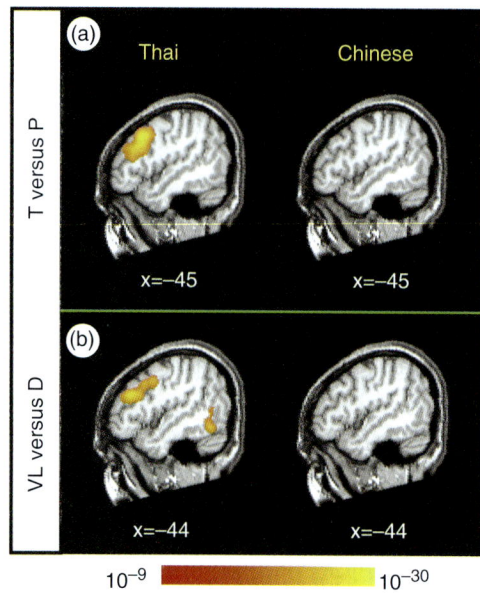

Figure 12.4 Cortical activation maps comparing discrimination of pitch and duration patterns in a speech relative to a non-speech condition for groups of Thai and Chinese listeners. (a) Spectral contrast between Thai tones (speech) and pitch (non-speech) and (b) temporal contrast between Thai vowel length (speech) and duration (non-speech). Left-sided activation foci in frontal and temporo-occipital regions occur in the Thai group only. T, tone; P, pitch; VL, vowel length; D, duration (Gandour *et al.* 2002a, b)

whole syllable itself. Selective attention to Mandarin tones elicited activation of a left-sided dorsal fronto-parietal, attention network, including a dorsolateral subregion of posterior prefrontal cortex. This cortical network in the left hemisphere is likely to reflect the engagement of attention-modulated, executive functions that are differentially sensitive to internal dimensions of a whole stimulus regardless of sensory modality or cognitive domain (Corbetta *et al.* 2000; Shaywitz *et al.* 2001; Corbetta & Shulman 2002).

12.9 Evidence that tonal categories are sufficient to account for left-hemisphere laterality effects

Because tones necessarily co-occur with real words in natural speech, it has been argued that cross-language differences in hemispheric laterality reflect nothing more than a lexical effect (Wong 2002). To isolate prelexical processing of tones, a recent study used a novel design in which hybrid stimuli were created by superimposing Thai tones onto Mandarin syllables (*tonal chimeras*) and Mandarin tones onto the same syllables (real words; Xu *et al.* 2006). Chinese and Thai speakers discriminated paired tonal contours in chimeric and Mandarin stimuli. None of the chimeric stimuli was identifiable as a Thai or Mandarin word. Thus, it was possible to compare Thai listeners' judgments of native tones in tonal chimeras with non-native tones in Mandarin words in the absence of lexical–semantic processing. In a comparison of native versus non-native tones, overlapping activity was identified in the left planum temporale. In this area, a double dissociation between language experience and neural representation of pitch occurred, such that stronger activity was elicited in response to native when compared with non-native tones (Figure 12.5). This finding suggests that cortical processing of pitch information can be shaped by language experience, and, moreover, that lateralized activation in left auditory cortex can be driven by higher-order abstract knowledge acquired through language experience. Both top-down and bottom-up processing are essential features of tonal processing. This reciprocity allows for modification of neural mechanisms involved in pitch processing based on language experience. It now appears that computations at a relatively early stage of acoustic-phonetic processing in auditory cortex can be modulated by the linguistic status of stimuli (Griffiths & Warren 2002).

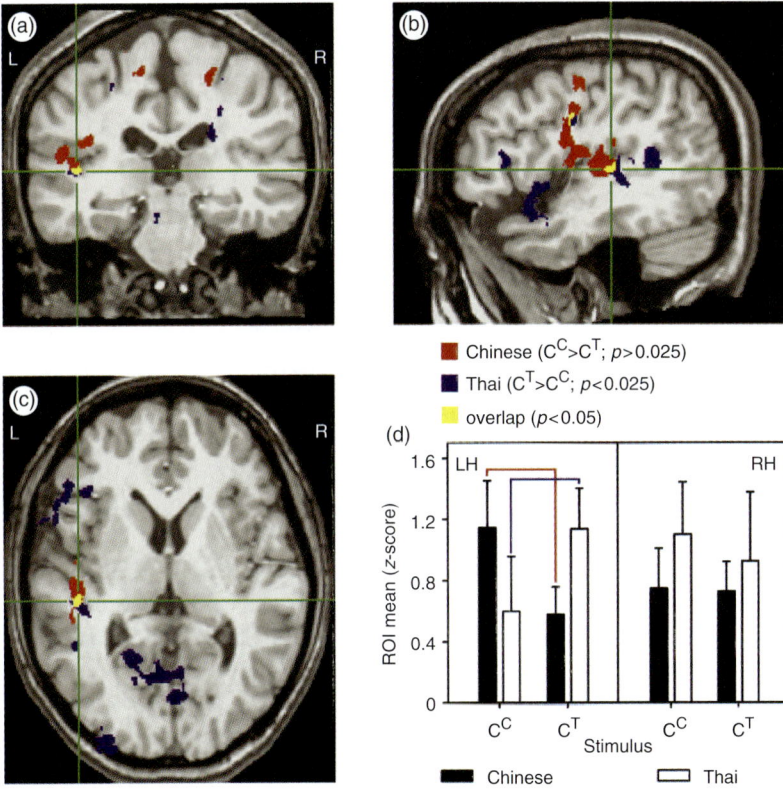

Figure 12.5 Cortical areas activated in response to discrimination of Chinese and Thai tones. A common focus of activation is indicated by the overlap (yellow) between Chinese and Thai groups in the functional activation maps. Green cross-hair lines mark the stereotactic centre coordinates for the overlapping region in the left planum temporale ((a) coronal section, $y = -25$; (b) sagittal section, $x = -44$; (c) axial section, $z = +7$). A double dissociation ((d) bar charts) between tonal processing and language experience reveals that for the Thai group, Thai tones elicit stronger activity relative to Chinese tones, whereas for the Chinese group, stronger activity is elicited by Chinese tones relative to Thai tones. C^C, Chinese tones superimposed on Chinese, i.e. Chinese words; C^T, Thai tones superimposed on Chinese syllable, i.e. *tonal chimeras*; L/LH, left hemisphere; R/RH, right hemisphere; ROI, region of interest (Xu et al. 2006).

12.10 Evidence for bilateral involvement in processing speech prosody regardless of the level of linguistic representation

In tone languages, pitch contours can be used to signal intonation as well as word meaning. A cross-language (Chinese and English) fMRI study (Gandour *et al.* 2004) was conducted to examine brain activity elicited by selective attention to Mandarin tone and intonation (statement versus question), as presented in three- and one-syllable utterance pairs. The Chinese group exhibited greater activity than the English in the left ventral aspects of the inferior parietal lobule across prosodic units and utterance lengths. It is possible that the 'categoricalness' or phonological relevance of the auditory stimuli triggers activation in this area (Jacquemot *et al.* 2003). In addition, only the Chinese group exhibited a leftward asymmetry in the intraparietal sulcus, anterior/posterior regions of the STG and frontopolar regions. However, both language groups showed activity within the STS and middle frontal gyrus (MFG) of the right hemisphere (Figure 12.6). The rightward asymmetry may reflect shared mechanisms underlying early attentional modulation in processing of complex pitch patterns irrespective of language experience (Zatorre *et al.* 1999). Albeit in the speech domain, this fronto-temporal network in the right hemisphere serves to maintain pitch information regardless of its linguistic relevance. Tone and intonation are therefore best thought of as being subserved by a mosaic of multiple local asymmetries, which allows for the possibility that different regions may be differentially weighted in laterality depending on language-, modality- and task-related features. We argue that left-hemisphere activity reflects higher-level processing of internal representations of Chinese tone and intonation, whereas right-hemisphere activity reflects lower-level, domain-independent pitch processing. Speech prosody perception is mediated primarily by the right hemisphere, but lateralized to task-dependent regions in the left hemisphere when language processing is required beyond the auditory analysis of the complex sound.

A cross-language study (Chinese and English) was also conducted to investigate the neural substrates underlying the discrimination of two *sentence-level* prosodic phenomena in Mandarin Chinese: contrastive (or emphatic) stress and intonation (Tong *et al.* 2005). Between-group comparisons revealed that the Chinese group exhibited significantly greater activity in the *left* supramarginal gyrus (SMG) and posterior middle temporal gyrus (MTG) relative to the English group for both tasks. The leftward asymmetry in the SMG and MTG, respectively, is consistent with the notion of a dorsal processing stream emanating from auditory cortex, projecting to the inferior parietal lobule, and ultimately to frontal lobe regions (Hickok & Poeppel 2004) and a ventral processing stream that projects to posterior inferior temporal lobe portions of the MTG and parts of the inferior temporal and fusiform gyri (Binder *et al.* 2000). The involvement of the posterior MTG in sentence comprehension is supported by voxel-based lesion–symptom mapping analysis (Dronkers & Ogar 2004). The left SMG serves as part of an auditory–motor integration circuit in speech perception, and supports phonological encoding–recoding processes in a variety of tasks. For both language groups, *rightward* asymmetries were observed in the middle portion of the MFG across tasks. The rightward asymmetry across tasks and language groups implicates more general auditory attention and working memory processes associated with pitch perception. Its activation is lateralized to the right hemisphere regardless of prosodic unit or language group. This area is not

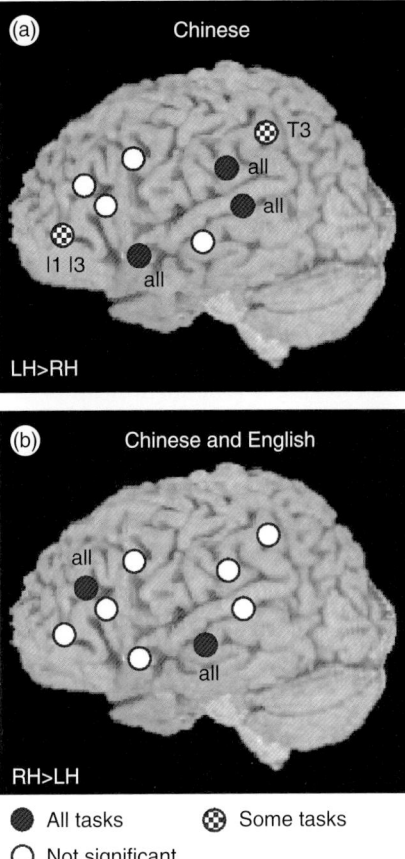

Figure 12.6 Laterality effects for ROIs in (a) the Chinese group only and in (b) both Chinese and English groups as rendered on a three-dimensional LH template for common reference. (a) In the Chinese group, the ventral aspects of the inferior parietal lobule and anterior and posterior STG are lateralized to the LH across tone and intonation tasks; the anterior MFG and intraparietal sulcus for a subset of tasks. (b) In both groups, the middle portions of the MFG and STS are lateralized to the RH across tasks. This right-sided fronto-temporal network subserves pitch processing regardless of its linguistic function. Other ROIs do not show laterality effects. MFG, middle frontal gyrus; STG, superior temporal gyrus (Gandour et al. 2004).

domain-specific since it is similarly recruited for extraction and maintenance of pitch information in processing music (Zatorre et al. 1994; Koelsch et al. 2002).

These findings from tone languages converge with imaging data on sentence intonation perception in non-tone languages (German and English). In degraded speech (prosodic information only), right-hemisphere regions are engaged predominantly in mediating slow modulation of pitch contours, whereas, in normal speech, lexical and syntactic processing elicits activity in left-hemisphere areas (Meyer et al. 2002, 2003). In high memory load tasks that result in recruitment of frontal lobe regions, a rightward asymmetry is found for prosodic stimuli and a leftward asymmetry for sentence processing (Plante et al. 2002).

12.11 Evidence from second-language learners for bilateral involvement in tonal processing

Another avenue for investigating cue- versus domain-specificity is the cortical processing of pitch in a tone language by second-language learners without any previous experience with lexical tones. As pointed out in this review, there is a lack of left-hemisphere dominance in the processing of lexical tone by English speakers who have had no prior experience with a tone language. In the case of lexical tone, there is nothing comparable in English phonology. Thus, a second-language learner of a tone language must develop novel processes for tone perception (Wang *et al*. 2003). The question then arises as to how learning a tone language as a second language affects cortical processing of pitch. The learning of Mandarin tones by adult native speakers of English was investigated by comparing cortical activation during a tone identification task before and after a two-week training procedure (Wang *et al*. 2003). Cortical effects of learning Chinese tones were associated with increased activation in the *left* posterior STG and adjacent regions, and recruitment of additional cortical areas within the *right* inferior frontal gyrus (IFG). These early cortical effects of learning lexical tones are interpreted to involve both the recruitment of pre-existing language-related areas (left STG), consistent with the findings of Golestani & Zatorre (2004) for learning of a non-pitch contrast, as well as the recruitment of additional cortical regions specialized for identification of lexical tone (right IFG). Whereas activation of the left STG is consistent with domain-specificity, the right IFG activation is consistent with cue-specificity. Native speakers of English appear to acquire a new language function, i.e. tonal identification, by enhancing existing pitch processing in the right IFG, a region that heretofore was not specialized for processing linguistic pitch on monosyllabic words. These findings suggest that cortical representations might be continuously modified as learners gain more experience with tone languages.

12.12 Evidence that knowledge of tonal categories may influence early stages of processing

While it is important to identify language-dependent processing systems at the cortical level, a complete understanding of the neural organization of language can only be achieved by viewing language processes as a set of computations or mappings between representations at different stages of processing (Hickok & Poeppel 2004). From the aforementioned haemodynamic imaging studies at the level of the cerebral cortex, it seems impossible to dismiss early processing stages as not relevant to language processing. Early stages of processing on the input side may perform computations on the acoustic data that are relevant to linguistic as well as non-linguistic auditory perception. However, to best characterize how pitch processing evolves in the time dimension, we need to turn our attention to auditory electrophysiological studies of tonal processing at the cortical level and even earlier stages of processing at the level of the brainstem.

In a study of the role of tone and segmental information in Cantonese word processing (Schirmer *et al*. 2005), the time course and amplitude of the N400 effect, a negativity that is associated with processing the semantic meaning of a word, were comparable for tone and segmental violations (cf. Brown-Schmidt & Canseco-Gonzalez 2004). The MMN, a cortical potential elicited by an odd-ball paradigm, reflects preattentive processing of auditory stimuli. Upon presentation of words with a similar tonal contour (falling) from native (Thai) and non-native (Mandarin) tone languages to Thai listeners (Sittiprapaporn *et al.* 2003), the MMN elicited by a native Thai word (/kʰa/) was greater than that elicited by a non-native Mandarin word (/ta/) and lateralized to the left auditory cortex. Both the Sittiprapaporn *et al.* (2003) and Schirmer *et al.* (2005) studies involve spoken word recognition instead of tonal processing *per se*. In a cross-language (Mandarin, English and Spanish) MEG study of spoken word recognition (Valaki *et al.* 2004), the Chinese group revealed a greater degree of late activity (greater than or equal to 200 ms) relative to the non-tone language groups in the right superior temporal and temporo-parietal regions. Since both phonological and semantic processes are involved in word recognition, we can only speculate that this increased RH activity reflects neural processes associated with the analysis of lexical tones.

The degree of linguistic specificity has yet to be determined for computations performed at the level of the auditory brainstem. The conventional wisdom appears to be that 'although there is considerable 'tuning' in the auditory system to the acoustic properties of speech, the processing operations conducted in the relay nuclei of the brainstem and thalamus are general to all sounds, and speech-specific operations probably do not begin until the signal reaches the cerebral cortex' (Scott & Johnsrude 2003, p. 100). In regard to tonal processing, the auditory brainstem provides a window for examining the effects of the linguistic status of pitch patterns at a stage of processing that is free of attention and memory demands.

To test whether early, preattentive stages of pitch processing at the brainstem level may be influenced by language experience, a recent study (Krishnan *et al.* 2005) investigated the human frequency following response (FFR), which reflects sustained phase-locked activity in a population of neural elements within the rostral brainstem. If based strictly on acoustic properties of the stimulus, FFRs in response to time-varying f_0 contours at the level of the brainstem would be expected to be homogeneous across listeners regardless of language experience. If, on the other hand, FFRs are sensitive to long-term language learning, they may be somewhat heterogeneous depending on how f_0 cues are used to signal pitch contrasts in the listener's native language. FFRs elicited by the four Mandarin tones were recorded from native speakers of Mandarin Chinese and English. Pitch strength and accuracy of pitch tracking were extracted from the FFRs using autocorrelation algorithms. In the autocorrelation functions, a peak at the fundamental period $1/f_0$ is observed for both groups, which means that phase-locked activity to the fundamental period is present regardless of language experience. However, the peak for the English group is smaller and broader relative to the Chinese group, suggesting that phase-locked activity is not as robust for English listeners. Auto-correlograms reveal a narrower band phase-locked interval for the Chinese group compared with the English, suggesting that phase-locked activity for the Chinese listeners is not only more robust, but also more accurate in tracking the f_0 contour (Figure 12.7). Both FFR pitch strength and pitch tracking were significantly greater for the Chinese group than for the English across all four Mandarin tones. These data suggest that language experience can induce neural plasticity at the brainstem level that may be enhancing or priming linguistically relevant features of the speech input. Moreover, the prominent phase-locked interval bands at $1/f_0$ were similar for stimuli that were spectrally different (speech versus non-speech) but produced equivalent pitch percepts in response to the Mandarin low falling–rising f_0 contour

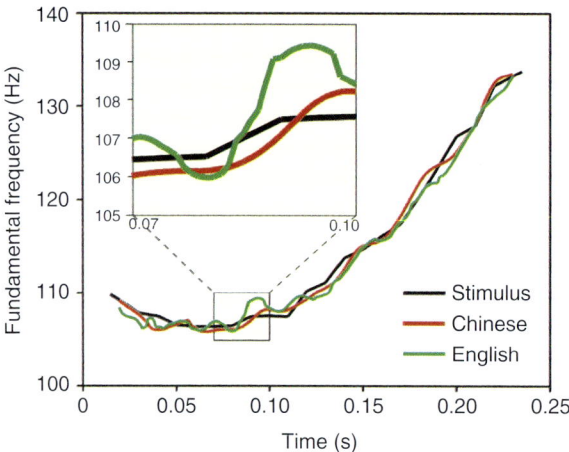

Figure 12.7 Grand-average f_0 contours of Mandarin tone 2 derived from the FFR waveforms of all subjects across both ears in the Chinese and English groups. The f_0 contour of the original speech stimulus is displayed in black. The enlarged inset shows that the f_0 contour derived from the FFR waveforms of the Chinese group more closely approximates that of the original stimulus (yi^2 'aunt') when compared with the English group. (Krishnan *et al.* 2005).

(Krishnan *et al.* 2004). We infer that experience-driven adaptive neural mechanisms are involved even at the level of the brainstem. They sharpen response properties of neurons tuned for processing pitch contours that are of special relevance to a particular language. Language-dependent operations may thus begin before the signal reaches the cerebral cortex.

12.13 Conclusion

What have we learned from reviewing this very dynamic and evolving field of research? It should be clear that a conceptualization of hemispheric specialization as being driven by just one simple dichotomy is inadequate. The data we have reviewed show that neural specializations are indeed predictable based on low-level features of the stimulus, and that they can be influenced by linguistic status. The body of evidence cannot be accounted for by claiming one or the other as the sole explanatory model. But how are they to be integrated? A complete account of language processing must allow for multiple dichotomies or scalar dimensions that either apply at different time intervals or interact within the same time interval at different cortical or subcortical levels of the brain. It is not our purpose in this discussion to describe a new model that will solve this problem, but rather to motivate researchers away from a 'zero-sum game' mentality.

Each of the models has its strengths and weaknesses. As we have seen, a strong form of either one is untenable. Indeed, several results cannot be predicted based on strong forms of either model. The cue-specific model not only predicts a large body of empirical evidence for both speech and other types of signals, but does so in a parsimonious way; it is best at explaining what is happening in the early stages of hemispheric processing, and is able to account for both speech and pitch processing. Its principal weakness is that it makes few predictions about later stages and hence does not directly take into account abstract knowledge. What needs to be done to improve it, therefore, is to add to it to take into account the influence of later stages. The domain-specific model as often presented, that is, as a monolithic view that speech is processed by a special mechanism unrelated to any other auditory process, is not very viable. But some account must be given of the many diverse phenomena we have discussed that indicate that higher-order knowledge does indeed influence patterns of neural processing. A less intransigent version of the domain-specific model would simply point out that abstract knowledge is relevant in explaining patterns of neural processing. Such a version can peacefully coexist with a modified version of the cue-specific hypothesis that focuses on early levels of processing and makes no claims about interaction occurring at later stages.

In the case of tonal processing, for example, it appears that a more complete account will emerge from consideration of general sensory-motor and cognitive processes in addition to those associated with linguistic knowledge. For example, the activation of frontal and temporal areas in the right hemisphere reflects their roles in mediating pitch processing irrespective of domain. We have also seen how various executive functions related to attention and working memory mediate prosody perception, in addition to stimulus characteristics, task demands and listeners' experience. All of these factors will need to be taken into account. Moreover, brain imaging should draw our attention to networks instead of isolated regions; in this respect, we look forward to (i) the further development of the idea that spatio-temporal patterns of neural interactions are key (McIntosh 2000) and (ii) the application of more sophisticated integrative computational models to imaging data (Horwitz *et al.* 2000).

There are several general principles that perhaps should be kept in mind in attempting to forge a solution to the dialectical issue before us. One such principle is that whenever one considers higher-order or abstract knowledge, one necessarily invokes memory mechanisms. There has been insufficient attention to the fact that any domain-specific effect is also a memory effect. As mentioned above, as soon as a sensory signal makes contact with a stored representation, there is going to be an interesting interaction based on the nature of that representation. Neurobiology would lead us to think that abstract memory representations are unlikely to be present in primary sensory cortices. Instead, abstract knowledge is probably present at multiple higher levels of cortical information processing. It is at these levels that many of the language-specific effects are likely to occur. In this respect, it is interesting to note that the effects of such learning can interact all the way down the processing stream, to include subcortical processes, as we have seen.

These kinds of interactive effects bring up a second major principle that is important in elucidating a more comprehensive model of hemispheric processing of speech, and that is the important role of top-down modulation. It is unlikely that early sensory cortices contain stored representations, but it is probable that the latter can influence early processing via top-down effects. Linguistic status, context effects, learning or attention, can modulate early processing. One promising avenue of investigation therefore would be to elucidate the efferent influences that come from amodal temporal, parietal or frontal cortices and interact with the neural computations occurring in earlier regions. A research effort devoted to understanding this specific sort of interaction is likely to prove fruitful.

In closing, we believe that the past 25 years have produced a very valuable body of evidence bearing on these issues, perhaps motivated by the creative tension inherent in the two contrasting models we have highlighted. If that is indeed the case, then these ideas have served their purpose. But we also believe that now is an appropriate time to move beyond the dichotomy. We look forward to the next 25 years.

We gratefully acknowledge research support from the Canadian Institutes of Health Research (R.J.Z.) and the US Institutes of Health (J.T.G.). We thank our former and present students and collaborators for their contributions to the work and ideas presented herein.

Acknowledgements

We wish to dedicate this paper to the memory of Peter D. Eimas (1934–2005), a devoted investigator and inspiring teacher who made significant contributions to the field of speech research, and whose thoughts influenced many of the ideas in the present text.

References

Bavelier, D., Corina, D. P. & Neville, H. J. 1998 Brain and language: a perspective from sign language. *Neuron* **21**, 275–278. (doi:10.1016/S0896-6273(00)80536-X)

Belin, P., Zilbovicius, M., Crozier, S., Thivard, L., Fontaine, A., Masure, M.-C. & Samson, Y. 1998 Lateralization of speech and auditory temporal processing. *J. Cogn. Neurosci.* **10**, 536–540. (doi:10.1162/089892998562834)

Binder, J., Frost, J., Hammeke, T., Cox, R., Rao, S. & Prieto, T. 1997 Human brain language areas identified by functional magnetic resonance imaging. *J. Neurosci.* **17**, 353–362.

Binder, J., Frost, J., Hammeke, T., Bellgowan, P., Springer, J., Kaufman, J. & Possing, J. 2000 Human temporal lobe activation by speech and nonspeech sounds. *Cereb. Cortex* **10**, 512–528. (doi:10.1093/cercor/10.5.512)

Blumstein, S. E. 1994 The neurobiology of the sound structure of language. In *The cognitive neurosciences* (ed. M. S. Gazzaniga), pp. 915–929. Cambridge, MA: MIT Press.

Boemio, A., Fromm, S., Braun, A. & Poeppel, D. 2005 Hierarchical and asymmetric temporal sensitivity in human auditory cortices. *Nat. Neurosci.* **8**, 389–395. (doi:10.1038/nn1409)

Brechmann, A. & Scheich, H. 2005 Hemispheric shifts of sound representation in auditory cortex with conceptual listening. *Cereb. Cortex* **15**, 578–587. (doi:10.1093/cercor/ bhh159)

Brown-Schmidt, S. & Canseco-Gonzalez, E. 2004 Who do you love, your mother or your horse? An event-related brain potential analysis of tone processing in Mandarin Chinese. *J. Psycholinguist. Res.* **33**, 103–135. (doi:10.1023/ B:JOPR.0000017223.98667.10)

Burns, E. M. & Ward, W. D. 1978 Categorical perception: phenomenon or epiphenomenon. Evidence from experiments in perception of melodic musical intervals. *J. Acoust. Soc. Am.* **63**, 456–468. (doi:10.1121/1.381737)

Corbetta, M. & Shulman, G. L. 2002 Control of goal-directed and stimulus-driven attention in the brain. *Nat. Rev. Neurosci.* **3**, 201–215. (doi:10.1038/nrn755)

Corbetta, M., Kincade, J. M., Ollinger, J. M., McAvoy, M. P. & Shulman, G. L. 2000 Voluntary orienting is dissociated from target detection in human posterior parietal cortex. *Nat. Neurosci.* **3**, 292–297. (doi:10.1038/73009)

Dehaene-Lambertz, G., Pallier, C., Serniclaes, W., Sprenger-Charolles, L., Jobert, A. & Dehaene, S. 2005 Neural correlates of switching from auditory to speech perception. *Neuroimage* **24**, 21–33. (doi:10.1016/j.neuroimage.2004. 09.039)

Diehl, R. L., Lotto, A. J. & Holt, L. L. 2004 Speech perception. *Annu. Rev. Psychol.* **55**, 149–179. (doi:10. 1146/annurev.psych.55.090902.142028)

Dronkers, N. & Ogar, J. 2004 Brain areas involved in speech production. *Brain* **127**, 1461–1462. (doi:10.1093/brain/ awh233)

Efron, R. 1963 Temporal perception, aphasia and deja vu. *Brain* **86**, 403–423. (doi:10.1093/brain/86.3.403)

Eimas, P. D. 1963 The relation between identification and discrimination along speech and non-speech continua. *Lang. Speech* **6**, 206–217.

Eimas, P., Siqueland, E., Jusczyk, P. & Vigorito, J. 1971 Speech perception in infants. *Science* **171**, 303–306. (doi:10.1126/science.171.3968.303)

Finney, E. M., Fine, I. & Dobkins, K. R. 2001 Visual stimuli activate auditory cortex in the deaf. *Nat. Neurosci.* **4**, 1171–1173. (doi:10.1038/nn763)

Gandour, J. 1998 Aphasia in tone languages. In *Aphasia in atypical populations* (eds P. Coppens, A. Basso & Y. Lebrun), pp. 117–141. Hillsdale, NJ: Lawrence Erlbaum.

Gandour, J. 2006a Tone: neurophonetics. In *Encyclopedia of language and linguistics* (ed. K. Brown), pp. 751–760, 2nd edn. Oxford, UK: Elsevier.

Gandour, J. 2006b Brain mapping of Chinese speech prosody. In *Handbook of East Asian psycholinguistics: Chinese* (eds P. Li, L. H. Tan, E. Bates & O. Tzeng), pp. 308–319. New York, NY: Cambridge University Press.

Gandour, J., Wong, D., Hsieh, L., Weinzapfel, B., Van Lancker, D. & Hutchins, G. D. 2000 A PET cross-linguistic study of tone perception. *J. Cogn. Neurosci.* **12**, 207–222. (doi:10.1162/089892900561841)

Gandour, J., Wong, D., Lowe, M., Dzemidzic, M., Satthamnuwong, N., Tong, Y & Li, X. 2002a A cross-linguistic FMRI study of spectral and temporal cues underlying phonological processing. *J. Cogn. Neurosci.* **14**, 1076–1087. (doi:10.1162/089892902320474526)

Gandour, J., Wong, D., Lowe, M., Dzemidzic, M., Satthamnuwong, N., Tong, Y & Lurito, J. 2002b Neural circuitry underlying perception of duration depends on language experience. *Brain Lang.* **83**, 268–290. (doi:10.1016/S0093-934X(02)00033-0)

Gandour, J., Dzemidzic, M., Wong, D., Lowe, M., Tong, Y, Hsieh, L., Satthamnuwong, N. & Lurito, J. 2003 Temporal integration of speech prosody is shaped by language experience: an fMRI study. *Brain Lang.* **84**, 318–336. (doi:10.1016/S0093-934X(02)00505-9)

Gandour, J., Tong, Y, Wong, D., Talavage, T, Dzemidzic, M., Xu, Y, Li, X. & Lowe, M. 2004 Hemispheric roles in the perception of speech prosody. *Neuroimage* **23**, 344–357. (doi:10.1016/j.neuroimage.2004.06.004)

Giraud, A.-L., Lorenzi, C, Ashburner, J., Wable, J., Johnsrude, I., Frackowiak, R. & Kleinschmidt, A. 2000 Representation of the temporal envelope of sounds in the human brain. *J. Neurophysiol.* **84**, 1588–1598.

Golestani, N. & Zatorre, R. J. 2004 Learning new sounds of speech: reallocation of neural substrates. *Neuroimage* **21**, 494–506. (doi:10.1016/j.neuroimage.2003.09.071)

Gougoux, F., Zatorre, R. J., Lassonde, M., Voss, P. & Lepore, F. 2005 Functional neuroimaging of sound localization with early blind persons. *PLoS Biol.* **3**, 324–333. (doi:10.1371/journal.pbio.0030027)

Griffiths, T. D. & Warren, J. D. 2002 The planum temporale as a computational hub. *Trends Neurosci.* **25**, 348–353. (doi:10.1016/S0166-2236(02)02191-4)

Griffiths, T. D., Johnsrude, I. S., Dean, J. L. & Green, G. G. R. 1999 A common neural substrate for the analysis of pitch and duration pattern in segmented sound? *NeuroReport* **10**, 3825–3830. (doi:10.1097/00001756-199912160-00019)

Hall, D. A., Johnsrude, I. S., Haggard, M. P., Palmer, A. R., Akeroyd, M. A. & Summerfield, A. Q. 2002 Spectral and temporal processing in human auditory cortex. *Cereb. Cortex* **12**, 140–149. (doi:10.1093/cercor/12.2.140)

Halpern, A. R. & Zatorre, R. J. 1999 When that tune runs through your head: a PET investigation of auditory imagery for familiar melodies. *Cereb. Cortex* **9**, 697–704. (doi:10.1093/cercor/9.7.697)

Halpern, A. R., Zatorre, R. J., Bouffard, M. & Johnson, J. A. 2004 Behavioral and neural correlates of perceived and imagined musical timbre. *Neuropsychologia* **42**, 1281–1292. (doi:10.1016/j.neuropsychologia.2003.12.017)

Hickok, G. & Poeppel, D. 2000 Towards a functional neuroanatomy of speech perception. *Trends Cogn. Sci.* **4**, 131–138. (doi:10.1016/S1364-6613(00)01463-7)

Hickok, G. & Poeppel, D. 2004 Dorsal and ventral streams: a framework for understanding aspects of the functional anatomy of language. *Cognition* **92**, 67–99. (doi:10.1016/ j. cognition.2003.10.011)

Hickok, G., Bellugi, U. & Klima, E. 1998 The neural organization of language: evidence from sign language aphasia. *Trends Cogn. Sci.* **2**, 129–136. (doi:10.1016/ S1364–6613(98)01154–1)

Horwitz, B., Friston, K. & Taylor, J. 2000 Neural modeling and functional brain imaging: an overview. *Neural Netw.* **13**, 829–846. (doi:10.1016/S0893–6080(00)00062–9)

Hsieh, L., Gandour, J., Wong, D. & Hutchins, G. D. 2001 Functional heterogeneity of inferior frontal gyms is shaped by linguistic experience. *Brain Lang.* **76**, 227–252. (doi:10. 1006/ brln.2000.2382)

Hugdahl, K., Bronnick, K., Kyllingsbaek, S., Law, I., Gade, A. & Paulson, O. 1999 Brain activation during dichotic presentations of consonant-vowel and musical instrument stimuli: a 15O-PET study. *Neuropsychologia* **37**, 431–440. (doi:10.1016/S0028–3932(98)00101–8)

Jacquemot, C, Pallier, C, LeBihan, D., Dehaene, S. & Dupoux, E. 2003 Phonological grammar shapes the auditory cortex: a functional magnetic resonance imaging study. *J. Neurosci.* **23**, 9541–9546.

Jamison, H. L., Watkins, K. E., Bishop, D. V. M. & Matthews, P. M. 2006 Hemispheric specialization for processing auditory nonspeech stimuli. *Cereb. Cortex.* **16**, 1266–1275. (doi:10.1093/cercor/bhj068)

Joanisse, M. F. & Gati, J. S. 2003 Overlapping neural regions for processing rapid temporal cues in speech and nonspeech signals. *NeuroImage* **19**, 64.

Johnsrude, I. S., Zatorre, R. J., Milner, B. A. & Evans, A. C. 1997 Left-hemisphere specialization for the processing of acoustic transients. *Neuroreport* **8**, 1761–1765. (doi:10. 1097/00001756–199705060–00038)

Johnsrude, I. S., Penhune, V. B. & Zatorre, R. J. 2000 Functional specificity in the right human auditory cortex for perceiving pitch direction. *Brain* **123**, 155–163. (doi: 10.1093/brain/123.1.155)

Klein, D., Milner, B. A., Zatorre, R. J., Zhao, V. & Nikelski, E. J. 1999 Cerebral organization in bilinguals: a PET study of Chinese-English verb generation. *Neuroreport* **10**, 2841–2846. (doi:10.1097/00001756–199909090–00026)

Klein, D., Zatorre, R., Milner, B. & Zhao, V. 2001 A cross-linguistic PET study of tone perception in Mandarin Chinese and English speakers. *NeuroImage* **13**, 646–653. (doi:10.1006/nimg.2000.0738)

Kluender, K. R., Diehl, R. L. & Killeen, P. R. 1987 Japanese quail can learn phonetic categories. *Science* **237**, 1195–1197. (doi:10.1126/science.3629235)

Koelsch, S., Gunter, T. C., von Cramon, D. Y., Zysset, S., Lohmann, G. & Friederici, A. D. 2002 Bach speaks: a cortical 'language-network' serves the processing of music. *NeuroImage* **17**, 956–966. (doi:10.1016/S1053–8119(02)91154–7)

Krishnan, A., Xu, Y., Gandour, J. T. & Cariani, P. A. 2004 Human frequency-following response: representation of pitch contours in Chinese tones. *Hear. Res.* **189**, 1–12. (doi:10.1016/S0378–5955(03)00402–7)

Krishnan, A., Xu, Y, Gandour, J. & Cariani, P. 2005 Encoding of pitch in the human brainstem is sensitive to language experience. *Brain Res. Cogn. Brain Res.* **25**, 161–168. (doi:10.1016/j.cogbrainres.2005.05.004)

Kuhl, P. K. 2000 A new view of language acquisition. *Proc. Natl Acad. Sci. USA* **97**, 11850–11857. (doi:10.1073/ pnas.97.22.11850)

Kuhl, P. K. & Miller, J. D. 1975 Speech perception by the chinchilla: voiced–voiceless distinction in alveolar plosive consonants. *Science* **190**, 69–72. (doi:10.1126/science. 1166301)

Kujala, T, Alho, K. & Näätänen, R. 2000 Cross-modal reorganization of human cortical functions. *Trends Neurosci.* **23**, 115–120. (doi:10.1016/S0166–2236(99) 01504–0)

Li, X., Gandour, J., Talavage, T., Wong, D., Dzemidzic, M., Lowe, M. & Tong, Y. 2003 Selective attention to lexical tones recruits left dorsal frontoparietal network. *Neuroreport* **14**, 2263–2266. (doi: 10.1097/00001756–200312020–00025)

Liberman, A. M. & Mattingly, I. G. 1985 The motor theory of speech perception revised. *Cognition* **21**, 1–36. (doi:10.1016/0010-0277(85)90021-6)

Liberman, A. M. & Whalen, D. H. 2000 On the relation of speech to language. *Trends Cogn. Sci.* **4**, 187–196. (doi:10.1016/S1364-6613(00)01471-6)

Liebenthal, E., Binder, J. R., Spitzer, S. M., Possing, E. T. & Medler, D. A. 2005 Neural substrates of phonemic perception. *Cereb. Cortex* **15**, 1621–1631. (doi:10.1093/cercor/bhi040)

Liégeois-Chauvel, C, de Graaf, J., Laguitton, V. & Chauvel, P. 1999 Specialization of left auditory cortex for speech perception in man depends on temporal coding. *Cereb. Cortex* **9**, 484–496. (doi: 10.1093/cercor/9.5.484)

MacSweeney, M., Woll, B., Campbell, R., McGuire, P. K., David, A. S., Williams, S. C. R., Suckling, J., Calvert, G. A. & Brammer, M. J. 2002 Neural systems underlying British sign language and audio-visual English processing in native users. *Brain* **125**, 1583–1593. (doi:10.1093/brain/awf153)

McIntosh, A. 2000 Towards a network theory of cognition. *Neural Netw.* **13**, 861–870. (doi:10.1016/S0893-6080 (00)00059-9)

Meyer, M., Alter, K., Friederici, A. D., Lohmann, G. & von Cramon, D. Y. 2002 FMRI reveals brain regions mediating slow prosodic modulations in spoken sentences. *Hum. Brain Mapp.* **17**, 73–88. (doi:10.1002/hbm.l0042)

Meyer, M., Alter, K. & Friederici, A. 2003 Functional MR imaging exposes differential brain responses to syntax and prosody during auditory sentence comprehension. *J. Neurolinguist.* **16**, 277–300. (doi: 10.1016/S0911-6044 (03)00026-5)

Miller, J. D., Wier, C. C, Pastore, R. E., Kelly, W J. & Dooling, R. J. 1976 Discrimination and labelling of noise-buzz sequences with varying noise-lead times: an example of categorical perception. *J. Acoust. Soc. Am.* **60**, 410–417. (doi:10.1121/1.381097)

Milner, B. A. 1962 Laterality effects in audition. In *Interhemispheric relations and cerebral dominance* (ed. V. Mountcastle), pp. 177–195. Baltimore, MD: Johns Hopkins Press.

Minagawa-Kawai, Y., Mori, K., Furuya, I., Hayashi, R. & Sato, Y. 2002 Assessing cerebral representations of short and long vowel categories by NIRS. *Neuroreport* **13**, 581–584. (doi:10.1097/00001756-200204160-00009)

Möttönen, R., Calvert, G. A., Jääskeläinen, I. P., Matthews, P. M., Thesen, T., Tuomainen, J. & Sams, M. 2006 Perceiving identical sounds as speech or non-speech modulates activity in the left posterior superior temporal sulcus. *NeuroImage* **30**, 563–569. (doi:10.1016/j.neuroimage.2005.10.002)

Näätänen, R. et al. 1997 Language-specific phoneme representations revealed by electric and magnetic brain responses. *Nature* **385**, 432–434. (doi:10.1038/385432a0)

Narain, C, Scott, S. K., Wise, R. J. S., Rosen, S., Leff, A., Iversen, S. D. & Matthews, P. M. 2003 Defining a left-lateralized response specific to intelligible speech using fMRI. *Cereb. Cortex* **13**, 1362–1368. (doi:10.1093/cercor/ bhgO83)

Nenonen, S., Shestakova, A., Huotilainen, M. & Naatanen, R. 2003 Linguistic relevance of duration within the native language determines the accuracy of speech-sound duration processing. *Brain Res. Cogn. Brain Res.* **16**, 492–95. (doi:10.1016/S0926-6410(03)00055-7)

Patel, A. & Balaban, E. 2001 Human pitch perception is reflected in the timing of stimulus-related cortical activity. *Nat. Neurosci.* **4**, 839–844. (doi:10.1038/90557)

Patterson, R. D., Uppenkamp, S., Johnsrude, I. S. & Griffiths, T. D. 2002 The processing of temporal pitch and melody information in auditory cortex. *Neuron* **36**, 767–776. (doi:10.1016/S0896-6273(02)01060-7)

Penhune, V. B., Zatorre, R. J. & Evans, A. C. 1998 Cerebellar contributions to motor timing: a PET study of auditory and visual rhythm reproduction. *J. Cogn. Neurosci.* **10**, 752–765. (doi:10.1162/089892998563149)

Perry, D. W., Zatorre, R. J., Petrides, M., Alivisatos, B., Meyer, E. & Evans, A. C. 1999 Localization of cerebral activity during simple singing. *NeuroReport* **10**, 3979–3984.

Petersen, S., Fox, P., Posner, M., Mintun, M. & Raichle, M. 1988 Positron emission tomographic studies of the cortical anatomy of single-word processing. *Nature* **331**, 585–589. (doi:10.1038/331585a0)

Petito, L. A., Zatorre, R. J., Gauna, K., Nikelski, E. J., Dostie, D. & Evans, A. C. 2000 Speech-like cerebral activity in profoundly deaf people processing signed languages: implications for the neural basis of human language. *Proc. Natl Acad. Sci. USA* **97**, 13 961–13 966. (doi:10.1073/pnas.97.25.13961)

Phillips, D. P. & Farmer, M. E. 1990 Acquired word deafness and the temporal grain of sound representation in the primary auditory cortex. *Behav. Brain Res.* **40**, 84–90. (doi:10.1016/0166-4328(90)90001-U)

Pisoni, D. B. 1977 Identification and discrimination of the relative onset time of two component tones: implications for voicing perception in stops. *J. Acoust. Soc. Am.* **61**, 1352–1361. (doi:10.1121/1.381409)

Plante, E., Creusere, M. & Sabin, C. 2002 Dissociating sentential prosody from sentence processing: activation interacts with task demands. *NeuroImage* **17**, 401–410. (doi:10.1006/nimg.2002.1182)

Poeppel, D. 2003 The analysis of speech in different temporal integration windows: cerebral lateralization as 'asymmetric sampling in time'. *Speech Commun.* **41**, 245–255. (doi:10.1016/S0167-6393(02)00107-3)

Remez, R. E., Rubin, P. E., Pisoni, D. B. & Carrell, T. D. 1981 Speech perception without traditional speech cues. *Science* **212**, 947–949. (doi:10.1126/science.7233191)

Robin, D. A., Tranel, D. & Damasio, H. 1990 Auditory perception of temporal and spectral events in patients with focal left and right cerebral lesions. *Brain Lang.* **39**, 539–555. (doi:10.1016/0093-934X(90)90161-9)

Sadato, N., Pascual-Leone, A., Grafman, J., Ibanez, V., Deiber, M. P., Dold, G. & Hallett, M. 1996 Activation of the primary visual cortex by Braille reading in blind subjects. *Nature* **380**, 526–528. (doi:10.1038/380526a0)

Schirmer, A., Tang, S. L., Penney, T. B., Gunter, T. C. & Chen, H. C. 2005 Brain responses to segmentally and tonally induced semantic violations in Cantonese. *J. Cogn. Neurosci.* **17**, 1–12. (doi: 10.1162/0898929052880057)

Schönwiesner, M., Rubsamen, R. & von Cramon, D. Y. 2005 Hemispheric asymmetry for spectral and temporal processing in the human antero-lateral auditory belt cortex. *Eur. J. Neurosci.* **22**, 1521–1528. (doi:10.1111/j.1460-9568. 2005.04315.x)

Scott, S. K. & Johnsrude, I. S. 2003 The neuroanatomical and functional organization of speech perception. *Trends Neurosd.* **26**, 100. (doi:10.1016/S0166-2236(02)00037-1)

Scott, S. K. & Wise, R. S. J. 2004 The functional neuroanatomy of prelexical processing in speech perception. *Cognition* **92**, 13–45. (doi:10.1016/j.cognition.2002. 12.002)

Scott, S. K., Blank, C. C., Rosen, S. & Wise, R. J. S. 2000 Identification of a pathway for intelligible speech in the left temporal lobe. *Brain* **123**, 2400–2406. (doi:10.1093/brain/ 123.12.2400)

Shaywitz, B. A. et al. 2001 The functional neural architecture of components of attention in language-processing tasks. *NeuroImage* **13**, 601–612. (doi: 10.1006/nimg.2000.0726)

Shtyrov, Y, Pihko, E. & Pulvermuller, F. 2005 Determinants of dominance: is language laterality explained by physical or linguistic features of speech? *NeuroImage* **27**, 37–47. (doi: 10.1016/j.neuroimage. 2005.02.003)

Sidtis, J. J. & Volpe, B. T. 1988 Selective loss of complex-pitch or speech discrimination after unilateral lesion. *Brain Lang.* **34**, 235–245. (doi:10.1016/0093-934X(88)90135-6)

Sittiprapaporn, W., Chindaduangratn, C., Tervaniemi, M. & Khotchabhakdi, N. 2003 Preattentive processing of lexical tone perception by the human brain as indexed by the mismatch negativity paradigm. *Ann. NY Acad. Sci.* **999**, 199–203. (doi:10.1196/annals.1284.029)

Swisher, L. & Hirsh, I. J. 1972 Brain damage and the ordering of two temporally successive stimuli. *Neuropsychologia* **10**, 137–152. (doi:10.1016/0028-3932(72)90053-X)

Tallal, P., Miller, S. & Fitch, R. 1993 Neurobiological basis of speech: a case for the preeminence of temporal processing. *Ann. NY Acad. Sci.* **682**, 27–7. (doi:10.1111/j.1749-6632.1993.tb22957.x)

Tallal, P., Miller, S. L., Bedi, G., Byma, G., Wang, X., Nagarajan, S. S., Schreiner, C., Jenkins, W. M. & Merzenich, M. M. 1996 Language comprehension in language-learning impaired children improved with acoustically modified speech. *Science* **271**, 81–84. (doi: 10.1126/science.271.5245.81)

Tervaniemi, M., Medvedev, S., Alho, K., Pakhomov, S., Roudas, M.j van Zuijen, T. & Näätänen, R. 2000 Lateralized automatic auditory processing of phonetic versus musical information: a PET study. *Hum. Brain Mapp.* **10**, 74–79. (doi:10.1002/(SICI)1097–0193(200006) 10:2 < 74::AID-HBM30 > 3.0.CO;2–2)

Tong, Y, Gandour, J., Talavage, T, Wong, D., Dzemidzic, M., Xu, Y, Li, X. & Lowe, M. 2005 Neural circuitry underlying sentence-level linguistic prosody. *NeuroImage* **28**, 417–428. (doi:10.1016/j.neuroimage.2005.06.002)

Valaki, C. E., Maestu, F., Simos, P. G., Zhang, W, Fernandez, A., Amo, C. M., Ortiz, T. M. & Papanicolaou, A. C. 2004 Cortical organization for receptive language functions in Chinese, English, and Spanish: a cross-linguistic MEG study. *Neuropsychologia* **42**, 967–979. (doi:10.1016/j.neuropsychologia.2003. 11.019)

von Steinbüchel, N. 1998 Temporal ranges of central nervous processing: clinical evidence. *Exp. Brain Res.* **123**, 220–233. (doi:10.1007/s002210050564)

Vouloumanos, A., Kiehl, K. A., Werker, J. F. & Liddle, P. F. 2001 Detection of sounds in the auditory stream: event-related fMRI evidence for differential activation to speech and nonspeech. *J. Cogn. Neurosci.* **13**, 994–1005. (doi:10. 1162/089892901753165890)

Wang, Y, Sereno, J. A., Jongman, A. & Hirsch, J. 2003 FMRI evidence for cortical modification during learning of Mandarin lexical tone. *J. Cogn. Neurosci.* **15**, 1019–1027. (doi:10.1162/0898 92903770007407)

Warrier, C. M. & Zatorre, R. J. 2004 Right temporal cortex is critical for utilization of melodic contextual cues in a pitch constancy task. *Brain* **127**, 1616–1625. (doi:10.1093/ brain/awh183)

Weeks, R., Horwitz, B., Aziz-Sultan, A., Tian, B., Wessinger, C., Cohen, L., Hallett, M. & Rauschecker, J. 2000 A positron emission tomographic study of auditory localization in the congenitally blind. *J. Neurosci.* **20**, 2664–2672.

Whalen, D. H. & Liberman, A. L. 1987 Speech perception takes precedence over nonspeech perception. *Science* **237**, 169–171. (doi:10.1126/science.3603014)

Wong, P. 2002 Hemispheric specialization of linguistic pitch patterns. *Brain Res. Bull.* **59**, 83–95.

Wong, P. C, Parsons, L. M., Martinez, M. & Diehl, R. L. 2004 The role of the insular cortex in pitch pattern perception: the effect of linguistic contexts. *J. Neurosci.* **24**, 9153–9160. (doi:10.1523/JNEUROSCI.2225–04.2004)

Xu, Y., Gandour, J., Talavage, T, Wong, D., Dzemidzic, M., Tong, Y., Li, X. & Lowe, M. 2006 Activation of the left planum temporale in pitch processing is shaped by language experience. *Hum. Brain Mapp.* **27**, 173–183. (doi:10.1002/hbm.20176)

Yip, M. 2003 *Tone*. New York, NY: Cambridge University Press.

Zaehle, T., Wustenberg, T., Meyer, M. & Jancke, L. 2004 Evidence for rapid auditory perception as the foundation of speech processing: a sparse temporal sampling fMRI study. *Eur. J. Neurosci.* **20**, 2447–2456. (doi:10.1111/ j.1460–9568.2004.03687.x)

Zatorre, R. J. 1983 Category–boundary effects and speeded sorting with a harmonic musical interval continuum. *J. Exp. Psychol. Hum. Percept. Perform.* **9**, 739–752. (doi:10.1037/0096–1523.9.5.739)

Zatorre, R. J. 1988 Pitch perception of complex tones and human temporal-lobe function. *J. Acoust. Soc. Am.* **84**, 566–572. (doi:10.1121/1.396834)

Zatorre, R. J. & Belin, P. 2001 Spectral and temporal processing in human auditory cortex. *Cereb. Cortex* **11**, 946–953. (doi:10.1093/cercor/11.10.946)

Zatorre, R. J. & Halpern, A. R. 1993 Effect of unilateral temporal-lobe excision on perception and imagery of songs. *Neuropsychologia* **31**, 221–232. (doi:10.1016/0028–3932(93)90086-F)

Zatorre, R. J. & Samson, S. 1991 Role of the right temporal neocortex in retention of pitch in auditory short-term memory. *Brain* **114**, 2403–2417. (doi:10.1093/brain/114. 6.2403)

Zatorre, R. J., Evans, A. C., Meyer, E. & Gjedde, A. 1992 Lateralization of phonetic and pitch processing in speech perception. *Science* **256**, 846–849. (doi:10.1126/science. 1589767) Zatorre, R. J., Evans, A. C. & Meyer, E. 1994 Neural mechanisms underlying melodic perception and memory for pitch. *J. Neurosci.* **14**, 1908–1919.

Zatorre, R. J., Meyer, E., Gjedde, A. & Evans, A. C. 1996 PET studies of phonetic processing of speech: review, replication, and re-analysis. *Cereb. Cortex* **6**, 21–30. (doi: 10.1093/cercor/6.1.21)

Zatorre, R. J., Mondor, T. A. & Evans, A. C. 1999 Auditory attention to space and frequency activates similar cerebral systems. *NeuroImage* **10**, 544–554. (doi:10.1006/nimg. 1999.0491)

Zatorre, R. J., Belin, P. & Penhune, V. B. 2002 Structure and function of auditory cortex: music and speech. *Trends Cogn. Sci.* **6**, 37–6. (doi:10.1016/S1364-6613(00)01816-7)

13

Language processing in the natural world
Michael K. Tanenhaus and Sarah Brown-Schmidt

The authors argue that a more complete understanding of how people produce and comprehend language will require investigating real-time spoken-language processing in natural tasks, including those that require goal-oriented unscripted conversation. One promising methodology for such studies is monitoring eye movements as speakers and listeners perform natural tasks. Three lines of research that adopt this approach are reviewed: (i) spoken word recognition in continuous speech, (ii) reference resolution in real-world contexts, and (iii) real-time language processing in interactive conversation. In each domain, results emerge that provide insights which would otherwise be difficult to obtain. These results extend and, in some cases, challenge standard assumptions about language processing.

Keywords: eye movements; language comprehension; spoken word recognition; conversation; parsing; speech perception

13.1 Introduction

As people perform everyday tasks involving vision, such as reading a newspaper, looking for the car keys or making a cup of tea, they frequently shift their attention to task-relevant regions of the visual world. These shifts of attention are accompanied by shifts in gaze, accomplished by ballistic eye movements known as *saccades*, which bring the attended region into the central area of the fovea, where visual acuity is greatest. The pattern and timing of saccades, and the resulting fixations, are one of the most widely used response measures in the brain and cognitive sciences, providing important insights into the functional and neural mechanisms underlying reading, attention, perception, memory (for reviews see Rayner (1998) and Liversedge and Findlay (2000)) and, most recently, visual behaviour in natural tasks (Land 2004; Hayhoe and Ballard 2005).

During the last decade, we have been using saccadic eye movements to investigate spoken-language processing in relatively natural tasks that combine perception and action (Tanenhaus *et al.* 1995; see Cooper (1974) for an important precursor). In these studies, participants' fixations are monitored, typically using a lightweight head-mounted eye tracker, as they follow instructions to manipulate objects or participate in a dialogue about a task-relevant workspace—the 'visual world'. These methods have made it possible to monitor real-time language comprehension at a grain fine enough to reveal subtle effects of sub-phonetic processing while using tasks as natural as unscripted interactive conversation. Before describing this work in more detail, we briefly sketch the motivation for studying language processing in the natural world.

Until recently, most psycholinguistic research on spoken-language comprehension could be divided into one of two traditions, each with its roots in seminal work from the 1960s (Clark 1992, 1996), and with its own characteristic theoretical concerns and dominant methodologies. The *language-as-product* tradition has its roots in George Miller's synthesis of the then-emerging information processing paradigm and Chomsky's theory of transformational grammar (Miller 1962; Miller and Chomsky 1963). The product tradition emphasizes the individual cognitive processes by which listeners recover linguistic representations—the 'products' of language comprehension. Psycholinguistic research within the product tradition typically examines moment-by-moment processes in real-time language processing, using fine-grained reaction time measures and carefully controlled stimuli.

The motivation for these measures comes from two observations. First, speech unfolds as a sequence of rapidly changing acoustic events. Second, experimental studies show that listeners make provisional commitments at multiple levels of representations as the input arrives (Marslen-Wilson 1973, 1975). Evaluating models of how linguistic representations are accessed, constructed and integrated given a continuously unfolding input requires data that can be obtained only by response measures that are closely time-locked to the input as it unfolds over time. We can illustrate this point by noting that one consequence of the combination of sequential input and time-locked processing is that the processing system is continuously faced with temporary ambiguity. For example, the initial portion of the spoken word *beaker* is temporarily consistent with many potential lexical candidates, e.g. *beaker, beet, beep, beetle* and *beagle*. Similarly, as the utterance, *Put the apple on the towel into the box* unfolds, the phrase, *on the towel*, is temporarily consistent with at least two mutually incompatible possibilities; *on the towel* could introduce a goal argument for the verb *put* (the location where the apple is to be put) or it could modify the theme argument, *the apple*, specifying the location of the theme (on the towel). A similar argument for the importance of time-locked response measures holds for studies of language production where the speaker must rapidly map thoughts onto sequentially produced linguistic forms (Levelt *et al*. 1999). The *language-as-action* tradition has its roots in work by the Oxford philosophers of language use, e.g. Grice (1957), Austin (1962) and Searle (1969), and work on conversational analysis, e.g. Schegloff and Sacks (1973). The action tradition focuses on how people use language to perform acts in conversation, the most basic form of language use. Psycholinguistic research within the action tradition typically focuses on interactive conversation involving two or more participants engaged in a cooperative task, typically with real-world referents and well-defined behavioural goals. One reason for the focus on these types of tasks and situations is that many aspects of utterances in a conversation can be understood only with respect to the context of the language use, which includes the time, place and participant's conversational goals, as well as the collaborative processes that are intrinsic to conversation. Moreover, many characteristic features of conversation emerge only when interlocutors have joint goals and when they participate in a dialogue both as a speaker and an addressee.

Table 13.1 illustrates some of these features using a fragment of dialogue from a study by Brown-Schmidt and colleagues (Brown-Schmidt *et al*. 2005; Brown-Schmidt and Tanenhaus 2008). Pairs of participants, separated by a curtain, worked together to arrange blocks in matching configurations and confirm those configurations. The excerpt contains many well-documented aspects of task-oriented dialogue, including fragments that can only be understood as combinations of utterances between two speakers, false starts, overlapping speech (marked by asterisks) and negotiated referential terms (e.g. *vertically* meaning up and down).

Detailed analyses of the participants' linguistic behaviour and actions in cooperative tasks have provided important insights into how interlocutors track information to achieve successful communication (Clark 1992, 1996). They also demonstrate that many

Table 13.1 Excerpt of dialogue taken from Brown-Schmidt *et al.* (2005).

Speaker	utterance
1	*ok, ok I got it* ele … ok
2	alright, *hold on*, I got another easy piece
1	*I got a* well wait I got a green piece right above that
2	above this piece?
1	well not exactly right above it
2	it can't be above it
1	it is to the … it' doesn't wanna fit in with the cardboard
2	it is to the right, right?
1	yup
2	w- how? *where*
1	*it is* kinda line up with the two holes
2	line 'em right next to each other?
1	yeah, vertically
2	vertically, meaning?
1	up and down
2	up and down

aspects of communication, establishing successful reference, for instance, are not simply an individual cognitive process; they are arrived at as the result of coordinated actions among two or more individuals across multiple linguistic exchanges (Clark and Wilkes-Gibbs 1986). However, because they interfere with interactive conversation, researchers in the action tradition have for the most part eschewed the time-locked response measures favoured by researchers in the product tradition (but see Marslen-Wilson *et al.* 1982; Brennan 2005). For example, some of the widely used experimental paradigms for examining real-time spoken-language comprehension require: (i) asking a participant to make a metalinguistic judgement while monitoring the speech input for a linguistic unit such as a phoneme, syllable or word, (ii) measuring a response to a visual target that is presented on a screen during a sentence, or (iii) monitoring EEG activity while the participant's head remains relatively still. None of these procedures can be used without disrupting interactive conversation. Thus, little is known about the moment-by-moment processes that underlie interactive language use.

13.2 Why study real-time language processing in natural tasks?

While it is tempting to view the product and action traditions as complementary, research in the product tradition examines the early perceptual and cognitive processes that build linguistic representations; research in the action tradition focuses on subsequent cognitive and social-cognitive processes that build upon these representations—we believe that this perspective is misguided. An increasing body of evidence in neuroscience demonstrates that even low-level perceptual processes are affected by task goals. Behavioural context, including attention and intention, affect basic perceptual processes in vision (Gandhi *et al.* 1998; Colby and Goldberg 1999). In addition, brain systems involved in perception and action are implicated in the earliest moments of language processing

(Pulvermüller *et al.* 2001). Thus, studies that examine sub-processes in isolation, without regard to other subsystems, and broader behavioural context, are likely to be misleading. Moreover, it is becoming clear that at least some aspects of conceptual representations are grounded in perception and action. The language used in interactive conversation is also dramatically different than the carefully scripted language that is studied in the product tradition. The characteristics of natural language illustrated in the excerpt from Brown-Schmidt *et al.* (2005) are ubiquitous, yet they are rarely studied outside of the action tradition. On the one hand, they raise important challenges for models of real-time language processing within the product tradition, which are primarily crafted to handle fluent, fully grammatical well-formed language. On the other hand, formulating and evaluating explicit mechanistic models of how and why these conversational phenomena arise requires data that necessitate real-time methods.

Moreover, the theoretical constructs developed within each tradition sometimes offer competing explanations for phenomena that have been the primary concern of the other tradition. For example, the product-based construct of *priming* provides an alternative mechanistic explanation for phenomena such as lexical and syntactic entrainment (the tendency for interlocutors to use the same words and/or the same syntactic structures). A priming account does not require appeal to the action-based claim that such processes reflect active construction of common ground between interlocutors (cf. Pickering and Garrod 2004). Likewise, the observation that speakers articulate lower frequency words more slowly and more carefully, which has been used within the action tradition to argue for speaker adaptation to the needs of the listener, has a plausible product-based mechanistic explanation – greater attentional resources are required to sequence and output lower frequency forms.

Conversely, the interactive nature of conversation may provide an explanation for why comprehension is so relentlessly continuous. Most work on comprehension within the product tradition takes as axiomatic the observation that language processing is continuous. If any explanation for *why* processing is incremental is offered, it is that incremental processing is necessitated by the demands of limited working memory, *viz.*, the system would be overloaded if it buffered a sequence of words rather than interpreting them immediately. However, working memory explanations are not particularly compelling. In fact, the first-generation models of language comprehension—models that were explicitly motivated by considerations of working memory limitations—assumed that comprehension was a form of sophisticated catch up in which the input was buffered long enough to accumulate enough input to reduce ambiguity (for discussion, see Tanenhaus (2004)). There is, however, a clear need for incremental comprehension in interactive conversation. Interlocutors, who are simultaneously playing the roles of speaker and addressee, need to plan and modify utterances in midstream in response to input from an interlocutor.

Finally, the action and product traditions often have different perspectives on constructs that are viewed as central within each tradition. Consider, for example, the notion of *context*. Within the product tradition, context is typically viewed as information that either enhances or instantiates a context-independent core representation or as a *correlated constraint* in which information from higher-level representations can, in principle, inform linguistic processing when the input to lower levels of representation is ambiguous. Specific debates about the role of context include whether, when and how: (i) lexical context affects sub-lexical processing, (ii) syntactic and semantic context affect lexical processing, and (iii) discourse and conversational context affect syntactic processing.

Each of these questions involves debates about the architecture of the processing system and the flow of information between different types of representations—classic information processing questions. In contrast, we have already noted that within the action tradition context includes the time, place and the participant's conversational goals, as well as the collaborative processes that are intrinsic to conversation. A central tenet is that utterances can only be understood relative to these factors. Although these notions can be conceptualized as a form of constraint, they are intrinsic to the comprehension process rather than a source of information that helps resolve ambiguity in the input. For these reasons, we believe that combining and integrating the product and action approaches is likely to prove fruitful by allowing researchers from each tradition to investigate phenomena that would otherwise prove intractable. Moreover, research that combines the two traditions is likely to deepen our understanding of language processing by opening up each tradition to empirical and theoretical challenges from the other tradition.

We now review three streams of research. The first two are by Tanenhaus and his collaborators. The third is a new line of research, which we have conducted jointly. First, we briefly discuss work examining how fine-grained acoustic information is used in spoken word recognition. We review this work to: (i) illustrate the sensitivity and temporal grain provided by eye movements, (ii) address some methodological concerns that arise from studying language processing in a circumscribed world, and (iii) highlight the importance of studying language processing as an integrated system. Second, we review studies demonstrating that real-world context, including intended actions, perceptually relevant properties of objects, and knowledge about the speaker's perspective combine to circumscribe the 'referential domain' within which a definite referring expression, such as *the empty bowl* is interpreted. Third, we present results from studies that begin to fully bridge the product and action traditions by examining real-time processing during unscripted conversation to explore how the participants in a task-oriented dialogue coordinate their referential domains.

13.3 Spoken word recognition in continuous speech

Since the seminal work of Marslen-Wilson and colleagues, an important goal of models of spoken word recognition has been to characterize how a target word is identified against the backdrop of alternatives or 'neighbours' that are temporarily consistent with the unfolding input (Marslen-Wilson and Welsh 1978). Some models, such as the neighbourhood activation model emphasize global similarity, without taking into account whether the overlap between the targets and potential competitors occurs early or late (Luce and Pisoni 1998). Some emphasize onset-based similarity by incorporating explicit bottom-up mismatch inhibition to strongly inhibit lexical candidates that have any mismatch with the input (Marslen-Wilson and Warren 1994; Norris *et al.* 2002). And some, such as the TRACE model (McClelland and Elman 1986) adopt a middle ground by avoiding explicit mismatch inhibition, but incorporating lateral inhibition at the lexical level. Similarity at any point can activate any word. However, there is an advantage for candidates overlapping at onset: since they become activated early in processing, they inhibit candidates that are activated later. Distinguishing among these alternative hypotheses about lexical neighbourhoods requires mapping out the time course of lexical processing.

Allopenna *et al.* (1998) examined the time course of activation for words that share initial phonemes with a target word (e.g. *beaker* and *beetle*), which we will refer to as 'cohort competitors' and which are predicted to compete by onset-similarity models, as well as words that rhyme with the target word (e.g. *beaker* and *speaker*), which are predicted to compete by global similarity models. Participants followed spoken instructions to move one of four objects displayed on a computer screen using the computer mouse (e.g. 'Look at the cross. Pick up the beaker. Now put it above the square'). Critical trials included cohort competitors (e.g. *beetle*) and/or rhyme competitors (*speaker*), and unrelated baseline items (e.g. *carriage*), as illustrated in Figure 13.1a. The assumption linking fixations to continuous word recognition processes is that as the instruction unfolds the probability that the listener's attention will shift to a potential referent of a referring expression increases with the activation (evidence for) of its lexical representation, with a saccadic eye movement typically following a shift in visual attention to the region in space where attention has moved. Since saccades are rapid, low cost, low-threshold responses, some saccades will be generated based on even small increases in activation, with the likelihood of a saccade increasing as activation increases. Thus, while each saccade is a discrete event, the probabilistic nature of saccades ensures that, with sufficient numbers of observations, the results will begin to approximate a continuous measure. For an insightful discussion, including the strengths and weaknesses of eye movements compared with

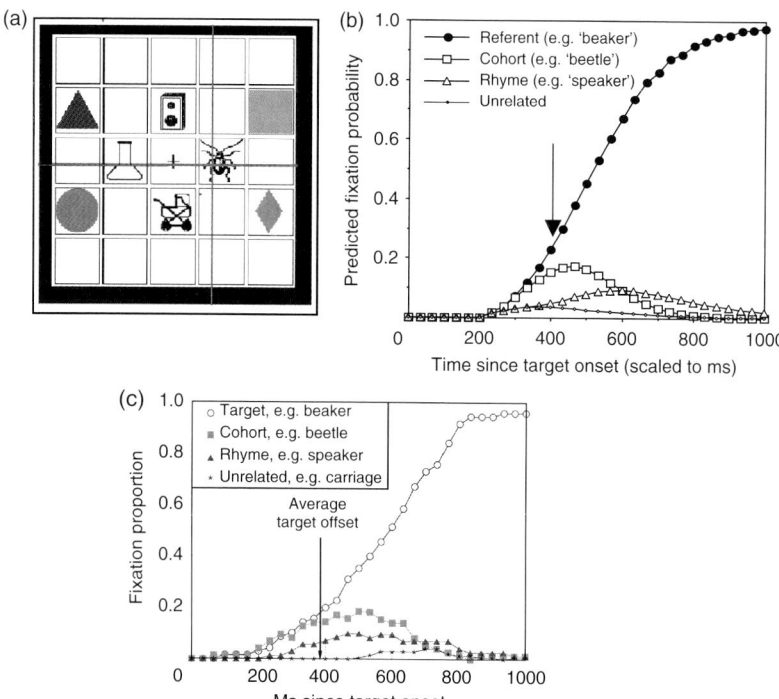

Figure 13.1 Shown are a sample display, simulations and data from Allopenna *et al.* (1998). (a) Sample display, (b) simulations of fixation proportions using TRACE and the linking hypothesis, and (c) the behavioural data. All figures are adapted from Allopenna *et al.* (1998).

a truly continuous measure, tracking the trajectories of hand movements, see Magnuson (2005) and Spivey *et al.* (2005).

Figure 13.1b shows the proportion of looks to each of the four pictures at each of 33 ms time slices, summed across trials and participants. The proportion of fixations maps onto phonetic similarity over time: targets and cohort competitor proportions increase and separate from the rhyme and unrelated baseline of approximately 200 ms after the onset of the target word (approximately the delay required to program and launch a saccade). As the input becomes more similar to the rhyme, looks to its referent increase compared with the unrelated baseline. At approximately 200 ms after the first acoustic/phonetic information that is more consistent with the target, fixation proportions to the cohort begin to drop off, returning to the unrelated baseline sooner than rhyme fixations. Simulations of these data using TRACE, and a formal linking hypothesis between activation in the model and likelihood of fixation, account for more than 90% of the variance in the time course of fixation proportions to the target and competitors.

These results suggest that the processing neighbourhood changes dynamically as a word unfolds. Early in processing, competition will be stronger for words with many cohort competitors compared with few cohort competitors; whereas, later in processing, competition will be stronger for words with a high density of globally similar competitors. Magnuson *et al.* (2007) report just this result using a display in which all of the pictures had unrelated names, and potential competitors were never pictured and never named. Magnuson *et al.* (2007) compared frequency-matched targets that differed in number of cohort competitors and neighbourhood density (words that differ from the target by one phoneme, or by adding or subtracting a phoneme). Even the earliest fixations to target words with many cohort competitors were delayed relative to fixations to targets with fewer cohort competitors. In contrast, density affected only later fixations to the target.

These results also address a potentially troubling methodological concern with the visual world approach. The use of a visual world with a limited set of pictured referents and potential actions creates a more restricted environment than language processing in many, if not most, contexts. Certainly, these closed-set characteristics impose more restrictions than most psycholinguistic tasks. In the Allopenna *et al.* (1998) paradigm, the potential response set on each trial was limited to four pictured items. If participants adopted a task-specific strategy, such as implicitly naming the pictures, then the unfolding input might be evaluated against these activated names, effectively bypassing the usual activation process. However, if this were the case, one would not expect to find effects of non-displayed competitors, as did Magnuson *et al.* (2007; for further discussion and other relevant data see Dahan and Tanenhaus (2004, 2005), Salverda and Altmann (2005) and Dahan *et al.* (2007)). Most crucially, the same linking hypothesis predicts the time course of looks to targets in experiments using displayed and non-displayed competitors (for specific examples, which compare displayed and non-displayed competitors, see Dahan *et al.* (2001*a, b*)).

In the procedure introduced by Allopenna *et al.* (1998), the time course of lexical processing is measured to words that are embedded in utterances. This is important because the prosodic environment in which a word occurs systematically affects its acoustic/phonetic realization. Recent studies of the processing of words such as *captain*, which begin with a phonetic sequence that is itself a word, e.g. *cap*, illustrate this point. One might think that the presence of embedded words would present a challenge to spoken word recognition. However, the language processing system exploits small systematic differences in vowel duration. In particular, the vowel in a monosyllabic word such as *cap* is typically longer than the same vowel in a polysyllabic word such as *captain*. (Davis *et al.* 2002; Salverda *et al.* 2003). However, the difference in vowel duration changes with the prosodic environment; it is smallest in the middle of a phrase and largest at the end of a phrase. Consequently, the extent to which an embedded word and its carrier compete for recognition varies with position in an utterance (Crosswhite *et al.* in preparation). More generally, prosodic factors will modulate the relative degree to which different members of a neighbourhood will be activated in different environments. A striking demonstration comes from a recent study from our laboratory (Salverda *et al.* 2007). Figure 13.2a shows a sample display with pictures of a cap, captain, cat and a picture with an unrelated name, a mirror, used with instructions in which *cap* is in either medial or final position. Figure 13.2b shows that in medial position *captain* is a stronger competitor than *cat*, whereas the opposite pattern is seen in utterance-final position.

Prosodic influences on processing neighbourhoods have broad implications for the architecture of word recognition system because pragmatic factors can have strong influences on prosody. For example, the duration of the first vowel in *captain* is similar to the typical duration of the vowel in *cap* when stress is being used to signal *contrast*, e.g. *The CAPtain was responsible for the accident*. It is possible then, that, during word recognition, the weighting of a sub-phonetic factor such as vowel duration might be modulated by a high-level property of the utterance. Evaluating this hypothesis requires examining the recognition of words embedded in an utterance at a fine temporal grain, and in a context rich enough to manipulate contrast. This can be accomplished with relatively minor extensions of the Allopenna *et al.* (1998) paradigm to create a richer discourse context. We now turn to studies that use real-world objects to focus on a particular type of context, the referential domain within which a linguistic expression is interpreted.

13.4 Referential domains: effects of action-based affordances

Many linguistic expressions can be understood only with respect to a circumscribed context or referential domain. Definite referring expressions are a paradigmatic example.

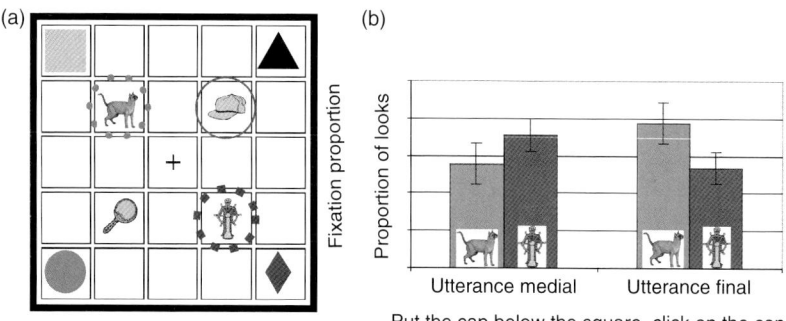

Figure 13.2 Sample display and proportion of looks to (a) the cohort competitor (picture of a cat) and (b) embedded carrier word competitor (picture of a captain) in utterance medial and utterance final positions.

Felicitous use of a definite noun phrase (NP) requires reference to, or introduction of, a *uniquely identifiable* entity (e.g. Roberts 2003). For example, imagine one is playing the role of sous-chef. If two bowls were in front of you on the kitchen counter, your cooking partner could not felicitously ask you *to pour the milk into the bowl*. Instead, he would have to use the indefinite phrase, *a bowl*. He could, however, say *the bowl*, if only one bowl was on the counter, and other bowls were on a visible shelf, or if there were several bowls on the counter and pouring the milk was embedded in a sequence of actions to add ingredients to a mixture in that particular bowl. The definite expression is felicitous in these situations because the satisfaction of uniqueness takes place with respect to a relevant context, or referential domain.

Recent research demonstrates that listeners dynamically update referential domains based on expectations driven by linguistic information in the utterance and the entities in the visual world (Eberhard *et al.* 1995; Altmann and Kamide 1999; Chambers *et al.* 2002). For example, in Eberhard *et al.* (1995), the participants touched one of four blocks that differed in marking, colour or shape. With instructions such as *Touch the starred yellow square*, the participants launched an eye movement to the target block on average 250 ms after the end of the word that uniquely specified the target with respect to the visual alternatives. In the example, the earliest possible point of disambiguation (POD) is after *starred* when only one of the blocks is starred, and after *square* when there are two starred yellow blocks. Similar results are obtained with more complex instructions and displays. With a display of seven miniature playing cards, including *two* five of hearts, Eberhard *et al.* (1995) used instructions such as, *Put the five of hearts that is below the eight of clubs above the three of diamonds*. To manipulate the POD, we varied whether or not the competitor five of hearts had a card above it, and if so, whether it differed in denomination or suit from the card above the target five. The following is a representative sequence of fixations. As the participant heard *the five of hearts*, she successively looked at each of the two potential referents. After hearing *below the*, she immediately looked at a 10 of clubs, which was above the (competitor) 5 that she had been fixating on. By the end *of clubs*, her eyes moved to interrogate the card above the other five, the eight of clubs, thus identifying that five as the target. The eye immediately shifted down to the target card and remained until the hand began to grasp the target, at which point gaze shifted to the three of diamonds.

In collaboration with Chambers and colleagues, we asked whether referential domains take into account the affordances of potential real-world referents with respect to the action evoked by the instruction. Chambers *et al.* (2002; Experiment 2) presented the participants with six objects in a workspace, as illustrated in Figure 13.3a. On test trials, the objects included a large and a small container, e.g. a large can and a small can. We manipulated whether the to-be-moved object, the cube in Figure 13.3a, could fit into both of the containers, as was the case for a small cube, or only fit into the larger container, as was the case for a large cube. Instructions for the display were: *Pick up the cube. Now put it inside a/the can*.

The size of the theme-object determined whether one or two of the potential goals (containers) were compatible referents. The instructions manipulated whether the goal was introduced with the definite article, *the*, which presupposes a unique referent, or the indefinite article, *a*, which implies that the addressee can choose from among more than one goal.

First, consider the predictions for the condition with the small cube. Here we would expect confusion when the definite article was used to introduce the goal because there is

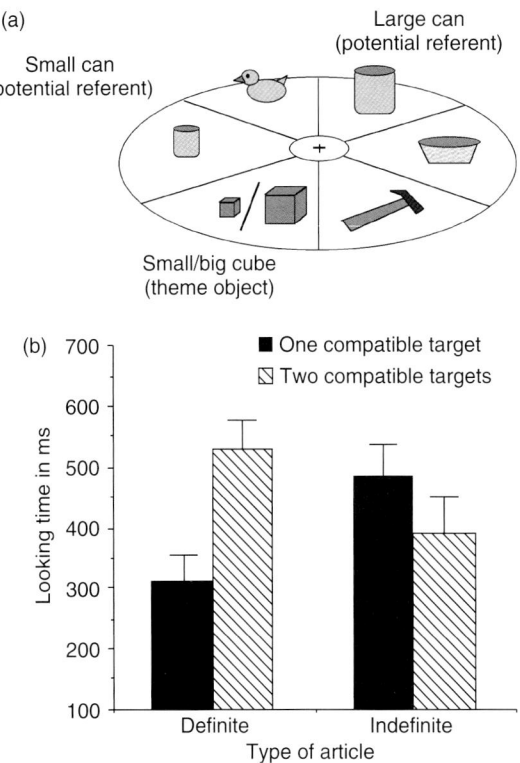

Figure 13.3 (a) Sample stimuli. The small cube will fit into both cans but the large cube will only fit into the big can. (b) The mean latency to launch an eye movement to the goal with definite and indefinite instructions and one and more than one compatible goal referents.

not a unique referent. In contrast, the indefinite article should be felicitous because there is more than one action-compatible goal referent. This is what we found: eye-movement latencies to fixate the goal chosen by the participant were slower in the definite condition compared with the indefinite condition. This confirms expectations derived from the standard view of how definite and indefinite articles are interpreted. Now consider predictions for the condition with the large cube, the theme-object that would fit into only one of the goal objects, i.e. the large can. If the referential domain consists of all of the objects in the visual world that meet the linguistic description in the utterance, that is both cans, then the pattern of results should be similar to that for the small cube. If, however, listeners dynamically update referential domains to include only those objects that afford the required action, i.e. containers that the object in hand would fit into, then only the large can is in the relevant referential domain. Therefore, use of a definite description, e.g. *the can* should be felicitous, because the cube could be put into only one can, whereas an indefinite description, e.g. *a can*, should be confusing.

Figure 13.3b shows the predicted interaction between definiteness and compatibility. Eye-movement latencies to the referent following the definite referring expressions were faster when there was only one compatible referent compared with when there were two compatible referents, whereas the opposite pattern occurred for the indefinite expressions.

Moreover, latencies for the one-referent compatible condition were comparable to control trials in which only a single object met the referential description in the instruction, e.g. trials with only a single large can. Thus, referential domains can be dynamically updated to take into account the real-world properties of potential referents with respect to a particular action.

To further evaluate the claim that intended actions were indeed constraining the referential domain, Chambers *et al.* (2002) conducted a second experiment in which the second instruction was modified to make it into a question, e.g. *Pick up the cube. Could you put it inside a/the can?* In order to prevent the participants from interpreting the question as an indirect request, the participant first answered the question. On about half of the trials when the participant answered 'yes', the experimenter subsequently asked the participant to perform the action. Unlike following a command, answering a question does not require the participant to perform an action that brings the affordance restrictions into play. Thus, the referential domain should now take into account all the potential referents that satisfy the linguistic description, not just those that would be compatible with the possible action mentioned in the question. If referential domains take into account behavioural goals then, under these conditions, definite expressions should be infelicitous regardless of compatibility, whereas indefinite expressions should always be felicitous. This is what we found. Time to answer the question was longer for questions with definite compared with indefinite referring expressions. Crucially, definiteness did not interact with compatibility (i.e. size of the theme-object). Moreover, compatibility had no effect on response times for questions with definite articles. These results demonstrate that referential domains are dynamically updated using information about available entities, properties of these entities and their compatibility with the action evoked by the utterance. This notion of referential domain is consistent with the rich view of context endorsed by researchers in the action tradition.

Assignment of reference necessarily involves mapping linguistic utterances onto entities in the world, or a conceptual model thereof. A crucial question, then, is whether these contextually defined referential domains influence core processes in language comprehension that many have argued operate without access to contextual information. In order to address this question, we examined whether action-based referential domains affect the earliest moments of syntactic ambiguity resolution.

Previously, we noted the temporary ambiguity in an utterance, such as, *Put the apple on the towel* ... Temporary 'attachment' ambiguities like these have long served as a primary empirical test bed for evaluating models of syntactic processing (Tanenhaus and Trueswell 1995). In many attachment ambiguities, the ambiguous phrase could either modify a definite NP or introduce a syntactic complement (argument) of a verb phrase. Under these conditions, the argument analysis is typically preferred. For instance, in Example 1, listeners will initially misinterpret the prepositional phrase, *on the towel*, as the goal argument of *put* rather than as an adjunct modifying the NP, *the apple*, resulting in temporary confusion.

Example 1 Put the apple on the towel into the box.

Crain and Steedman (1985) noted that one use of modification is to differentiate an intended referent from other alternatives (also see Altmann and Steedman, 1998). For example, it would be odd for Example 1 to be uttered in a context in which there was only one perceptually salient apple, whereas it would be natural in contexts with more than one apple. In the latter context, the modifying phrase, *on the towel*, provides information about which of the apples is intended. They proposed that listeners might initially prefer the modification analysis to the argument analysis in situations that provided the appropriate referential context. They also argued that referential fit to the context, rather than syntactic complexity, was the primary factor controlling syntactic preferences (also see Altmann and Steedman 1988).

Tanenhaus *et al.* (1995) and Spivey *et al.* (2002) compared the processing of temporarily ambiguous sentences such as *Put the apple on the towel in the box* and unambiguous control sentences, such as *Put the apple that's on the towel in the box*, in contexts such as the ones illustrated in Figure 13.4.

The objects were placed on a table in front of the participants. Eye movements were monitored as they followed the spoken instruction. The results, which are presented in Figure 13.4c, provided clear evidence for immediate use of the visual context. In the one-referent context, the participants frequently looked at the false (competitor) goal, indicating that they initially misinterpreted the prepositional phrase, *on the towel*, as introducing the goal. Looks to the competitor goal were dramatically reduced in the two-referent context. Crucially, the participants were no more likely to look at the competitor goal with ambiguous instructions compared to the unambiguous baseline (also see Trueswell *et al.* 1999).

Clearly, then, referential context can modulate syntactic preferences from the earliest moments of syntactic ambiguity resolution. But is the relevant referential domain defined by all of the salient entities that meet the referential description in the utterance or can it be dynamically updated based on real-world constraints, including action-based

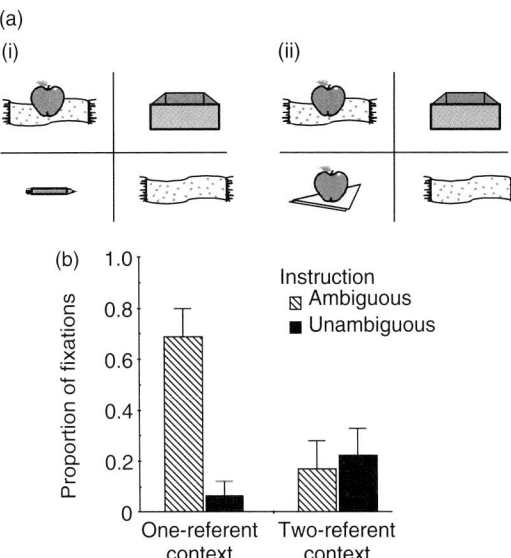

Figure 13.4 (a) Sample stimuli for (i) one-referent (pencil) and (ii) two-referent (apple on napkin) conditions. (b) The proportion of looks to the competitor goal (the towel) for instructions with locally ambiguous and unambiguous prepositional phrases in one-referent and two-referent contexts.

affordances of objects? Chambers *et al.* (2004) addressed this question using temporarily ambiguous instructions such as, *Pour the egg in the bowl over the flour*, and unambiguous instructions such as, *Pour the egg that's in the bowl over the flour*, with displays such as the one illustrated in Figure 13.5.

The display for test trials included the goal (the flour), a competitor goal (the bowl), the referent (the egg in the bowl), and a competitor referent (the egg in the glass). The referent was always compatible with the action evoked by the instruction, e.g. the egg in the bowl was liquid and therefore could be poured. We manipulated whether the competitor referent was also compatible with the action evoked by the instruction, e.g. one can pour a liquid egg, but not a solid egg. In the compatible condition, the other potential referent, the egg in the glass, was also in liquid form. In the incompatible condition, it was an egg in a shell. The crucial result was the time spent looking at the competitor goal, which is presented in Figure 13.5c.

When both potential referents matched the verb (e.g. the condition with two liquid eggs, as in Figure 13.5a), there were few looks to the false goal (e.g. the bowl) and no differences between the ambiguous and unambiguous instructions. Thus, the prepositional phrase was correctly interpreted as a modifier, replicating the pattern found by Spivey *et al.* (2002) for two-referent contexts. However, when the competitor was incompatible, as in Figure 13.5c (e.g. the condition where there was a liquid egg and a solid egg), we see the same data pattern as Spivey *et al.* (2002) found with one-referent contexts. Participants were more likely to look to the competitor goal (the bowl) with the ambiguous instruction than with the unambiguous instruction. Thus, listeners misinterpreted the ambiguous

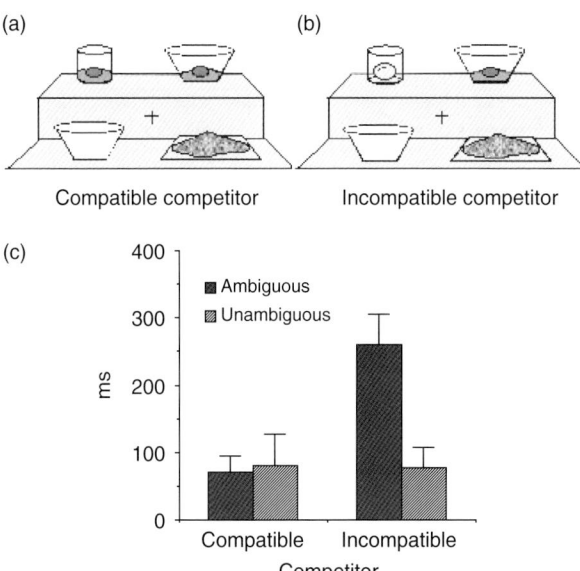

Figure 13.5 Sample stimuli for trials with (a) action-compatible competitor (two liquid eggs) and (b) action-incompatible competitor (one solid egg), (c) The mean proportion of time spent looking at the competitor goal (the empty bowl) for instructions with locally ambiguous and unambiguous prepositional phrases with action-compatible and action-incompatible competitors.

prepositional phrase as introducing a goal only when a single potential referent (the liquid egg) was compatible with a pouring action.

Unlike the Spivey *et al.* (2002) study, which used the verb *put*, all of the relevant affordances were related to properties that might be plausibly be attributed to the lexical semantics of the verb. For example, *pour* requires its theme to have the appropriate liquidity for pouring. There is precedent going back to Chomsky (1965) for incorporating a subset of semantic features, so-called 'selectional restrictions', into lexical representations when those features have syntactic or morphological reflexes in at least some languages. Thus, it could be argued that only real-world properties referred to with selectional restrictions can influence syntactic processing.

Chambers *et al.* (2004) addressed this issue in a second experiment. The critical instructions contained *put* (e.g. *Put the whistle (that's) on the folder in the box*), a verb that obligatorily requires a goal argument. Figure 13.6a shows a corresponding display, containing potential referents that are whistles, one of which is attached to a loop of string. Importantly, *put* does not constrain which whistle could be used in the action described by the instruction.

The compatibility of the referential competitor was manipulated by varying whether or not the participants were provided with an instrument. The experimenter handed the instrument to the participant without naming it. For example, before the participants were given the instruction described earlier, they might be given a small hook. Critically, this hook could not be used to pick up the competitor whistle without a string. Thus, upon

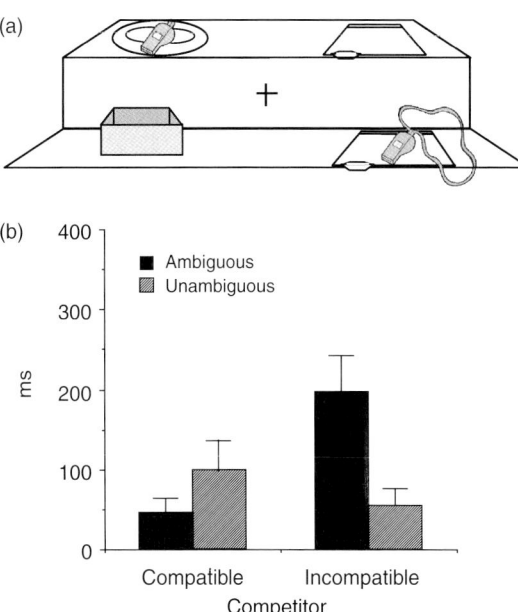

Figure 13.6 (a) Sample stimuli. Both whistles can be moved by hand, but only the whistle with the string attached can be picked up with a hook, (b) The proportion of time spent looking at the competitor goal when the presence or absence of an instrument makes the competitor action-compatible or action-incompatible.

hearing *put the* ... the competitor could be excluded from the referential domain based on the affordances of the object with respect to the intended action, i.e. using the hook to move an object. If so, the participants should misinterpret *on the folder* as the goal only when ambiguous instructions are used *and* when an instrument is provided. If, however, the relevant referential domain is defined using only linguistic information, then a goal misanalysis should occur regardless of whether an instrument is supplied beforehand. Figure 13.6b shows the mean time spent fixating the false goal object within the 2500 ms after the first prepositional phrase. The false goal is most often fixated when ambiguous instructions are used and the competitor cannot afford the evoked action. The remaining conditions all show fewer fixations to the false goal.

Thus, the syntactic role assigned to a temporarily ambiguous phrase varies according to the number of possible referents that can afford the action evoked by the unfolding instruction. The same results hold regardless of whether the constraints are introduced linguistically by the verb, or non-linguistically by the presence of a task-relevant instrument. Thus, the referential domain for an initial syntactic decision was influenced by the listener's consideration of how to execute an action—an information source that cannot be isolated within the linguistic system. This action itself can be partially determined by situation-specific factors such as the presence of a relevant instrument. The syntactic role assigned to an unfolding phrase in turn depends on whether these factors jointly determine a unique referent without additional information. These results add to the growing body of literature indicating that multiple constraints affect even the earliest moments of syntactic processing. They are incompatible with the claim that the language processing system includes subsystems (modules) that are informationally encapsulated, and thus isolated from high-level expectations (Fodor 1983; Coltheart 1999).

So far, we have described experiments that extend investigations of real-time language processing into more natural real-world tasks. However, it was not clear that this approach could be used to study unscripted interactive conversation. We now turn to some work in progress, which shows that such studies are both tractable and informative.

13.5 Real-time language processing in interactive conversation

Since Stalnaker's pioneering work on mutual knowledge (Stalnaker 1978, 2002), formal theories of discourse in computational linguistics, pragmatics and semantics have assumed that keeping track of what is known, and not known, to the individual participants in a discourse is fundamental for coordinating information flow (Clark 1992, 1996; Brennan and Hulteen 1995). For example, a speaker making a statement (such as *The coffee's ready*) is expected to contribute information that is not known to the listener, a choice that involves judgments about the information states of the participants. If the addressee is already holding a fresh cup of coffee, this is not likely to be an informative contribution. The speaker's statement reflects what the speaker takes to be not yet commonly known at that point in the discourse. Asking a question (*Is the coffee ready?*) would similarly seem to reflect both the speaker's assessment of the addressee's information state—i.e. that the addressee is in a position to provide the answer—and the speaker's own state of ignorance or uncertainty.

However, conversational partners may not continuously update their mental representations of each other's knowledge. Building, maintaining and updating a model of a conversational partner's beliefs could be memory intensive (Keysar et al. 1998). In addition, many conversational situations are constrained enough that an individual participant's perspective will provide a sufficient approximation of the knowledge and beliefs shared between interlocutors. Moreover, information about another's beliefs can be uncertain at best. For these reasons, Keysar and colleagues propose that whereas considerations of an interlocutor's perspective might control language performance at a macro level, the moment-by-moment processes that accomplish production and comprehension could take place relatively egocentrically. Indirect supporting evidence comes from a growing number of studies demonstrating that speakers typically do not adapt the form of their utterances to avoid constructions that are difficult for listeners (Brown and Dell 1987; Bard et al. 2000; Ferreira and Dell 2000; Keysar and Barr 2005; but see Metzing and Brennan 2003). More direct evidence comes from studies showing that addressees often fail to reliably distinguish their own knowledge from that of their interlocutor when interpreting a partner's spoken instructions (Keysar et al. 2000, 2003). For example, in Keysar et al. (2000), participants were seated on opposite sides of a vertical grid of squares, some of which contained objects. Most of the objects were in 'common' ground because they were visible from both sides of the display, but a few objects in the grid were hidden from the director's view, and thus were in the matcher's privileged ground. On critical trials, the director, a confederate, referred to a target object in common ground using an expression that could also refer to a hidden object in privileged ground, which was always the more prototypical referent for the expression. Matchers initially preferred to look at the hidden object, and on some trials even picked it up and began to move it. Subsequent studies that equate the typicality of potential referents in common and privileged ground also find intrusions of information from privileged ground. However, under these conditions, addressees seem to make partial use of common ground from the earliest moments of processing (Nadig and Sedivy 2002; Hanna et al. 2003).

Thus, whether and, if so, when interlocutors seek and use information about each other's probable intentions, commitments and probable knowledge remain open questions. The answers will determine which classes of theoretical constructs can be imported from semantics, pragmatics and computational linguistics. They will also determine the extent to which production and comprehension can be viewed as encapsulated from social/cognitive processes. However, the standard approaches that have been used to address these questions have serious problems. One problem is that those studies that have shown the strongest support for use of common ground may invite the subject to adopt the speaker's perspective, e.g. by drawing attention to the mismatch in perspective by mislabelling objects. The second problem is that using a confederate to generate instructions eliminates many of the natural collaborative processes that occur in interactive conversation and dramatically changes the form of the language. More seriously, in studies with confederates, all of the instructions are simple declarative commands, which carry the presupposition that there is a unique action that the addressee can perform, in effect attributing omniscience to the speaker. In addition, asking a participant to follow instructions from a director may create a weak version of the suspension of skepticism that can occur in situations where there is an authority giving directions (e.g. an experimenter, a health professional, etc.). The addressee aims to do what he is told on the assumption that the person generating the instruction has the relevant knowledge.

The solution, which is to examine interactive conversation in unscripted joint tasks, is rife with methodological challenges. The experimenter gives up a substantial degree of control because trials cannot be scripted in advance. Rather, they have to emerge from the conversational interaction. Thus, tasks have to be carefully crafted to generate appropriate trials and baseline control conditions. Data analysis is time consuming because the conversation and state of the workspace have to be transcribed in order to identify relevant trials for subsequent data analysis. In addition, as we have seen, the form of the language differs from the sanitized language used in the typical psycholinguistic experiment with pre-recorded materials. With these challenges in mind, we conducted a series of preliminary investigations to determine the feasibility of monitoring real-time language processing with naive participants during task-oriented interactive dialogue.

Our approach adopts a 'targeted language games' methodology. Pairs of naive participants complete a type of language game—a referential communication task (Krauss and Weinheimer 1966)—while gaze and speech are monitored. These language games are 'targeted' in that the parameters of the game are carefully designed to elicit specific types of utterances, without explicitly restricting what the participants say. The task is structured so that the conditions of interest that would be manipulated in a typical within-subjects design, including control conditions, emerge during the game. Each utterance is naturally produced assuring that it is contextually appropriate. Conversations are lengthy enough to generate sufficient trials in the conditions of interest to approximate a standard factorial design. We then compare characteristics of these utterances and the corresponding eye movements across conditions. In what follows, we test the validity of the methodology by replicating some standard findings. We then extend the methodology to investigate aspects of language production and comprehension that have been difficult or impossible to study using standard techniques. In doing so, we make novel observations that clarify how interlocutors generate and comprehend referential expressions and how and when they take into account their partner's perspective—an ability that comprises part of our theory of mind (see Keysar *et al.* 2003).

(a) Language interpretation

In our initial experiments, pairs of participants, separated by a curtain, worked together to arrange blocks in matching configurations and confirm those configurations (Brown-Schmidt *et al.* 2005; Brown-Schmidt and Tanenhaus 2008). The characteristics of the blocks afforded comparison with findings from scripted experiments investigating language-driven eye movements, specifically those demonstrating POD effects during reference resolution. We investigated: (i) whether these effects could be observed in a more complex domain during unrestricted conversation and (ii) under what conditions the effects would be eliminated, indicating that factors outside of the speech itself might be operating to circumscribe the referential domain. Figure 13.7 presents a schematic of the experimental setup. We divided participants' boards into five physically distinct sub-areas, within which the blocks were arranged. Most of the blocks were of assorted shapes (square or rectangle) and colours (red, blue, green, yellow, white or black). The configuration of the blocks was such that their colour, size and orientation would encourage the use of complex NPs. Other blocks contained pictures that could be easily described by naming the picture (e.g. 'the candle'). We included pairs whose names were cohort competitors, e.g. *clown* and *cloud*.

Partners were highly engaged and worked closely with one another to complete the task. Each pair worked through the game board in a different way, and none used overt

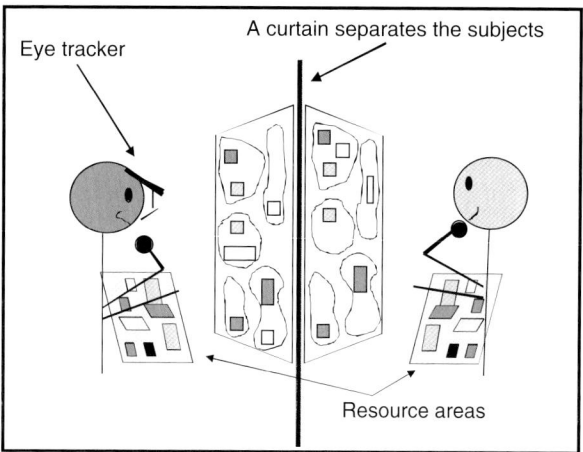

Figure 13.7 Schematic of the setup used in the referential communication task. Shaded squares and rectangles represent blocks and unshaded squares and rectangles represent stickers (which will eventually be replaced with blocks). The scene pictured is midway through the task, so some portions of the partners' boards match, while other regions are not completed yet.

strategies, such as establishing a grid system. All pairs made frequent references to the game pieces. Referential expressions were indefinite NPs, definite NPs and pronouns. We designed the task to focus exclusively on the interpretation of definite references, and this is what we turn to now.

The POD for each of the non eye-tracked partner's definite references to coloured blocks was defined as the onset of the word in the NP that uniquely identified a referent, given the visual context at the time. Just over half of these NPs (53%) contained a linguistic POD. The remaining 47% were technically ambiguous with respect to the sub-area that the referent was located in (e.g. *the red one* uttered in a context of multiple red blocks). Eye movements elicited by NPs with a unique linguistic POD were analysed separately from those that were never fully disambiguated linguistically. The eye-tracking analysis was restricted to cases where at least one competitor block was present. Eye movements elicited by disambiguated NPs are pictured in Figure 13.8a. Before the POD, listeners showed a preference to look at the target block. Within 200 ms of the onset of the word in the utterance that uniquely specified the referent (POD), looks to targets rose substantially. This POD effect for looks to the target is similar to that seen by Eberhard *et al.* (1995), demonstrating that we were successful in using a more natural task to investigate online language processing. The persistent target bias and lack of a significant increase in looks to competitors are probably due to additional pragmatic constraints that we will discuss shortly.

For ambiguous utterances (Figure 13.8b), fixations were primarily restricted to the referent, and there were very few requests for clarification. Thus, the speaker's underspecified referential expressions did not confuse listeners, indicating that referential domains of the speaker and the listener were closely coordinated. These results suggest that: (i) speakers systematically use less specific utterances when the referential domain has been otherwise constrained, (ii) the attentional states of speakers and addressees

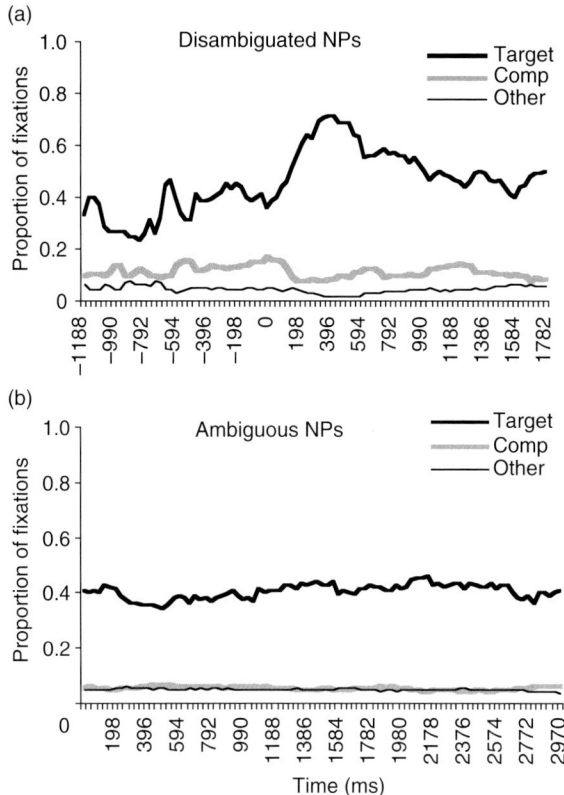

Figure 13.8 (a) The proportion of fixations to targets, competitors, and other blocks by time (ms) for linguistically disambiguated definite NPs. The graph is centred by item with 0ms = POD onset. (b) The proportion of fixations for the linguistically ambiguous definite NPs.

become closely coordinated, and (iii) utterances are interpreted with respect to referential domains circumscribed by contextual constraints. In order to identify what factors led speakers to choose underspecified referring expressions, and enabled addressees to understand them, we performed a detailed analysis of all of the definite references, focusing on factors that seemed likely to be influencing the generation and comprehension of referential expressions. We hypothesized that speakers would choose to make a referential expression more specific when the intended referent and at least one competitor block were each salient. We focused on recency, proximity and compatibility with task constraints—factors similar to those identified by Beun and Cremers (1998), who employed a task in which the participants, separated by a screen, worked together in a mutually co-present visual space to build a structure out of blocks.

(i) Recency
We assumed that recency would influence the salience of a referent, with the most recently mentioned entities being more salient than other (non-focused) entities. Thus, how recently the target block was last mentioned should predict the degree of specification,

with references to the most recently mentioned block of a type, resulting in ambiguous referring expressions. For example, if *the green block* was uttered in the context of a set of 10 blocks, 2 of which were green, recency would predict that the referent should be the green block that was most recently mentioned.

(ii) Proximity

We examined the proximity of each block to the last mentioned block, because partners seemed to adopt a strategy of focusing their conversation on small regions within each sub-area. Table 13.2 presents a segment of discourse where the referent of an otherwise ambiguous NP is constrained by proximity. The underlined referring expression is ambiguous given the visual context; there are approximately three green blocks up and to the left of the previously focused block (the one referred to in the NP as *this green piece*). In this case, the listener does not have difficulty dealing with the ambiguity because he considers only the block closest to the last mentioned block.

(iii) Task compatibility

Task compatibility refers to constraints on block placement due to the size and shape of the board, as well as the idiosyncratic systems that partners used to complete the task. In the second exchange in Table 13.2, compatibility circumscribes the referential domain as the partners strive to determine where the clown block should be placed. Again, the underlined referring expression is ambiguous given the visual context. While the general task is to make their boards match, the current sub-task is to place the clown piece (which they call *the green piece with the clown*). In order to complete this sub-task, Speaker 2 asks whether the clown should be lined up with the target, *thuuh square*. The listener does not have difficulty dealing with this ambiguous reference because, although there are a number of blocks one could line up with *the green piece with the clown*, only one is task relevant. Given the location of all the blocks in the relevant sub-area, the target block is the easiest

Table 13.2 Excerpts from dialogue illustrating proximity and task compatibility constraints.

Speaker	utterance
proximity constraint	
2	ok, so it is four down, you're gonna go over four, and then you're gonna put the piece right there
1	ok... how many spaces do you have between this green piece and *the one to the left of it, vertically up?*
task compatibility	
1	ok, you're gonna line it up. it is gonna go <pause> one row *above* the green one, directly next to it
2	can't fit it
1	cardboard?
2	can't yup, cardboard
1	well, take it two back
2	the only way I can do it is if I move, alright, should the green piece with the clown be directly lined up with *thuuh square?*

block to line up with the clown. The competitor blocks are inaccessible due to the position of the other blocks or the design of the board.

For all ambiguous and disambiguated trials, each coloured block in the relevant sub-area was coded for recency (number of turns since last mention), proximity (ranked proximity to last mentioned item) and task constraints (whether or not the task predicted a reference to that block). Target blocks were more recently mentioned and more proximal than competitor blocks, and better fit the task constraints, establishing the validity of these constructs. However, recency, proximity and task compatibility of the target blocks did not predict speaker ambiguity. Ambiguity was, however, determined by the proximity and task constraints associated with the *competitor* blocks. When a competitor block was proximate and fit the task constraints, speakers were more likely to linguistically disambiguate their referential expression. A logistic regression model supported these observations: ambiguity was significantly predicted by a model that included task and proximity effects, with no independent contribution of recency.

These results suggest that the relevant referential domain for the speakers and addressees were restricted to a small task-relevant area of the board. Striking support comes from an analysis of trials in which there was a cohort competitor for the referent in the addressee's referential domain. Brown-Schmidt *et al.* (2005) found that looks to cohort competitors were no more probable than looks to competitors with unrelated names. This is not simply a null effect. Owing to the length of the experiment, the participants occasionally needed to take a bathroom break. Following the break, the eye tracker had to be recalibrated. The experimenter did so by instructing the participant to look at some of the blocks. The referential domain now consists of the entire display because there is no constraining conversational or task-based goal. When the intended referent had a cohort competitor, the participant frequently looked at the competitor, showing the classic cohort effect. A follow-up experiment replicated this finding, focusing exclusively on cohort trials that were in and outside the context of the conversation, by systematically incorporating simulated calibration checks (Brown-Schmidt and Tanenhaus 2008).

In summary, these results demonstrate that it is possible to study real-time language processing in a complex domain during unrestricted conversation. When a linguistic expression is temporarily ambiguous between two or more potential referents, reference resolution is closely time-locked to the word in the utterance that disambiguates the referent, replicating effects found in controlled experiments with less complex displays and pre-scripted utterances. Most importantly, our results provide a striking demonstration that participants in a task-based or 'practical dialogue' (Allen *et al.* 2001), closely coordinate referential domains as the conversation develops.

(b) Language Production

In conversation, speakers often update messages on the fly based on new insights, new information and feedback from addressees, all of which can be concurrent with the speaker's planning and production of utterances. Thus, message formulation and utterance planning are interwoven in time and must communicate with one another at a relatively fine temporal grain. During the last two decades, detailed models of utterance planning have been developed to explain a growing body of evidence about how speakers retrieve lexical concepts, build syntactic structures and translate these structures into linguistic forms (Dell 1986; Levelt 1989; Bock 1995; Levelt *et al.* 1999; Indefrey and Levelt 2004). However, much less is known about how speakers plan and update the non-linguistic thoughts (messages) that are translated into utterances during language production, or about how message formulation and utterance planning are coordinated (but cf. Griffin and Bock 2000; Bock *et al.* 2003).

We created situations in which the speaker, while in the process of planning or producing an utterance, encounters new information that requires revising the message. If eye movements can be used to infer when the speaker first encounters that information, then the timing between the uptake of the new information and the form of the utterance might shed light on the interface between message formulation and utterance planning. In one study (Brown-Schmidt and Tanenhaus 2006), we explored this interface by exploiting properties of scalar adjectives.

Speakers typically use a scalar adjective, such as *big* or *small*, only when the relevant referential domain contains both the intended referent and an object of the same semantic type that differs along the scale referred to by the adjective (Sedivy 2003). Pairs of naive participants took turns describing pictures of everyday objects situated in scenes with multiple distractor objects. For example, on one trial the target was a picture of a large horse. In a scene with multiple *unrelated* objects, we would expect the speaker to refer to the target as *the horse*. Sometimes, however, our displays also contained a *contrast* object that differed from the target only in size (in this example, a small horse). On these trials, we expected that speakers would use a scalar adjective to describe the target, as in *the large horse*.

This paradigm allowed us to control what speakers referred to (the target was predetermined by us and highlighted on the screen), but the choice of how to refer to that entity, including whether a scalar adjective was used, was entirely up to the speaker. When a contrast object was present, the speaker's first fixation to the contrast should provide an estimate of when the speaker first encountered information that size must be included in the message.

Speakers tailored their messages to the referential context, rarely using a size adjective when there was not a contrast in the display; when there was a size contrast, approximately three quarters of utterances included a size adjective. These proportions are consistent with those found by Sedivy and colleagues (Gregory *et al.* 2003; Sedivy 2003) in experiments that used simpler scenes with fewer objects.

Speakers typically gazed first at the highlighted target and then looked around the scene, sometimes fixating the contrast, and then returned to the target before speaking. When speakers looked at the contrast, over 80% of the referring expressions included a size modifier compared with less than 20% on trials when the contrast was not fixated. Thus, fixation on the contrast indexes whether size was included in the message. The position of the size adjective was variable (e.g. *the big horse*, versus *the horse... big one*). We reasoned that timing of the first fixation on the contrast would be linked to the planning of the message elements underlying the size adjective. Consistent with this prediction, we found that earlier size adjectives were associated with earlier first fixations on the contrast. For example, first fixations on the contrast for utterances like *the horse* (pause) *big one were* delayed by more than 800 ms compared to first fixations for utterances like *the big horse*.

Having demonstrated that one can monitor real-time processes in both comprehension and production using unscripted interactive language, we can now examine how and when interlocutors take account of each other's knowledge, intentions and beliefs. In recent work, we examined how having information about what an addressee knows shapes the form of the speaker's utterances, and how the addressee's knowledge guides the interpretation of these utterances (Brown-Schmidt *et al.* 2008).

If speakers can distinguish their own knowledge from an estimate of their partner's knowledge during utterance formulation, they should use declarative and interrogative forms in complementary situations, asking questions when the addressee knows more, and vice-versa for declaratives. Declarative questions, which are marked with special intonation, share features of declarative statements and interrogatives; consistent use of these forms requires even more fine-grained distinctions between the relative knowledge of speaker and addressee. Unlike typical declaratives, which contribute information to the discourse, declarative questions do not commit the speaker to the propositional content of what is said (see Gunlogson 2001, unpublished data), and unlike interrogatives, declarative questions do not preserve the speaker's neutrality. For example, a common use of declarative questions is to indicate surprise or skepticism, as seen in:

Example 2 A: *This is a painting by Chuck Close.*

              B: *That's a painting?*

Appropriate use of different utterance forms would seem to crucially depend on the ability of a speaker to distinguish, if even at a coarse grain, information that is shared with a specific interlocutor from information that is not shared. Successful communication may also require estimating what private information an interlocutor might have. To examine these issues, we designed a targeted language game that elicited spontaneously produced questions and statements. Some task-relevant information was mutually known to the participants, and some was privately known. As in Keysar *et al.* (2000, 2003), the distinction between mutual and private knowledge was initially established by whether an object was visually available to one or both partners.

Pairs of naive participants rearranged game-cards as they sat on either side of a gameboard made up of cubbyholes. A card with a picture of an identical animal figure on either side stood in each cubbyhole (Figure 13.9). Some cubbyholes were blocked-off on one side making them visible to only the eye-tracked partner or only the non eye-tracked partner; the remaining cubbyholes were visible to both partners. Each card featured a clip art picture of an animal (pig, cow or horse), with an accessory (glasses, shoes or a hat).

At the beginning of the task, the cards were randomly arranged. The participants' task was to rearrange the cards such that no two adjacent squares matched with respect to type of animal or type of accessory (e.g. neither two horses nor two animals wearing hats could be adjacent). Participants were required to avoid matches in adjacent squares, some of which were hidden from each participant. The task therefore required extensive interaction.

We report an analysis of a subset of utterances produced by four of the twelve pairs tested by Brown-Schmidt, *et al.*, (2008). We examined *responses to questions*, *declarative questions*, simple *declaratives*, and *wh-questions* that inquired about or made a statement about the identity of a card. Example 3 shows excerpts illustrating examples of each utterance type.

328   Michael K. Tanenhaus and Sarah Brown-Schmidt

Example 3

3.1 *(partner: What's below that?)… It is a horse with a hat*
3.2 *oh ok… You have a pig?*
3.3 *And then I got a pig with shoes next to that.*
3.4 *What's under the cow with the hat?*

Production analyses compared responses to questions (3.1), and declarative questions (3.2). These constructions share a similar grammatical form, but differ in communicative intent. *Comprehension* analyses compared declaratives (3.3) and wh-questions (3.4). An addressee who is sensitive to information state might take an interrogative form at the beginning of a question (e.g. *What*) as a cue that the speaker would be asking about cubbyholes that were in the addressee's private ground. In contrast, a declarative would indicate that the information was in common ground or the speaker's private ground.

Figure 13.10a shows the distribution of referent types for declaratives that were used either to respond to a question, or to ask a question. The referent of the utterance (e.g. the referent of *a horse with a hat* in 3.1 and *a pig* in 3.2) was identified and categorized in

Figure 13.9 Schematic of part of the game board for the 'questions' experiment from the perspective of one of the participants. The animals in grey squares are in that participant's privileged ground. The animals in the white squares are in common ground, that is, visible to both participants. The black squares contain animals that are only visible to the participant's partner.

terms of whether it was visible to just the speaker (private), just the addressee (hidden) or to both (shared). When speakers responded to a question, most of the time they referred to entities that only they could see (e.g. private), and to a lesser extent hidden and shared entities. In contrast, when asking a declarative question, the referent was likely to be either private or shared. Figure 13.10b shows the distribution of referent types for declaratives and wh-questions, analysed from the perspective of the addressee. When interpreting a question, the referent (e.g. the object of *What* in 3.4) is almost always in one of the addressee's private cubbyholes, indicating that the speaker asked about something she did not have direct information about. In contrast, when interpreting a declarative, the referent is most likely to be hidden from the addressee, and to a lesser extent shared or private.

We also examined addressee's eye movements as they heard wh-questions and declaratives (eye movement data are for all 12 pairs of participants). During wh-questions, approximately 47% of fixations were to private entities, and 43% to shared entities. In contrast, during declaratives, only approximately 29% of fixations were to private cubbyholes compared with 47% to shared cubbyholes. Most strikingly, as shown in Figure 13.11, addressee looks to objects in private ground increase during interpretation of wh-questions, rising significantly during the third analysis region. Following declaratives, looks to private entities decrease significantly across the three regions. Clearly, then, interlocutors in goal-oriented communicative tasks can use representations of their partner's perspective to guide real-time reference resolution.

13.6 Conclusions and implications

The studies that we have reviewed demonstrate the feasibility of examining real-time language processing in natural tasks. We showed that monitoring eye movements can be used to examine spoken word recognition in continuous speech, and why doing so is important.

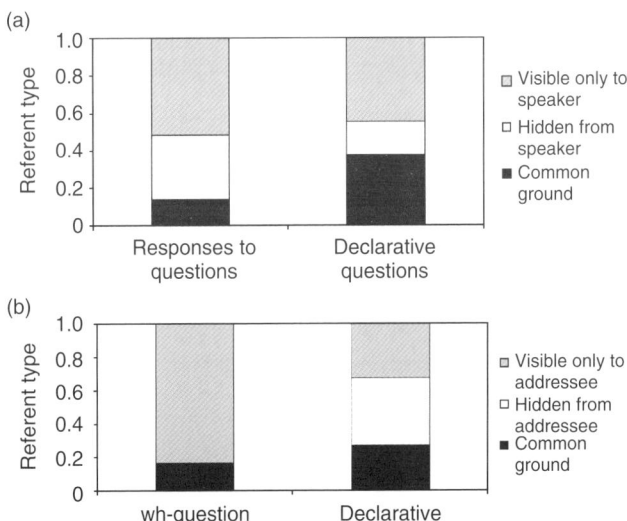

Figure 13.10 The proportion of referents for responses to questions, declarative questions, wh-questions, and declaratives, categorized by mutuality of the referent, from the viewpoint of (a) the speaker and (b) the addressee.

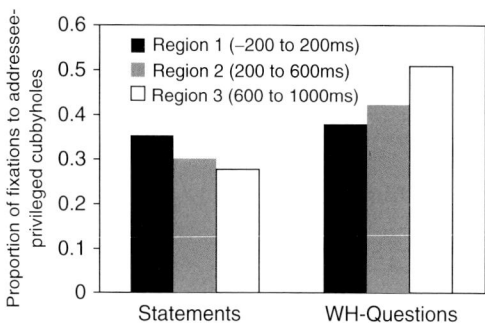

Figure 13.11 Fixations to privileged ground cubbyholes during interpretation of wh-questions and statements (consecutive, 400 ms analysis regions; baseline region began 200 ms before utterance onset).

We then showed that actions, intentions and real-world knowledge circumscribe referential domains, and that these context-specific domains affect processes such as syntactic ambiguity resolution. Example 4 illustrates how this notion of a context-specific referential domain alters the standard view obtained from studies of decontextualized language.

Example 4 After putting the pencil below the big apple, James put the apple on top of the towel.

A traditional account would go something like this. When the listener encounters the scalar adjective *big*, interpretation is delayed because a scalar dimension can only be interpreted with respect to the noun it modifies (e.g. compare a big building and a big pencil). As *apple* is heard, lexical access activates the apple concept, a prototypical red apple. The apple concept is then modified resulting in a representation of a *big apple*. When *apple* is encountered in the second clause, lexical access again results in activation of a prototypical *apple* concept. Because *apple* was introduced by a definite article, this representation would need to be compared with the memory representation of the *big apple* to decide whether the two co-refer (see Tanenhaus *et al.* (1985) for an outline of such an approach).

This account of moment-by-moment processing seems reasonable when we focus on the processing of just the linguistic forms. However, let us reconsider how the real-time interpretation of Example 4 proceeds in the context illustrated in Figure 13.12, taking into account results we have reviewed. At *big*, the listener's attention will be drawn to the larger of the two apples because a scalar adjective signals a contrast among two or more entities of the same semantic type. Thus, *apple* will be immediately interpreted as the misshapen (green) apple, even though a more prototypical (red) apple is present the display. And, when *the apple* is encountered in the second clause, the red apple would be ignored in favour of the large green apple.

This account is incompatible with the view that the initial stages of language processing create context-independent representations. However, it is compatible with increasing evidence throughout the brain and cognitive sciences that (i) behavioural context, including attention and intention affect even basic perceptual processes (Gandhi *et al.* 1998; Colby and Goldberg 1999) and (ii) brain systems involved in perception and action are implicated in the earliest moments of language processing (Pulvermüller *et al.* 2001).

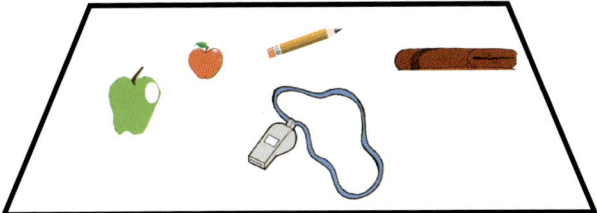

Figure 13.12 Hypothetical context for utterance *After putting the pencil below the big apple, James put the apple on top of the towel* to illustrate the implausibility of standard assumptions about context-independent comprehension. Note that the small (red) apple is intended to be a more prototypical apple than the large (green) apple.

An important goal for future research will be integrating action-based notions of linguistic context with perceptual and action-based accounts of perception and cognition (cf. Barsalou 1999; Glenberg and Robertson 2000; Spivey and Dale 2004).

Finally, we showed that it is possible to examine real-time processing in unscripted interactive conversation at the same temporal grain as in tightly controlled studies with scripted utterances. When we do so, we find that interlocutors make immediate use of information about each other's probable perspective, knowledge, and intentions that is closely tied to the pragmatic constraints on different utterance types. This result is compatible with the growing evidence that social pragmatic cues such as joint attention and intentionality are critical in early language development (Bloom 1997; Sabbagh and Baldwin 2001), as well as evidence showing that non-linguistic gestures contribute to the understanding of speech (Goldin-Meadow 1999; McNeill 2000). By using natural tasks, and moving towards a methodological and theoretical union of the action and product traditions, work in language processing can more fruitfully identify points of contact with these areas of research. Most generally, investigating real-time language processing in natural tasks reveals a system in which multiple types of representations are integrated remarkably quickly, and a system in which processing language and using language for action are closely intertwined.

Acknowledgements

The research presented here has been generously supported by NIH grants HD 27206 and DC 005071. The first author would like to thank the numerous students and colleagues whose work is presented here.

References

Allen, J. F., Byron, D. K., Dzikovska, M., Ferguson, G., Galescu, L. & Stent, A. 2001 Towards conversational human–computer interaction. *AIMag.* **22**, 27–35.
Allopenna, P. D., Magnuson, J. S. & Tanenhaus, M. K. 1998 Tracking the time course of spoken word recognition: evidence for continuous mapping models. *J. Mem. Lang.* **38**, 419–439. (doi:10.1006/jmla.1997.2558)

Altmann, G. T. M. & Kamide, Y. 1999 Incremental interpretation at verbs: restricting the domain of subsequent reference. *Cognition* **73**, 247–264. (doi:10.1016/ S0010–0277(99)00059–1)

Altmann, G. T. M. & Steedman, M. J. 1988 Interaction with context during human sentence processing. *Cognition* **30**, 191–238. (doi:10.1016/0010–0277(88)90020–0)

Austin, J. L. 1962 *How to do things with words*. Cambridge, MA: Harvard University Press.

Bard, E. G., Anderson, A. H., Sotillo, C., Aylett, M., Doherty-Sneddon, G. & Newlands, A. 2000 Controlling the intelligibility of referring expressions in dialogue. *J. Mem. Lang.* **42**, 1–22. (doi:10.1006/jmla.1999.2667)

Barsalou, L. 1999 Language comprehension: archival memory or preparation for situated action? *Discourse Process.* **28**, 61–80.

Beun, R.-J. & Cremers, A. H. M. 1998 Object reference in a shared domain of conversation. *Pragm. Cogn.* **6**, 121–151.

Bock, J. K. 1995 Sentence production: from mind to mouth. In *Handbook of perception and cognition*, vol. 11 (eds J. Miller & P. Eimas). Speech, language, and communication, pp. 181–216. New York, NY: Academic Press.

Bock, J. K., Irwin, D. E., Davidson, D. J. & Levelt, W. J. M. 2003 Minding the clock. *J. Mem. Lang.* **48**, 653–685. (doi:10.1016/S0749–596X(03)00007-X)

Bloom, P. 1997 Intentionality and word learning. *Trends Cogn. Sci.* **1**, 9–12. (doi:10.1016/ S1364–6613(97)01006–1)

Brennan, S. E. 2005 How conversation is shaped by visual and spoken evidence. In *Approaches to studying world-situated language use*: *Bridging the language-as-product and language-as-action traditions* (eds J. C. Trueswell & M. K. Tanenhaus), pp. 95–129. Cambridge, MA: The MIT Press.

Brennan, S. E. & Hulteen, E. 1995 Interaction and feedback in a spoken language system: a theoretical framework. *Knowledge-Based Syst.* **8**, 143–151. (doi:10.1016/0950–7051(95)98376-H)

Brown, P. & Dell, G. 1987 Adapting production to comprehension: the explicit mention of instruments. *Cogn. Psychol.* **19**, 441–472. (doi:10.1016/0010–0285(87)90015–6)

Brown-Schmidt, S. & Tanenhaus, M. K. 2006 Watching the eyes when talking about size: an investigation of message formulation and utterance planning. *J. Mem. Lang.* **54**, 592–609. (doi:10.1016/j.jml.2005.12.008)

Brown-Schmidt S. & Tanenhaus, M. K. 2008. Real-time investigation of referential domains in unscripted conversation: a targeted language game approach. *Cogn. Sci.* **32**, 643–684.

Brown-Schmidt, S., Campana, E. & Tanenhaus, M. K. 2005 Real-time reference resolution in a referential communication task. In *Approaches to studying world-situated language*: *Bridging the language-as-action and language-as-product traditions* (eds J. C. Trueswell & M. K. Tanenhaus), pp. 153–172. Cambridge, MA: The MIT Press.

Brown-Schmidt, S., Gunlogson, C. & Tanenhaus, M. K. (2008). Addressees distinguish shared from private information when interpreting questions during interactive conversation. *Cognition.* **107**, 1122–1134.

Chambers, C. G., Tanenhaus, M. K., Eberhard, K. M., Filip, H. & Carlson, G. N. 2002 Circumscribing referential domains in real-time sentence comprehension. *J. Mem. Lang.* **47**, 30–49. (doi:10.1006/jmla.2001.2832)

Chambers, C. G., Tanenhaus, M. K. & Magnuson, J. S. 2004 Action-based affordances and syntactic ambiguity resolution. *J. Exp. Psychol. Learn. Mem. Cogn.* **30**, 687–696. (doi:10.1037/0278–7393.30.3.687)

Chomsky, N. 1965 *Aspects of the theory of syntax*. Cambridge, MA: The MIT Press.

Clark, H. H. 1992 *Arenas of language use*. Chicago, IL: University of Chicago Press. Clark, H. H. 1996 *Using language*. Cambridge, UK: Cambridge University Press.

Clark, H. H. & Wilkes-Gibbs, D. 1986 Referring as a collaborative process. *Cognition* **22**, 1–39. (doi:10.1016/ 0010–0277(86)90010–7)

Colby, C. L. & Goldberg, M. E. 1999 Space and attention in parietal cortex. *Annu. Rev. Neurosci.* **22**, 97–136. (doi:10. 1146/annurev.neuro.22.1.319)

Coltheart, M. 1999 Modularity and cognition. *Trends Cogn. Sci.* **3**, 115–120. (doi:10.1016/ S1364–6613(99)01289–9)

Cooper, R. M. 1974 The control of eye fixation by the meaning of spoken language: a new methodology for the real-time investigation of speech perception, memory, and language processing. *Cogn. Psychol.* **6**, 84–107. (doi:10. 1016/0010–0285(74)90005-X)

Crain, S. & Steedman, M. 1985 On not being led up the garden path: the use of context by the psychological parser. In *Natural language parsing: psychological, computational, and theoretical perspectives* (eds D. Dowty, L. Karttunen & A. Zwicky), pp. 320–358. Cambridge, UK: Cambridge University Press.

Crosswhite, K., Masharov, M., McDonough, J. M. & Tanenhaus, M. K. In preparation. Phonetic cues to word length in the online processing of onset-embedded word pairs.

Dahan, D. & Tanenhaus, M. K. 2004 Continuous mapping from sound to meaning in spoken-language comprehension: evidence from immediate effects of verb-based constraints. *J. Exp. Psychol. Learn. Mem. Cogn.* **30**, 498–513. (doi:10.1037/0278–7393.30.2.498)

Dahan, D. & Tanenhaus, M. K. 2005 Looking at the rope when looking for the snake: conceptually mediated eye movements during spoken-word recognition. *Psychol. Bull. Rev.* **12**, 455–459.

Dahan, D., Magnuson, J. S. & Tanenhaus, M. K. 2001a Time course of frequency effects in spoken word recognition: evidence from eye movements. *Cogn. Psychol.* **42**, 317–367. (doi:10.1006/cogp.2001.0750)

Dahan, D., Magnuson, J. S., Tanenhaus, M. K. & Hogan, E. 2001b Subcategorical mismatches and the time course of lexical access: evidence for lexical competition. *Lang. Cogn. Process.* **16**, 507–534. (doi:10.1080/01690960143000074)

Dahan, D., Tanenhaus, M. K. & Salverda, A. P. 2007 How visual information influences phonetically-driven saccades to pictures: effects of preview and position in display.In *Eye movements: a window on mind and brain* (eds R. P. G. van Gompel, M. H. Fischer, W. S. Murray & R. L. Hill), pp. 471–486. Oxford, UK: Elsevier.

Davis, M. H., Marslen-Wilson, W. D. & Gaskell, M. G. 2002 Leading up the lexical garden-path: segmentation and ambiguity in spoken word recognition. *J. Exp. Psychol. Hum. Percept. Perform.* **28**, 218–244. (doi:10.1037//0096–1523.28.1.218)

Dell, G. S. 1986 A spreading activation theory of retrieval in language production. *Psychol. Rev.* **93**, 283–321. (doi:10. 1037/0033–295X.93.3.283)

Eberhard, K. M., Spivey-Knowlton, M. J., Sedivy, J. C. & Tanenhaus, M. K. 1995 Eye-movements as a window into spoken language comprehension in natural contexts. *J. Psychol. Res.* **24**, 409–436. (doi:10.1007/BF02143160)

Ferreira, V. S. & Dell, G. S. 2000 The effect of ambiguity and lexical availability on syntactic and lexical production. *Cogn. Psychol.* **40**, 296–340. (doi:10.1006/cogp. 1999.0730)

Fodor, J. A. 1983 *Modularity of mind*. Cambridge, MA: Bradford Books.

Gandhi, S. P., Heeger, M. J. & Boyton, G. M. 1998 Spatial attention affects brain activity in human primary visual cortex. *Proc. NatlAcad. Sci. USA* **96**, 3314–3319. (doi:10. 1073/pnas.96.6.3314)

Glenberg, A. M. & Robertson, D. A. 2000 Symbol grounding and meaning: a comparison of high-dimensional and embodied theories of meaning. *J. Mem. Lang.* **43**, 379^01. (doi:10.1006/jmla.2000.2714)

Goldin-Meadow, S. 1999 The role of gesture in communication and thinking. *Trends Cogn. Sci.* **3**, 419–429. (doi:10. 1016/S1364–6613(99)01397–2)

Gregory, M. L., Joshi, A., Grodner, D. & Sedivy, J. C. 2003 Adjectives and processing effort: so, uh, what are we doing during disfluencies? Paper presented at the *16th Annual CUNY Sentence Processing Conf.*, March, Cambridge, MA. Grice, H. P. 1957 Meaning. *Phil. Rev.* **66**, 377–388. (doi:10. 2307/2182440)

Griffin, Z. M. & Bock, J. K. 2000 What they eyes say about speaking. *Psychol. Sci.* **11**, 274–279. (doi:10.1111/1467–9280.00255)

Hanna, J. E., Tanenhaus, M. K. & Trueswell, J. C. 2003 The effects of common ground and perspective on domains of referential interpretation. *J. Mem. Lang.* **49**, 43–61. (doi:10.1016/S0749–596X(03)00022–6)

Hayhoe, M. & Ballard, D. 2005 Eye movements in natural behavior. *Trends Cogn. Sci.* **9**, 188–194. (doi:10.1016/j.tics. 2005.02.009)

Indefrey, P. & Levelt, W. J. M. 2004 The spatial and temporal signatures of word production components. *Cognition* **92**, 101–144. (doi:10.1016/j.cognition.2002.06.001)

Keysar, B. & Barr, D. J. 2005 Coordination of action and belief in communication. In *Approaches to studying world situated language use: Bridging the language-as-product and language-as-action traditions* (eds J. C. Trueswell & M. K. Tanenhaus), pp. 71–94. Cambridge, MA: The MIT Press.

Keysar, B., Barr, D. J., Balin, J. A. & Paek, T. S. 1998 Definite reference and mutual knowledge: process models of common ground in comprehension. *J. Mem. Lang.* **39**, 1–20. (doi:10.1006/jmla.1998.2563)

Keysar, B., Barr, D. J., Balin, J. A. & Brauner, J. S. 2000 Taking perspective in conversation: the role of mutual knowledge in comprehension. *Psychol. Sci.* **11**, 32–38. (doi:10.1111/1467–9280.00211)

Keysar, B., Lin, S. & Barr, D. J. 2003 Limits on theory of mind use in adults. *Cognition* **89**, 25–41. (doi:10.1016/ S0010–0277(03)00064–7)

Krauss, R. M. & Weinheimer, S. 1966 Concurrent feedback, confirmation, and the encoding of referents in verbal communication. *J. Person. Social Psychol.* **4**, 343–346. (doi:10.1037/h0023705)

Land, M. 2004 Eye movements in daily life. In *The visual neurosciences*, vol. 2 (eds L. Chalupa & J. Werner), pp. 1357–1368. Cambridge, MA: The MIT Press. Levelt, W. J. M. 1989 *Speaking: from intention to articulation*. Cambridge, MA: The MIT Press.

Levelt, W. J. M., Roelofs, A. P. A. & Meyer, A. S. 1999 A theory of lexical access in speech production. *Behav. Brain Sci.* **22**, 1–37. (doi:10.1017/S0140525X99001776)

Liversedge, S. P. & Findlay, J. M. 2000 Saccadic eye movements and cognition. *Trends Cogn. Sci.* **4**, 6–14. (doi:10.1016/S1364–6613(99)01418–7)

Luce, P. A. & Pisoni, D. B. 1998 Recognizing spoken words: the neighborhood activation model. *Ear Hear.* **19**, 1–36.

Magnuson, J. S. 2005 Moving hand reveals dynamics of thought. *Proc. Natl Acad. Sci. USA* **102**, 9995–9996. (doi:10.1073/pnas.0504413102)

Magnuson, J. S., Dixon, J. A., Tanenhaus, M. K. & Aslin, R. N. 2007 The dynamics of lexical competition during spoken word recognition. *Cogn. Sci.* **31**, 1–24. (doi:10. 1207/s 15516709cog000_90)

Marslen-Wilson, W. D. 1973 Linguistic structure and speech shadowing at very short latencies. *Nature* **244**, 522–523. (doi:10.1038/244522a0)

Marslen-Wilson, W. D. 1975 Sentence perception as an interactive parallel process. *Science* **189**, 226–228. (doi:10. 1126/science.189.4198.226)

Marslen-Wilson, W. & Warren, P. 1994 Levels of perceptual representation and process in lexical access. *Psychol. Rev.* **101**, 653–675. (doi:10.1037/0033–295X. 101.4.653)

Marslen-Wilson, W. D. & Welsh, A. 1978 Processing interactions and lexical access during word recognition in continuous speech. *Cogn. Psychol.* **10**, 29–63. (doi:10. 1016/0010–0285(78)90018-X)

Marslen-Wilson, W., Levy, E. & Tyler, L. K. 1982 Producing interpretable discourse: the establishment and maintenance of reference. In *Speech, place and action* (eds R. J. Jarvella & W Klein), pp. 339–378. New York, NY: Wiley.

McClelland, J. L. & Elman, J. L. 1986 The TRACE model of speech perception. *Cogn. Psychol.* **18**, 1–86. (doi:10.1016/ 0010–0285(86)90015–0)

McNeill, D. (ed.) 2000 *Language and gesture*, Cambridge, UK: Cambridge University Press.

Metzing, C. & Brennan, S. E. 2003 When conceptual pacts are broken: partner-specific effects on the comprehension of referring expressions. *J. Mem. Lang.* **49**, 201–213. (doi: 10.1016/S0749–596X(03)00028–7)

Miller, G. A. 1962 Some psychological studies of grammar. *Am. Psychol.* **17**, 748–762. (doi:10.1037/h0044708)

Miller, G. A. & Chomsky, N. 1963 Finitary models of language users. In *Handbook of mathematical psychology* (eds R. D. Luce, R. R. Bush & E. Galanter), pp. 421–491. New York, NY: Wiley.

Nadig, A. & Sedivy, J. 2002 Evidence for perspective-taking constraints in children's on-line reference resolution. *Psychol. Sci.* **13**, 329–336. (doi:10.1111/j.0956–7976. 2002.00460.x)

Norris, D., McQueen, J. M. & Cutler, A. 2002 Bias effects in facilitatory phonological priming. *Mem. Cogn.* **30**, 399–411.

Pickering, M. J. & Garrod, S. C. 2004 Towards a mechanistic theory of dialog. *Behav. Brain Sci.* **7**, 169–190.

Pulvermüller, F., Härle, M. & Hummel, F. 2001 Walking or talking? Behavioral and neurophysiological correlates of action verb processing. *Brain Lang.* **78**, 143–168. (doi:10. 1006/brln.2000.2390)

Rayner, K. 1998 Eye movements in reading and information processing: twenty years of research. *Psychol. Bull.* **124**, 372–422. (doi: 10.1037/0033–2909.124.3.372)

Roberts, C. 2003 Uniqueness in definite noun phrases. *Linguist. Philos.* **26**, 287–350. (doi:10.1023/A: 1024157 132393)

Sabbagh, M. A. & Baldwin, D. A. 2001 Learning words from knowledgeable versus ignorant speakers: links between preschoolers' theory of mind and semantic development. *ChildDev.* **72**, 1054–1070. (doi: 10.1111/1467–8624.00334)

Salverda, A. P. & Altmann, G. T. M. 2005 Cross-talk between language and vision: interference of visually-cued eye movements by spoken language. Poster presented at the *Architectures and Mechanisms in Language Processing (AMLaP) Conf.*, September 2005, Gent.

Salverda, A. P., Dahan, D. & McQueen, J. M. 2003 The role of prosodic boundaries in the resolution of lexical embedding in speech comprehension. *Cognition* **90**, 51–89. (doi:10.1016/S0010–0277(03)00139–2)

Salverda, A. P., Dahan, D., Tanenhaus, M. K., Crosswhite, K., Masharov, M. & McDonough, J. M. 2007 Effects of prosodically modulated sub-phonetic variation on lexical competition. *Cognition.* **105**, 466–476. (doi:10.1016/j.cognition.2006.10.008)

Schegloff, E. A. & Sacks, H. 1973 Opening up closings. *Semiotica* **8**, 289–327.

Searle, J. R. 1969 *Speech acts. An essay in the philosophy of language*. Cambridge, UK: Cambridge University Press.

Sedivy, J. C. 2003 Pragmatic versus form-based accounts of referential contrast: evidence for effects of informativity expectations. *J. Psycholinguis. Res.* **32**, 3–23. (doi:10.1023/ A:1021928914454)

Spivey, M. J. & Dale, R. 2004 On the continuity of mind: toward a dynamical account of cognition. In *The psychology of learning and motivation*, vol. **45** (ed. B. Ross), pp. 87–142. Amsterdam, The Netherlands: Elsevier.

Spivey, M. J., Tanenhaus, M. K., Eberhard, K. M. & Sedivy, J. C. 2002 Eye movements and spoken language comprehension: effects of visual context on syntactic ambiguity resolution. *Cogn. Psychol.* **45**, 447–481. (doi:10.1016/S0010–0285(02)00503–0)

Spivey, M. J., Grosjean, M. & Knoblich, G. 2005 Continuous attraction toward phonological competitors. *Proc. Natl Acad. Sci. USA* **102**, 10 393–10 398. (doi:10.1073/pnas. 0503903102)

Stalnaker, R. C. 1978 Assertion. In *Syntax and semantics*: *pragmatics*, vol. **9** (ed. P. Cole), pp. 315–332. New York, NY: Academic Press.

Stalnaker, R. C. 2002 Common ground. *Linguist. Philos.* **25**, 701–721. (doi:10.1023/A:1020867916902)

Tanenhaus, M. K. 2004 On-line sentence processing: past, present and, future. In *On-line sentence processing*: *ERPS, eye movements and beyond* (eds M. Carreiras & C. Clifton), pp. 371–392. London, UK: Psychology Press.

Tanenhaus, M. K. & Trueswell, J. C. 1995 Sentence comprehension. In *Speech, language, and communication* (eds J. Miller & P. Eimas), pp. 217–262. San Diego, CA: Academic Press.

Tanenhaus, M. K., Carlson, G. & Seidenberg, M. S. 1985 Do listeners compute linguistic representations? In *Natural language parsing*: *psychological, computational, and theoretical perspectives* (eds D. Dowty, L. Kartunnen & A. Zwicky), pp. 359–408. Cambridge, UK: Cambridge University Press.

Tanenhaus, M. K., Spivey-Knowlton, M. J., Eberhard, K. M. & Sedivy, J. C. 1995 Integration of visual and linguistic information in spoken language comprehension. *Science* **268**, 1632–1634. (doi: 10.1126/science.7777863)

Trueswell, J. C., Sekerina, I., Hill, N. & Logrip, M. 1999 The kindergarten-path effect: studying on-line sentence processing in young children. *Cognition* **73**, 89–134. (doi: 10. 1016/S0010–0277(99)00032–3)

Index

acoustic and auditory phonetics
   adaptive design of speech sound systems 79–99, 151
   auditory enhancement hypothesis 95–6
   dispersion theory: vowels 91–5
   dispersion theory and quantal theory 96–8
   quantal theory: consonants 89–91
   quantal theory: vowels 85–9
   restricted character of speech sound systems 82–5
   source-filter theory 79–82
   vocal-tract cavity properties and formant frequencies 82–4
acoustic features of sounds 277–82
acoustic signals 79
acoustic specification of speech segment 144
acoustic trauma 17, 23
action-based affordances 312–19
affine-scaling transform 178
algorithmic level 6–7, 250
alveolars 265
ambiguities 208–9, 210, 211, 212, 322, 324–5
   referential 238–9
   temporal 315–17, 319
American 108, 109
amplitude modulation 59, 154
amplitude reduction 265
anaesthesia 41
analysis-by-synthesis 7, 262–6
anterior cingulate cortex (ACC) 201, 203–4, 217
anterior stream 144
antinodes 82–3, 84, 95
aperiodic sources 80
aphasia 281
articulation 81, 89–90, 135, 144
aspiration 80
asynchronies 136, 161, 260
audio-visual (AV) integration 261, 264–5
auditory adaptation to glottal pulse rate and vocal tract length 176–9
auditory core 173–4
auditory cortex 32, 140, 172–3
auditory distance 97
auditory enhancement hypothesis 95–6
auditory evoked field (AEF) 181
auditory filters 47–8, 53, 97, 177
auditory grouping 158–64
   harmonicity 158–60

   onset-time 160–1
   spatial direction 161–4
auditory image model (AIM) 176–9, 181
auditory nerve (AN) 9, 30, 33
   fibres 15–17, 18, 20, 23, 25, 27–9, 39
   representation in 26–7
   temporal envelope of speech in 31–3
   tuning 11–13, 21
auditory primal sketch 253
auditory processes and speech sound analysis 47–71
   internal representation of sounds, calculation of 69–70
   masking, across-channel processes in 56–60
   pitch perception 63–5
   temporal analysis 65–9
   timbre perception 60–2
   see also frequency selectivity
auditory scene analysis 153–5
autism spectrum disorder 4, 111, 115–18, 119, 123, 125

background sounds *see* listening to speech with background sounds
bandwidth 82
Bark scale 92
basilar membrane 11, 13, 16, 21
bats 39
Bayesian methodology 263
Bayes's rule 264
behavioural scores 199
belt regions of cortex 173–4
best frequency 11–13, 15–18, 22–6, 29–31, 33–7, 39
bilabials 265–6
bilateral involvement 292–3, 294
bilingual children 120–1, 125
binaural masking level difference 162
binding 135–7
biology and culture, interaction between 106
birds, perception by 34, 109
brain damage 197, 199, 200
broadband Gaussian noise 13
broadband phase-locking 18
broadband sounds and within-channel temporal analysis 66–7
Broca's area 119, 145–6, 173, 205, 257
Broca-Wernicke-Lichtheim model 193–4, 200

Cantonese 90, 294
categorical perception 91
cats 20–2, 34, 35–6
centre frequency 82
cepstral coefficients 156
characteristic frequency (CF) 52–3
chinchillas 12, 91–2, 277
chopper neurones 27–31
cochlear implants 155
cochlear nucleus 27–31, 32–3, 39, 172
cognitive enhancement 110
cohort competitors 310
communication sounds 176
co-modulation masking release 56–8
complementarity in speech stream 4, 139–40, 142, 144
compound-action-potential tuning curve 21
comprehension analyses 328
compression, fast-acting 177
computational level of description 6, 250–1
concepts 269–70
conjoint allocation 157
consonants 10–11, 26, 31–2, 33, 37, 38, 122
    adaptive design of speech sound inventories 84, 85
    alveolar 121
    audio-visual speech processing 145
    bilingual infants 121
    continuant 89
    dental 121
    labiodental 137
    lexical representation and distinctive features 267
    nasal 89
    neural signatures of phonetic learning 106
    oral 89
    posterior superior temporal sulcus 143
    quantal theory 89–91
    source-filter theory 80–1
    timbre perception 62
    visible speech stream 138, 139
    voiceless 161
    see also fricatives; glides; stop consonants
consonant-vowel-consonant syllables 37
consonant-vowel syllables 23–6
context 308–9
    effects and sensory cortical response 285–6
    specific models 329–30
continuous speech and spoken word recognition 309–12
coronal palatalization 266–7
correlated mode 4, 140, 142, 144, 145
cortex 32
    anterior cingulate 201, 203–4, 217
    auditory 32, 34, 140, 172–3
    orofacial motor 175
    see also left inferior frontal cortex
cortical activation, time course of 141–2

cortical representation of temporal and spectral parameters 33–9
cue-specific model 275–7, 282, 285, 287, 288, 294, 296–7
CW 239, 240

deafness 138, 152
    prelingual 140
    pure word 281–2
    see also hearing impairment
decision device 69
declaratives 327–9, 330
decompositional morphemic substrate 7
definite articles 314–15
derivation 196
detection 152
developmental disabilities 125
difference filter 61–2
diffusion tensor imaging (DTI) 195, 205, 214, 223
Direct Realism 118–19
disambiguated trials 325
discriminability 24, 26
discrimination 22–3, 64–5, 91
disjoint allocation 157
dispersion theory 3, 84–5, 91–5, 96–8
dissimilarity 24–5
distinctive features 250–4, 266–8
domain-specific model 275–7, 285, 288, 293, 294, 297
dominance principle 63–4
dominance ratings 208
dorsal route 198
Dutch 90, 237–8, 239
dynamic ranges 40

electrical brain signals *see* fractionation of spoken language processing measuring electrical and magnetic brain signals
electroencephalography (EEG) 7–8, 103–6, 125, 224, 241, 307
    event-related brain potentials 233, 236
    fronto-temporal brain systems 195
English 90, 91, 105, 108, 121, 269
    audio-visual speech processing 137–8
    bilateral involvement 292–3
    fronto-temporal brain systems 198
    left-hemisphere involvement 288
    lexical representation and distinctive features 267
    second-language and bilateral involvement 294
    tonal category knowledge 295–6
equivalent rectangular bandwidth (ERB) 51–2
    see also ERB_N-number scale
ERB_N-number scale 52, 54–5, 57–61, 65, 70, 92
event-related brain potentials (ERP) 16, 224–9, 237–9, 240, 241
    analysis-by-synthesis-internal forward models 265

early language acquisition 103–4, 106–11, 113–15, 121
late anterior negativity 226–7
measures of early language processing and autism spectrum disorder 115–18
mirror neurons and shared brain systems 119
P600/SPS 227–9
time course of cortical activation 141
evoked potentials 141, 175, 286
excitation patterns 2, 52–6, 58–9
eye movements *see* saccadic eye movements

facial expression 140
fasciculus 194, 195
ferrets 34, 38
filter function 134
Flemish 269
foreign language *see* non-native language
formants 29, 156
  frequencies 10–11, 16, 19, 82–4
forward model 7
Fourier transform 14–15, 16, 37, 251
fractional anisotrophy (FA) 214–15
fractionation of spoken language processing measuring electrical and magnetic brain signals 223–42
  auditory N400/N200 229–31
  event-related brain potentials 224–9
  lexical selection versus integration 233–5
  N400 neuronal generators 235–6
  semantic and referential processing 236–40
  spoken word recognition, aspects of 232–3
French 121, 184
frequency following response (FFR) 295–6
frequency-modulated sine waves 154
frequency selectivity 2–3, 47–56
  auditory filter 47–8
  excitation pattern of vowel sound 54–6
  impaired hearing 56
  masking patterns and excitation patterns 52–4
  notched-noise method 50–2
  psychophysical tuning curves 48–50
frequency spectra 9
frequency tuning 40
frication 80, 81
fricatives 24, 89, 90
frontal region 174
fronto-temporal brain systems 193–217
  functional connectivity in intact and damaged brain: syntax and semantics 209–15
  functional connectivity in intact and damaged brain: words and morphemes 204–7
  functional organization: words and morphemes 198–204
  functional organization in intact and damaged brain: syntax and semantics 207–9

functional anatomic background 255–7
functional imaging of auditory processing 171–86
  auditory cortex and projections in macaque 172–3
  auditory image model and auditory adaptation to glottal pulse rate and vocal tract length 176–9
  communication sounds 176
  functional magnetic resonance imaging 179
  Heschl's gyrus 179–81
  magnetoencephalography 181–3
  PAC 179–81
  perception and production, links between in humans 173–5
  planum polare 180–1
  regular-interval sounds 179
  subcortical auditory system in humans 171–2
  voice-specific and speech-specific processing 183–6
functional magnetic resonance imaging (fMRI) 7, 179, 186, 205, 223, 241
  audio-visual and visual speech 140
  bilateral involvement 292–3
  early language acquisition 103–5, 119, 125
  fronto-temporal brain systems 195, 201
  functional organization in intact and damaged brain 207–8
  general auditory processing 182
  hemispheric specialization 278–80, 284
  multi-time resolution processing 260, 262
  visible speech stream 139
  voice-specific and speech-specific processing in brain 183, 185

gain functions 12–13
gammatone filterbank 10
gaze
  direction 140
  following 110
  shifting 305
gender 135
general purpose associative mechanism 145
German 110–12, 114, 293
gestures 118, 186, 331
glides 80, 89
glimpsing 158, 159
global dispersion 96
global similarity models 309, 310
glottal pulse rate (GPR) 175, 176–9, 181, 183
glottal repetition rate (voice pitch) 10
glottis 79, 82
guinea pigs 21

Haas effect 163–4
hair cells 11, 70

half-wave resonators 84
harmonicity 154, 158–60, 162–3
harmonics 63, 64, 119, 161
head action kinematics 134
hearing by eye 137–8
hearing impairment 11, 56, 152
Helmholtz resonators 84, 96
hemispheric differences 6
hemispheric specialization 277–82, 284–5
Heschl's gyrus (HG) 175, 179–81, 182, 283
hidden Markov models 263
hierarchical linear growth curve modelling 108
hierarchy of processing 181
Hilbert transform 251
Hindi 90
'how' stream 4–5, 144–5
hyper-speech 95
hypo-speech 95
hypothesize-and-test approach 250, 263

ideal binary mask 157
immediacy principle 241
implementational level 6–7, 250
indefinite articles 314–15
inferior colliculus (IC) 31–3, 39, 172
inferior frontal gyrus (IFG) 173, 175, 258, 294
  see also right inferior frontal gyrus
inferior frontal regions 141
inferior parietal lobule (IPL) 144, 208, 213, 214, 217, 293
  left 201, 210
inferior temporal gyrus (ITG) 258
inflection 196, 200
inhibition, loss of 41
integration 232, 233–5
intelligence quotient 117, 119
intentionality 331
interactive conversation and real-time language processing 319–29
inter-aural time differences (ITDs) 162–3, 164
intermediate representations 155–6
internal forward models 262–6
internal representation of sounds calculation of 69–70
intonation 3, 292, 293
invariance 251
Italian 94, 97

jabberwocky sentences 114–15
Japanese 184
jaw 83, 92, 95, 96
joint attention 331

kinematics 134, 135, 139, 142
known words 117–18, 119, 121, 125
Korean 267

LACC 202, 204, 205, 206
landmarks approach 251, 255
language-as-action tradition 306, 308–9, 315, 316–17
language-as-product tradition 305–6, 308–9
language interpretation 321–5
language production 325–9
larynx 92
  lowering 96
late anterior negativity (LAN) 113, 226–7, 239
lateral inhibition 309
left hemisphere (LH) 195, 196, 204–6, 209, 211–13, 216
  bilateral involvement 293
  laterality effects 290–1
  and linguistic tone 288–90
left inferior frontal cortex (LIFC) 195–7, 201–3, 206–7, 210, 212–13
left inferior parietal lobule (LIPL) 201, 210
left middle frontal gyrus (LMFG) 212
left middle temporal gyrus (LMTG) 201, 204–5, 206, 208, 210, 212, 214, 217
left superior temporal gyrus (LSTG) 200, 202, 212
lexical entrainment 308
lexical representation 266–8
lexical selection 233–5
lexicons, early 112–13
LIFG 217
  functional connectivity: syntax and semantics 209, 210, 211, 212, 213, 214
  functional connectivity : words and morphemes 204, 205, 206
  functional organization: syntax and semantics 207, 208, 209
  functional organization: words and morphemes 199, 201
linear frequency scale 65
linguistic tone 288–90
lips 82–3, 84, 92, 142
liquids 80
listener-oriented constraints 84, 85, 87–8, 95
listening to speech with background sounds 5, 58, 151–65
  auditory scene analysis 153–5
  intermediate representations 155–6
  multiple sound sources, perception of 151–2
  noise 152–3
  see also separation of speech signals
local focalization 96
logarithmic frequency scale 65
Logogen model 235
loudness integration 261

M350 236
macaque monkeys 172–3, 184, 194
MacArthur Bates Communicative Development Inventories (CDI) 108

McGurk effect 133–4, 136, 137, 138, 141, 144, 265
magnetic brain signals *see* fractionation of spoken language processing measuring electrical and magnetic brain signals
magnetic evoked potentials 286
magnetic resonance imaging (MRI) 105
  fronto-temporal brain systems 199
  structural 124
  see also functional magnetic resonance imaging
magnetoencephalography (MEG) 5, 175, 181–3, 186, 224, 240, 241
  early language acquisition 103–6, 110–11, 119–20, 122, 124, 125
  event-related brain potentials 229, 233, 235–6
  fronto-temporal brain systems 195
  hemispheric specialization 285
  time course of cortical activation 141, 142
  tonal category knowledge 295
Mandarin Chinese 6, 107, 109–10, 288, 289–91
  bilateral involvement 292–3
  second-language and bilateral involvement 294
  tonal category knowledge 295–6
marmoset monkeys 34, 37, 181
masking 48
  across-channel processes 56–60
  energetic 157
  experiments 2
  informational 157
  patterns 52–4
  upward spread 52
medial geniculate body (MGB) 32, 172
Mel scale 92
middle frontal gyrus (MFG) 292, 293
middle temporal gyrus (MTG) 196, 203–5, 208–14, 217, 253
  bilateral 201
  functional anatomic background 257
  multi-time resolution processing 258
  posterior 292
  see also right middle temporal gyrus
mid-palate 96
mirror-neurones 118–20, 124, 125, 145–6
mismatch inhibition 309
Mismatch Negativity (MMN) 107–8, 110, 116, 121, 285, 294–5
mispronunciations 16
missing fundamental phenomenon 63
modification analysis 316
modulation discrimination interference 59–60
moment-by-moment processing 330
monkeys 21, 35, 37
  see also macaque; marmoset
monosyllabic words 16, 162
morphemes
  derivational 198
  fronto-temporal brain systems 196

functional connectivity in intact and damaged brain 204–7
functional organization 198–204
grammatical 198
inflectional 198, 200
past tense 198
plural 198
semantic 198
morphological processes 216
morphological violations 113
motherese (infant-directed speech) 116–17, 122
Motor theory 118–19, 276
mouth 138, 142
multiple sound sources, perception of 151–2
multi-time resolution process 250, 251, 253, 254, 255, 257–62
multiunit recording 40
mutual knowledge 327

N1 response 180–1, 265
N100 180–1
N200 229–31, 233, 241
N400 113–14, 229–31, 239–40, 241
  event-related brain potentials 225–6, 232–4
  neuronal generators 235–6
  tonal category knowledge 294
narrowband sounds and within-channel temporal analysis 67–8
nasality 81
nasals 24
native language
  auditory scene analysis 154
  phonemes 125
  phonetic cues 124
  skill 4
Native Language Magnet Expanded (NLM-e) 113, 122–4, 125
Native Language Neural Commitment (NLNC) hypothesis 106–7, 108, 122
near-infrared spectroscopy (NIRS) 103–6, 117, 125
neighbourhood activation model 309
neural activity pattern (NAP) 177
neural commitment process 124
neural mechanisms for audio-visual and visual speech 140–1
neural representation of spectral and temporal information 9–41
  analysis of neural representation 13–15
  auditory nerve, representation in 26–7
  cat model, usefulness of for human ear 20–2
  changes in neural representation of spectral shape in cochlear nucleus 27–31
  consonant-vowel (CV) syllables responses to 23–6
  cortical representation of temporal and spectral parameters 33–9
  discrimination data, inferences from 22–3

neural representation of spectral and temporal information *(Continued)*
  neurones, how speech is represented by 39–40
  rate representation of spectral shape 18–20
  temporal envelope of speech in auditory nerve and inferior colliculus 31–3
  tonotopic representation of vowels 15–18
  tuning in auditory nerve: tonotopic representation 11–13
neural specializations for speech and pitch 275–98
  acoustic features of sounds and hemispheric specialization 277–82
  bilateral involvement 292–3
  context effects and sensory cortical response 285–6
  left-hemisphere involvement and linguistic tone 288–90
  left-hemisphere laterality effects 290–1
  linguistic features of stimulus and hemispheric specialization 284–5
  right auditory cortex processing 282–3
  second-language learners and bilateral involvement in tonal processing 294
  sign language studies 286–7
  tonal categories 290–1, 294–6
  tonal languages 288
neural substrates *see* phonetic and word learning and neural substrates
neural subsystems 27
neuro-biological context 5–6
neurobiology and linguistics 249–70
  algorithmic level of description 250
  analysis-by-synthesis 250, 251, 253, 254, 262–6
  computational level of description 250–1
  distinctive features 250–4, 266–8
  functional anatomic background 255–7
  hypothesize-and-test methods 250
  implementational level of description 250
  lexical representation 266–8
  multi-time resolution processing 250, 251, 253, 254, 255, 257–62
neuro-imaging context 5–6
neurones 2, 39–40
  primary-like 27, 29–31
  see also mirror neurones
neuroscience techniques 4, 103–6
nodes 82–3, 84, 96
  see also antinodes
nonlinear device 69
non-native language 113
  patterns 125
  phonetic cues 124
  skill 4
non-speech 286
  signals 116–17, 119–20
  sounds 185, 277–80
  vocalizations 184
non-words 16, 200, 201, 202

notched-noise method 50–2, 53
noun phrases (NPs) 196, 207, 237–8, 321–4
nouns 198

object 113
oddball responses 34
off-frequency listening (off-place listening) 49
offset asynchrony 161
Old and New heuristic 157
one-referent context 315, 316–17
onset-based similarity 309, 310
onset-time 154, 160–1, 162–3
oral (front) cavity 83–4, 85, 86–7
ordinate 10, 177
orofacial motor cortex 175
oscillogram 31–2

P1m 182
P2 response 265
P600 113–14
P600/SPS 227–9, 241
PAC 175, 179–81, 182, 186
parabelt regions 173–4
past tense 199–200
pattern-detection strategies 111
peak latency 265
perceptual constancy 251
periodic sources 80
peri-stimulus time (PST) histogram 13–14
pharyngeal (back) cavity 83–4, 85–6, 87
pharynx 96
phase 1 in development 106, 122–3, 125
phase 2 in development 106, 122–4, 125
phase 3 in development 123–4
phase 4 in development 123–4
phase locking 2, 24–6, 97
phonemes 38
  event-related potential measures of language processing 116
  hemispheric specialization 277
  infants' early lexicons 112–13
  native 107–9
  neural signatures of phonetic learning 107
  neural signatures of word learning 111
  non-native 107–9
  second language 109–10
phonemically equivalent class (PEC) 137–8
phonemic level 137
phonetic learning 122
phonetic patterns 116
phonetic and word learning and neural substrates 103–25
  bilingual infants 120–1
  event-related brain potentials measures of early language processing and autism spectrum disorder 115–18

lexicons, early 112–13
mirror neurones and shared brain systems 118–20
neural signatures of early sentence-processing 113–15
neural signatures of phonetic learning in typically developing children 106–9
neural signatures of word learning 110–12
neuroscience techniques 103–6
second language, learning from exposure to 109–10
theoretical model and future research 121–5
phonological message 176
phonological mismatch negativity (PMN) 230
phonological primal sketch (PPS) 7, 253, 255, 263
phonotactic patterns 124
photographs of speech 143–4
physio-physiological interaction 204
pitch 5
  general auditory processes 176
  low 63
  onset response (POR) 182
  patterns 3
  perception 63–5
  see also neural specializations for speech and pitch
planum polare 180–1
point of disambiguation (POD) 313, 321–3
positron emission tomography (PET) 195, 223, 278
posterior lateral superior temporal (PLST) 175
posterior stream 145
power-spectrum model of masking 48
precedence effect 164
prepositional phrases 196
preserved phase 68
primary auditory cortex 34
primary-like neurones 27, 29–31
PRIMER 113
priming 199–200, 203, 308
private knowledge 327
processing of audio-visual speech: empirical and neural bases 133–46
  binding 135–7
  complementarity and redundancy in speech stream 139–40
  cortical activation, time course of 141–2
  hearing by eye 137–8
  neural mechanisms 140–1
  photographs of speech 143–4
  posterior superior temporal sulcus (pSTS) 142–3
  source-filter model 134–5
  visible speech stream 138–9
production analyses 328
profile analysis 58–9
pronunciations, correct 16
proximity 324, 325
psychophysical tuning curves 48–50, 52
psycho-physiological interaction 204
psycho-physio-physiological interaction 204
PT 181

pulse rate 176
pulse-resonance sound 176
pure word deafness 281–2

quails 277
quantal theory 3, 84–5, 96–8
  consonants 89–91
  vowels 85–9
quarter-wave resonators 83, 85, 96

RACC 204–5, 206
radiation characteristic 80
rate 37–9
  locked representation 24
  place profile 18
rats 37
real-time language processing in interactive conversation 319–29
  language interpretation 321–5
real-time language processing in natural tasks 307–9
real words 200, 201, 202
recency 323–4, 325
recognition point (RP) 232, 234, 235
redundancy in speech stream 139–40
referential ambiguity 238–9
referential communicative task 8
referential domains 312–19
referential processing 236
regular-interval sounds 179
relative amplitude 82
residue 63
resonance 82–3, 84, 176
  frequency 85–6, 87–8, 89, 95
resonators
  half-wave 84
  Helmholtz 84, 96
  quarter-wave 83, 85, 96
responses to questions 327–8
retroflex stops 90
reverse correlation 12
RIFC 205, 206, 207, 212
right auditory cortex processing of pitch information 282–3
right hemisphere (RH) 195, 205, 206, 209, 212, 213, 216, 293
right inferior frontal gyrus (RIFG) 204, 205, 206, 211, 212–13, 214
right middle temporal gyrus (RMTG) 204–5, 206
right superior temporal gyrus (RSTG) 202, 208, 212, 214
RINS 206
RIPL 212
RI sound (RIS) 182

saccadic eye movement 305, 310–11, 313, 314, 316, 322, 329
scalar adjectives 326, 330

scale 37–9
schizophrenia 136
second language
　and bilateral involvement in tonal
　　processing 294
　learning from exposure to 109–10
　see also bilingual
segmental contrasts 139
segmentation 251
selectional restrictions 318
selective listening hypothesis 29
semantically anomalous words 113–15, 125
semantics
　functional connectivity in intact and damaged
　　brain 209–15
　functional organization in intact and damaged
　　brain 207–9
　processing 236–40
sensory cortical response 285–6
sentences 125
　mirror neurones and shared brain systems 119
　processing, early 113–15
separation of speech signals 156–64
　see also auditory grouping
sequential grouping 158
seven-vowel system 97
shared brain systems 118–20
signal intensity 198–9
signal-to-masker ratio 48, 50
signal-to-noise ratio 48, 133, 152
sign language studies 286–7
silent speech reading 137, 140, 141–2, 143
simultaneous grouping 158
sine-wave consonant-vowel syllables 185
sine-wave speech 153, 155, 161, 284
sliding temporal integrator 3, 70
smoothing device 69
social agent 124
social gating hypothesis 109–10
social interaction 109, 122–3
songbirds 34
sound patterns and objects association between 124
source-filter theory 3, 79–82, 134–5
Spanish 90, 91, 94, 107, 108, 109–10, 121, 195
spatial direction 161–4
spatio-temporal coding 98
specialization 268
specific language impairment 281
spectral changes 66
spectral energy 60
spectral information see neural representation of
　spectral and temporal information
spectral peaks 97
spectral representation 10
spectral salience 98
spectral shape 58, 61

spectral structure 61
spectral template-matching 151
spectrogram 31–2, 69
　log-amplitude 157
　separation of speech signals 156, 159
spectrotemporal excitation pattern (STEP) 3, 70–1
spectrotemporal receptive fields (STRFs) 35, 37, 38,
　254, 256, 259
Speech Intelligibility Index 152
speech-recognition algorithms 151
speech-specific processing 183–6
speech stream, complementarity and redundancy in
　139–40
Speech Transmission Index 152
speed of processing 7–8, 223
spike trains 40
spoken word recognition in continuous speech
　309–12
spondees 162
SQUID sensors 105
SR of fibres 23, 28–9
stabilized auditory image (SAI) 177
statistical learning 111
stimulus, linguistic features of 284–5
stimulus onset asynchrony (SOA) 260
stop consonants 33, 35
　adaptive design of speech sound inventories 89
　hemispheric specialization 277
　separation of speech signals 161
　source-filter theory 80
　velar 90
　vowel stimuli 23–4
strobed temporal integration (STI) 177–8
subcortical auditory system in humans 171–2
subject 113
superior olivary complex (SOC) 32, 172
superior temporal gyrus (STG) 181, 196, 201,
　203–5, 209–14, 217
　analysis-by-synthesis-internal forward
　　models 264
　anterior 208
　bilateral 201, 292, 293
　functional anatomic background 255, 256–7
　hemispheric specialization 279, 281
　left 200
　left anterior 210
　multi-time resolution processing 258, 262
　posterior 253
　second-language and bilateral involvement 294
　see also right
superior temporal regions 141
superior temporal sulcus (STS) 217, 253
　analysis-by-synthesis-internal forward models 264
　bilateral involvement 292, 293
　functional anatomic background 255, 256–7
　functional imaging 173

hemispheric specialization 281, 284
multi-time resolution processing 262
posterior 140–1, 142–4, 145
voice-specific and speech-specific processing in brain 183–4, 185
super-speech-readers 138
suppression 2, 17, 53–4
decreased 18
supralaryngeal regions of vocal tract 79–80
supramarginal gyrus (SMG) 292
syllables 124
general auditory processes 175
initial stops 90
mirror neurones and shared brain systems 119
stressed 16
sylvian parieto-temporal area (SPT) 258
syntactically anomalous words 113–15, 125
syntactic entrainment 308
syntactic processes 216
syntactic and semantic processing 7
syntax
functional connectivity in intact and damaged brain 209–15
functional organization in intact and damaged brain 207–9

talker-oriented constraints 84, 85, 88, 95
targeted language games 321, 327
task compatibility 324–5
temporal analysis 65–9
temporal coding 97–8
temporal envelope 39
temporal information *see* neural representation of spectral and temporal information
temporal integration 259–62, 265
temporal modulation transfer function (TMTF) 66–7
temporal region 174
temporal sampling rate 139
temporal window shape 69
temporary ambiguity 315–17, 319
Thai 6, 91, 288, 289–91, 295
thalamus 32
thresholds 40
timbre perception 60–2
time
locked processing 306–7
see also temporal analysis
tonal categories 290–1, 294–6
tonal chimeras 290
tonal languages 288, 294
tonal processing 294
tone 289, 292, 293
tongue 92, 96, 138, 142
body 83–4, 95
root 84, 96

tonotopic representation 11–13, 15–18, 34
tonotopic speech signals 2
TRACE model 235, 309, 310–11
transduction mechanism 177
transient source 80
transitional probabilities between segments and syllables 124
troughs/minima in speech spectrum 29
tuning 11–13, 16
two-referent context 316–17
two-tone suppression 20, 177
two-tube model 86–7

UCLA Phonological Segment Inventory Database 88, 89, 90
unambiguous instructions 316–17
unknown words 117–18, 121

velars 265
ventriloquism illusion 136
verbal working memory 227
verb phrases 196, 207
verbs 113, 198
visemes 137
visible kinematics 139
visible speech processing 143–4
visible speech stream 138–9
vision, influence of 4
visual distractor task 136
visual kinematics (mouth and lips) 135
visual world paradigm 8
vocabulary explosion 110
vocal folds vibration 134
vocal-tract 79–81, 83, 96
cavity properties and formant frequencies 82–4
length (VTL) 5, 134, 175, 176–9, 183
voice
bar 90
onset time (VOT) 33, 35–6, 37, 90–2
specific processing 183–6
voicing 81
vowel-like sounds 89
vowels 3, 10, 13, 17–19, 21–3, 27–9, 32–3, 38–9, 122
adaptive design of speech sound inventories 84, 87–8, 95
anterior 95
audio-visual speech processing 145
auditory scene analysis 152, 156
bilingual infants 121
central 93–4
dispersion theory 91–5
English point 137
frequency selectivity 56
general auditory processes 176, 177, 178
high 94, 95, 96, 97

vowels (Continued)
  high back 95
  length 289
  lexical representation and distinctive features 267
  lower 95
  masking 58–9
  mid-back 94, 97
  mid-front 94, 97
  neural signatures of phonetic learning 106
  non-nasalized 95
  percept 5–6
  posterior superior temporal sulcus 143
  quantal 85–9, 97–8
  separation of speech signals 158–61, 162, 163
  seven-vowel system 97
  sounds, excitation pattern of 54–6
  source-filter theory 80–1
  timbre perception 60–2
  tonotopic representation 15–18
  visible speech stream 138, 139

vocal-tract cavity properties and formant frequencies 82–3, 84
Welsh 269
Wernicke's area 200, 205
West African xylophone (amadinda) music 163
'what' (ventral) pathway 4–5, 144–5, 256–7
white matter tracts 214–15
white noise 61, 66–7
wh-questions 327–9, 330
word-category constraints 226–7
word learning *see* phonetic and word learning and neural substrates
words
  functional connectivity in intact and damaged brain 204–7
  functional organization 198–204
  second language 109–10
working memory 239

Xhosa language 154